Neural Networks for Signal Processing IX—Proceedings of the 1999 IEEE Signal Processing Society Workshop

IEEE Catalog Number	98TH8468
ISBN	0-7803-5673-X (softbound)
ISSN	1089-3555

Additional copies of this publication are available from

IEEE Operations Center
P.O. Box 1331
445 Hoes Lane
Piscataway, NJ 08855-1331 USA

1-800-678-IEEE
1-732-981-1393
1-732-981-9667 (FAX)
email: customer.services@ieee.org

NEURAL NETWORKS FOR SIGNAL PROCESSING IX

PROCEEDINGS
OF THE 1999 IEEE
SIGNAL PROCESSING
SOCIETY WORKSHOP

Ninth in a Series of Workshops
Organized by the IEEE Signal Processing Society
Neural Networks Technical Committee

Edited by

Yu-Hen Hu
Jan Larsen
Elizabeth Wilson
Scott Douglas

Published under the sponsorship of the IEEE Signal Processing Society
with support from the THOR Center for Neuroinformatics, Denmark

The Institute of Electrical and Electronics Engineers, Inc.
New York, New York

PREFACE

This book contains refereed papers presented at the Ninth IEEE Workshop on Neural Networks for Signal Processing (NNSP'99) held at Madison Concourse Hotel and Governor's Club, Madison, Wisconsin, August 23–25, 1999. The Neural Networks for Signal Processing Technical Committee of the IEEE Signal Processing Society sponsored NNSP'99 with co-sponsorship from THOR Center for Neuroinformatics, Denmark.

In recent years the field of neural networks has matured considerably in both methodology and real-world application domains going far beyond the "standard" backpropagation network and academic examples. Contemporary neural networks for signal processing research shares many ideas of adaptive signal/image processing, machine learning, and advanced statistics. We consider the field of neural networks as the open-minded playground which combines many approaches for solving complex real-world signal processing problems.

This year the workshop features a special session arranged by Klaus-Robert Muller on Support Vector Machines which is an interesting new paradigm.

Our sincere thanks goes to Professor Thomas Huang, University of Illinois at Urbana-Champaign, Dr. Vladimir Vapnik, AT&T Labs Research, and Professor Grace Wahba, University of Wisconsin for accepting to give keynote speeches at the workshop. Also we would like to thank the Technical Committee for taking time to provide quality reviews, and above all, thanks to all the author's who made this workshop possible and interesting. Special thanks go to Ms. Jean Stover for careful handling of the workshop registrations.

Continuing the tradition of paperless and easy communication we provided a re-designed workshop webpage (http://eivind.imm.dtu.dk/nnsp99) which, among other features, included web-based submission, review and registration.

Yu-Hen Hu, *University of Wisconsin*
Jan Larsen, *Technical University of Denmark*
Elizabeth Wilson, *Raytheon Company*
Scott Douglas, *Southern Methodist University*

NNSP'99 Technical Committee

Tulay ADALI	University of Maryland, Baltimore County
Amir ASSADI	University of Wisconsin - Madison
Andrew BACK	The Institute of Physical and Chemical Research (RIKEN)
Herve BOURLARD	Institut Dalle Molle d'Intelligence Artificielle Perceptive
Andrzej CICHOCKI	The Institute of Physical and Chemical Research (RIKEN)
Anthony G. CONSTANTINIDES	Imperial College
Bert DE VRIES	David Sarnoff Research Center
Scott C. DOUGLAS	Southern Methodist University
Kevin R. FARRELL	T-NETIX, Inc.
Hsin-Chia FU	National Chiao-Tung University
Ling GUAN	University of Sydney
Jenq-Neng HWANG	University of Washington
Shigeru KATAGIRI	NTT Communication Science Laboratories, ATR
Fa-Long LUO	ReSound Corporation
David MILLER	Pennsylvania State University
Nelson MORGAN	UC Berkeley
Klaus-Robert MULLER	GMD First
Mahesan NIRANJAN	University of Sheffield
Dragan OBRADOVIC	Siemens AG
Volker TRESP	Siemens AG, Central Research
Marc VAN HULLE	K.U. Leuven

CONTENTS

SUPPORT VECTOR MACHINES

ALGORITHMS AND ARCHITECTURES

APPLICATIONS

SUPPORT VECTOR MACHINES

USING CLASS-CENTER VECTORS TO BUILD SUPPORT VECTOR MACHINES

Xuegong Zhang
Dept. of Automation, Tsinghua University, Beijing 100084, China
xzhang@simba.au.tsinghua.edu.cn

ABSTRACT

A support vector machine builds the final classification function on only a small part of the training samples (the support vectors). It is believed that all the information about classification in the training set can be represented by these samples. However, this is actually not always true when the training set is polluted by noises (training data are not i.i.d.). In this paper, we present a different way for the problem, which applies the idea of capacity control in SVM but tries to make the machine less sensitive to noises and outliners. The new method can be called *Central Support Vector Machine* or *CSVM*, for it uses the class centers in building the support vector machine.

1. INTRODUCTION

Support vector machine or SVM is a new pattern recognition technique developed by Dr. Vapnik and his research group[1,2,3]. It is derived from the idea of (generalized) optimal hyperplane with maximum margin between the two classes. Although intuitively simple, this idea actually implements the SRM or structural risk minimization principle in statistical learning theory[1]. Maximizing the margin plays the role of capacity control so that the learned machine will not only have small empirical risks, but also have good generalization performance.

Solving the prime optimization problem of optimal hyperplane via its dual form is another important idea in SVM. By this method, the complexity of the optimization problem depends on the size of the training set but not on the dimensionality, which makes solving large dimensional problems possible. And the final classification function turns to be only depending on a small part of the training samples, which are called support vectors or SVs. SVM is first derived from linear cases, but since only the inner-product calculation between samples is involved in both the training and the recognition procedures, it can be extended to various nonlinear cases by defining different types of kernel functions. In fact, traditional polynomial, radial basis function and multi-layer perceptron classifiers can all find their equivalents in SVM by defining corresponding kernel functions.

However, we should notice that all the discussions about SVM are based on the basic precondition that all samples in the training set are independent and identically distributed (i.i.d.) according to the (unknown) dependency to be estimated[1]. But in many practical engineering applications, the observation is often affected by noises, and this precondition may not be well satisfied. If some of the training samples are actually not i.i.d. observations of the system, but are affected by some random noises, is SVM still a proper choice for the classification problem?

Intuitively we can say the answer is always yes. Since the classifier obtained by SVM depends on only a small part of the samples, it is very easy for it to become sensitive to noises or outliners in the training set. As noted in [3], the outliners tend to be support vectors with large Lagrangian coefficients. Although [3] and [4] had discussed method for removing these outliners from the samples, but the method is rather empirical, and there is still no general way to distinguish outliners from those really informative samples.

In fact, the most attractive characteristic of SVM is not the idea of support vectors, but the idea of capacity control which implements SRM principle, the idea of solving the prime problem with its dual form and the idea of kernels for nonlinear transformations. The feature that the final classifier depends on only the support vectors is good for its implementation, but is actually not what we are pursuing. In this paper, we present a new method which follows those important ideas of SVM but try to prevent the classifier from becoming too sensitive to outliners in the training set by building the classifier on both the class-center vectors and the support vectors. We call the method *Central Support Vector Machine* or *CSVM*. Our work is still quite preliminary but we hope the idea can lead to solutions that suit better for some engineering applications where one can only guarantee that the large part of observations are identically distributed.

2. AVERAGE: the Simple yet Effective Tool to Handle Noises

As noted in [3], classification functions obtained by minimizing mean square error (MSE) are not sensitive to outliners in the training set, while the decision functions obtained by SVM are very sensitive to outliners close to the decision boundaries. This can be seen clearly from Figure 1, the example given by [3].

It is true that if all the samples in the training set are drawn i.i.d. from the underlying distribution, the decision boundary can be decided by only those samples closest to the boundary, i.e. those support vectors. However, as shown in Figure 1, the decision boundary will deviate from the optimal boundary when outliners exist in the training sets.

The reason that the classical MSE result is not affected by outliners is because it is an "averaged" algorithm. A specific sample in the training set only contributes little to the final result. In fact, if it can be assumed that the large part of the training samples are not affected by noises, or if we can assume that although all samples are affected by noises, the noises are random (with zero mean) relative to the underlying system, then by taking averages on the samples the effect of the noises can be eliminate by a great deal. This reflects the technique of stacking for depressing random noises in signal processing, in which the signal-to-noise ratio can be improved N times if the N observations of the same signal are stacked.

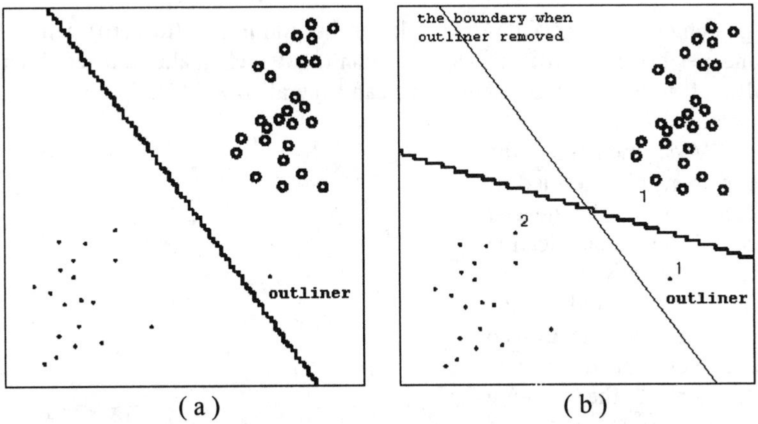

Figure 1. Decision functions by MSE(a) and SVM(b).
It can be seen that the MSE result is not affected by the outliner in the training set but the SVM result is affected severely. (from [3])

This consideration leads to our idea of Central Support Vector Machines. We want to apply the principle of capacity control as what SVM does to guarantee good generalization ability, and we want to build the classifier with the classification information contained in the averages or the means of the training samples but not only those in some specific samples. Also we also try to follow the strategy of dual optimization and kernels in SVM to make the algorithm practical and powerful for different cases. We hope that the idea of extracting or learning the *averaged* information from the noisy training set can lead to some general learning principle for noisy cases. The specific algorithm presented in the following sections is only the first trial we did on this idea.

3. CENTRAL SUPPORT VECTOR MACHINES: the Basic Idea

Let's follow the major procedures of the derivation of SVM. Consider the two-class linear classifier problem. The task is to find some "optimal" separating hyperplane

$$\mathbf{w} \cdot \mathbf{x} + b = 0 \tag{1}$$

according to the training sample set

$$(\mathbf{x}_i, y_i), \quad i = 1, \ldots, n, \quad \mathbf{x}_i \in R^d, y_i \in \{+1, -1\}. \tag{2}$$

The decision function is then

$$f(\mathbf{x}) = \text{sgn}[\mathbf{w} \cdot \mathbf{x} + b] \ . \tag{3}$$

First we consider linear separable cases. According to [1], for the linear classifier to be optimal, all samples in (2) should be correctly classified by (3) and the separation margin should be maximized. The former requirement can be written as:

$$y_i(\mathbf{w} \cdot \mathbf{x}_i + b) \geq \varepsilon > 0, \quad i = 1,...,n, \tag{4}$$

which guarantees that the empirical risk can be minimized (to zero), while the latter one performs the role of capacity control, which makes the confidence interval small so that the learned machine can generalize well[1].

In SVM methods, the separation margin is defined as the distances from the nearest samples to the classification boundary. Actually it is this definition that decides the famous feature that the trained SVM classifier depends on only a small part of the training samples (the SVs). To avoid this, we define another kind of margin, that is the distances from the two class centers to the separation boundary. Figure 2 illustrates this idea. First we define the mean of all training samples of a class as the class-center. Denote the center of class +1 as $\mathbf{x}^+$, and the center of class -1 as $\mathbf{x}^-$. It is easy to check that the distances from the class-centers to the classification boundary are:

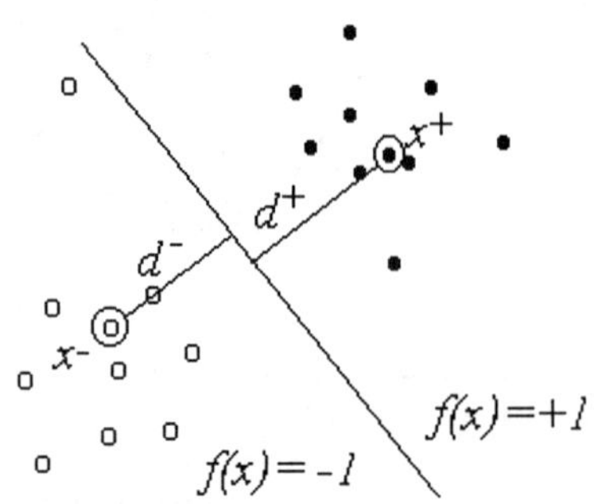

Figure 2. Central separation margin

$$d^+ = \frac{|\mathbf{w} \cdot \mathbf{x}^+ + b|}{\|\mathbf{w}\|} = \frac{y^+(\mathbf{w} \cdot \mathbf{x}^+ + b)}{\|\mathbf{w}\|} \tag{5a}$$

$$d^- = \frac{|\mathbf{w} \cdot \mathbf{x}^- + b|}{\|\mathbf{w}\|} = \frac{y^-(\mathbf{w} \cdot \mathbf{x}^- + b)}{\|\mathbf{w}\|} \tag{5b}$$

where $y^+ = 1, y^- = -1$ are the class labels of $\mathbf{x}^+$ and $\mathbf{x}^-$. Denote the number of samples of the two classes as n^+ and n^-, then we can formulate our version of margin as:

$$d = d^+ + d^- = \frac{\sum_{i=1}^{n} l_i y_i (\mathbf{w} \cdot \mathbf{x}_i)}{\|\mathbf{w}\|} \tag{6}$$

where $l_i = 1/n^+$ for samples in class +1 and $l_i = 1/n^-$ for samples in class -1. We call the margin defined by (6) the *central separation margin*, and call the classifier to maximize this margin as *Central Support Vector Machine* or *CSVM*.

Maximizing (6) directly will not be an easy problem, so as in the derivation of the SVM algorithm, we introduce some constraint to the scale of the separation function so that

$$\sum_{i=1}^{n} l_i y_i (\mathbf{w} \cdot \mathbf{x}_i) = 1 \cdot \tag{7}$$

Under this constraint, maximizing the margin of (6) is equivalent to minimizing $\|\mathbf{w}\|$. Remembering the requirements for our optimal hyperplane, we arrive at the prime optimization problem of our CSVM:

$$\text{minimize} \qquad (\mathbf{w}) = \frac{1}{2}\|\mathbf{w}\|^2 = \frac{1}{2}(\mathbf{w} \cdot \mathbf{w}) \tag{8}$$

subject to constraints (4) and (7).

which in form is similar to the prime problem of SVM with one more constraints.

4. THE DUAL PROBLEM AND NONSEPARABLE CASES

The Lagrangian function of our prime problem is

$$L_P = \frac{1}{2}(\mathbf{w} \cdot \mathbf{w}) - \sum_{i=1}^{n} \alpha_i \{y_i[(\mathbf{w} \cdot \mathbf{x}_i) + b] - \varepsilon\} - \beta\left\{\sum_{i=1}^{n} l_i y_i (\mathbf{w} \cdot \mathbf{x}_i) - 1\right\} \tag{9}$$

Following the same strategy of SVM, we can get the dual form of the problem:

$$\text{maximize } L_D = \sum_{i=1}^{n} \varepsilon\alpha_i + \beta - \frac{1}{2}\sum_{i,j=1}^{n} (\alpha_i + \beta l_i)(\alpha_j + \beta l_j) y_i y_j (\mathbf{x}_i \cdot \mathbf{x}_j) \tag{10}$$

$$\text{subject to:} \qquad \sum_{i=1}^{n} y_i \alpha_i = 0, \tag{11a}$$

$$\text{and} \qquad \alpha_i \geq 0, \quad i = 1, \ldots n \tag{11b}$$

which is a quadratic programming problem of α_i and β. And the solution weight vector **w** is the linear combination of the training samples:

$$\mathbf{w}^* = \sum_{i=1}^{n} (\alpha_i^* + \beta^* l_i) y_i \mathbf{x}_i = \sum_{i=1}^{n} \alpha_i^* y_i \mathbf{x}_i + \beta^* (\mathbf{x}^+ - \mathbf{x}^-) \cdot \tag{12}$$

And the threshold b in (3) can be decided from any sample for which the equality in (4) holds true or be calculated from the two class centers.

Similar as in the SVM case, here we can also see from the Kühn-Tucker condition that only the α_i^* corresponding to those samples that the equality holds in (4) are non-zero. These samples are also the ones that are nearest to the separation boundary, corresponding to the support vectors in SVM. From (12) we see that the final classifier is decided by both the support vectors and the two class-center vectors.

In the above discussion, we did not consider the choice of ε, which should be small positive number. In SVM $\varepsilon = 1$, but here it is hard to predefine this parameter because of the existence of constraint (7), for if we choose ε to be too large, condition (4) may be violated for some samples, and if we choose ε to be too small, the equality in (4) may be hold for none of the samples so that the decision boundary will be decided only by the two class-center vectors. In the latter case, the classification function will be a hyperplane orthogonal to the line crossing the two class-centers and any other information about the distribution of samples besides the means are not taken into account. Obviously this will not be an optimal classifier unless the samples in each class are distributed in a hypersphere or the directions of the largest variances of the two classes happen to coincide with the line crossing the two class-centers.

This problem will be solved when we consider the nonseparable cases. Again following the ideas in SVM, we introduce slack variables into condition (4) to include the samples that violate this condition:

$$y_i(\mathbf{w}\cdot\mathbf{x}_i + b) + \xi_i \geq \varepsilon > 0, \quad \xi_i \geq 0, \quad i = 1,\ldots,n, \qquad (13)$$

and define the new objective function as:

$$\text{minimize} \qquad (\mathbf{w}, \xi) = \frac{1}{2}(\mathbf{w}\cdot\mathbf{w}) + C\left(\sum_{i=1}^{n}\xi_i\right). \qquad (14)$$

For this new problem, the dual problem turns to be same as defined by (10) and (11) except that condition (11b) becomes:

$$0 \leq \alpha_i \leq C, \quad i = 1,\ldots n. \qquad (15)$$

Parameter C controls penalty on the errors and also controls the trade-off between the classification information contained in the means of the samples and in the individual samples near the boundary.

5. FURTHER CONSIDERATIONS AND EXPERIMENTS

As we can see from (5), the final decision function can be considered as the combination of two parts, one is the SVM part and the other part is decided by the two class-center vectors. This fact leads to a more simple and direct modification to SVM. That is, after we get the weight of a SVM, we can bias the obtained hyperplane by adding the difference vector between the two class-centers with certain factor, e.g.

$$\mathbf{w}^{new} = (1-\lambda)\mathbf{w}^{svm} + \lambda(\mathbf{x}^{+} - \mathbf{x}^{-}), \qquad (16)$$

where $\mathbf{w}^{svm}$ is the weight obtained by SVM and $0 \leq \lambda \leq 1$ is a constant controlling the balance between the classification information contained in the support vectors and in the class-centers. In practice λ can be selected according to the *a priori* estimation about the chance that the training samples be polluted by noises. If $\lambda = 0$ then we are using standard SVM, and if $\lambda = 1$ then the classifier is degenerated to classical minimum-distance classifier.

Figure 3 is an experimental result testing this idea. In this example, Fig.3(b) is the result obtained by SVM when the training set is noiseless. If an outliner exists in the training set, the standard SVM boundary will no longer be optimal, as shown in Fig.3(a). Fig.3(c) shows the result obtained by modifying the SVM weight by (16) with $\lambda = 0.5$. We can see that although the outliner exists in the training set, the result is almost the same as that obtained when the training set is clean. Note that here we did not attempt to identify and remove the outliner, but simply made the outliners less effective to the result by taking the class-center information into account.

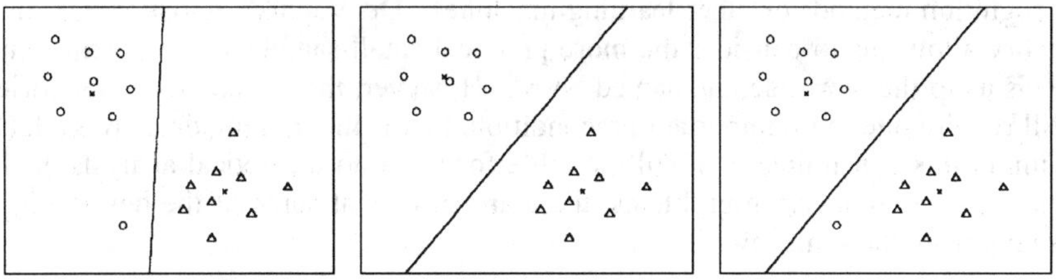

(a) (b) (c)

Figure 3. An experiment example about CSVM.
Circles and triangles denote the training samples of two classes, and x denotes the two class-centers. (a) is the standard SVM result at the existence of an outliner in the first class; (b) is the standard SVM result if the training set does not contain this outliner; and (c) is the result obtained by CSVM when the outliner exists.

From (10) and (12) we can see that just as the SVM case, here the problem also only involves in inner-product calculation, which makes it possible for using kernels $K(\mathbf{x}, \mathbf{x}')$ in place of the inner-products to implement nonlinear transformations. However, (12) shows that although it is true that only some of the α_i^* are none-zero, we will have to calculate the inner-product of the new sample with all training samples when applying the decision function, unless it holds true that

$$K(\mathbf{x}^+, \mathbf{x}') = K\left((\frac{1}{n^+} \sum_{y_i=+1} \mathbf{x}_i), \mathbf{x}' \right), \quad K(\mathbf{x}^-, \mathbf{x}') = K\left((\frac{1}{n^-} \sum_{y_i=-1} \mathbf{x}_i), \mathbf{x}' \right), \qquad (17)$$

which means that the image of the class-center vectors after the transformation are still class-centers in the new space, in which case we can only calculated the inner-products with the support vectors and the two class-center vectors.

6. DISCUSSIONS AND CONCLUSIONS

Support vector machine is a new technique for pattern recognition which can guarantee good generalization. But the characteristic that its final decision function depends on only the support vectors makes it easy to becoming sensitive to noises and outliners in the training set. In this paper, we present a method named central support vector machine or CSVM which tries to make the algorithm less sensitive to noises and outliners by taking class-center information into account. It follows the major ideas in SVM so we hope the property of good generalization ability can be kept in our new method. We believe that CSVM can be more robust in many practical engineering applications where the observations are severely affected by noises. The current study of CSVM is rather abecedarian and some further justifications are needed. However, experiments on toy data already show some usefulness of the idea.

In classical statistics, we can only apply asymptotic analysis on pattern recognition methods or other learning machines. Dr. Vapnik's statistical learning theory allows us to consider the more practical small-sample-size problems, and leads us to the new machine named SVM. However, these methods and theories still require one more important precondition, that is the training data are i.i.d. (If without this constraints, is it still possible for us to do theoretical analysis?) In our real-life engineering problems, there are cases that some of the observations are not from the same law.

Although there is still no sound theory for this kind of cases, we believe that there should be many *practical* ways to solve the problem. The idea presented in this paper is only one example. An optional idea may be first partitioning the training set into several clusters before training a support vector machine, then using the cluster center to represent all the samples in that cluster for the training. In this way, not only the effect of outliners can be eliminated, but also the computation labor of the training procedure can be reduced greatly, for the size of the quadratic programming problem depends on the number of training samples. We call this idea Clustered SVM or Reduced SVM, since the number of training samples will be reduced much. The primary study on this idea will be presented in a later publication of us. Similar with the condition of (17), one should be careful when choosing clustering algorithms and the kernels so that the cluster centers in the original space are still centers of the corresponding clusters in the transformed space, otherwise the partition will have to be done in the transformed space which one may even not explicitly know.

Acknowledgement

This paper is supported by National Natural Science Foundation of China, project No. 69885004. Special thanks to the website <http://svm.first.gmd.de> for the useful information and literatures.

References:

[1] V. N. Vapnik, **The Nature of Statistical Learning Theory**, Springer-Verlag, NY, 1995

[2] C. Cortes, V. Vapnik, "Support vector networks", **Machine Learning**, vol. 20, pp.273-297, 1995

[3] B. Boser, I. Guyon, V. Vapnik, "A training algorithm for optimal margin classifiers", presented at the 5th Annual Workshop on Computational Learning Theory, ACM Press, Pittsburgh, 1992

[4] I. Guyon, Matic, V. Vapnik, "Discovering informative patterns and data cleaning", in U.M. Fayyad et.al. (eds.), **Advances in Knowledge Discovery**

and Data Mining, MIT Press, pp. 181-203, 1996

[5] C.J.C. Burges, "A tutorial on support vector machines for pattern recognition", **Data Mining and Knowledge Discovery**, vol.2 no.2, pp.1-43, 1998

[6] Steve Gunn, **Support Vector Machines for Classification and Regression**, ISIS Technical Report, University of Southampton, May, 1998

MARGIN-LIKE QUANTITIES AND GENERALIZED APPROXIMATE CROSS VALIDATION FOR SUPPORT VECTOR MACHINES

Grace Wahba, Yi Lin and Hao Zhang
Department of Statistics
University of Wisconsin
Madison WI 53706 USA
Phone: (608) 262-2598
Fax: (608) 262-0032
E-mail: wahba, yilin, hzhang@stat.wisc.edu
Web: www.stat.wisc.edu

Abstract. We examine Support Vector Machines from the point of view of solutions to variational problems in a reproducing kernel Hilbert space. We discuss the Generalized Comparative Kullback-Leibler Distance as a target for choosing tuning parameters in SVM's, and we propose that the Generalized Approximate Cross Validation estimate of them is a reasonable proxy for this target. We indicate an interesting relationship between the GACV and the SVM margin.

INTRODUCTION

It is well known that the popular support vector machine (SVM) paradigm, which has proved highly successful in a number of classification studies, can be cast as a variational/regularization problem in a reproducing kernel Hilbert space (RKHS), see [10], [15], [8], [13], [3], the papers and references in [14], and elsewhere. In this note, which constitutes part of a sequel to [16], we look at the SVM paradigm from the point of view of a regularization problem. This allows comparison with penalized likelihood and other regularization problems, but that is not done here. However, there is a long literature on choosing smoothing and tuning parameters in regularization problems. Early papers include [5] for choosing a smoothing parameter in spline smoothing, more recent work includes [18] and [11], for penalized likelihood estimates with Bernoulli data. In those works minimizing the Comparative Kullback-Leibler Distance is the target criteria for choosing tuning parameters, and the Generalized Approximate Cross Validation (GACV) is proposed and studied

0-7803-5673-X/99/$10.00 © 1999 IEEE

as the method.

After describing the variational problem, we review the Generalized Comparative Kullback-Leibler Distance (GCKL) as an optimization criteria for the SVM paradigm, and observe that it is trivially a simple upper bound on the expected misclassification rate. Next we revisit the GACV as a proxy for the GCKL proposed in [16] and the argument that it is a reasonable estimate of the GCKL. We found that it is not necessary to do the randomization of the GACV in [16], because it can be replaced by an equally justifiable approximation which is readily computed exactly, along with the SVM solution to the dual mathematical programming problem. This estimate turns out interestingly, but not surprisingly to be simply related to what several authors have identified as the (observed) VC dimension of the estimated SVM. This work is an abbreviated version of [17], where more detail and the results of some preliminary simulation studies may be found.

THE SVM VARIATIONAL PROBLEM

Let $\mathcal{T}$ be an index set, $t \in \mathcal{T}$. Usually $\mathcal{T} = E^d$, Euclidean d-space, but not necessarily. Let $K(s,t), s,t \in \mathcal{T}$, be a positive definite function on $\mathcal{T} \otimes \mathcal{T}$, and let $\mathcal{H}_K$ be the RKHS with reproducing kernel K. See [15], [16], [11] for more on RKHS. RK's which are tensor sums and products of RK's are discussed there and elsewhere. K may contain one or more tuning parameters, to be chosen. A variety of RK's with success in practical applications have been proposed by various authors, see e. g. the Publications list at `http://svm.first.gmd.de/`. Recently [13] interestingly observed how different scales may be accomodated using RKHS methods.

We are given a training set $\{y_i, t_i\}$, where the attribute vector $t_i \in \mathcal{T}$, and $y_i = \pm 1$ according as an example with attribute vector t_i is in category $\mathcal{A}$ or $\mathcal{B}$. The classical SVM paradigm is equivalent to: find f_λ of the form $const + h$, where $h \in \mathcal{H}_K$ to minimize

$$\frac{1}{n}\sum_{i=1}^{n}(1 - y_i f_i)_+ + \lambda \|h\|^2_{\mathcal{H}_K}, \tag{1}$$

here $f_i = f(t_i)$, and $(\tau)_+ = \tau, \tau > 0; = 0$ otherwise. Once the minimizer, call it f_λ is found, then the decision rule for a new example with attribute vector t is: $\mathcal{A}$ if $f_\lambda(t) > 0$, $\mathcal{B}$ if $f_\lambda(t) < 0$.

We will assume for simplicity that K is strictly positive definite on $\mathcal{T} \otimes \mathcal{T}$, although this is not necessary. The minimizer of (1) is known to be in the span $\{K(\cdot, t_i), i = 1, \cdots n\}$, of representers of evaluation in $\mathcal{H}_K$. The function $K(\cdot, t_i)$ is $K(s, t_i)$ considered as a function of s with t_i fixed. The famous 'reproducing' property gives the inner product in $\mathcal{H}_K$ of two representers as $< K(\cdot, t_i), K(\cdot, t_j) >_{\mathcal{H}_K} = K(t_i, t_j)$. Thus, if $h(\cdot) = \sum_{i=1}^n c_i K(\cdot, t_i)$, then $\|h\|^2_{\mathcal{H}_K} = \sum_{i,j=1}^n c_i c_j K(t_i, t_j)$. Letting $e = (1, \cdots, 1)'$, $c = (c_1, \cdots, c_n)'$, $(f(t_1), \cdots f(t_n))' = (f_1, \cdots, f_n)'$, and with some abuse of notation, letting

$f = (f_1, \cdots, f_n)'$ and K now be the $n \times n$ matrix with ijth entry $K(t_i, t_j)$, and noting that $f(t) = d + \sum_{i=1}^{n} c_i K(t, t_i)$ for some c, d, we have

$$f = Kc + ed \tag{2}$$

and the variational problem (1) becomes: find (c, d) to minimize

$$\frac{1}{n}\sum_{i=1}^{n}(1 - y_i f_i)_+ + \lambda c' K c. \tag{3}$$

THE DUAL PROBLEM

Let Y be the $n \times n$ diagonal matrix with y_i in the iith position, and let $H = \frac{1}{2n\lambda} YKY$. By going to the dual form of (3), it can be shown that $c = \frac{1}{2n\lambda} Y\alpha$, where α is the solution to the problem

$$maxL = -\frac{1}{2}\alpha' H \alpha + e'\alpha \tag{4}$$

$$\text{subject to} \left\{ \begin{array}{c} 0 \le \alpha \le 1 \\ e'Y\alpha = y'\alpha = 0 \end{array} \right.$$

Assiming that there is an i for which $0 < \alpha_i < 1$, it can also be shown that $d = 1/y_i - \sum_{j=1}^{n} c_j K(t_i, t_j)$.

For future reference we review the relation between the (hard) margin (γ) of the support vector machine classifier and $\sum_{y_i f_{\lambda i} \le 1} \alpha_{\lambda i}$. In the situation where we can separate the training set points perfectly, γ is given by

$$\gamma^2 = 2n\lambda \left(\sum_{y_i f_{\lambda i} \le 1} \alpha_{\lambda i} \right)^{-1}.$$

See [4], [1]. (Notice the notation is a bit different from ours in these papers.) By definition the margin of the (hard margin) support vector machine classifier is $\gamma = \frac{1}{\|h\|_{\mathcal{H}_K}} = (c'Kc)^{-1/2}$. This equality can be seen from the following: In the perfectly separable case, where all members of the training set are classified correctly, $\alpha_{\lambda i}$ is the solution of the problem below:

$$maxL = -\frac{1}{2}\alpha' H \alpha + e'\alpha$$

subject to $\alpha_i \ge 0$ and $y'\alpha = 0$.

Introducing the Lagrangian multipliers $\xi = (\xi_1, \cdots, \xi_n)'$ and β for the constraints, the Lagrangian for this problem is

$$L_P = -\frac{1}{2}\alpha' H \alpha + e'\alpha - \beta y'\alpha - \xi'\alpha$$

and $\alpha_{\lambda i}$ satisfies the Kuhn-Tucker conditions:

$$\begin{aligned} \frac{\partial}{\partial \alpha} L_P = -H\alpha + e - \beta y - \xi &= 0 \\ \alpha_i &\geq 0, \; i = 1, 2, ..., n \\ y'\alpha &= 0 \\ \xi_i &\geq 0, \; i = 1, 2, ..., n \\ \xi_i \alpha_i &= 0, \; i = 1, 2, ..., n \end{aligned}$$

From these and the relation that $c = Y\alpha_\lambda/(2n\lambda)$, it is easy to get

$$\begin{aligned} c'Kc &= \frac{1}{2n\lambda} \alpha'_\lambda H \alpha_\lambda \\ &= \frac{1}{2n\lambda} [\alpha'_\lambda e - \beta \alpha'_\lambda y - \alpha'_\lambda \xi] \\ &= \frac{1}{2n\lambda} [\alpha'_\lambda e] \end{aligned}$$

Since $\alpha_{\lambda i} = 0$ if $y_i f_i > 1$, we finally get

$$\gamma^2 = (c'Kc)^{-1} = 2n\lambda \left[\sum_{y_i f_{\lambda i} \leq 1} \alpha_{\lambda i} \right]^{-1}.$$

THE GENERALIZED COMPARATIVE KULLBACK-LIEBLER DISTANCE

Suppose unobserved y_i's will be generated according to an (unknown) probability model with $p(t) = p_{true}(t)$ being the probability that an instance with attribute vector t is in class $\mathcal{A}$. Let y_j be an (unobserved) value of y associated with t_j. Given f_λ, define the generalized comparative Kullback-Liebler distance (GCKL distance) with respect to g as

$$GCKL(p_{true}, f_\lambda) \doteq GCKL(\lambda) = E_{true} \frac{1}{n} \sum_{j=1}^{n} g(y_j f_{\lambda j}). \tag{5}$$

Here f_λ is considered fixed and the expectation is taken over future, unobserved y_j. If $g(\tau) = ln(1 + e^{-\tau})$, then $GCKL(\lambda)$ reduces to the usual CKL for Bernoulli data [1] averaged over the attribute vectors of the training set. More details may be found in [16]. If $g(\tau) = [-\tau]_*$, then

$$\begin{aligned} E_{true}[-y_j f_{\lambda j}]_* &= p_{[true]j}[-f_{\lambda j}]_* + (1 - p_{[true]j})[f_{\lambda j}]_* && (6) \\ &= p_{[true]j}, \;\; f_{\lambda j} < 0 && (7) \\ &= (1 - p_{[true]j}), \;\; f_{\lambda j} > 0, && (8) \end{aligned}$$

[1] The usual CKL (comparative Kullback-Liebler distance) is the Kullback-Liebler distance plus a term which depends only on $p_{[true]}$.

where $p_{[true]j} = p_{[true]}(t_j)$, so that the $GCKL(\lambda)$ is the expected misclassification rate for f_λ on unobserved instances if they have the same distribution of t_j as the training set. Similarly, if $g(\tau) = (1-\tau)_+$, then

$$
\begin{aligned}
E_{true}(1 - y_j f_{\lambda j})_+ &= p_{[true]j}(1 - f_{\lambda j}), \quad f_{\lambda j} < -1 && (9)\\
&= 1 + (1 - 2p_{[true]j}) f_{\lambda j}, \quad -1 \le f_{\lambda j} \le 1 && (10)\\
&= (1 - p_{[true]j})(1 + f_{\lambda j}), \quad f_{\lambda j} > 1. && (11)
\end{aligned}
$$

Note that $[-y_i f_i]_* \le (1 - y_i f_i)_+$, so that the $GCKL$ for $(1 - y_i f_i)_+$ *is an upper bound for the expected misclassification rate.* - see Figure 1.

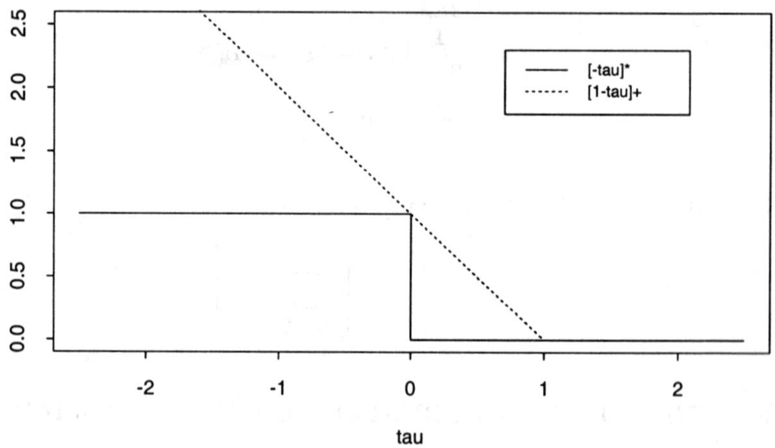

Figure 1: Comparison of $g(\tau) = (1-\tau)_+$ and $g(\tau) = [-\tau]_*$.

LEAVING OUT ONE AND THE GACV

Recently there has been much interest in choosing λ (or its equivalent, referred to in the literature as $\frac{1}{2nC}$), as well as other parameters inside K. See for example [2], [6], [9], surely not a complete list. Important references in the statistics literature that are related include [7], [19]. X. [11] consider in detail the case $g(\tau) = ln(1 + e^{-\tau})$. We now obtain the GACV estimate for λ and other tuning parameters.

Let $f_\lambda^{[-i]}$ be the solution to the variational problem: find f of the form $f = const + h$ with $h \in \mathcal{H}_K$ to minimize

$$\frac{1}{n} \sum_{\substack{j=1 \\ j \ne i}}^{n} g(y_j f_j) + \lambda \|h\|^2_{\mathcal{H}_K} \qquad (12)$$

Then the leaving-out-one function $V_0(\lambda)$ is defined as

$$V_0(\lambda) = \frac{1}{n}\sum_{i=1}^{n} g(y_i f_{\lambda i}^{[-i]}). \tag{13}$$

Since $f_{\lambda i}^{[-i]}$ does not depend on y_i but is (presumably) on average close to $f_{\lambda i}$, we may consider $V_0(\lambda)$ a proxy for $GCKL(\lambda)$, albeit one that is not generally feasible to compute in large data sets. Now let

$$V_0(\lambda) = OBS(\lambda) + D(\lambda), \tag{14}$$

where $OBS(\lambda)$ is the observed match of f_λ to the data,

$$OBS(\lambda) = \frac{1}{n}\sum_{i=1}^{n} g(y_i f_{\lambda i}) \tag{15}$$

and

$$D(\lambda) = \frac{1}{n}\sum_{i=1}^{n}[g(y_i f_{\lambda i}^{[-i]}) - g(y_i f_{\lambda i})]. \tag{16}$$

Using a first order Taylor series expansion gives

$$D(\lambda) \approx -\frac{1}{n}\sum_{i=1}^{n} \frac{\partial g}{\partial f_{\lambda i}}(f_{\lambda i} - f_{\lambda i}^{[-i]}). \tag{17}$$

Next we let $\mu(f)$ be a 'prediction' of y given f. Here we let

$$\mu_i = \mu(f_i) = \sum_{y\in\{+1,-1\}} \frac{\partial}{\partial f_i} g(y_i f_i). \tag{18}$$

When $g(\tau) = ln(1 + e^{-\tau})$ then $\mu(f) = 2p - 1 = E\{y|p\}$. For $g(\tau) = (1 - \tau)_+, \mu(f) = -1, f < -1; \mu(f) = 0, -1 \le f \le 1$ and $\mu(f) = 1$ for $f > 1$.

Letting $\mu_{\lambda i} = \mu(f_{\lambda i})$ and $\mu_{\lambda i}^{[-i]} = \mu(f_{\lambda i}^{[-i]})$, we may write (ignoring, for the moment, the possibility of dividing by 0),

$$D(\lambda) \approx -\frac{1}{n}\sum_{i=1}^{n} \frac{\partial g}{\partial f_{\lambda i}} \frac{(f_{\lambda i} - f_{\lambda i}^{[-i]})}{(y_i - \mu_{\lambda i}^{[-i]})}(y_i - \mu_{\lambda i}^{[-i]}) \tag{19}$$

This is equation (1.40) in [16]. We now provide somewhat different arguments than in [16] to obtain a similar result, which, however is easily computed as soon as the dual variational problem is solved.

Let $f_\lambda[i, x]$ be the solution of the variational problem 1 [2] given the data $\{y_1, \cdots, y_{i-1}, x, y_{i+1}, \cdots, y_n\}$. Note that the variational problem does not require that $x = \pm 1$. Thus $f_\lambda[i, y_i](t_i) \equiv f_{\lambda i}$. To simplify the notation,

[2] d is not always uniquely determined, this however does not appear to be a problem in practice, and we shall ignore it.

let $f_\lambda[i,x](t_i) = f_{\lambda i}[i,x] = f_{\lambda i}[x]$. In [16] it is shown, via a generalized leaving-out-one lemma, that $\mu(f)$ as we have defined it has the property that $f_{\lambda i}^{[-i]} = f_\lambda[i, \mu_{\lambda i}^{[-i]}](t_i)$. Letting $\mu_{\lambda i}^{[-i]} = x$, this justifies the aproximation

$$\frac{f_{\lambda i} - f_{\lambda i}^{[-i]}}{y_i - \mu_{\lambda i}^{[-i]}} \equiv \frac{f_{\lambda i}[y_i] - f_{\lambda i}[x]}{y_i - x} \approx \frac{\partial f_{\lambda i}}{\partial y_i}. \tag{20}$$

Furthermore, $\mu_{\lambda i}^{[-i]} \equiv \mu(f_{\lambda i}^{[-i]}) = \mu(f_{\lambda i}) = \mu_{\lambda i}$ whenever $f_{\lambda i}^{[-i]}$ and $f_{\lambda i}$ are both in the interval $(-\infty, -1)$, or $[-1, 1]$, or $(1, \infty)$, which can be expected to happen with few exceptions. Thus, we make the further approximation $(y_i - \mu_{\lambda i}^{[-i]}) \approx (y_i - \mu_{\lambda i})$, and we replace (19) by

$$D(\lambda) \approx -\frac{1}{n} \sum_{i=1}^{n} \frac{\partial g}{\partial f_{\lambda i}} \frac{\partial f_{\lambda i}}{\partial y_i} (y_i - \mu_{\lambda i}). \tag{21}$$

Now, for $g(\tau) = (1-\tau)_+$

$$\begin{aligned} \frac{\partial g}{\partial f_{\lambda i}}(y_i - \mu_{\lambda i}) &= -2, \; y_i f_{\lambda i} < -1 \\ &= -1, \; y_i f_{\lambda i} \in [-1, 1] \\ &= 0, \; y_i f_{\lambda i} > 1, \end{aligned}$$

giving finally

$$D(\lambda) \approx \frac{1}{n} \sum_{y_i f_{\lambda i} < -1} 2 \frac{\partial f_{\lambda i}}{\partial y_i} + \frac{1}{n} \sum_{y_i f_{\lambda i} \in [-1,1]} \frac{\partial f_{\lambda i}}{\partial y_i}. \tag{22}$$

It is not hard to see how $\frac{\partial f_{\lambda i}}{\partial y_i}$ should be interpreted. Fixing λ and solving the variational problem for f_λ we obtain $\alpha = \alpha_\lambda$, $c = c_\lambda = \frac{1}{2n\lambda} Y \alpha_\lambda$ and for the moment letting f_λ be the column vector with ith component $f_{\lambda i}$, we have $f_\lambda = K c_\lambda + ed = \frac{1}{2n\lambda} K Y \alpha_\lambda + ed$. From this we may write

$$\frac{\partial f_{\lambda i}}{\partial y_i} = K(t_i, t_i) \frac{\alpha_{\lambda i}}{2n\lambda} \equiv \|K(\cdot, t_i)\|_{\mathcal{H}_K}^2 \frac{\alpha_{\lambda i}}{2n\lambda}. \tag{23}$$

The resulting $GACV(\lambda)$, which is believed to be a reasonable proxy for $GCKL(\lambda)$, is, finally

$$GACV(\lambda) = \frac{1}{n} \sum_{i=1}^{n} (1 - y_i f_{\lambda i})_+ + \hat{D}(\lambda), \tag{24}$$

where

$$\hat{D}(\lambda) = \frac{1}{n} \left[2 \sum_{y_i f_{\lambda i} < -1} \frac{\alpha_{\lambda i}}{2n\lambda} \cdot \|K(\cdot, t_i)\|_{\mathcal{H}_K}^2 + \sum_{y_i f_{\lambda i} \in [-1,1]} \frac{\alpha_{\lambda i}}{2n\lambda} \cdot \|K(\cdot, t_i)\|_{\mathcal{H}_K}^2 \right]. \tag{25}$$

If $K = K_\theta$, where θ are some parameters inside K to which the result is sensitive, then we may let $GACV(\lambda) = GACV(\lambda, \theta)$. Note the relationship between $\hat{D}$ and $\sum_{y_i f_{\lambda i} \le 1} \alpha_{\lambda i}$ and the margin γ. If $K(\cdot,\cdot)$ is a radial basis function then $||K(\cdot, t_i)||^2_{\mathcal{H}_K} = K(0,0)$. Furthermore $||K(\cdot, t_i) - K(\cdot, t_j)||^2_{\mathcal{H}_K}$ is bounded above by $2K(0,0)$. If all members of the training set are classified correctly then $y_i f_i > 0$ and the sum following the 2 in (25) does not appear and $\hat{D}(\lambda) = K(0,0)/n\gamma^2$.

We note that [12] have obtained a different approximation for $f_{\lambda i} - f_{\lambda i}^{[-i]}$.

ACKNOWLEDGMENTS

The authors thank Fangyu Gao and David Callan for important suggestions in this project. This work was partly supported by NSF under Grant DMS-9704758 and NIH under Grant R01 EY09946.

REFERENCES

[1] P. Bartlett and J. Shawe-Taylor, "Generalization performance of support vector machines and other pattern classifiers," in B. Schoelkopf, C. Burges and A. Smola, eds., **Advances in Kernel Methods-Support Vector Learning**, MIT Press, pp. 43–54, 1999.

[2] C. Burges, "A tutorial on support vector machines for pattern recognition," **Data Mining and Knowledge Discovery**, vol. 2, pp. 121–167, 1998.

[3] S. Chen, D. Donoho and M. Saunders, "Atomic decomposition by basis pursuit," Tech. Rep. 479, Dept of Statistics, Stanford University, Stanford CA, 1995.

[4] C. Cortes and V. Vapnik, "Support vector networks," **Machine Learning**, vol. 20, pp. 1–25, 1995.

[5] P. Craven and G. Wahba, "Smoothing noisy data with spline functions: estimating the correct degree of smoothing by the method of generalized cross-validation," **Numer. Math.**, vol. 31, pp. 377–403, 1979.

[6] N. Cristianini, C. Campbell and J. Shawe-Taylor, "Dynamically adapting kernels in support vector machines," Tech. Rep. NC2-TR-1998-017, Royal Holloway University of London, Surrey England, 1998.

[7] B. Efron and R. Tibshirani, "Improvements on cross-validation: the .632+ bootstrap method," **J. Amer. Statist. Assoc**, vol. 92, pp. 548–560, 1997.

[8] F. Girosi, "An equivalence between sparse approximation and support vector machines," Tech. Rep. A. I. 1606, MIT artificial Intelligence Laboratory, Boston MA, 1997.

[9] M. Kearns, A. Ng, Y. Mansour and D. Ron, "An experimantal and theoretical comparison of model selection methods," **Machine Learning**, vol. xx, pp. xx, to appear.

[10] G. Kimeldorf and G. Wahba, "Some results on Tchebycheffian spline functions," **J. Math. Anal. Applic.**, vol. 33, pp. 82–95, 1971.

[11] X. Lin, G. Wahba, D. Xiang, F. Gao, R. Klein and B. Klein, "Smoothing spline ANOVA models for large data sets with Bernoulli observations and the

randomized GACV," Tech. Rep. 998, Department of Statistics, University of Wisconsin, Madison WI, 1998.

[12] M. Opper and O. Winther, "Gaussian process classification and SVM: Mean field results and leave-out-one estimator," in **Advances in Large Margin Classifiers**, MIT Press, 1999.

[13] T. Poggio and F. Girosi, "A sparse representation for function approximation," **Neural Computation**, vol. 10, pp. 1445–1454, 1998.

[14] B. Schoelkopf, C. Burges and A. Smola, **Advances in Kernel Methods-Support Vector Learning**, MIT Press, 1999.

[15] G. Wahba, **Spline Models for Observational Data**, SIAM, 1990, CBMS-NSF Regional Conference Series in Applied Mathematics, v. 59.

[16] G. Wahba, "Support vector machines, reproducing kernel Hilbert spaces and the randomized GACV," in B. Schoelkopf, C. Burges and A. Smola, eds., **Advances in Kernel Methods-Support Vector Learning**, MIT Press, pp. 69–88, 1999.

[17] G. Wahba, Y. Lin and H. Zhang, "Generalized approximate cross validation for support vector machines, or, another way to look at margin-like quantities," Tech. Rep. 1006, Department of Statistics, University of Wisconsin, Madison WI, 1999.

[18] D. Xiang and G. Wahba, "A generalized approximate cross validation for smoothing splines with non-Gaussian data," **Statistica Sinica**, vol. 6, pp. 675–692, 1996.

[19] J. Ye and W. Wong, "Evaluation of highly complex modeling procedures with Binomial and Poisson data," manuscript, University of Chicago School of Business, 1997.

SUPANOVA
A Sparse, Transparent Modelling Approach

Steve R. Gunn
ISIS Research Group
Dept. of Electronics and Computer Science
University of Southampton, U.K.
Email: S.R.Gunn@ecs.soton.ac.uk
Tel.: +44 1703 592338
Fax: +44 1703 594498
Web: www.isis.ecs.soton.ac.uk

Martin Brown
Unilever Research Port Sunlight
Quarry Road East
Wirral, U.K.
Email: Martin.Q.Brown@unilever.com

Abstract. Traditional neural networks produce opaque models that are difficult to interpret. This work describes a transparent, non-linear, modelling approach that enables constructed models to be visualised, enhancing their validation and interpretation. The technique combines the representational advantage of a sparse ANOVA decomposition, with the good generalisation ability of a Support Vector Machine.

INTRODUCTION

The problem of empirical data modelling is germane to many applications. In empirical data modelling a process of induction is used to build up a model of the system from examples. Ultimately the quantity and quality of the observations govern the performance of the model. By its observational nature, data obtained is finite and sampled; typically this sampling is non-uniform and due to the high dimensional nature of the problem, the data will form only a sparse distribution in the input space. Consequently, the problem is nearly always ill-posed. To address the ill-posed nature of the problem it is necessary to convert the problem to one that is well-posed. For a problem to be well-posed, a unique solution must exist that varies continuously with the data. Conversion to a well-posed problem is typically achieved with some form of capacity control, which aims to balance the fitting of the data with constraints on the model flexibility, producing a robust model that generalises successfully. Previous approaches to restoring the well posedness have included regularisation methods. In this paper, the method chosen is based around support vector methods introduced by Vapnik [11,12]. Girosi [5] and Smola [8] has shown that these methods can be placed in a regularisation framework, which guarantees the well posedness. Support Vector Machines (SVMs) employ the method of Structural Risk Minimisation to enable good generalisation for small sample sizes. However, the solution of SVMs is a summation of Kernels, and consequently the solution is opaque.

0-7803-5673-X/99/$10.00 © 1999 IEEE

may not be strictly well-posed. Other strategies for sparse selection have been proposed in the literature, notably Wahba [13] using the basis function energy.

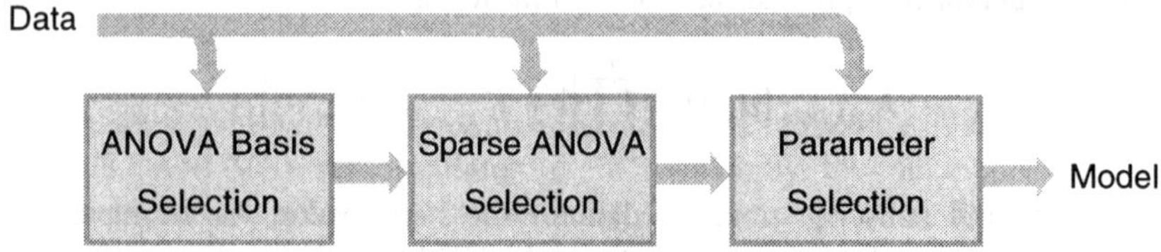

Figure 1. SUPANOVA Technique.

Stage I : Basis Selection

The initial goal is to obtain a complete ANOVA basis that is a suitable representation for the model. This is a difficult modelling task, which is addressed by using the kernel-based method of SVMs to address the curse of dimensionality. Splines are an attractive choice for modelling [13] due to their ability to approximate arbitrary functions. Accordingly, an SVM with an ANOVA spline kernel, equation (4), is used to approximate the data. This produces a complete ANOVA basis from which the 2^n sub-models can be extracted. The actual model will be composed of a product of piecewise cubic splines with a finite number of knots located at a subset of the data points. The flexibility of the model goes against traditional thinking in statistical modelling, but the over-fitting problem is resisted by the inclusion of the capacity control of the SVM. The capacity control parameter C can be determined using 8-fold cross validation, combined with an automatic procedure, which searches for a local minimum of the validation error.

Another attractive feature of the SVM approach to regression is that it will accommodate a variety of loss functions [9, 10]. In this paper, we employ two, an ε-insensitive and a quadratic loss function. The quadratic loss function gives a solution which is identical to a Gaussian process, and as such this technique may also be employed within a Gaussian process framework. The optimal regression function is given by the minimum of the functional,

$$\Phi(\mathbf{w},\xi^*,\xi)=\frac{1}{2}\|\mathbf{w}\|^2+C\left(\sum_{i=1}^{l}\xi_i+\sum_{i=1}^{l}\xi_i^*\right), \tag{6}$$

where C is the smoothing parameter determined by cross-validation, and ξ, ξ^* are slack variables representing upper and lower errors on the outputs of the model. The resultant optimisation problem for a quadratic loss function is given by,

$$\bar{\beta}=\arg\min_{\beta}\frac{1}{2}\sum_{i=1}^{l}\sum_{j=1}^{l}\beta_i\beta_j K(x_i,x_j)-\sum_{i=1}^{l}\beta_i y_i+\frac{1}{2C}\sum_{i=1}^{l}\beta_i^2, \tag{7}$$

and for an ε-insensitive loss function is given by,

$$\overline{\alpha},\overline{\alpha}^* = \arg\min_{\alpha,\alpha^*}\left\{\begin{array}{l}\frac{1}{2}\sum_{i=1}^{l}\sum_{j=1}^{l}(\alpha_i-\alpha_i^*)(\alpha_j-\alpha_j^*)K(x_i,x_j)\\ -\sum_{i=1}^{l}(\alpha_i-\alpha_i^*)y_i+\sum_{i=1}^{l}(\alpha_i+\alpha_i^*)\varepsilon\end{array}\right\}, \quad (8)$$

with constraints,

$$\begin{array}{ll}0\leq\alpha_i\leq C, & i=1,\ldots,l\\ 0\leq\alpha_i^*\leq C, & i=1,\ldots,l\end{array}. \quad (9)$$

The ANOVA spline kernel has an implicit bias, and hence the equality constraint is not required for either of the two loss functions [5]. Letting $\overline{\beta}_i=(\overline{\alpha}_i-\overline{\alpha}_i^*)$, the model resulting from the solution of either (7) or (8) is given by,

$$m(z)=\sum_{\mathrm{SVs}}\overline{\beta}_i K(x_i,z). \quad (10)$$

In the quadratic case, there will be no sparseness in the support vector expansion and hence all data points will be support vectors. This technique produces a complete ANOVA expansion from which the 2^n components of equation (1) can be extracted, by expanding the kernel in the manner of equation (5). However, some terms in the ANOVA expansion will typically be negligible and the goal is to extract the significant terms, to reflect the underlying model structure. The second stage uses a technique to select the most significant terms from the ANOVA expansion to provide a parsimonious representation.

Stage II : Sparse Basis Selection

The complete ANOVA representation is difficult to interpret. By introducing a technique that extracts a subset of terms, it is possible to provide a model, which is interpretable whilst retaining good approximating capability. The individual ANOVA terms can be extracted from the SVM solution to the first stage by the appropriate sub-kernel. For example, the univariate terms are given by,

$$m_j(z)=\sum_{SVs}\beta_i K_{spline}({}_j x_i,{}_j z), \quad (11)$$

and the bi-variate terms are given by,

$$m_{j,k}(z)=\sum_{SVs}\beta_i K_{spline}({}_j x_i,{}_j z)K_{spline}({}_k x_i,{}_k z), \quad (12)$$

where β_i are the Lagrange multipliers obtained from the ANOVA kernel solution. In this stage we borrow a technique from the wavelet community [3], where it is desirable to select a sparse basis from an overdetermined one. A commonly employed method for determining a sparse representation is to trade off the error in approximation with the sparseness of the representation,

$$\min Error + \lambda Sparsity\ , \tag{13}$$

where λ controls this trade-off. To enforce continuity the technique retains the loss function that was employed in the basis selection stage. The optimisation problem is given by,

$$\min_{c} L(y, \Phi c) + \lambda S(c) \quad \text{subject to} \quad c_i \geq 0, \lambda \geq 0 \tag{14}$$

where L is the loss function, Φ is the ANOVA basis obtained from the first stage, S is the sparseness measure and c is the coefficient vector. The goal in selecting a sparse representation is to minimise the number of non-zero coefficients, c_i. Here non-negativity of the sub-kernel multipliers, c_i, is enforced since the requirement is to simply scale the basis functions (not to invert them). A practical measure of sparseness is a p-norm,

$$S(c) = \|c\|_p\ . \tag{15}$$

As p increases the solution becomes less sparse and the computational complexity of the resulting optimisation problem is relaxed. Ideally a value of p=0, which measures the number of terms in the expansion is attractive. This case is referred to as atomic decomposition [3] and results in a hard combinatorial optimisation problem. Alternatively choosing a value of p=2 produces a straightforward optimisation problem. This case is referred to as the method of frames or ridge regression [3], but crucially the sparseness within the basis is now lost. A good compromise occurs when p=1 producing a sparse solution, with a practical implementation. This norm has been employed in basis-pursuit de-noising [3] to achieve a very similar goal. It is also advantageous in that it produces a Quadratic Program (QP) which can be solved using the same optimiser that is used to optimise the SVM. In the case of an ε-insensitive loss function, the problem simplifies to a Linear Program (LP). Finally, the sparseness parameter, λ, can be selected by choosing a set of values, and selecting the one which gives the nearest loss to the error obtained on the validation sets in the first stage. The solution for a quadratic loss function is given by,

$$\min_{c} \|y - \Phi c\|_2^2 + \lambda \|c\|_1 \quad \text{subject to} \quad c_i \geq 0, \tag{16}$$

$$\min_{c} c^T \Phi^T \Phi c + \left(\lambda \vec{1} - 2 y^T \Phi\right) c \quad \text{subject to} \quad c_i \geq 0, \tag{17}$$

and for an ε-insensitive loss function by,

$$\min_{c} \|y - \Phi c\|_{1,\varepsilon} + \lambda \|c\|_1 \quad \text{subject to} \quad c_i \geq 0, \tag{18}$$

$$\min_{\mathbf{c}} \begin{bmatrix} \lambda\vec{1} \\ \vec{1} \\ \vec{1} \end{bmatrix}^T \begin{bmatrix} c \\ m \\ n \end{bmatrix} \quad \text{subject to} \quad \begin{bmatrix} \Phi & I & -I \\ -\Phi & -I & I \\ -I & 0 & 0 \\ 0 & -I & 0 \\ 0 & 0 & -I \end{bmatrix} \begin{bmatrix} c \\ m \\ n \end{bmatrix} \leq \begin{bmatrix} \varepsilon + y \\ \varepsilon - y \\ 0 \\ 0 \\ 0 \end{bmatrix}. \quad (19)$$

Stage III : Parameter Selection

The first two stages together can be considered as a sparse kernel selection method. In this final stage the sparse kernel representation of stage II is deployed within an SVM to produce a sparse ANOVA model.

MPG DATA MODELLING

The performance of the SUPANOVA approach is demonstrated by application to the problem of modelling automobile miles per gallon data [1]. The MPG data set contains the miles travelled, per gallon of fuel consumed, for various different cars. The input variables measure six characteristics of a car; the number of cylinders (discrete), displacement, horsepower, weight, acceleration and model year (discrete). The goal is to discover a relationship between the MPG and the cars' characteristics. After removing a small number of entries with missing values from the original data set, the experiments were performed using 392 examples, 352 for training and 40 for estimating the generalisation performance. The SUPANOVA algorithm was executed 50 times using quadratic and ε-insensitive loss functions, with the whole data set "randomly" partitioned into the training and test sets.

Figure 2 illustrates one of the 100 models, obtained from the SUPANOVA technique. It can be seen that it has selected 8 interaction terms (bias, 3 univariate, 3 bivariate and one trivariate) from a possible 64 terms. Table 1 demonstrates that the difference in the mean of the estimated generalisation error between a full ANOVA model and a parsimonious ANOVA model is negligible. However, it also demonstrates that the parsimonious kernel has a lower variance and hence suggest it is more robust. These results were corroborated by the quadratic loss function results. Comparing the two different loss functions shows that, for this particular data-set, there is very little performance difference. Inspection of the ANOVA terms selected by the 100 models shows a high consistency, and confirms the robustness of the technique.

Bias 30.4

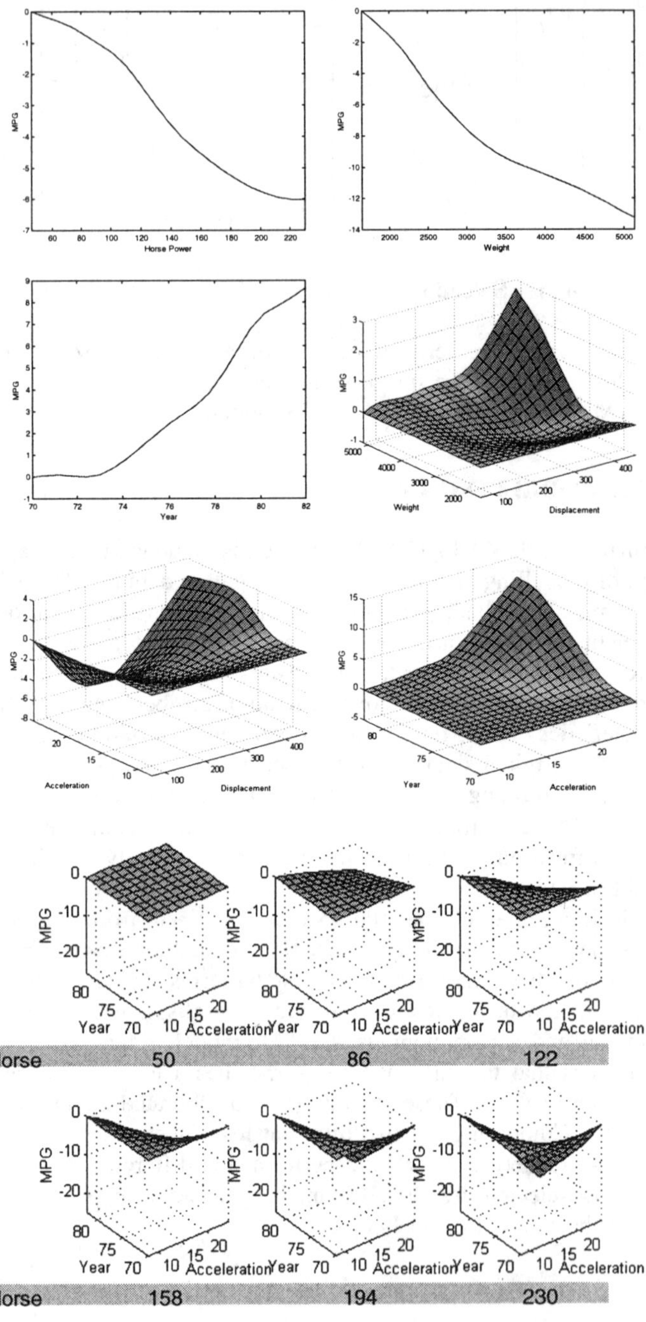

Figure 2. ε-Insensitive AMPG model (1 of 50).

Loss Function		Estimated Generalisation Error					
Training	Testing	Stage I		Stage III		Linear Model	
		Mean	Variance	Mean	Variance	Mean	Variance
Quadratic	Quadratic	6.97	7.39	7.08	6.19	11.4	11.0
ε Insensitive	ε Insensitive	0.48	0.04	0.49	0.03	1.80	0.11
ε Insensitive	Quadratic	7.07	6.52	7.13	6.04	11.72	10.94

Table 1. SUPANOVA Results for MPG Data Set. (ε=2.5)

Future work will investigate the nature of the regularisation term within the SVM regression framework. The regularisation is effectively zeroth order and this may be inappropriate in some circumstances. The dimension of the optimisation problem for the first and last stages is dependent upon the data size. However, the second stage is computed in the feature space and hence the dimension of the optimisation problem is exponentially related to the input dimension. This curse of dimensionality can be addressed by considering a restricted ANOVA expansion, e.g. consider only univariates and bivariates. This will reduce the complexity of the second stage to polynomial in the input dimension. Finally, the technique can just as easily be applied to the SVM classification scenario.

ACKNOWLEDGEMENTS

This work was funded by the UK Engineering and Science Research Council via grant GR/K55110.

REFERENCES

[1] Automobile Dataset, University of California Machine Learning Repository, http://www.ics.uci.edu/~mlearn/MLSummary.html.

[2] R. E. Bellman. *Adaptive Control Processes.* Princeton University Press, Princeton, NJ, 1961.

[3] S. S. Chen. Basis Pursuit. Ph.D. Thesis, Dept. of Statistics, Stanford University, 1995.

[4] J. H. Friedman. Multivariate Adaptive Regression Splines. *The Annals of Statistics*, **19**(1): 1-141, 1991.

[5] F. Girosi. An Equivalence between Sparse Approximation and Support Vector Machines. Technical Report AIM-1606, Artificial Intelligence

Laboratory, Massachusetts Institute of Technology (MIT), Cambridge, Massachusetts, 1997.

[6] S. R. Gunn, M. Brown and K. M. Bossley. Network Performance Assessment for Neurofuzzy Data Modelling. *Lecture Notes in Computer Science*, **1280**:313-323, 1997.

[7] T. Kavli and E. Weyer. On {ASMOD} - an algorithm for building multivariable spline models. *Neural Networks Engineering in Control*, Springer-Verlag, 83-104, 1995.

[8] A. Smola, B. Schölkopf and K.-R. Müller. The connection between regularization operators and support vector kernels. *Neural Networks*, 11:637-649, 1998.

[9] A. Smola, B. Schölkopf and K.-R. Müller. General Cost Functions for Support Vector Regression, *in Proceedings of the Ninth Australian Conference on Neural Networks*, T. Downs and M. Frean and M. Gallagher (eds.), University of Queensland, 79-83, 1998.

[10] A. Smola, B. Schölkopf and K.-R. Müller. Convex cost functions for support vector regression, in L. Niklasson, M. Bod\'en, and T. Ziemke, editors, *Proceedings of the 8th ICANN*, Perspectives in Neural Computing, Berlin, 99 -- 104, 1998. Springer Verlag.

[11] V. Vapnik. *The Nature of Statistical Learning Theory*. Springer-Verlag, New York, 1995.

[12] V. Vapnik, S. Golowich and A. Smola. 1997. Support Vector Method for Function Approximation, Regression Estimation, and Signal Processing. In: M. Mozer, M. Jordan, and T. Petsche (eds.): *Neural Information Processing Systems*, Vol. 9. MIT Press, Cambridge, MA, 1997.

[13] G. Wahba. *Spline Models for Observational Data*. Series in Applied Mathematics, Vol. 59, SIAM, Philadelphia, 1990.

SEQUENTIAL SUPPORT VECTOR MACHINES

Nando de Freitas Marta Milo Philip Clarkson
Mahesan Niranjan Andrew Gee

Cambridge University Engineering Department
Cambridge CB2 1PZ England
Tel: +44 1223 332754
Fax: +44 1223 332662
Email: {jfgf,mm300,prc14,niranjan,ahg}@eng.cam.ac.uk

Abstract. In this paper, we derive an algorithm to train support vector machines sequentially. The algorithm makes use of the Kalman filter and is optimal in a minimum variance framework. It extends the support vector machine paradigm to applications involving real-time and non-stationary signal processing. It also provides a computationally efficient alternative to the problem of quadratic optimisation.

INTRODUCTION

Sequential learning and inference methods are important in many applications involving real-time signal processing, where data arrival is inherently sequential. Furthermore, one might wish to adopt a sequential processing strategy to deal with non-stationarity in signals, so that information from the recent past is given greater weighting than information from the distant past. Increased computational simplicity due to not having to store all the data might also constitute an additional motivating factor for sequential methods. The latter point is of great relevance to the support vector machine (SVM) paradigm.

The theory of support vector machines was first introduced by Vapnik and was developed from the theory of structural risk minimisation [21]. SVMs learn the boundary regions between samples belonging to two classes by mapping the input samples into a high dimensional space, and seeking a separating hyper-plane in this space. The separating hyper-plane is chosen in such a way as to maximise its distance from the closest training samples (a quantity referred to as the *margin*).

The appeal of SVMs is twofold. Firstly they do not require fine tuning of many parameters, and secondly they exhibit a great ability to generalise. In many problems SVMs have been shown to provide better performance than more traditional techniques, such as highly tuned neural networks. In recent

0-7803-5673-X/99/$10.00 © 1999 IEEE

years they have been used in multiple applications (see [3]), from machine vision to text classification. However, there has been no serious attempt to develop efficient sequential support vector algorithms. SVMs are, therefore, limited to tasks where the data has to be processed in batches. Furthermore, the standard way of training SVMs involves storing all the data and solving an extremely computationally demanding quadratic programming problem. In this paper, we surmount these limitations by introducing an optimal algorithm to train SVMs sequentially.

SUPPORT VECTOR MACHINES

This section introduces the theory behind SVMs. Lack of space prohibits a more detailed discussion, but interested readers are referred to [21] for an in depth discussion or to [3] for an excellent short tutorial.

The Linear Case

Suppose we have a set of training data $\{\mathbf{x}_{1:N} \triangleq \{\mathbf{x}_1, \mathbf{x}_2, \ldots, \mathbf{x}_N\} | \mathbf{x}_i \in \mathbb{R}^d \ \forall i\}$ and a set of corresponding labels $\{\mathbf{y}_{1:N} \triangleq \{\mathbf{y}_1, \mathbf{y}_2, \cdots, \mathbf{y}_N\} | \mathbf{y}_i \in \{-1, 1\} \ \forall i\}$ that indicates which of two classes each sample belongs to. Then the hyper-plane $(\mathbf{w} \cdot \mathbf{x}) + \mathbf{b}$ separates the data if and only if

$$\mathbf{y}_i((\mathbf{w} \cdot \mathbf{x}_i) + \mathbf{b}) \geq 1 \quad \forall \quad i \tag{1}$$

To find the optimal separating hyper-plane, we need to find the plane which maximises the distance between the hyper-plane and the closest positive and negative samples. This distance is given by:

$$d(\mathbf{w}, \mathbf{b}) = \frac{1}{||\mathbf{w}||} - \frac{-1}{||\mathbf{w}||} = \frac{2}{||\mathbf{w}||} \tag{2}$$

Therefore our problem is equivalent to minimising $||\mathbf{w}||^2/2$ subject to the constraints expressed in (1). By forming the Lagrangian, this can be translated into a problem of minimising:

$$L_p = \frac{1}{2}||\mathbf{w}||^2 + \sum_i^N \alpha_i - \sum_i^N \alpha_i \mathbf{y}_i (\mathbf{x}_i \cdot \mathbf{w} + \mathbf{b}) \tag{3}$$

with respect to $\mathbf{w}$ and $\mathbf{b}$. Setting the derivative of L_p to zero gives:

$$\mathbf{w} = \sum_i^N \alpha_i \mathbf{y}_i \mathbf{x}_i \qquad \text{and} \qquad \sum_i^N \alpha_i \mathbf{y}_i = 0 \tag{4}$$

The α_i are the Lagrange multipliers; there is one Lagrange multiplier for each training sample. The training samples for which the Lagrange multiplier is non-zero are called *support vectors*, and are such that the equality in equation (1) holds. The samples with Lagrange multipliers of zero value may be removed from the training set without affecting the position of the final hyper-plane. This is a well understood quadratic programming problem, and software packages exist which can find a solution [8]. Such solvers are non-trivial, however, especially in cases involving large training sets [16, 18]. Note that we can dispense with the second condition of equation (4) by setting one of the inputs to 1 and extending the dimension of $\mathbf{w}$.

The Non-Linear Case

The classification framework outlined above is limited to linear separating hyper-planes. SVMs circumvent this problem by mapping the sample points into a higher dimensional space using a non-linear mapping chosen in advance. That is, we choose a map $\Phi : \mathbb{R}^n \mapsto \mathcal{H}$ where the dimension of $\mathcal{H}$ is greater than n. We then seek a separating hyper-plane in the higher dimensional space: this is equivalent to a non-linear separating surface in $\mathbb{R}^n$.

The data only ever appears in our training problem (equations (3) to (4)) in the form of dot products, so in the higher dimensional space we are only dealing with the data in the form $\Phi(\mathbf{x}_i) \cdot \Phi(\mathbf{x}_j)$. If the dimensionality of $\mathcal{H}$ is very large, then this could be difficult, or very computationally expensive to compute. However, if we have a *kernel function* such that $K(\mathbf{x}_i, \mathbf{x}_j) = \Phi(\mathbf{x}_i) \cdot \Phi(\mathbf{x}_j)$, then we can use this in place of $\mathbf{x}_i \cdot \mathbf{x}_j$ everywhere in the optimisation problem, and never need to know explicitly what Φ is. For instance, when a new data sample arrives, the SVM output is given by:

$$\begin{aligned} f(\mathbf{x}_j) &= \mathbf{w} \cdot \Phi(\mathbf{x}_j) + \mathbf{b} \\ &= \sum_{i}^{Ns} \alpha_i \mathbf{y}_i \Phi(\mathbf{x}_i) \cdot \Phi(\mathbf{x}_j) + b \\ &= \sum_{i}^{Ns} \alpha_i \mathbf{y}_i K(\mathbf{x}_i, \mathbf{x}_j) + \mathbf{b} \end{aligned} \tag{5}$$

where N_s denotes the number of support vectors $\mathbf{x}_i$ obtained in the training phase. This trick has been well understood for many decades [14]. Some examples of kernel functions are the polynomial kernel $K(\mathbf{x}_i, \mathbf{x}_j) = (\mathbf{x}_i \cdot \mathbf{x}_j + 1)^p$, the sigmoidal kernel $K(\mathbf{x}_i, \mathbf{x}_j) = \tanh(\kappa \mathbf{x}_i \cdot \mathbf{x}_j - \delta)$ and the Gaussian radial basis function (RBF) kernel $K(\mathbf{x}_i, \mathbf{x}_j) = e^{-\|\mathbf{x}_i - \mathbf{x}_j\|^2 / 2\sigma^2}$. In our experiments, we have adopted RBF kernels.

SEQUENTIAL SUPPORT VECTOR MACHINES

In our sequential SVM paradigm, we re-express equation (5) in terms of a moving window over the data and compute the SVM output each time a new

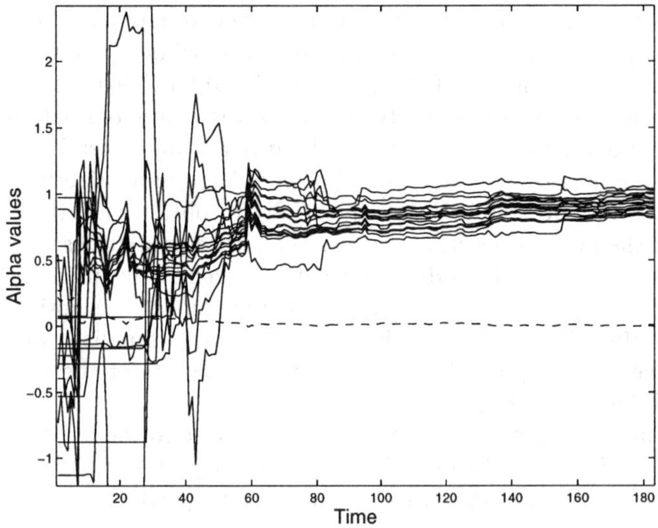

Figure 3: Convergence of the sequential SVM parameters.

sequential Monte Carlo methods [6, 7]. As discussed earlier, it will be possible to improve the sequential SVM (SVEKF) algorithm by adopting noise estimation and sampling strategies. We intend to address these issues in future work. The important thing to note is that the SVEKF algorithm works well on a real data set and that it is very computationally efficient.

Strike price	2925	3025	3125	3225	3325
Trivial	0.0783	0.0611	0.0524	0.0339	0.0205
BS	0.0761	0.0598	0.0534	0.0377	0.0262
SVEKF	0.0572	0.0509	0.0509	0.0302	0.0204
MLP-EKFE	0.0404	0.0366	0.0394	0.0283	0.0150
MLP-HySIR	0.0394	0.0362	0.0404	0.0273	0.0151

Table 1: One-step-ahead prediction errors on call options.

CONCLUSIONS

In this paper, we derived an algorithm to train SVMs sequentially. This extends the SVM application domain to encompass on-line, non-stationary problems. It also becomes possible to introduce specifications for the noise in the data (Gaussian in this paper). In addition, the new algorithm provides a computationally efficient alternative to the quadratic programming problem. From our experimental results on a real data set, we believe that the method can be improved by adopting a sequential sampling strategy and estimating the noise covariances R and Q. In future work, we intend to devise algorithms for estimating the kernels, apply sequential Monte Carlo methods to incorporate any type of noise distributions, estimate the noise parameters and test the algorithm on additional data sets.

ACKNOWLEDGEMENTS

Nando de Freitas is financially supported by two University of the Witwatersrand Merit Scholarships, a Foundation for Research Development Scholarship (South Africa), an ORS Award (UK) and a Trinity College External Research Studentship (Cambridge).

REFERENCES

[1] C Andrieu, J F G de Freitas, and A Doucet. Robust full Bayesian learning for neural networks. Technical Report CUED/F-INFENG/TR 343, Cambridge University, http://svr-www.eng.cam.ac.uk/, April 1999.

[2] C Andrieu, J F G de Freitas, and A Doucet. Sequential MCMC for Bayesian model selection. In *IEEE Higher Order Statistics Workshop*, Ceasarea, Israel, 1999.

[3] Christopher Burges. A Tutorial on Support Vector Machines for Pattern Recognition. *Data Mining and Knowledge Discovery*, 2(2), 1998.

[4] C Cortes and V Vapnik. Support vector networks. *Machine Learning*, 20:273–297, 1995.

[5] J F G de Freitas, M Niranjan, and A H Gee. The EM algorithm and neural networks for nonlinear state space estimation. Technical Report CUED/F-INFENG/TR 313, Cambridge University, http://svr-www.eng.cam.ac.uk/~jfgf, 1998.

[6] J F G de Freitas, M Niranjan, and A H Gee. Hierarchichal Bayesian models for regularisation in sequential learning. To appear in Neural Computation, 1999.

[7] J F G de Freitas, M Niranjan, A H Gee, and A Doucet. Sequential Monte Carlo methods to train neural network models. To appear in Neural Computation, 1999.

[8] R Fletcher. *Practical Methods of Optimization.* John Wiley and Sons, 1987.

[9] P J Green. Reversible jump Markov chain Monte Carlo computation and Bayesian model determination. *Biometrika*, 82:711–732, 1995.

[10] J C Hull. *Options, Futures, and Other Derivative.* Prentice Hall, third edition, 1997.

[11] J M Hutchinson, A W Lo, and T Poggio. A nonparametric approach to pricing and hedging derivative securities via learning networks. *The Journal of Finance*, 49(3):851–889, 1994.

[12] R E Kalman and R S Bucy. New results in linear filtering and prediction theory. *Transactions of the ASME (Jounal of Basic Engineering)*, 83D:95–108, 1961.

[13] G Kitagawa. Monte Carlo filter and smoother for non-Gaussian nonlinear state space models. *Journal of Computational and Graphical Statistics*, 5:1–25, 1996.

[14] J Mercer. Functions of positive and negative type and their connection with the theory of integral equations. *Philosophical Transactions of the Royal Society*, A:415–446, 1909.

[15] M Niranjan. Sequential tracking in pricing financial options using model based and neural network approaches. In M C Mozer, M I Jordan, and T Petsche, editors, *Advances in Neural Information Processing Systems*, pages 960–966, 1996.

[16] Edgar Osuna. *Support Vector Machines: Training and Applications.* PhD thesis, Department of Electrical Engineering and Computer Science, Massachusetts Institute of Technology, May 1998.

[17] J Platt. A resource allocating network for function interpolation. *Neural Computation*, 3:213–225, 1991.

[18] J Platt. Fast training of support vector machines using sequential minimal optimization. In B Schölkopf, C J C Burges, and A J Smola, editors, *Advances in Kernel Methods – Support Vector Learning*, pages 185–208, MIT Press, 1999.

[19] S Roweis and Z Ghahramani. A unifying review of linear Gaussian models. *Neural Computation*, 11(2):305–345, 1999.

[20] T.-T. Frieß, N Cristianini, and C Campbell. The kernel adatron: A fast and simple learning procedure for support vector machines. In J Shavlik, editor, *Fifteenth International Conference on Machine Learning*, Morgan Kaufmann, San Francisco, 1998.

[21] V Vapnik. *The Nature of Statistical Learning Theory.* Springer-Verlag, New York, 1995.

[22] M West and J Harrison. *Bayesian Forecasting and Dynamic Linear Models.* Springer, 1996.

FISHER DISCRIMINANT ANALYSIS WITH KERNELS

Sebastian Mika[†], Gunnar Rätsch[†], Jason Weston,[‡]
Bernhard Schölkopf[†], and Klaus-Robert Müller[†]

†GMD FIRST, Rudower Chaussee 5, 12489 Berlin, Germany
‡Royal Holloway, University of London, Egham, Surrey, UK
{mika, raetsch, bs, klaus}@first.gmd.de, jasonw@dcs.rhbnc.ac.uk

Abstract. A non–linear classification technique based on Fisher's discriminant is proposed. The main ingredient is the kernel trick which allows the efficient computation of Fisher discriminant in feature space. The linear classification in feature space corresponds to a (powerful) non–linear decision function in input space. Large scale simulations demonstrate the competitiveness of our approach.

DISCRIMINANT ANALYSIS

In classification and other data analytic tasks it is often necessary to utilize pre–processing on the data before applying the algorithm at hand and it is common to first extract features suitable for the task to solve.

Feature extraction for classification differs significantly from feature extraction for describing data. For example PCA finds directions which have minimal reconstruction error by describing as much variance of the data as possible with m orthogonal directions. Considering the first directions they need not (and in practice often will not) reveal the class structure that we need for proper classification. Discriminant analysis addresses the following question: Given a data set with two classes, say, which is the best feature or feature set (either linear or non–linear) to discriminate the two classes? Classical approaches tackle this question by starting with the (theoretically) optimal Bayes classifier and, by assuming normal distributions for the classes, standard algorithms like quadratic or linear discriminant analysis, among them the famous Fisher discriminant, can be derived (e.g. [5]). Of course any other model different from a Gaussian for the class distributions could be assumed, this, however, often sacrifices the simple closed form solution. Several modifications towards more general features have been proposed (e.g. [6]); for an introduction and review on existing methods see e.g. [3, 5, 8, 11].

In this work we propose to use the kernel idea [1], originally applied in Support Vector Machines [19, 14]), kernel PCA [16] and other kernel based algorithms (cf. [14]) to define a non–linear generalization of Fisher's discriminant. Our method uses kernel feature spaces yielding a highly flexible

0-7803-5673-X/99/$10.00 © 1999 IEEE

algorithm which turns out to be competitive with Support Vector Machines.

Note that there exists a variety of methods called *Kernel Discriminant Analysis* [8]. Most of them aim at replacing the parametric estimate of class conditional distributions by a non–parametric kernel estimate. Even if our approach might be viewed in this way too, it is important to note that it goes one step further by interpreting the kernel as a dot–product in another space. This allows a theoretically sound interpretation together with an appealing closed form solution.

In the following we will first review Fisher's discriminant, apply the kernel trick, then report classification results and finally present our conclusions. In this paper we will focus on two–class problems only and discriminants linear in the feature space.

FISHER'S LINEAR DISCRIMINANT

Let $\mathcal{X}_1 = \{\boldsymbol{x}_1^1, \ldots, \boldsymbol{x}_{\ell_1}^1\}$ and $\mathcal{X}_2 = \{\boldsymbol{x}_1^2, \ldots, \boldsymbol{x}_{\ell_2}^2\}$ be samples from two different classes and with some abuse of notation $\mathcal{X} = \mathcal{X}_1 \cup \mathcal{X}_2 = \{\boldsymbol{x}_1, \ldots, \boldsymbol{x}_\ell\}$. Fisher's linear discriminant is given by the vector $\boldsymbol{w}$ which maximizes

$$J(\boldsymbol{w}) = \frac{\boldsymbol{w}^T S_B \boldsymbol{w}}{\boldsymbol{w}^T S_W \boldsymbol{w}} \tag{1}$$

where

$$S_B := (\boldsymbol{m}_1 - \boldsymbol{m}_2)(\boldsymbol{m}_1 - \boldsymbol{m}_2)^T \text{ and} \tag{2}$$

$$S_W := \sum_{i=1,2} \sum_{\boldsymbol{x} \in \mathcal{X}_i} (\boldsymbol{x} - \boldsymbol{m}_i)(\boldsymbol{x} - \boldsymbol{m}_i)^T \tag{3}$$

are the between and within class scatter matrices respectively and $\boldsymbol{m}_i$ is defined by $\boldsymbol{m}_i := \frac{1}{\ell_i} \sum_{j=1}^{\ell_i} \boldsymbol{x}_j^i$. The intuition behind maximizing $J(\boldsymbol{w})$ is to find a direction which maximizes the projected class means (the numerator) while minimizing the classes variance in this direction (the denominator). But there is also a well known statistical way to motivate (1):

Connection to the optimal linear Bayes classifier: The optimal Bayes classifier compares the a posteriori probabilities of all classes and assigns a pattern to the class with the maximal probability (e.g. [5]). However, the a-posteriori probabilities are usually unknown and have to be estimated from a finite sample. For most classes of distributions this is a difficult task and often it is impossible to get a closed form estimate. However, by assuming normal distributions for all classes, one arrives at quadratic discriminant analysis (which essentially measures the Mahalanobis distance of a pattern towards the class center). Simplifying the problem even further and assuming equal covariance structure for all classes, quadratic discriminant analysis becomes linear. For two–class problems it is easy to show that the vector $\boldsymbol{w}$ maximizing (1) is in the same direction as the discriminant in the corresponding Bayes optimal classifier. Although relying on heavy assumptions which are not true in many applications, Fisher's linear discriminant has proven very powerful. One reason is certainly that a linear model is rather robust against noise and most likely will not overfit. Crucial, however, is the estimation of the scatter matrices, which might be highly biased. Using simple "plug-in" estimates as in (2) when the number of samples is small compared to the dimensionality will result in a high variability. Different ways to deal with such a situation by regularization have been proposed (e.g. [4, 7]) and we will return to this topic later.

FISHER'S DISCRIMINANT IN THE FEATURE SPACE

Clearly, for most real-world data a linear discriminant is not complex enough. To increase the expressiveness of the discriminant we could either try to use more sophisticated distributions in modeling the optimal Bayes classifier or look for non-linear directions (or both). As pointed out before, assuming general distributions will cause trouble. Here we restrict ourselves to finding non-linear directions by first mapping the data non-linearly into some feature space $\mathcal{F}$ and computing Fisher's linear discriminant there, thus thus implicitly yielding a non-linear discriminant in input space.

Let Φ be a non-linear mapping to some feature space $\mathcal{F}$. To find the linear discriminant in $\mathcal{F}$ we need to maximize

$$J(\boldsymbol{w}) = \frac{\boldsymbol{w}^T S_B^\Phi \boldsymbol{w}}{\boldsymbol{w}^T S_W^\Phi \boldsymbol{w}} \tag{4}$$

where now $\boldsymbol{w} \in \mathcal{F}$ and S_B^Φ and S_W^Φ are the corresponding matrices in $\mathcal{F}$, i.e.

$$\begin{aligned} S_B^\Phi &:= (\boldsymbol{m}_1^\Phi - \boldsymbol{m}_2^\Phi)(\boldsymbol{m}_1^\Phi - \boldsymbol{m}_2^\Phi)^T \text{ and} \\ S_W^\Phi &:= \sum_{i=1,2} \sum_{\boldsymbol{x} \in \mathcal{X}_i} (\Phi(\boldsymbol{x}) - \boldsymbol{m}_i^\Phi)(\Phi(\boldsymbol{x}) - \boldsymbol{m}_i^\Phi)^T \end{aligned}$$

with $\boldsymbol{m}_i^\Phi := \frac{1}{\ell_i} \sum_{j=1}^{\ell_i} \Phi(\boldsymbol{x}_j^i)$.

Introducing kernel functions: Clearly, if $\mathcal{F}$ is very high– or even infinitely dimensional this will be impossible to solve directly. To overcome this limitation we use the same trick as in Kernel PCA [16] or Support Vector Machines. Instead of mapping the data explicitely we seek a formulation of the algorithm which uses only dot-products $(\Phi(\boldsymbol{x}) \cdot \Phi(\boldsymbol{y}))$ of the training patterns. As we are then able to compute these dot-products efficiently we can solve the original problem without ever mapping explicitely to $\mathcal{F}$. This can be achieved using Mercer kernels (e.g. [12]): these kernels $\mathrm{k}(\boldsymbol{x}, \boldsymbol{y})$ compute a dot-product in some feature space $\mathcal{F}$, i.e. $\mathrm{k}(\boldsymbol{x}, \boldsymbol{y}) = (\Phi(\boldsymbol{x}) \cdot \Phi(\boldsymbol{y}))$. Possible choices for k which have proven useful e.g. in Support Vector machines or Kernel PCA are Gaussian RBF, $\mathrm{k}(\boldsymbol{x}, \boldsymbol{y}) = \exp(-\|\boldsymbol{x} - \boldsymbol{y}\|^2/c)$, or polynomial kernels, $\mathrm{k}(\boldsymbol{x}, \boldsymbol{y}) = (\boldsymbol{x} \cdot \boldsymbol{y})^d$, for some positive constants c and d respectively.

To find Fisher's discriminant in the feature space $\mathcal{F}$, we first need a formulation of (4) in terms of only *dot products* of input patterns which we then replace by some kernel function. From the theory of reproducing kernels we know that any solution $\boldsymbol{w} \in \mathcal{F}$ must lie in the span of *all* training samples in $\mathcal{F}$. Therefore we can find an expansion for $\boldsymbol{w}$ of the form

$$\boldsymbol{w} = \sum_{i=1}^{\ell} \alpha_i \Phi(\boldsymbol{x}_i) \tag{5}$$

Using the expansion (5) and the definition of $\boldsymbol{m}_i^\Phi$ we write

$$\begin{aligned} \boldsymbol{w}^T \boldsymbol{m}_i^\Phi &= \frac{1}{\ell_i} \sum_{j=1}^{\ell} \sum_{k=1}^{\ell_i} \alpha_j \, \mathrm{k}(\boldsymbol{x}_j, \boldsymbol{x}_k^i) \\ &= \boldsymbol{\alpha}^T \boldsymbol{M}_i \end{aligned} \tag{6}$$

where we defined $(\boldsymbol{M}_i)_j := \frac{1}{\ell_i} \sum_{k=1}^{\ell_i} \mathrm{k}(\boldsymbol{x}_j, \boldsymbol{x}_k^i)$ and replaced the dot products by the kernel function. Now consider the numerator of (4). Be using the definition of S_B^Φ and (6) it can be rewritten as

$$\boldsymbol{w}^T S_B^\Phi \boldsymbol{w} = \boldsymbol{\alpha}^T M \boldsymbol{\alpha} \tag{7}$$

where $M := (\boldsymbol{M}_1 - \boldsymbol{M}_2)(\boldsymbol{M}_1 - \boldsymbol{M}_2)^T$. Considering the denominator, using (5), the definition of $\boldsymbol{m}_i^\Phi$ and a similar transformation as in (7) we find:

$$\boldsymbol{w}^T S_W^\Phi \boldsymbol{w} = \boldsymbol{\alpha}^T N \boldsymbol{\alpha} \tag{8}$$

where we set $N := \sum_{j=1,2} K_j (I - \mathbf{1}_{\ell_j}) K_j^T$, K_j is a $\ell \times \ell_j$ matrix with $(K_j)_{nm} := \mathrm{k}(\boldsymbol{x}_n, \boldsymbol{x}_m^j)$ (this is the kernel matrix for class j), I is the identity and $\mathbf{1}_{\ell_j}$ the matrix with all entries $1/\ell_j$.

Combining (7) and (8) we can find Fisher's linear discriminant in $\mathcal{F}$ by maximizing

$$J(\boldsymbol{\alpha}) = \frac{\boldsymbol{\alpha}^T M \boldsymbol{\alpha}}{\boldsymbol{\alpha}^T N \boldsymbol{\alpha}}. \tag{9}$$

This problem can be solved (analogously to the algorithm in the input space) by finding the leading eigenvector of $N^{-1}M$. We will call this approach (non–linear) *Kernel Fisher Discriminant* (KFD). The projection of a new pattern $\boldsymbol{x}$ onto $\boldsymbol{w}$ is given by

$$(\boldsymbol{w} \cdot \Phi(\boldsymbol{x})) = \sum_{i=1}^{\ell} \alpha_i \, \mathrm{k}(\boldsymbol{x}_i, \boldsymbol{x}). \tag{10}$$

Numerical issues and regularization: Obviously, the proposed setting is ill–posed: we are estimating ℓ dimensional covariance structures from ℓ samples. Besides numerical problems which cause the matrix N not to be positive, we need a way of capacity control in $\mathcal{F}$. To this end, we simply add a multiple of the identity matrix to N, i.e. replace N by N_μ where

$$N_\mu := N + \mu I. \tag{11}$$

This can be viewed in different ways: (i) it clearly makes the problem numerically more stable, as for μ large enough N_μ will become positive definite; (ii) it can be seen in analogy to [4], decreasing the bias in sample based estimation of eigenvalues; (iii) it imposes a regularization on $||\boldsymbol{\alpha}||^2$ (remember that we are maximizing (9)), favoring solutions with small expansion coefficients. Although the real influence in this setting of the regularization is not yet fully understood, it shows connections to those used in Support Vector Machines (see also [14]). Furthermore, one might use other regularization type additives to N, e.g. penalizing $||\boldsymbol{w}||^2$ in analogy to SVM (by adding the full kernel matrix $K_{ij} = \mathrm{k}(\boldsymbol{x}_i, \boldsymbol{x}_j)$).

EXPERIMENTS

Figure 1 shows an illustrative comparison of the feature found by KFD and the first and second (non–linear) feature found by Kernel PCA [16] on a toy data set. For both we used a polynomial kernel of degree two and for KFD the regularized within class scatter (11) where $\mu = 10^{-3}$. Depicted are the two classes (crosses and dots), the feature value (indicated by grey level) and contour lines of identical feature value. Each class consists of two noisy parabolic shapes mirrored at the x and y axis respectively. We see, that the KFD feature discriminates the two classes in a nearly optimal way, whereas the Kernel PCA features, albeit describing interesting properties of the data set, do not separate the two classes well (although higher order Kernel PCA features might be discriminating, too).

To evaluate the performance of our new approach we performed an extensive comparison to other state-of-the-art classifiers. The experimental setup was chosen in analogy to [10] and we compared the Kernel Fisher Discriminant to AdaBoost, regularized AdaBoost (also [10]) and Support Vector Machines (with Gaussian kernel). For KFD we used Gaussian kernels, too, and the regularized within-class scatter from (11). After the optimal direction $\boldsymbol{w} \in \mathcal{F}$ was found, we computed projections onto it by using (10). To estimate an optimal threshold on the extracted feature, one may use any classification technique, e.g. as simple as fitting a sigmoid [9]. Here we used a linear Support Vector Machine (which is optimized by gradient descent as we

Figure 1: Comparison of feature found by KFD (left) and those found by Kernel PCA: first (middle) and second (right); details see text.

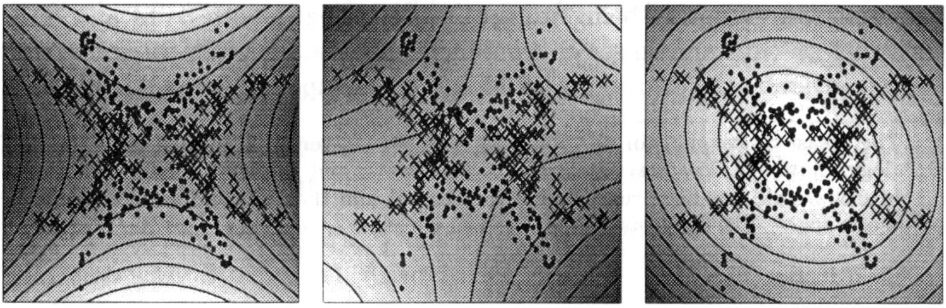

Parameters of the LTE are fully specified by the Wiener filter solution when a priori knowledge of the channel is available. In practice, the channel statistics are unknown and the parameters are estimated by the least-mean-square (LMS) algorithm. The LTE is easy to train but suffers from poor performance in severe channel conditions compared to nonlinear equalizers based on neural networks, such as multilayer perceptrons [10] or radial basis functions (RBFs)[4]. Unfortunately, neural networks typically require a prohibitive amount of training data before exhibiting good generalization performance. This is a serious drawback in case of time-varying channels because the training sequence needs to be periodically transmitted. In order to limit the side information required for training, there is a need for constructing nonlinear equalizers that generalize well using a restricted amount of training data. A similar problem arises in machine learning [12], [13]. A classifier that generalizes well in a high-dimensional space is the so-called *optimal hyperplane.* It provides the largest distance or *margin* from the separating hyperplane to the closest training vector. The basic idea behind the Support Vector Machines (SVMs) is first to transform the original data onto a high-dimensional space by some fixed a priori mapping, and then find the optimal hyperplane in this feature space [2], [6]. This paper investigates the application of Support Vector machines (SVMs) with Gaussian kernels for the equalization of communication systems corrupted with additive white Gaussian noise, intersymbol and co-channel interference.

SYSTEM DESCRIPTION

The discrete-time model for the digital communication system considered here is depicted in Fig.1. In this model [3], $H_0(z)$ represents the transfer function of the channel of interest. In addition, there exists p interfering co-channels with transfer functions $H_i(z)$, $i = 1 \cdots p$. Channels are traditionally modeled by finite impulse response filters

$$H_i(z) = \sum_{j=0}^{n_i} h_{ij} z^{-j}, \quad i = 0 \cdots p$$

where, n_i and h_{ij} is the filter length and the jth impulse response component for the ith channel, respectively. It is further assumed that the transmitted binary symbols $s_i = \pm 1$ are independently, identically distributed random variables and e is an additive white Gaussian noise with zero mean and variance σ_e^2. At time k, the channel observation $x(k)$ is the sum of the signal of interest $\hat{y}(k)$, the interfering signal $u(k)$ and the noise $e(k)$. The signal to noise ratio (SNR), signal to interference ratio (SIR) and signal to interference noise ratio are usually defined as

$$\text{SNR} = \sigma_s^2/\sigma_e^2 \qquad \text{SIR} = \sigma_s^2/\sigma_{co}^2 \qquad \text{SINR} = \sigma_s^2/(\sigma_e^2 + \sigma_{co}^2)$$

where σ_s^2 and σ_{co}^2 are the signal power and the co-channel signal power, respectively.
The task of the equalizer is then to estimate the transmitted symbol $s_0(k)$ from the observation vector $\mathbf{x} = (x(k), x(k-1) \cdots x(k-m+1))$. For training the equalizer, a desired signal $s_0(k-d)$ including the equalizer delay d is available at each time step k of the training sequence. As an example, a channel corrupted with a single interfering co-channel is considered in Fig. 2.

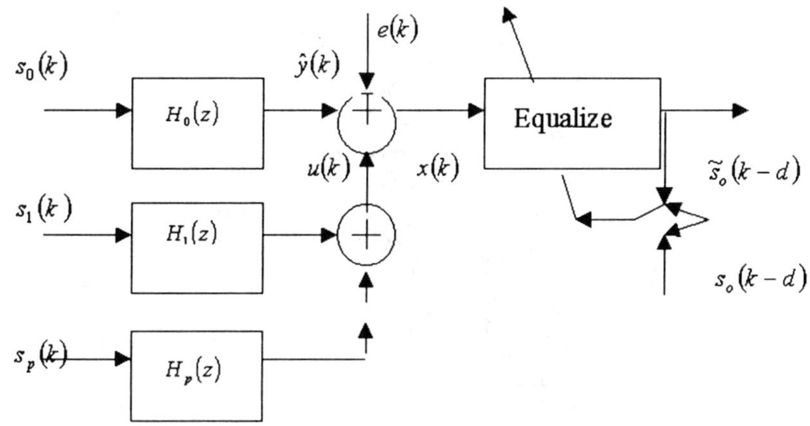

Figure 1: Discrete-time model of data transmission system corrupted with co-channel interference. At time k, the channel observation $x(k)$ is the sum of the signal $\hat{y}(k)$, the interfering signal $u(k)$ and the noise $e(k)$.

RBF EQUALIZERS

A radial basis function (RBF) network is a two-layer network with a single hidden layer. Each neuron i within the hidden layer computes the Euclidean distance between a center vector $\mathbf{c_i}$ and the input vector $\mathbf{x} = (x(k), x(k-1) \cdots x(k-m+1))$. The result is then passed through a Gaussian function $\phi_i = \exp(-||\mathbf{x}-\mathbf{c_i}||^2/\sigma_i^2)$ of width σ_i and the output is obtained by a weighted linear combination $f(\mathbf{x}) = \sum_i w_i\phi_i$. Provided a priori knowledge of the channel, the parameters $\mathbf{c_i}$, σ_i and w_i can be chosen so that the RBF network implements the optimal Bayesian equalizer [4]. Unfortunately, this requires a correct estimate of the channel order. In addition, the resulting RBF equalizer can have a prohibitive number of neurons for large m, as shown in [3]. The performance of the RBF equalizer depends to a great extent on locating the centers at the desired channel states. If this is achieved, the full effects of the co-channel interference can be taken into account and the

RBF equalizer can discriminate the signal of interest from the interfering signals and the noise [3]. Center locations are usually estimated by using supervised or unsupervised clustering algorithms [4] which typically requires a prohibitive amount of training data before exhibiting good generalization performance.

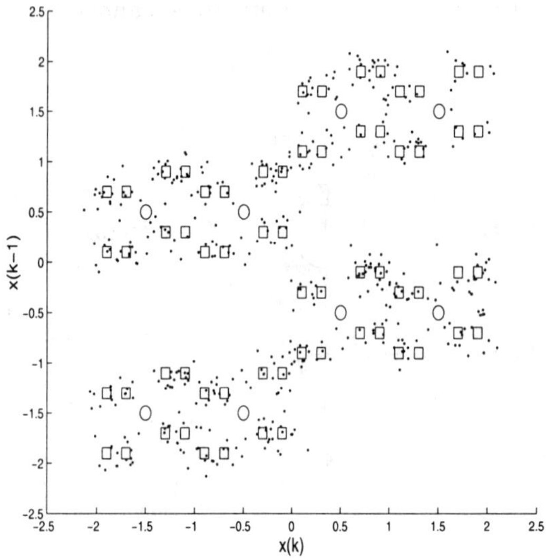

Figure 2: Example of channel output vectors and signal states for a channel $H_0(z) = 0.5 + z^{-1}$ corrupted by a single interfering co-channel $H_1(z) = 0.1 + 0.3z^{-1}$. The signal to noise ratio is 20 dB. Data points represent 500 observation vectors $(x(k), x(k-1))$, $k = 1 \cdots 500$. The locations of the signal states for channel H_0 are indicated by circles. The interfering co-channel H_1 causes an increase in the number of signal states, represented by squares. For lower signal to interference ratios, the distribution of the observation vectors spreads more widely around the signal states [3].

SVM EQUALIZERS

Let us consider a training set of n labeled observations $(\mathbf{x}_i, y_i)$, $i = 1 \cdots n$, where $\mathbf{x}_i \in R^d$ and $y_i \in \{-1, +1\}$. If the training set is linearly separable, then there exist a weight vector $\mathbf{w}$ and a bias b such that $y_i = \text{sgn}(\mathbf{w}.\mathbf{x}_i + b)$ for all i, where $\text{sgn}(a) = +1$ if $a \geq 0$ and -1 otherwise. The optimal hyperplane is the one that maximizes the margin or, equivalently minimizes the norm of $\mathbf{w}$ subject to $y_i(\mathbf{w}.\mathbf{x}_i + b) \geq 1$ for all i, see e.g. [12]. The solution of this optimization problem is given by the saddle point of the Lagrangian

$$\mathcal{L}_P = \frac{1}{2}||\mathbf{w}||^2 - \sum_i \alpha_i y_i(\mathbf{w}.\mathbf{x}_i + b) + \sum_i \alpha_i \quad (1)$$

where the $\alpha_i \geq 0$ are the Lagrange multipliers taking into account the inequality constraints. At the saddle point, the solution should satisfy the Karush-Kuhn-Tucker (KKT) conditions

$$\begin{aligned} \frac{\partial \mathcal{L}_P}{\partial \mathbf{w}} &= 0 \Rightarrow \mathbf{w} = \sum_i \alpha_i y_i \mathbf{x}_i \\ \frac{\partial \mathcal{L}_P}{\partial b} &= 0 \Rightarrow \sum_i \alpha_i y_i = 0 \end{aligned} \tag{2}$$

The weight vector $\mathbf{w}$ given by (2) is a linear combination of the vectors of the training set. Those vectors $\mathbf{x}_i$, for which α_i are nonzero, are called support vectors. Putting expressions of $\mathbf{w}$ into (1), and taking into account the other KKT condition, one obtains

$$\mathcal{L}_D = \sum_i \alpha_i - \frac{1}{2} \sum_i \sum_j \alpha_i \alpha_j y_i y_j (\mathbf{x}_i . \mathbf{x}_j) \tag{3}$$

which now has to be maximized with respect to the α_i under the constraints $\alpha_i \geq 0$. Maximizing (3) is a quadratic programming problem with linear constraints.

The optimal hyperplane is only suitable for the linearly separable case. Possible classification errors are taking into account in the optimization problem by introducing slack variables $\xi_i \geq 0$. The optimization problem now consists of finding the minimum of $\frac{1}{2}||\mathbf{w}||^2 + C \sum_i \xi_i$ subject to

$$y_i(\mathbf{w}.\mathbf{x}_i + b) \geq 1 - \xi_i$$

where C is a parameter to be chosen by the user, a larger value corresponding to assigning a higher penalty to errors. It can be shown that the quadratic programming problem (3) remains unchanged except for the constraints which now become $0 \leq \alpha_i \leq C$ (see [6] for details). The basic idea behind the Support Vector Machines (SVMs) is first to transform the original data onto a high-dimensional space by some fixed a priori mapping, and then find the optimal hyperplane in this feature space [2], [6]. The mapping Φ onto the feature space is obtained by a given kernel representation $K(\mathbf{x}_i, \mathbf{x}_j)$ of the inner product $\Phi(\mathbf{x}_i).\Phi(\mathbf{x}_j)$ [1]. In practice, we used a Gaussian kernel $K(\mathbf{x}_i, \mathbf{x}_j) = \exp(-||\mathbf{x_i} - \mathbf{x_j}||^2/\sigma^2)$. For the optimization problem, we used an active set strategy (also known as a projection method) [11].

SIMULATION STUDY

The channel used for our simulations has the form [5]

$$H_0(z) = -0.2052 - 0.5131z^{-1} + 0.7183z^{-2} + 0.3695z^{-3} + 0.2052z^{-4}$$

and the co-channel transfer function considered was [3]

$$H_1(z) = \lambda(0.6 + 0.8z^{-1})$$

for $\lambda > 0$. This way, the SIR and SINR can be modified by varying the factor λ. The bit error rate (BER) performance of the different equalizers was estimated using 10^5 to 10^7 samples, depending on the SINR value. The delay of the equalizers was set to $d = 2$. Two experimental conditions were used. 1) The SIR was set to 24 dB and the noise power was changed to produce different SINR ratios. 2) The SNR was set to 24 dB by fixing the noise power and the interfering signal power was changed by choosing different values for λ.

The performance obtained using the Wiener filter is the best that a linear transversal equalizer can attain. The BER performance of the Wiener filter is shown in Fig. 3 as a function of the order of the filter for different SINR values and for the two experimental conditions described above (SIR=24dB and SNR=24dB). These results indicate that the BER performance does not improve when the filter order is higher than 5. This is known as the problem of noise enhancement [3]. In the following, the order of the Wiener filter was set to $m = 5$.

We now compare the performance of the Wiener filter with the one of RBF and SVM equalizers under a variety of SINR conditions. The order of the equalizer was set to $m = 5$ for the Wiener filter and $m = 4$ for the nonlinear equalizers. The amount of training data (640 samples) is small compared to the number of desired signal states, equal to 256. It should be noted that the optimal Bayesian solution would be very costly (an RBF network with 8192 centers), and was not investigated in our study. Instead, the number of centers of the RBF equalizer was set to the number of desired signal states. Training the RBF equalizer was done in the following way. The center locations were first determined by a k-means clustering procedure [7] and the Gaussian widths were set as in [3]. Then, the weights were trained with the LMS algorithm. Parameters for the SVM equalizer were determined empirically. We experimented with various C and σ^2 but we only report here the results for $C = \infty$ and $\sigma^2 = 0.5$ since they performed best overall on the data tested. The BER performance for linear and nonlinear equalizers is reported in Fig. 4 for the two experimental conditions. These results show that both non-linear equalizers outperform the Wiener filter, particularly for high SINRs. In addition, the BER value attained with an SVM equalizer is lower to that of an RBF equalizer, particularily for high SINRs. For lower values of SINR, the larger C value was penalizing too much the errors. The

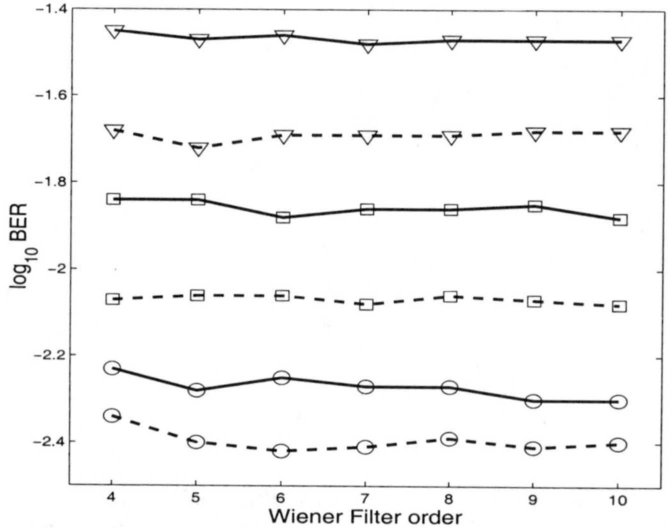

Figure 3: Performance in terms of Bit Error Rate (BER) for the Wiener filter as a function of the filter order for different values of SINR = 18 dB (circles), 14 dB (squares) and 10 dB (triangles). These performance have been estimated on the two experimental conditions : constant SIR=24dB (full lines) and constant SNR=24dB (dashed lines).

value for C could be chosen optimally by estimating both the empirical risk and the VC dimension on the training set [13].

The BER performance for the RBF and SVM equalizers is reported in Fig. 5 as a function of the length of the training sequence for the two experimental conditions. It shows that the SVM equalizer trained with 100 samples yields better performance than the RBF equalizer trained with 400 samples. Therefore, in case of time-varying channels, it is envisaged that the length of the training sequence which needs to be periodically transmitted would be reduced by SVM equalizers.

CONCLUSION

A comparative study of BER performance among three techniques for channel equalization in hostile environments has been given. Our simulations showed the superiority in overcoming co-channel interference of SVM equalizers over RBF and linear equalizers. SVMs provide a better way of choosing the number and locations of RBF centers using a limited amount of training data as was the case in our equalization problem. Our future work will focus on extending the current study to more general channel models and to online training algorithms such as, for instance, the recently proposed kernel adatron algorithm [9].

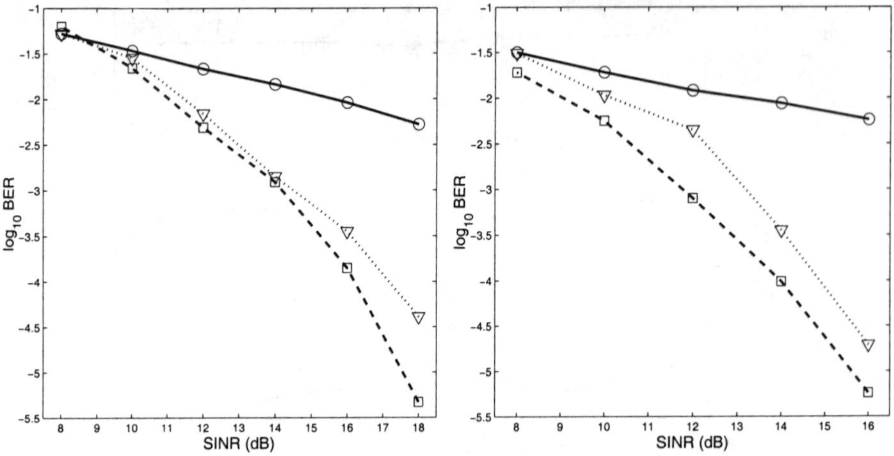

Figure 4: BER performance for the Wiener filter (full lines), the RBF equalizer (dotted lines) and the SVM equalizer (dashed lines) for different values of SINR. Figures on the left and right are for the two experimental conditions SIR=24dB and SNR=24dB, respectively.

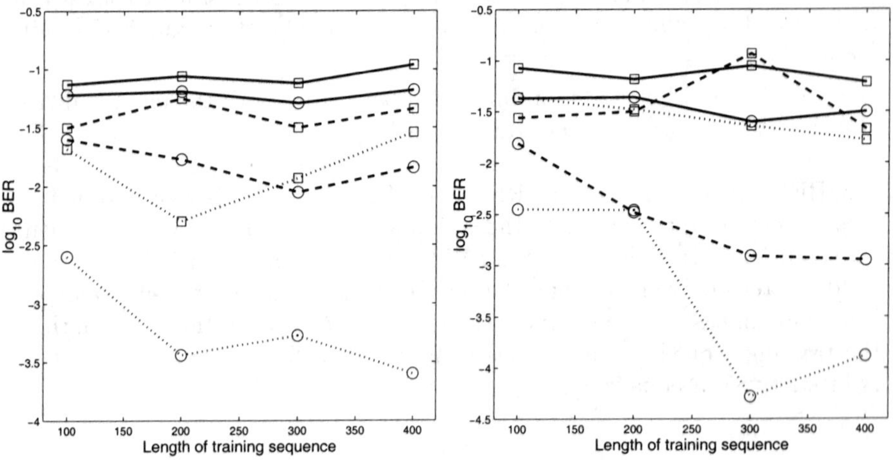

Figure 5: BER performance for the RBF equalizer (squares) and the SVM equalizer (circles) as a function of the length of the training sequence for different values of SINR equal to 8 dB (full lines), 12 dB (dashed lines) and 16 dB (dotted lines). Figures on the left and right are for the two experimental conditions SIR=24dB and SNR=24dB, respectively.

Acknowledgement

The authors wish to thank the reviewers for their valuable comments on the manuscript.

REFERENCES

[1] M. Aizerman, B. E.M. and L. Rozonoer, "Theoretical foundations of the potential function method in pattern recognition learning," **Automation and remote control**, vol. 25, pp. 821–837, 1964.

[2] B. Boser, I. Guyon and V. Vapnik, "A training algorithm for optimal margin classifiers," in **Fith Annual Workshop on Computational Learning Theory, Pittsburgh ACM,**, 1992, pp. 144–152.

[3] S. Chen and B. Mulgrew, "Overcoming co-channel interference using an adaptive radial basis function equaliser," in **EURASIP Signal Processing**, 1992, vol. 28, pp. 91–107.

[4] S. Chen, B. Mulgrew and P. M. Grant, "A clustering technique for digital communications channel equalization using radial basis function networks," **IEEE Transactions on Neural Networks**, vol. 4, no. 4, pp. 570–578, 1993.

[5] S. Chen, B. Mulgrew and S. McLaughlin, "Adaptive bayesian equalizer with decision feedback," **IEEE Transactions on Signal Processing**, vol. 41, no. 9, pp. 2918–2927, 1993.

[6] C. Cortes and V. Vapnik, "Support vector networks," **Machine Learning**, vol. 20, pp. 1–25, 1995.

[7] R. O. Duda and P. E. Hart, **Pattern Classification and Scene Analysis**, John Wiley & Sons, 1973.

[8] G. Forney, "Maximum-likelihood sequence estimation of digital sequences in the presence of intersymbol interference," **IEEE Trans. Inform. Theory**, vol. IT-18, no. 3, pp. 363–378, 1972.

[9] T.-T. Friess, N. Cristianini and C. Campbell, "The kernel adatron algorithm: a fast and simple learning procedure for support vector machines," in **15th Intl. Conf. Machine Learning, Morgan Kaufman Publishers**, 1998.

[10] G. Gibson, S. Siu and C. Cowan, "The application of nonlinear structures to the reconstruction of binary signals," **IEEE Transactions on Signal processing**, vol. 39, no. 8, pp. 1877–1884, 1991.

[11] P. E. Gill, W. Murray and M. H. Wright, **Practical Optimization**, Academic Press, 1981.

[12] V. Vapnik, **The Nature of Statistical Learning Theory**, New York: Springer-Verlag, 1995.

[13] V. Vapnik, **Statistical learning theory**, John Wiley & Sons, Inc, 1998.

A GAUSSIANITY MEASURE FOR BLIND SOURCE SEPARATION INSENSITIVE TO THE SIGN OF KURTOSIS

Hsiao-Chun Wu, Jose C. Principe
Computational NeuroEngineering Laboratory
Department of Electrical and Computer Engineering
NEB 486, New Engineering Building
University of Florida
Gainesville, FL 32611
Tel: (352) 392-2662
Fax: (352) 392-0044
wu@cnel.ufl.edu, principe@cnel.ufl.edu

ABSTRACT

Various existing criteria to characterize the statistical independence are applied in blind source separation and independent component analysis. However, almost all of them are based on parametric models. The distribution model mismatch between the output PDF (Probability Density Functions) and the chosen underlying distribution model is a serious problem in blind signal processing. Nonparametric PDF estimates like the Parzen window applied to the popular Kullback-Leibler divergence produce computational difficulties. Hence we propose a new measure, the *Quadratic Gaussianity Measure*, which is associated with the Euclidean distance between the marginal probability density function and the Gaussian distribution. We show that it outperforms other Gaussianity measures in signal processing applications, such as standardized kurtosis tests because our novel Gaussianity measure is robust to changes in the distribution form.

INTRODUCTION

Information theory has been widely applied in communication engineering and digital signal processing. Since statistical independence among different sources or adjacent communication symbols is the most reasonable assumption for source separation and blind equalization problems, the Kullback-Leibler divergence $K(p||q)$

$$K(p||q) = E\left\{\log\left[\frac{p(x)}{q(x)}\right]\right\} = \int q(x)\log\left[\frac{p(x)}{q(x)}\right]dx \ , \qquad (1)$$

0-7803-5673-X/99/$10.00 © 1999 IEEE

where x is an N-dimensional random variable vector, has been applied as an independence measure [1]. However, algorithms based on Kullback-Leibler divergence involve numerical integration which poses computational difficulties [2]. Another method of estimating $K(p||q)$ replaces the analytic form in Equation (1) by a sample mean approximation, but it may cause a large statistical bias when the sample size T is not sufficiently large [3].

To overcome such a practical difficulty Fisher [4] and Xu [5] proposed alternative measures of entropy and mutual information based on the Euclidean distance between the probabilistic functionals such as

$$J(f(x), g(x)) = \int\int_{S_x}\ldots\int[f(x) - g(x)]^2 dx \quad , \tag{2}$$

where S_x is the region of support. In Fisher's quadratic entropy difference measure, $g(x) = 1$ and for Xu's quadratic mutual information $g(x) = \prod_i f_i(x_i)$ where $f(x)$ is the joint distribution of multi-dimensional random variable $x = [x_i]$ and $f_i(x_i)$ is the marginal distribution of x_i. The non-parametric Parzen window [6] was applied to estimate $f(x)$ in [4, 5]. The computational complexity is $O(N^2)$ where N is the size of the random variable vector x, so there are practical difficulties when N gets large.

Let us restrict our discussion to the adaptation of a linear demixing system. The blind source separation/extraction from an instantaneous or convolutive mixture, as well as the blind equalization of non-Gaussian i.i.d. signals can be unified in the framework of Figure 1. The common goal for both multi-channel blind source separation and single-channel blind equalization is to extract the original signal s(t) up to some undetermined delay q and scaling factor α ($\alpha s(t-q)$) by linearly transforming the mixed or channel-distorted $x(t)$ through a rank-one recovery system W such that $y(t) = Wx(t)$. If the problem is blind source separation, the vector $s(t) = [s_1(t), s_2(t),\ldots]^T$ is composed of multiple independent sources. If the problem is blind channel equalization, the vector is composed by the delays of $s(t) = [s(t), s(t-1),\ldots]$.

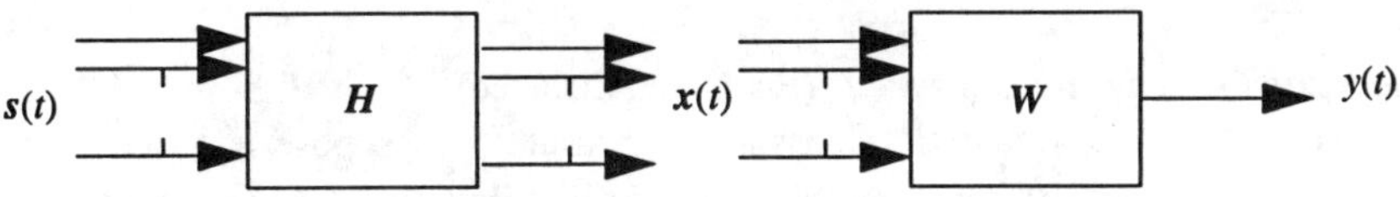

Figure 1. A source extraction or blind equalization system.

STANDARDIZED KURTOSES

The eigenspread criterion to extract sources one-by-one among selected frames can be considered as a threshold on the second-order energy ratio [7]. Although a large and time-varying energy discrepancy is very common for some nonstationary natural signals such as speech, it is neither an universal characteristic nor a plausible technique for blind separation of a convolutive mixture. Hence, a novel way to extract a single source from the linear mixture using statistical information is necessary.

Donoho established measures associated with the so-called Gaussianity of the random variables [8]. The linear combination of random variables always increase the Gaussianity or entropy of the distribution. Then the goal of blind source separation or blind equalization must be simply to decrease the Gaussianity or entropy of the output [8]. Donoho [8] and Wiggin [9] proposed the *standardized* ρ^{th} *cumulant* as the objective function, defined as

$$C_s(\rho, x) = C^{\rho}{}_x / {C^2{}_x}^{\rho/2}, \tag{3}$$

where $C^{\rho}{}_x$ is the ρth order cumulant of the data $\boldsymbol{x}$, and proved that we may extract a source out form the linear mixture by maximizing or minimizing (3). Another objective function is the "*Gray norm*" of the separated or deconvolved sequence [8, 9, 10], which can be expressed as

$$N_{Gray}(\rho, x) = \frac{E\{|x|^{\rho}\}}{\sqrt{E\{|x|^2\}}^{\rho}}. \tag{4}$$

The Gray norm or standardized kurtosis has been widely utilized for blind source separation [11, 13] and blind channel equalization [12].

QUADRATIC GAUSSIANITY MEASURES

Gray norms or standardized cumulants require a minimum of a priori knowledge about the sources, namely, the sign of the sources' kurtoses [11, 13]. We will propose here a novel measure of Gaussianity to extract sources *without any a priori requirements*, i.e. independently of the sign of the kurtosis. Our novel measures are described as follows:

Quadratic Gaussianity measure I: (Euclidean distance)

The first novel quadratic Gaussianity measure we propose is based on the Euclidean distance between a marginal output distribution $f_{y_i}(y_i)$ and its corresponding Gaussian distribution $f_G(y_i)$ with the same variance. The Euclidean distance between these two probabilistic functionals is defined by

$$G_E = \int_{-\infty}^{\infty} [f_{y_i}(y_i) - f_G(y_i)]^2 dy_i \ . \tag{5}$$

The measure $G_E(g_{i1}, g_{i2})$ of (5) will include the trivial zero solution, hence an alternative measure $G_N(g_{i1}, g_{i2})$ with appropriate normalization (the sample standard deviation) was developed [14]. G_N becomes scale invariant.

Quadratic Gaussianity measure II: (Absolute cosine)

The other novel quadratic Gaussianity measure we propose is based on the square cosine value between a marginal output functional $f_{y_i}(y_i)$ and its corresponding Gaussian functional $f_G(y_i)$ with the same variance. The inverse of the square cosine value between these two probabilistic functionals is defined by

$$G_A = \frac{\int_{-\infty}^{\infty} f_{y_i}(y_i)^2 dy_i \int_{-\infty}^{\infty} f_G^2(y_i) dy_i}{\left[\int_{-\infty}^{\infty} f_{y_i}(y_i) f_G(y_i) dy_i\right]^2} \quad . \tag{6}$$

Each of the integrals in (5) and (6) will be estimated with the information potential described in [6],

$$\begin{cases} H(\{a_i\}) = H_{R2}(Y|\{a_i\}) = -\log V(\{a_i\}) \\ \\ V(\{a_i\}) = \dfrac{1}{N^2} \sum_{i=1}^{N} \sum_{j=1}^{N} G(a_i - a_j, 2\sigma^2 I) \end{cases}$$

where G(.,.) is a multidimensional Gaussian kernel. We will utilize an example to illustrate the difference among all the measures defined here and the existing Gaussianity measures, e.g. the Gray norms $N_{Gray}(\rho, y_i)$ in Equation (4) and the standardized cumulants $C_s(\rho, y_i)$ in Equation (3). For a linear channel, the input vector is $\boldsymbol{x} = \boldsymbol{Hs}$, where $\boldsymbol{s}$ consists of independent source $[s_1, s_2, s_3,...]^T$ or original symbols, $\boldsymbol{H}$ is the channel or mixing system, and $\boldsymbol{x}$ are the observations. Now we assume the output y_i is the projection $\boldsymbol{w}_i^T\boldsymbol{x}$ of the N-dimensional input vector $\boldsymbol{x} = [x_1, x_2,..., x_N]^T$ onto a separating or extracting column vector $\boldsymbol{w}_i = [w_{i1}, w_{i2},..., w_{iN}]^T$. We define the product $\boldsymbol{w}_i^T\boldsymbol{H} = \boldsymbol{G} = [g_{i1}, g_{i2},..., g_{iN}]$ as the "*separation/extraction performance vector*". The goal of our extracting system is to maximize the *signal-to-interference ratio*, denoted as $P_{dB} = 20 \log\left(\sum_{p \neq j}^{N} (g_{ij}^2/g_{ip}^2)\right)$, for some specific j, namely, $N_{Gray}(4, y_i)$, $N_{Gray}(1, y_i)$.

In this example we consider a mixture of two sources, a sub-Gaussian source (Laplacian stochastic process) and a super-Gaussian source (uniform stochastic process) each with 600 samples. The various measures are functions of the mixing coefficients (g_{i1}, g_{i2}) through the relationship $y_i = g_{i1}s_1 + g_{i2}s_2$. Figure 2 depicts the measures $G_N(g_{i1}, g_{i2})$, $G_A(g_{i1}, g_{i2})$ of (6), $C_s(4, y_i) = N_{Gray}(4, y_i)$-3 of (3), and 1/

$[N_{Gray}(1, y_i)]^2$. From Figure 2a through Figure 2d, we can observe the following facts:

The maxima of (5) and (6) are always two line segments excluding the origin, i.e., $g_{i1}= 0$ or $g_{i2}= 0$ except $g_{i1} = g_{i2} = 0$. In these conditions, maximization will always separate one sole sub-Gaussian or super-Gaussian source. Consequently, with the new quadratic Gaussianity measures $G_N(g_{i1}, g_{i2})$ or $G_A(g_{i1}, g_{i2})$ *there is no need to know the kurtosis sign.* This is unlike the Gray norms or the standard cumulants, where we have to decide between a minimization or maximization operation depending upon the sign of the kurtosis.

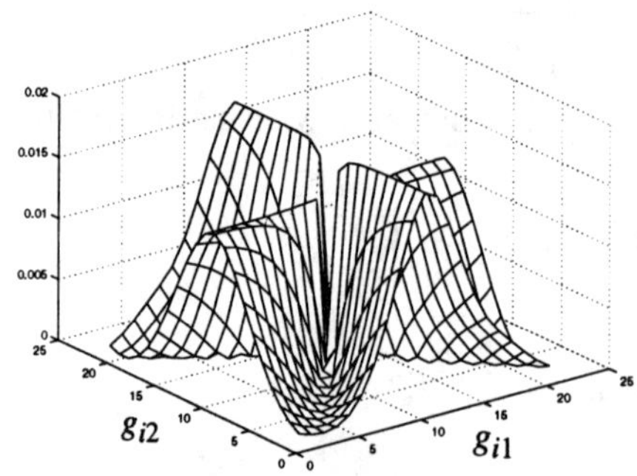

The plot of measure $G_N(g_{i1}, g_{i2})$

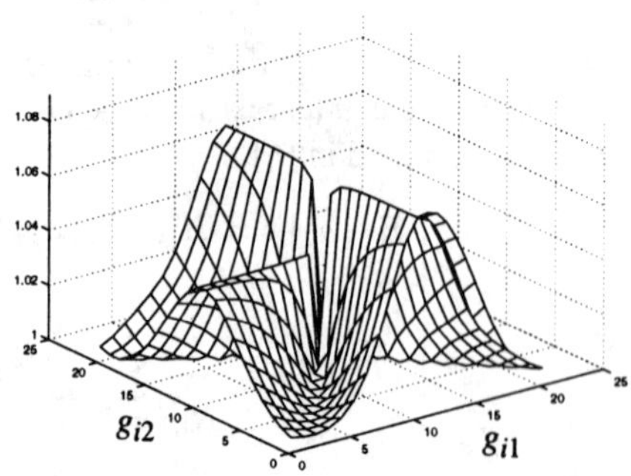

The plot of measure $G_A(g_{i1}, g_{i2})$

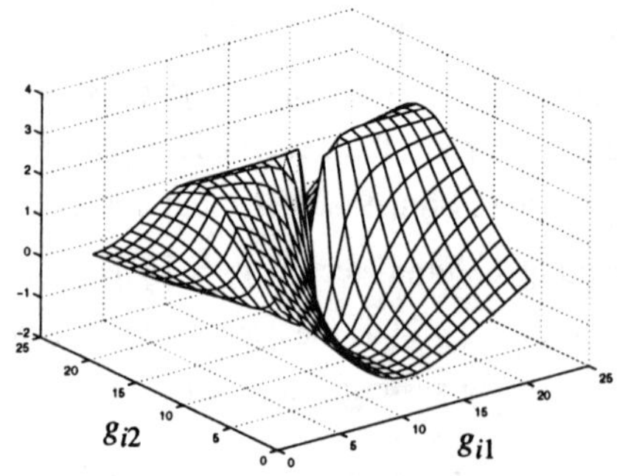

The plot of measure $N_{Gray}(4, y_i)$-3

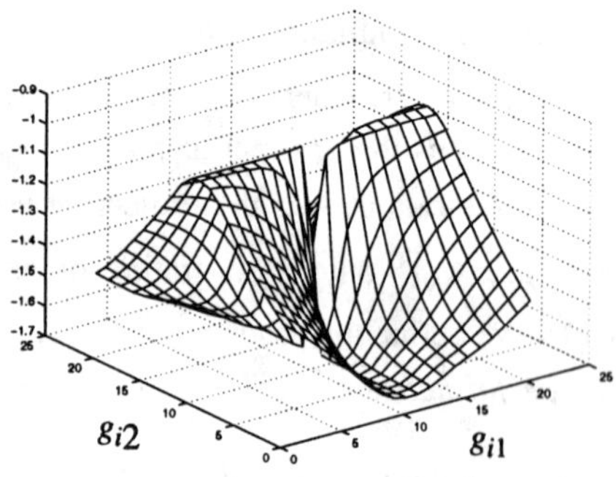

The plot of measure $1/[N_{Gray}(1, y_i)]^2$

Figure 2. The measures $G_N(g_{i1}, g_{i2})$, $G_A(g_{i1}, g_{i2})$, $N_{Gray}(4, y_i)$-3 and $1/[N_{Gray}(1, y_i)]^2$ for the linear combination of a sub-Gaussian source with a super-Gaussian source.

SIMULATION AND CONCLUSION

The new measures can be widely applied in blind source separation and blind channel equalization [14]. We just discuss the source separation here due to page limitations. We will present results using G_N only because this learning rule is much simpler to implement. Here we would like to show three source separation simulations and compare the results among different criteria. All simulations are based on artificial mixtures.

Instantaneous Mixture

We randomly choose a static mixing channel $\boldsymbol{H} = \begin{bmatrix} 0.5077 & 0.5913 \\ 1.6924 & -0.6436 \end{bmatrix}$. Table 1 gives the signal to interference ratio of the extraction for nonstationary sources such as speech. If we estimate the Gray norm using the whole data set (window size T = 38,196 samples) and maximize we obtain the best overall performance. However, if we use the more realistic method of a short sliding window (T = 300 samples), the optimization of our new measure (normalization of (5)) gives much better results. We believe that this is due to the better estimation of the time-varying statistics obtained with the information potential (notice that it uses the interactions among pairs of samples which increases the number of estimations from N to Nx(N-1)/2).

Table 1: Separation performance by optimization of the two measures (two nonstationary sources, speech)

Objective functions	$1/[N_{Gray}(1, y_i)]^2$, $T = 38196$	$1/[N_{Gray}(1, y_i)]^2$, $T = 300$	G_N, $T = 300$
P_{dB}	44.5019	1.5542	41.8162

Another comparison between these two measures is presented in Table 2. The sources are now four non-Gaussian stationary processes (2 sub and 2 super-Gaussians) and the mixing channel is $\boldsymbol{H} = \begin{bmatrix} 0.2944 & -0.6918 & -1.4410 & 0.8156 \\ -1.3362 & 0.8580 & 0.5711 & 0.7119 \\ 0.7143 & 1.2540 & -0.3999 & 1.2902 \\ 1.6236 & -1.5937 & 0.6900 & 0.6686 \end{bmatrix}$. We simulate source separation by either training with a small portion of data (T = 300 samples) or training with a sliding window of the same size over the full data. The two measures perform more evenly for stationary signals than for nonstationary signals since the statistics are consistent for the whole data set. However, our proposed measure G_N still yields slightly better performance.

Table 2: Separation performance by optimization of the two measures (four stationary sources including super/sub-Gaussian random processes)

Objective functions	$1/[N_{Gray}(1, y_i)]^2$ T=300	$1/[N_{Gray}(1, y_i)]^2$ T=300, sliding window	G_N T=300	G_N T=300, sliding window
P_{dB}	13.7894	28.8579	17.3165	34.9422

Convolutive Mixture

In this experiment we choose the same speech sources as above. The mixing channel impulse responses are $H_{11}(z) = H_{22}(z) = 1$, $H_{12}(z) = 0.9 + 0.8z^{-1} + 0.1z^{-2}$, and $H_{21}(z) = -0.7 + 0.4z^{-1} + 0.3z^{-2} + 0.2z^{-3} + 0.1z^{-4}$. We mix them in the computer and separate these two mixed signals using our G_N maximization method and Gray norm ($1/[N_{Gray}(1, y_i)]^2$) maximization method. We utilize a feedforward structure with FIRs of order 5, and a non-overlapping sliding windows (window size T = 700). The mixed signals and the separated signals are depicted in Figure 3. We will compute the *SIR*s as above. The SIR using Gray norm maximization is 0.87 *dB*, while using our G_N maximization is 15.88 *dB*. We can observe in the Figure that the output using the G_N maximization is much closer to Source 2 while the output using the Gray norm maximization is still scrambled. The listening test corroborates the visual analysis. This verifies again that even for convolutive mixtures our G_N maximization method is more robust for nonstationary signals than Gray norm maximization method.

From these three simulations, we can conclude that the optimization of our measure will outperform the Gray norms for blind signal processing. Further applications of these new measures will be provided in future work. These applications include the blind equalization of short-term fading channels, whose model can be characterized as short finite-support impulse responses.

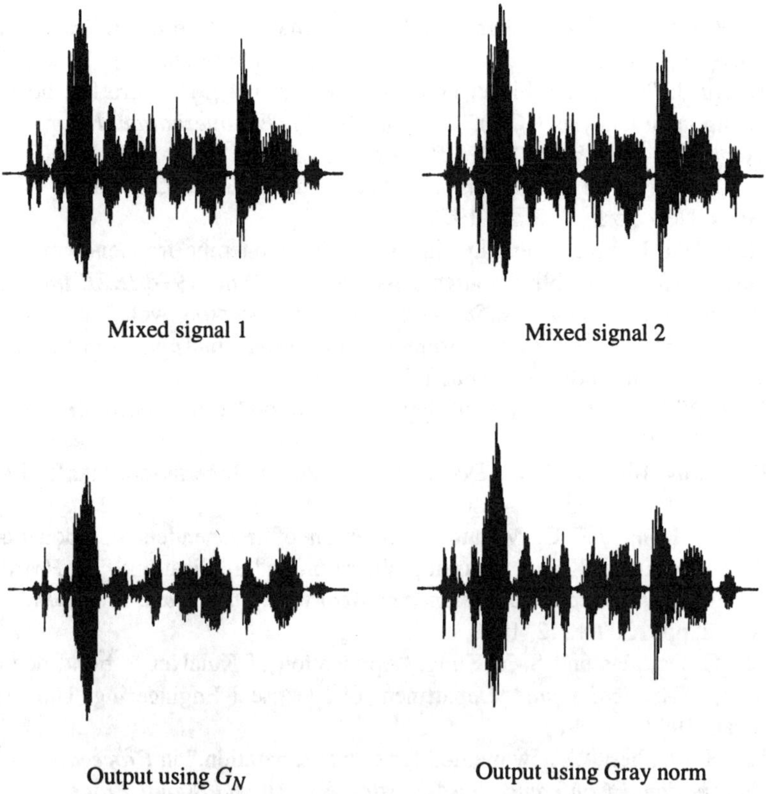

Figure 3. The mixed speech, separated speech for a convolutive mixture.

Acknowledgments:
This work was partially supported by NSF grant ECS-9900394.

REFERENCES

[1] A. Papoulis, *Probability, random variables, and stochastic processes*, New York, McGraw-Hill, 1991.

[2] D. T. Pham, "Blind separation of instantaneous mixture of sources via an independent component analysis," *IEEE Transactions on Signal Processing*, vol. 44, no. 11, pp. 2768-2779, November 1996.

[3] P. Viola, N. Schraudolph and T. Sejnowski, "Empirical entropy manipulation for real-world problems," in Proceeding of *Neural Information Processing Systems Conference*, *NIPS95*, p. 851-857, 1995.

[4] J. W. Fisher and J. C. Principe, "Entropy manipulation of arbitrary nonlinear mappings," in *Proc. Neural Network for Signal Processing*, pp. 14-23, 1997.

[5] D. Xu, J. Principe, J. Fisher and H.-C. Wu, "A novel measure for independent component analysis (ICA)," in *Proc. 1998 IEEE International Conference on Acoustics, Speech and Signal Processing*, vol. 2, p. 1161-1164.

[6] R. Duda, *Pattern classification and scene analysis*, p. 85-95, John Wiley and Sons, New York, U. S. A., 1973.

[7] H.-C. Wu, J. Principe and D. Xu, "Exploring the tempo-frequency micro-structure of speech for blind source separation," in *Proc. 1998 IEEE International Conference on Acoustics, Speech and Signal Processing*, vol. 2, p. 1145-1148.

[8] D. Donoho, "On minimum entropy deconvolution," in *Applied time-series analysis II*, pp. 565-609, Academic Press, 1981.

[9] R. A. Wiggins, "Minimum entropy deconvolution," *Geoexploration*, vol. 16, pp. 21-35, 1978.

[10] W. Gray, *Variable Norm Deconvolution*, Ph. D. dissertation, Stanford University, 1979.

[11] S. Y. Kung and C. Mejuto, "Extraction of independent components from hybrid mixture: Kuicnet learning algorithm and applications," in *Proceedings of IEEE International Conference on Acoustics, Speech and Signal Processing*, vol. 2, pp. 1209-1212, 1998.

[12] S. C. Douglas and S.-Y. Kung, "Application of KuicNet to blind deconvolution," *Technical report*, Department of Electrical Engineering, University of Utah, 1998.

[13] R. H. Lambert, "A new method for source separation," in *Proceedings of IEEE International Conference on Acoustics, Speech and Signal Processing*, vol. 3, pp. 2116-2119, 1995.

[14] H.-C. Wu, *Blind source separation using information measures in the time and frequency domain*, Ph. D. dissertation, University of Florida, 1999.

ROBUST MACHINE FAULT DETECTION WITH INDEPENDENT COMPONENT ANALYSIS AND SUPPORT VECTOR DATA DESCRIPTION

Alexander Ypma, David M. J. Tax and Robert P.W. Duin
Pattern Recognition Group
Dept. of Applied Physics, Delft University of Technology
Lorentzweg 1, 2628 CJ Delft, The Netherlands
Phone: +31-15-2781845
Fax: +31-15-2786740
Email: {ypma,davidt}@ph.tn.tudelft.nl

Abstract.

We propose a novel approach to fault detection in rotating mechanical machines: fusion of multichannel measurements of machine vibration using Independent Component Analysis (ICA), followed by a description of the admissible domain (part of the feature space indicative of normal machine operation) with a Support Vector Domain Description (SVDD) method. The SVDD-method enables the determination of an arbitrary shaped region that comprises a target class of a dataset. In this particular application, it provides a way to quantify the compactness of the admissible class in relation to data preprocessing. Application to monitoring of a submersible pump indicates that combination of measurement channels with ICA gives improved results in fault detection, without requiring detailed prior knowledge on origin and type of the failure.

INTRODUCTION

Monitoring of mechanical machinery involves the analysis of an ensemble of vibration measurements, yielding a combined view of the periodic force variations within the machine. Fault development or wear is usually accompanied by increased harmonics of the machine running frequency, increased (amount of) sidebands around structure-related fault frequencies and a shift in the amplitude distribution of the measurement signals. Several authors proposed the use of higher-order statistics in this problem [3, 1, 7] to exploit nongaussian and nonlinear behaviour in a machine (e.g. in the presence of bearing and gearbox failure and cyclostationarity).

0-7803-5673-X/99/$10.00

In this paper we use Independent Component Analysis for the combination of several sensor measurements in the time domain (*sensor fusion*). The aim is to find the manifestation of a vibrating (fault) source by using the spatial redundancy in the multichannel measurement time series. This could provide a fault detection procedure that is less sensitive to careful sensor choice and specific details of the sensor mounting process.

In machine condition monitoring it is not possible to collect an exhaustive set of measurements from all possible failure scenarios. Moreover, since a machine may be used under very different operating modes (running speed, load, type of lubricants) and environmental conditions (indoor or outdoor, placed near interfering machinery), the admissible behaviour (usually called: the machine fingerprint) actually comprises a *set* of fingerprints. In order to perform reliable fault detection, significant deviations from this normal description should be detected. We propose a new method for description of a dataset (the target class), that enables to determine the rejection rate of the target class a priori. The method is based on the support vector method proposed by Vapnik [12].

SENSOR FUSION WITH INDEPENDENT COMPONENT ANALYSIS

In Independent Component Analysis one tries to retrieve an ensemble of independent source signals $\mathbf{s}$ out of an ensemble of linear mixtures $\mathbf{x}$ by determining the inverse of the mixing matrix A (up to permutation and scaling) [5]

$$\mathbf{y} = W\mathbf{x}, \quad W = \Lambda P A^{-1} \tag{1}$$

Several solutions to this *blind signal separation* problem have been developed, usually based on the notion of minimization of the mutual information between the reconstructed components $\mathbf{y} = \hat{\mathbf{s}}$ (which is sufficient for obtaining independency) [6]. There are a few assumptions involved here: the model assumes linear instantaneous mixing (which is expected to be correct for the pump under investigation), whereas only one of the sources is allowed to be Gaussian and noise is not explicitly taken into account. Lately, attempts have been made to take temporal coherence into account [2, 4, 13], leading to algorithms in which second-order statistics suffice for separation (if sources have different spectral characteristics).

The linear approach of (1) often leads to algorithms that maximize nongaussianity of the recovered components, i.e. the distribution of the projection of the data onto a vector in the ICA-basis should be as far from Gaussian as possible. If it cannot be assumed that independent sources underly a dataset, ICA may therefore be regarded as a form of exploratory projection pursuit. In the application at hand (machine diagnostics) it is known that incipient faults often manifest themselves as impulses in the time signal, e.g.

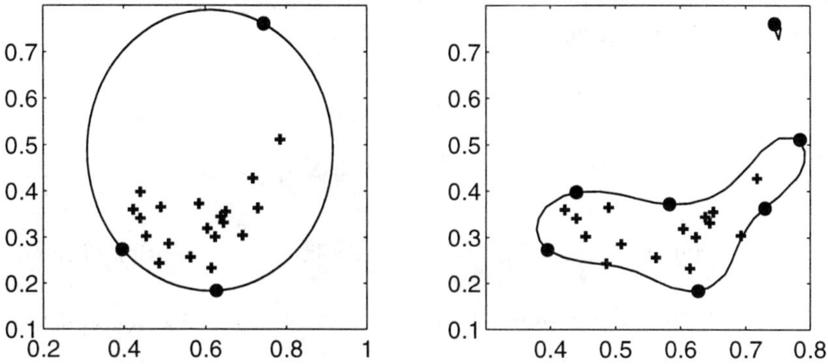

Figure 1: Data description of a small data set, (left) normal spherical description, (right) description using a Gaussian kernel.

it has been reported that bearing failure and loose foundation give rise to nongaussianity (increased kurtosis) of vibration measurements.

We performed the ICA-based sensor fusion in the experiments of section 4 using the *fast-ica*-algorithm developed by Hyvärinen [5]. This algorithm works with both sub- and supergaussian sources and shows cubic convergence. More specifically, we used the deflation approach with cubic nonlinearity. If a component couldn't be found because the algorithm didn't converge in 100 cycles, the multichannel time series at hand was disregarded in feature extraction (see section 4).

Note that we consider the vibration signal as a stationary signal on the scale of several shaft rotations: ICA basis vectors will be tuned to the average signal content within a few rotations.

DATA DOMAIN DESCRIPTION WITH SUPPORT VECTORS

For description of the domain of a dataset one can try to capture it with a sphere with minimum volume. Inspired by the Support Vector Method by Vapnik (see [12], or for a more simple introduction [9]) one can extend this idea to determine an arbitrary shaped region in the original feature space. We will refer to the method as the Support Vector Data Description (SVDD) method [10].

Assume we a have data set containing N data objects, $\{x_i, i = 1, .., N\}$ and the sphere is described by center a and radius R. We now try to minimize an error function containing the volume of the sphere. The constraints that objects are within the sphere are imposed by applying Lagrange multipliers:

$$L(R, a, \alpha_i) = R^2 - \sum_i \alpha_i \{R^2 - (x_i^2 - 2ax_i + a^2)\} \qquad (2)$$

with Lagrange multipliers $\alpha_i \geq 0$. This function has to be minimized with respect to R and a and maximized with respect to α_i.

Setting the partial derivatives of L to R and a to zero, gives:

$$\begin{aligned} \sum_i \alpha_i &= 1 \\ a &= \frac{\sum_i \alpha_i x_i}{\sum_i \alpha_i} = \sum_i \alpha_i x_i \end{aligned} \tag{3}$$

This shows that the center of the sphere a is a linear combination of the data objects x_i.

Resubstituting these values in the Lagrangian gives to maximize with respect to α_i:

$$L = \sum_i \alpha_i (x_i \cdot x_i) - \sum_{i,j} \alpha_i \alpha_j (x_i \cdot x_j) \tag{4}$$

with $\alpha_i \geq 0$, $\sum_i \alpha_i = 1$.

This function should be maximized with respect to α_i. In practice this means that a large fraction of the α_i become zero. For a small fraction $\alpha_i > 0$ and the corresponding objects are called Support Objects. We see that the center of the sphere depends just on the few support objects, objects with $\alpha_i = 0$ can be disregarded.

Object z is accepted when:

$$\begin{aligned} &(z-a)(z-a)^T = (z - \sum_i \alpha_i x_i)(z - \sum_i \alpha_i x_i) \\ &\leq R^2 \end{aligned} \tag{5}$$

In general this does not give a very tight description. Analogous to the method of Vapnik [12], we can replace the inner products $(x \cdot y)$ in equations (4) and in (5) by kernel functions $K(x, y)$ which gives a much more flexible method. When we replace the inner products by Gaussian kernels for instance, we obtain:

$$(x \cdot y) \to K(x, y) = \exp(-(x-y)^2/s^2) \tag{6}$$

Equation (4) now changes into:

$$L = 1 - \sum_i {\alpha_i}^2 - \sum_{i \neq j} \alpha_i \alpha_j K(x_i, x_j) \tag{7}$$

and and the formula to check if a new object z is within the sphere (equation (5)) becomes:

$$1 - 2\sum_i \alpha_i K(z, x_i) + \sum_{i,j} \alpha_i \alpha_j K(x_i, x_j) \leq R^2 \tag{8}$$

We obtain a more flexible description than the rigid sphere description. In figure 1 both methods are shown applied on the same two dimensional data set. The sphere description on the left includes all objects, but is by no means very tight. It includes large areas of the feature space where no target patterns are present. In the right figure the data description using Gaussian kernels is shown, and it clearly gives a superior description. No empty areas are included, what minimizes the change of accepting outlier patterns.

This Gaussian kernel contains one extra free parameter, the width parameter s in the kernel (equation (6)). As shown in [10] this parameter can be set by setting a priori the maximal allowed rejection rate of the target set, i.e. the error on the target set. This error can be estimated by the number of support vectors:

$$E[P(\text{error})] = \frac{\#SV}{N} \tag{9}$$

where $\#SV$ is the number of support vectors. We can regulate the number of support vectors by changing the width parameter s and therefore we can also set the error on the target set on the prespecified value. Unfortunately the parameter s is incorporated in the kernel function K and therefore this parameter cannot be optimized directly during the maximization of L (see [8] for direct control of the number of support vectors).

Note we cannot set a priori restrictions on the error on the outlier class. In general we only have a good representation of the target class and the outlier class is per definition everything else.

EXPERIMENTS

An in-house test rig (figure 2) consists of a small submersible pump in a water basin, which can be made to run at several speeds (from 46 to 54 Hz) and several loads (by closing a membrane controlling the water flow). A number of faults were induced to this pump (loose foundation, imbalance, failure in the outer race of the uppermost ball bearing), and vibration was measured with three accelerometers (yielding 5 measurement channels, one of the sensors is triaxial). Both normal and faulty behaviour was measured at several speeds and loads.

Feature extraction

We compared several methods for feature extraction from vibration data. It is well known that faults in rotating machines will be visible in the acceleration spectrum as increased harmonics of running speed or presence of sidebands around characteristic (structure-related) frequencies. Due to overlap in series of harmonic components and noise, high spectral resolution may be required for adequate fault identification. This may lead to difficulties because of the curse of dimensionality: one needs large sample sizes in high-dimensional spaces in order to avoid overfitting of the train set. Hence we

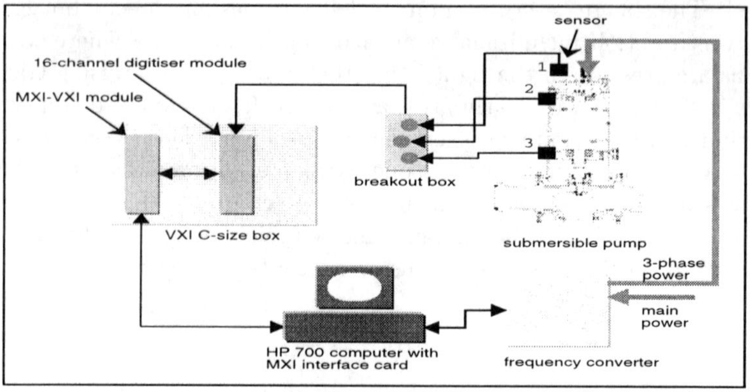

Figure 2: Experimental setup with submersible pump

focused on relatively low feature dimensionality (64 spectral bins). A previous study [11] showed that features based on autoregressive modelling (order 64) and MUSIC spectrum estimation (64 spectral bins) performed very well in this application. We compared this to feature vectors obtained with a 64-bins envelope spectrum, i.e. the spectrum of an envelope detected (or *demodulated*) time signal. No bandpass-filtering was employed here. As a reference, a high-resolution (512 spectral bins) normalized power spectrum was obtained from the measurements. For further details on experimental setup and feature extraction for machine diagnostics, see [11, 13].

We compared our ICA combination approach with taking a linear combination of multichannel measurements based on variance maximization (Principal Component Analysis). Alternatively, we used the SVDD with datasets that were generated with feature extraction from measurement signals without applying ICA or PCA. Here, features from all measurement channels were included into one dataset.

Detection Results

The SVDD is applied to several datasets, differing in sensor combination method and feature extraction method. Since test objects from the outlier class are available (i.e. the fault class comprising imbalance, loose foundation and bearing failure), the rejection performance on the outlier set can also be measured.

In all experiments we have used the SVDD with a Gaussian kernel. For each of the feature sets we have optimized the width parameter s in the SVDD such that 1%, 5%, 10%, 25% and 50% of the target objects will be rejected, so for each data set and each target error another width parameter s is obtained. For each feature set this gives a acceptance-rejection curve for the target and the outlier class.

In figure 3 the performances of the different features are compared with the 512 bin power spectrum with no ICA applied. The high resolution power

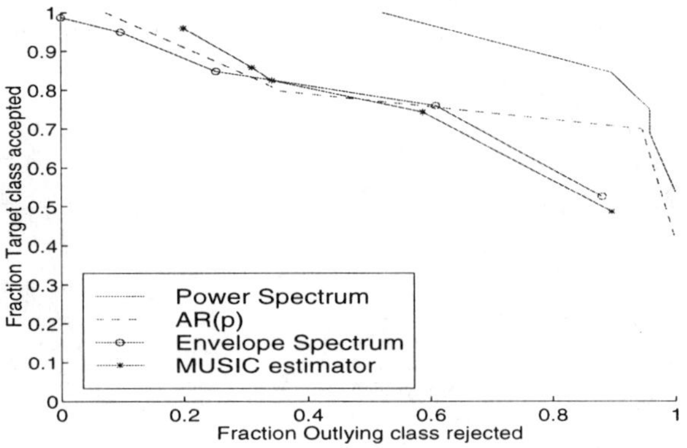

Figure 3: Acceptance/rejection performance of the SVDD on all types of features with all sensor measurements collected.

spectrum clearly outperforms all other methods. When all of the target class objects is accepted, over 50% of the outlier class is rejected. When about 10% of the target class is rejected, almost 80% of the outlier class can be rejected. This shows that a large fraction of the abnormal patterns can be distinguished from the normal class, but some overlap exist. Performance of the SVDD on the AR-model features, envelope spectrum features and the MUSIC estimator are clearly inferior. Especially when large target acceptance rates are required, large fractions of the outlier class are accepted. Only the AR-model features approximates the performance of the 512 bin power spectrum for smaller target acceptance rates.

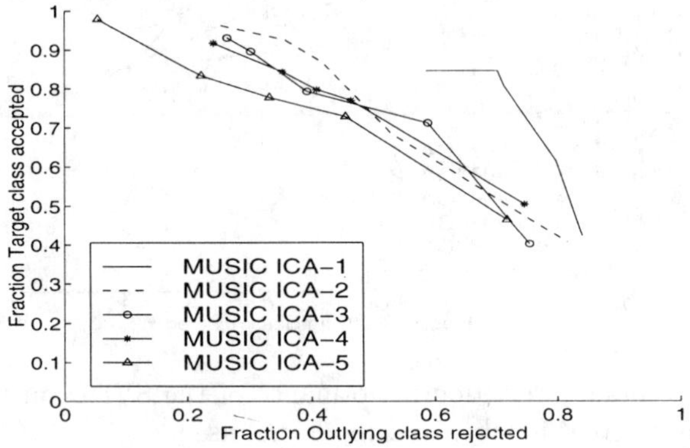

Figure 4: Acceptance/rejection performance of the SVDD on the MUSIC features with different number of components used.

In figure 4 the SVDD is applied to the MUSIC estimator data on which an ICA is performed. Results vary with the number of independent components that are included into the dataset. Performance is optimal when just one component is used, using more than one component results in (comparable) worsened performance for all situations. Inclusion of all 5 independent components gives the worst results. Clearly the first ICA-component contains almost all useful information to distinguish between normal and abnormal behaviour.

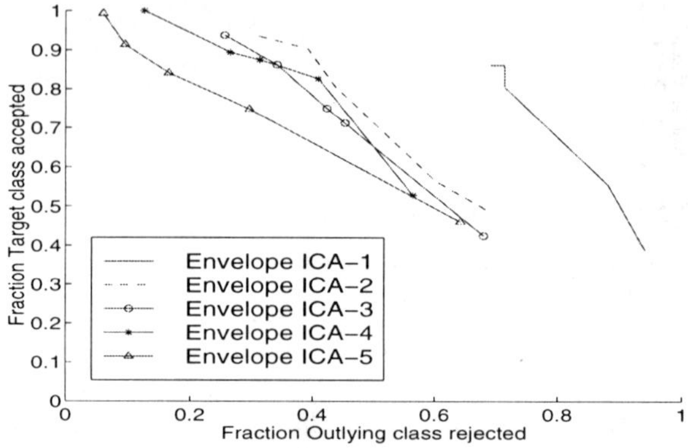

Figure 5: Acceptance/rejection performance of the SVDD on the envelope features with different number of components used.

In figure 5 can observe the same trend for the envelope features. Again, the first independent component yields most information for fault detection, including multiple components gives significantly worse results. The envelope features also perform slightly better than the MUSIC frequency features, but still do not reach the performance obtained with the power spectrum.

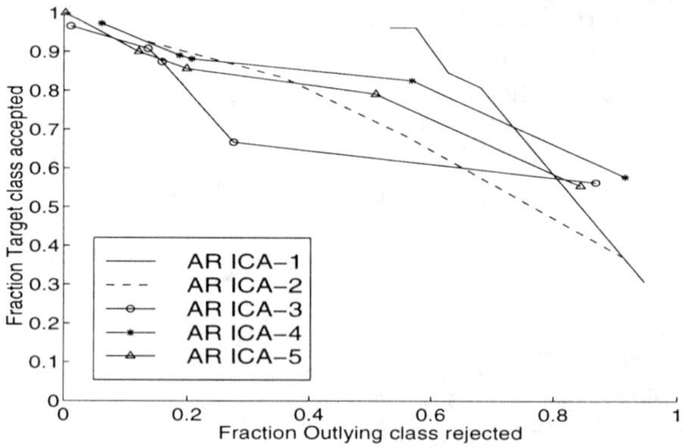

Figure 6: Acceptance/rejection performance of the SVDD on the envelope features with different number of components used.

In figure 6 the performance is shown for the AR-model features. Again the first component gives best performance, but for smaller target class acceptance rates, inclusion of multiple components is useful. In this case it is less clear how much useful information is retained when taking only the first component into account. Although in the non-ICA case the AR model features better distinguish between target and outlier class (see figure 3) than envelope spectrum features, in the ICA case the AR features perform worse than envelope features (compare figures 5 and 6).

Last, AR-model features were computed on data with PCA preprocessing

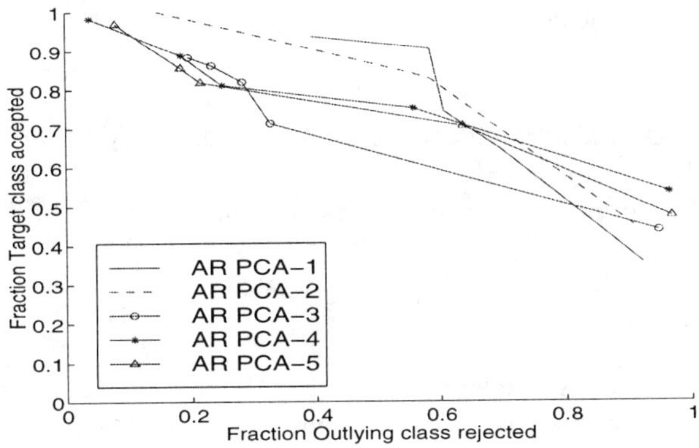

Figure 7: Acceptance/rejection performance of the SVDD on the envelope features with different number of components used.

(figure 7). Taking just the first principal component is not enough, only for very large target acceptance rates it outperforms the sets using more components. When larger fractions of the outlier class should be rejected, information from more components should be used.

The final observation is that performance is best using envelope spectrum features on ICA-combined measurement channels.

CONCLUSION

For robust fault detection in rotating mechanical machinery it may be useful to use spatial redundancy in measurements that are obtained simultaneously with multiple spatially distributed vibration sensors. In order to quantify this effect, we used a data domain description method to capture the normal behaviour of a machine and analyzed the trade-off between acceptance of data from normal behaviour and rejection of data from fault situations. For three typical features (AR-modelling, MUSIC spectrum estimation and envelope spectrum estimation) performance on ICA-combined measurement channels was usually better than linear combination using Principal Components Analysis. Moreover, if we fuse all channels into one combined channel with ICA, detection results are superior to the results obtained with datasets from all original (not fused) measurement channels. Previous results [11] showed that careful choice of sensor channel to include in a dataset may enhance performance, but this requires prior knowledge on the nature and spatial origin of a failure. Using ICA for sensor fusion may overcome this limitation, while retaining high detection accuracy.

ACKNOWLEDGMENTS

This work was partly supported by the Dutch Technology Foundation (STW) and the Dutch Organisation for Scientific Research (NWO). We would like to thank Ronald Ligteringen, SKF Condition Monitoring B. V. and Landustrie Sneek B. V. for cooperation in measurements and setup of the test rig.

REFERENCES

[1] R. W. Barker, G. Klutke and M. Hinich, "Monitoring rotating tool wear using higher-order spectral features," **Journal of engineering for industry**, vol. 115, pp. 23 – 29, February 1993.

[2] A. Belouchrani, K. Abed-Meraim, J.-F. Cardoso and E. Moulines, "A blind source separation technique using second-order statistics," **IEEE Transactions on signal processing**, vol. 45, no. 2, pp. 434 – 444, 1997.

[3] B. Boashash, E. J. Powers and A. M. Zoubir, eds., **Higher-Order Statistical Signal Processing**, Melbourne, 1995, LONGMAN (Wiley).

[4] A. Hyvärinen, "Independent component analysis of time-dependent stochastic processes," in **Proceedings of the 8th International Conference on Artificial Neural Networks (ICANN98)**, Springer-Verlag, 1998, vol. 1, pp. 135–140.

[5] A. Hyvärinen and E. Oja, "A fast fixed-point algorithm for independent component analysis," **Neural computation**, vol. 9, no. 7, pp. 1483–1492, 1997.

[6] T.-W. Lee, M. Girolami, A. J. Bell and T. J. Sejnowski, "A unifying information-theoretic framework for independent component analysis," **International journal on mathematical and computer modeling**, 1998.

[7] A. Murray and J. Penman, "Extracting useful higher order features for condition monitoring using artificial neural networks," **IEEE Transactions on signal processing**, vol. 45, no. 11, pp. 2821 – 2828, November 1997.

[8] B. Scholkopf, P. Bartlett, A. Smola and R. Williamson, "Shrinking the tube: A new support vector regression algorithm," M. S. Kearns, S. A. Solla, and D. A. Cohn, editors, Advances in Neural Information Processing Systems 11, MIT Press, Cambridge, MA.

[9] D. Tax, D. de Ridder and R. Duin, "Support vector classifiers: a first look." in **Proceedings of ASCI'97**, ASCI, 1997.

[10] D. Tax and R. Duin, "Data domain description using support vectors," in Verleysen, M., ed., **Proceedings of the European Symposium on Artificial Neural Networks 1999**, Brussels: D-Facto, April 1999, pp. 251 – 256.

[11] D. M. Tax, A. Ypma and R. P. W. Duin, "Support vector data description applied to machine vibration analysis," To appear in the Proceedings of ASCI'99.

[12] V. Vapnik, **The nature of statistical learning theory**, Springer-Verlag, New York, 1995.

[13] A. Ypma and P. Pajunen, "Rotating machine vibration analysis with second-order independent component analysis," in **Proceedings of the First International Workshop on Independent Component Analysis and Signal Separation, ICA'99**, January 1999, pp. 37 – 42.

VIEW-BASED 3D OBJECT RECOGNITION WITH SUPPORT VECTOR MACHINES

Danny Roobaert [†‡] & Marc M. Van Hulle[‡]

[†] Computer Vision and Active Perception laboratory
Royal Institute of Technology, S-100 44 Stockholm, SWEDEN
Tel.: + 46 8 790 6209, Fax: + 46 8 723 0302
E-mail: roobaert@nada.kth.se

[‡] Laboratory for Neuro- and Psychophysiology
K. U. Leuven, Herestraat, B-3000 Leuven, BELGIUM
Tel.: + 32 16 34 59 61, Fax: + 32 16 34 59 93
E-mail: marc@neuro.kuleuven.ac.be

Abstract. Support Vector Machines have demonstrated excellent results in pattern recognition tasks and 3D object recognition. In this contribution, we confirm some of the results in 3D object recognition and compare it to other object recognition systems. We use different pixel-level representations to perform the experiments, while we extend the setting to the more challenging and practical case when only a limited number of views of the object are presented during training. We report high correct classification of unseen views, especially considering that no domain knowledge is including into the proposed system. Finally, we suggest an active learning algorithm to reduce further the required number of training views.

INTRODUCTION

Humans are able to recognize everyday 3D objects when shown previously only one - or at most a few - views of the object. In contrast, artificial systems must either been shown many views of an object (*e.g.* [8]) or either a lot of knowledge of object structure must be build-in into the system to attain a reasonable performance (say $> 95\%$). Generic 3D object recognition systems capable of recognizing a reasonably large set ($n > 25$) of objects from only a few views, has not been reported yet.

Recently a new learning algorithm was proposed by V. Vapnik and co-workers, based on Statistical Learning Theory [13]. This algorithm, named the Support Vector Machine (SVM), has produced excellent performances

0-7803-5673-X/99/$10.00 © 1999 IEEE

in a number of difficult learning tasks. In particular, on the handwritten digit recognition problem, SVMs have shown to largely outperform all other learning algorithms, if one excludes the influence of domain-knowledge [5][12]. SVMs also have generated high performance on 3D object recognition tasks [1, 9]. However, in both these cases, only comparisons with neural networks were performed. In addition always many views (resp. 89 and 36) are presented to the system during training. Despite the latter fact does not come near to human performance, it is also questionable in which extend the generalization ability of the SVM approach is exploited by the presentation of so many views.

Hence, in this contribution, we will study 3D object recognition with SVMs and compare the method with two other methods in the case only a limited number of views of the objects are presented during training. Since SVMs are designed to optimally cope with small sample sizes, we will test a key aspect of SVMs in a real application. Note also that we will not preprocess the images nor will we incorporate any other domain knowledge in the system.

RECOGNITION SYSTEM

Support Vector Machines

Most classical neural network algorithms, require an *ad hoc* choice of the system's generalization ability, *e.g.* the designer of the system has to predefine, guided by some heuristic, the network's architecture and/or the network parameter(s) controlling the complexity of the network. In contrast, the SVM approach proposes a learning algorithm to control the generalization ability of the system automatically. Using the result that the system's generalization error is bounded by the Vapnik-Chervonenkis (VC) dimension [13], the SVM algorithm searches the minimal VC-dimension for the system given the training data.

In the case of a two-class pattern recognition problem, the VC-dimension reduction principle, is implemented as a search for the Optimal Hyper-Plane (OHP) that separates the data. Vapnik shows that hyper-planes that maximize the (orthogonal) distance between the two (or more) closest points to the hyper-plane, effectively have a minimal VC-dimension. In brief (we refer to the literature for a full exposition [13]), given training data $(\mathbf{x}_1, y_1), ..., (\mathbf{x}_l, y_l)$, $\mathbf{x}_i \in \Re^m, y_i \in \{-1, 1\}$ the OHP, $\mathbf{w} \cdot \mathbf{x} + b = 0$, can be found by minimizing:

$$\|\mathbf{w}\|^2 + C \sum_{i=1}^{l} \xi_i$$

constrained by:

$$\xi_i \geq 0 \quad \text{and} \quad y_i((\mathbf{x}_i \cdot \mathbf{w} + b) \geq (1 - \xi_i) \qquad \text{for } i = 1, \ldots, l$$

The ξ_i are slack variables which are introduced to account for non-separable data [4]. The (only) parameter C quantifies the trade-off between training error and system capacity (in all experiments C was set very high in order to minimize training error). The solution $\mathbf{w}^*$ of this quadratic programming problem, can be written as a linear combination of a few $\mathbf{x}_i$. These $\mathbf{x}_i$ are defined as support vectors and determine the OHP. Intuitively, the support vectors correspond to the previously mentioned closest points to the OHP.

Linear SVMs can be generalized to non-linear SVMs, by first mapping the input $\phi(\mathbf{x})$ to a high-dimensional feature space (say with dimension M) and to search the OHP in that new space. However, it has been discovered that this mapping must not be computed explicitly. Since only scalar products are calculated in the input space, this scalar product can be replaced by a kernel $K(\mathbf{x},\mathbf{y}) = \sum_{j=1}^{M} \lambda_j \phi_j(\mathbf{x})\phi_j(\mathbf{y})$ (under certain conditions) to realize this mapping implicitly [2]. We will use $K(\mathbf{x},\mathbf{y}) = (\mathbf{x}\cdot\mathbf{y}+1)^p$. Using this kernel method, one is able to classify data in a high-dimensional feature space, while requiring only computations in the (low-dimensional) input space. That overcomes what is known as the curse of dimensionality.

Multi-class pattern recognition systems can be obtained by combining two-class SVMs. In the 1-against-1 scheme, $\frac{n(n-1)}{2}$ two-class machines are constructed (with n the number of objects). Each machine is trained as a classifier for one class against another class. In order to classify test data, pair-wise competition between all the machines is performed and in analogy with a tennis tournament, each winner competes against another winner until a single winner remains. This final winner determines the class of the test data. In an alternative 1-against-all scheme, only n two-class machines are constructed. Each machine is trained for one class against all other classes. The disadvantage of this scheme is that some test data might not be classified in a single class. In addition, experiments show lower performances for this scheme, even excluding the ambiguities. Hence, in this contribution, we report on experiments with the 1-against-1 scheme only.

Dataset and experimental set-up

We use the COIL image library of Columbia University [8] which contains pictures of 100 every day objects (see Fig. 1). This database consists of 7200 color images: 72 different views of every object. The images were obtained by placing the objects on a turntable and taking a view every 5°. The views were consecutively clip out from the background using a rectangular bounding box and illumination normalization was performed.

Experiments are performed on the raw color-images of the object. In order to estimate the influence of the cues "color" and "shape" in the images, two additional representations were created as a reference. We describe these representations in more detail. For the images called "Shape&Color", the original color images were sub-sampled to 16x16 pixels. Taking into account RGB-composition of individual pixels, input vectors are of dimension $m = 768$. For the representation "Color only", the average color of an image was

Figure 1: (Top row) A few of the objects in the COIL Columbia Image Library (all in view= 30°). (Bottom row) The views that make up the training set for an object, in the case of four training views per object.

calculated and the average R, G and B values were fed to the system as a $m = 3$ dimensional (floating point) vector. For the representation "Shape only", the original images were sub-sampled to 32x32 grey-level images. The color-pixels are reduced to grey as follows: grey-value = (integer) $\frac{R+G+B}{3}$. The resulting $m = 1024$ dimensional vector were fed into the system.

For the experiments, the SVM implementation [11] was used. For the high-dimensional representations "Shape&Color" and "Shape only", linear SVMs were used, since the dimension of the input space is higher than the number of training examples for a single machine. We checked and confirmed the validity of this choice by experimenting with SVMs with kernel K and $p > 1$. This gave resulted in a lower performance. Considering the low dimensionality of the input vector in the representation "Color only", we trained SVMs with the kernel K and $p = 2$. This gave experimentally better results than linear SVMs. In the case of 36 views, $p = 3$ gave the best result. The regularization parameter C was set very high *i.e.* an arbitrary number much higher than any λ_j [10].

RESULTS

In the experiment, the number of training views is step-wise reduced from 36 views per object (10° intervals) to 2 views per object (180° intervals). In Fig. 2 (left panel) we plot the generalization ability of the recognition system on the remaining views in the database (*i.e.* respectively 36 to 70 views per object). All performances mentioned are averages over the $n = 30$ objects in the set.

Concretely, we confirm the results reported in [9], *i.e.* 100% accuracy in the case of 36 training views per object. If the number of training views per object is lower than 18 however, faultless performance cannot be obtained anymore. To predict scaling properties of the object recognition system for larger sets of objects, different object set sizes n were evaluated. Performances are plotted in Fig. 2 (right panel) in the rather difficult setting of only 4 training views per object. Note that performance decrease less than linear, with increasing number of objects to be discriminated. Remark that

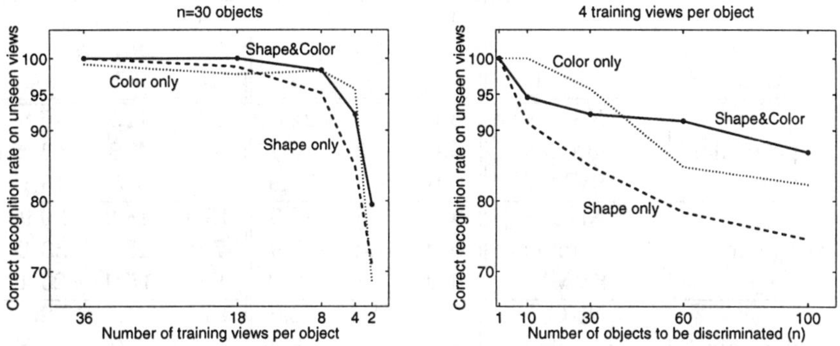

Figure 2: Performance of the recognition system with different input representations. (Left panel) The behavior when the number of training views is reduced and $n = 30$ objects have to be discriminated. (Right panel) The scaling behavior of the recognition system in the (hard) case when only 4 training views per object are presented during training.

the quite primitive representation "Color only" has good results for small sets of objects and hence appears to be a strong cue in this case. For larger sets, the full representation "Shape&Color" gives the best results.

Interpreting the produced support vectors (see Fig. 3), one can remark that the "Color only" representation requires less support vectors than the "Shape&Color" representation (in the case $n = 30$). One can remark as well that (circle) symmetric objects require fewer support vectors than a-symmetric objects, an observation which is in correspondence with intuition.

Comparison with other object recognition systems

In order to assess the obtained results, we compare our approach with two other approaches. The first approach is a simple image matching procedure. Starting from the distances between the training images and the test image, we use a Nearest Neighbor Classifier (NNC) to classify the test image. Using an Euclidean distance measure, the results obtained are detailed in Table 1.

The second method is the 3D object recognition system constructed at Columbia University by Murase and Nayar [6]. This system includes also a NNC classifier, but incorporates knowledge about the database into the system. It assumes that the 3D rotation of the objects in the database can be modeled with a 1-dimensional spline in a eigenspace. This eigenspace is spanned by the eigenvectors obtained by Principle Components Analysis on all the training images. Performance details are reported in Table 1. These results were obtained using the SLAM software developed by Nene S., Nayar S. and Murase H. at Columbia University. One can observe that the performance of both approaches are similar. Except for the representation "Color only" or when only very few (2) training images per object are shown, the SVM performs better.

representation	method	($n = 30$ objects) views/obj. 36	8	2	(4 views/obj.) n objects 10	30	100
Shape&Color	NNC	**0.0**	3.7	22.5	8.2	11.2	16.8
	Columbia	**0.0**	**1.2**	27.9	**4.8**	**7.8**	**12.4**
	SVM	**0.0**	1.7	**20.5**	5.4	**7.8**	13.1
Shape only	NNC	**0.0**	7.5	29.5	13.5	18.2	29.9
	Columbia	**0.0**	**4.4**	32.9	**7.9**	15.4	**23.0**
	SVM	**0.0**	4.8	**29.0**	9.0	**15.1**	25.4
Color only	NNC	1.9	11.6	32.0	18.7	21.5	32.9
	Columbia	1.0	2.0	38.0	5.9	5.9	22.5
	SVM	**0.8**	**1.7**	**31.5**	**0**	**4.3**	**17.7**

Table 1: 3D object recognition error rates with three different approaches: Nearest Neighbor Classifier (NNC), the Columbia 3D object recognition system and the SVM-based system. In the left panel, the error rates for a database of 30 objects are reported, in the case the number of training images per object are varied. In the right panel, 4 training views per object are presented during training, while the size of the set of objects to be discriminated is varied.

Active learning

The expected error on unseen views is bounded by the average compression factor *i.e.* the ratio of training vectors to the number of support vectors produced [13]:

$$E[\text{test-error rate}] \leq E\left[\frac{\text{nr. of support vectors}}{\text{nr. of training vectors}}\right]$$

Since acquiring different views of objects in the real world requires physical movements, it might be (time and/or energy) expensive to acquire views. In order to minimize the number of training views needed, we suggest following exploratory object recognition algorithm. First, the SVM-based system is trained with only a few views per object. Next, the system acquires additional views only from these objects which required a large number of support vectors during training. Thirdly, these additional views are used to retrain the system. The system repeats the previous steps until the generation of new support vectors is smaller than a certain threshold. In a real-world setting, this behavior would resemble the sort of active, exploratory learning performed by humans.

We made a simulation of a single iteration in this algorithm. Starting from 4 (resp. 8) views per object, we doubled the number of training views only for these objects that had a compression ratio worse than $\frac{3}{4}$ and consecutively retrained the system. The resulting performances were similar as if all 8 (resp. 16) views were presented, but in average 0.13 to 3.3 less views per object were needed.

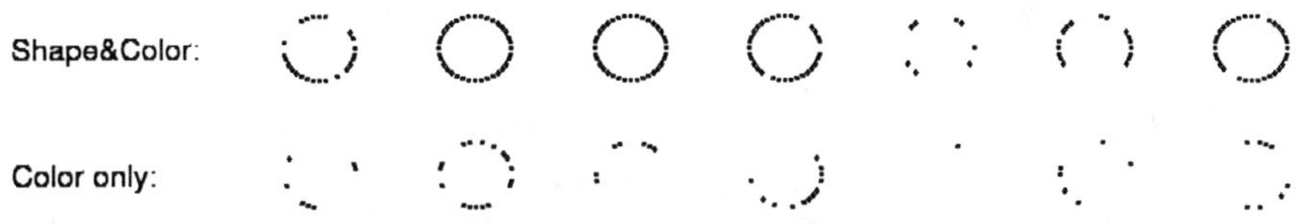

Figure 3: The support vectors of the 7 objects in Fig. 1 in a training set of 30 objects when 36 views per object are presented. The dots represent support vectors (*i.e.* support views) and are drawn counter-clockwise with view= 0° to the right. (Top row) Support vectors when views are represented as "Shape&Color". (Bottom row) Support vectors when views are represented as "Color only".

CONCLUSION

We demonstrated that the SVM approach holds promise for 3D object recognition systems. Direct application of this learning algorithm on a standard 3D object recognition database produces a performance that is comparable to the known 3D object recognition system developed by Murase and Nayar at Columbia University. The latter system incorporates knowledge of the the object database, while in contrast the SVM-based system does not incorporate any domain knowledge. Consequently, the SVM-based system promises also to perform well on other object recognition problems that require other types of invariances apart of the 3D rotational invariance that is required for this database. In addition, the SVM algorithm comes with a way to predict classification error. This is exploited in this contribution in the suggested active learning algorithm. This algorithm can reduce the number of views needed during training.

Considering that no a-priori knowledge is included in the proposed SVM-based system, the incorporation of domain knowledge, such as image preprocessing, 3D geometrical properties and/or object-specific features, can still potentially improve the performance, as is shown by the Columbia University system. Finally, the support vector approach could also provide a computational theory of a 2D view-based model of 3D object recognition as proposed by [3]. In the latter case, the support vectors correspond to the 2D views that characterize the object.

Acknowledgments

D.R. wishes to thank Jan-Olof Eklundh for continuous encouragement and discussions and Peter Nillius for help with running the SLAM software. D.R. is a Marie Curie research fellow and is supported by a Training and Mobility of Researchers grant of the European Commission (ERBFMBICT961126). M.M.V.H. is a research associate of the Fund for Scientific Research - Flanders (Belgium) and is supported by research grants received from the Fund for Scientific Research (G.0185.96), the National Lottery (9.0185.96), and the Flemish Regional Ministry of Education (GOA 95/99-06).

REFERENCES

[1] Blanz, V., Schölkopf, B., Bülthoff, H., Burges, C., Vapnik, V., Vetter, T. (1996) Comparison of view-based object recognition algorithms using realistic 3D models, *Proc. Int. Conf. Art. Neural Netw. 1996*, 251-256

[2] Boser, B., Guyon, I., Vapnik, V. (1992). A training algorithm for optimal margin classifiers, *5th Ann. Workshop on Computational Learning Theory*, Pittsburgh ACM, 144-152.

[3] Bülthoff, H., Edelman, S. (1992) Psychophysical support for a two-dimensional view-interpolation theory of object recognition, *Proc. National Academy of Sciences* **89**, 60-64

[4] Cortes, C., Vapnik, V. (1995) Support-Vector Networks, *Machine Learning* **20**, 1-25

[5] LeCun, Y., Jackel, L., Bottou, L., Brunot, A., Cortes, C., Denker, J., Drucker, H., Guyon, I., Müller, U., Sckinger, E., Simard, P., Vapnik, V. (1995) Comparison of learning algorithms for handwritten digit recognition, *Proc. Int. Conf. on Artificial Neural Networks 1995*, 53-60

[6] Murase, H., Nayar, S. (1995) Visual Learning and Recognition of 3-D Objects from Appearance, *Int. Journal of Computer Vision* **14**, 5-24

[7] Nayar, S. K., Nene, S. A., Murase, H. (1996) Real-Time 100 Object Recognition System, Techn. Rep. CUCS-019-95, *Proc. ARPA Image Understanding Workshop*, San Fransisco, February 1996

[8] Nene, S. A., Nayar, S. K., Murase, H. (1996) Columbia Object Image Library (COIL-100), Techn. Rep. No. CUCS-006-96, dept. Comp. Science, Columbia University

[9] Pontil, M., Verri, A. (1998) Support Vector Machines for 3D Object Recognition, *IEEE Trans. on Pattern Analysis & Machine Intelligence* **20**, 637-646

[10] Pontil, M., Verri, A. (1998) Properties of Support Vector Machines, *Neural Computation* **10**, 955-974

[11] Saunders, C., Stitson, M.O., Weston, J., Bottou, L., Schölkopf, B., Smola, A. (1998) Support Vector Machine - Reference Manual, CSD-TR-98-03, Royal Holloway Techn. Rep.

[12] Schölkopf, B., Burges, C., Vapnik, V. (1995) Extracting support data for a given task. In: U. M. Fayyad and R. Uthurusamy (eds.), *Proc. 1st. Int. Conf. Knowl. Disc. & Data Mining*, AAAI Press, Menlo Park, CA

[13] Vapnik, V. N. (1998) *Statistical Learning Theory*, in series adaptive and Learning Systems for Signal Processing, Communications and Control, John Wiley & Sons, New-York

ALGORITHMS AND ARCHITECTURES

GRADIENT BASED ADAPTIVE REGULARIZATION

Robert Eigenmann and Josef A. Nossek
Institute for Network Theory and Circuit Design
Munich University of Technology, D-80333 Munich, Germany
Phone: ++49 89 289 28510, Fax: ++49 89 289 28504
email: eigenmann@ei.tum.de, nossek@ei.tum.de

Abstract. A technique to optimize regularization parameters for a given supervised training problem is presented. A training database is applied to minimize a regularized cost function, and a validation database is used to estimate and optimize generalization properties by means of a modification of regularization. The performance is validated for a vowel classification task and compared to other approaches.

INTRODUCTION

In supervised training, finding the optimal complexity of a model is of major concern in the model design. Given only a very few number of training examples, an overly complex model tends to memorize the training set rather then representing the systematic structure of the data. A less complex model can represent the data only in a coarse approximation. In both cases, the model performs poorly on unknown data, expressed by a high generalization error [4].

Optimizing the model complexity therefore requires an estimation of generalization ability, achieved by a partitioning of the given data into a training set and a validation set. The model parameters are optimized in order to fit the training data, whereas the model complexity is optimized in order to minimize the validation error.

Besides growing [3, 2] and pruning techniques [6, 9, 10] which modify the number of model parameters, regularization affects the effective model complexity, encouraging smooth mappings by adding a penalty term to the error function [1, 5]. The necessary amount of regularization depends on the data structure and therefore has to be adapted during the training process. In [7] an adaptive regularization method was presented, based on the computation of the inverse Hessian of the loss function. In this work we propose an alternative, where the gradient information is sufficient to adapt the proportion of regularization.

0-7803-5673-X/99/$10.00 © 1999 IEEE

TRAINING AND ADAPTIVE REGULARIZATION

Suppose a given nonlinear and differentiable model with adaptive parameters $\boldsymbol{w}$ and an error function $E(\boldsymbol{w}, \mathcal{T})$ to be minimized in a supervised training process for a given database $\mathcal{T}$. Preventing the model from overfitting the given data, a model based regularizer term $R(\boldsymbol{w}, \boldsymbol{r})$ is added to penalize immoderate curvature mappings. The amount of regularization is parameterized with vector $\boldsymbol{r}$ and limits the effective model complexity.

Usually, the regularizer is composed of a weighted sum of penalty terms

$$R(\boldsymbol{w}, \boldsymbol{r}) = \sum_i r_i \, p(\boldsymbol{w}_i),$$

each regarding a subset of adaptive parameters $\boldsymbol{w}_i \subset \boldsymbol{w}$ and contributing with a positive weight $r_i \geq 0$ to the augmented cost function

$$C(\boldsymbol{w}, \boldsymbol{r}, \mathcal{T}) = E(\boldsymbol{w}, \mathcal{T}) + R(\boldsymbol{w}, \boldsymbol{r}). \tag{1}$$

Let $\hat{\boldsymbol{w}}(\boldsymbol{r})$ be the minimum of the cost function (1) for a given regularization weight $\boldsymbol{r}$:

$$\hat{\boldsymbol{w}}(\boldsymbol{r}) = \arg\min_{\boldsymbol{w}} C(\boldsymbol{w}, \boldsymbol{r}, \mathcal{T}). \tag{2}$$

Further let $E(\hat{\boldsymbol{w}}(\boldsymbol{r}), \mathcal{V})$ be the estimated generalization error for this solution, based on a validation dataset $\mathcal{V}$. The subject of adaptive regularization is to find optimal regularization parameters $\boldsymbol{r}^*$, minimizing the validation error in the manifold $\hat{\boldsymbol{w}}(\boldsymbol{r})$:

$$\boldsymbol{r}^* = \arg\min_{\boldsymbol{r}} E(\hat{\boldsymbol{w}}(\boldsymbol{r}), \mathcal{V}) \quad ST \ \boldsymbol{r} \geq \boldsymbol{0}. \tag{3}$$

With this, the optimal model parameters are given by

$$\boldsymbol{w}^* = \arg\min_{\boldsymbol{w}} C(\boldsymbol{w}, \boldsymbol{r}^*, \mathcal{T}). \tag{4}$$

We suppose an iterative update of parameters $\boldsymbol{w}$ and $\boldsymbol{r}$ for the combined optimization problem (2) and (3) with the following heuristic: Let (2) be optimized using an update rule

$$\boldsymbol{w}^{new} = \boldsymbol{w}^{old} - \mu \, \boldsymbol{s}(\boldsymbol{r}), \tag{5}$$

with an appropriate step size μ and a search direction

$$\boldsymbol{s}(\boldsymbol{r}) = \frac{\partial}{\partial \boldsymbol{w}} C(\boldsymbol{w}, \boldsymbol{r}, \mathcal{T}). \tag{6}$$

To avoid the effect of overfitting, this update (6) must also decrease the validation error. A necessary condition to avoid overfitting is that the search direction $\boldsymbol{s}(\boldsymbol{r})$ points in the direction of the validation error gradient. Therefore, we suggest minimizing the divergence of the gradient vectors form loss function (1) and validation error $E(\boldsymbol{w}, \mathcal{V})$ in a least squares sense w.r.t. the regularization parameters $\boldsymbol{r}$

$$\min_{\boldsymbol{r}} \left\| \frac{\partial}{\partial \boldsymbol{w}} C(\boldsymbol{w}, \boldsymbol{r}, \mathcal{T}) - \frac{\partial}{\partial \boldsymbol{w}} E(\boldsymbol{w}, \mathcal{V}) \right\|_2^2 . \tag{7}$$

Optimal regularization parameters are determined for each update (5), constituting the program for adaptive regularization

$$\boldsymbol{r}^{new} = \arg\min_{\boldsymbol{r}} \|\boldsymbol{g}_T - \boldsymbol{g}_V + \boldsymbol{g}_R(\boldsymbol{r})\|_2^2 , \tag{8}$$

$$\boldsymbol{w}^{new} = \boldsymbol{w}^{old} - \mu\ \boldsymbol{s}(\boldsymbol{r}^{new}), \tag{9}$$

with the substitution

$$g_T := \frac{\partial}{\partial \boldsymbol{w}} E(\boldsymbol{w}, \mathcal{T}), \quad g_V := \frac{\partial}{\partial \boldsymbol{w}} E(\boldsymbol{w}, \mathcal{V}), \quad g_R(\boldsymbol{r}) := \frac{\partial}{\partial \boldsymbol{w}} R(\boldsymbol{w}, \boldsymbol{r}).$$

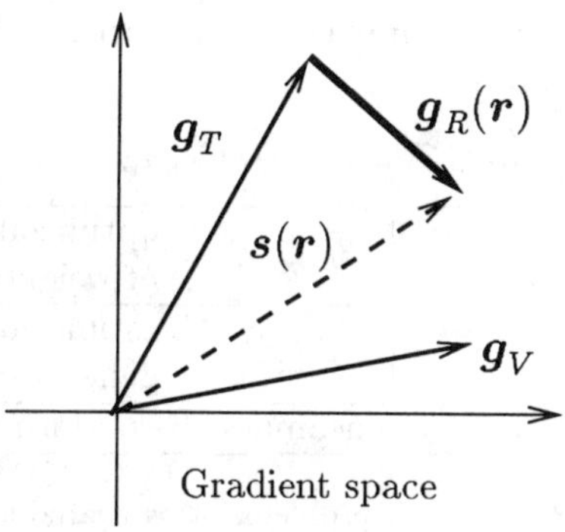

Figure 1: Key motivation of the gradient based adaptive regularization method.

Update (8), (9) is repeated, until $\boldsymbol{r}$ converges to an equilibrium $\hat{\boldsymbol{r}}$, i.e.

$$\|\boldsymbol{r}^{new} - \boldsymbol{r}^{old}\| \leq \delta \quad \Longrightarrow \quad \text{Stop adaptive regularization, } \hat{\boldsymbol{r}} = \boldsymbol{r}^{new}.$$

for a small value δ. At this point, the cost function (1) is finally optimized with constant regularization parameters $\hat{\boldsymbol{r}}$ and the complete available training data

$$\mathcal{T}^{new} = \mathcal{T} \cup \mathcal{V}.$$

THE CHOICE OF VALIDATION DATA

One critical aspect of validation is to appropriately choose the training and validation set from an available training database $\mathcal{X}$. A common technique (see e.g [7]) is to split $\mathcal{X}$ in two disjoint sets $\mathcal{T}$ and $\mathcal{V}$, whereas the choice of the split ratio is a nontrivial task.

$$\mathcal{T} \subset \mathcal{X}, \quad \mathcal{V} \subset \mathcal{X}, \quad \mathcal{T} \cup \mathcal{V} = \mathcal{X}, \quad \mathcal{T} \cap \mathcal{V} = \emptyset \tag{10}$$

According to this technique, we constitute the training set to be a subset $\mathcal{T} \subset \mathcal{X}$, e.g. half of the available training data. Optimization of the cost function with this limited representation of data will favor an overfitting of the training set $\mathcal{T}$. This overfitting effect is compensated by the regularizer, based on the estimation of generalization. The validation database $\mathcal{V}$ should represent the underlying data structure as well as possible, to give a reliable estimation of the generalization ability of the model. Therefore we propose to use all available training data for validation (including $\mathcal{T}$).

$$\mathcal{T} \subset \mathcal{X}, \quad \mathcal{V} = \mathcal{X}. \tag{11}$$

Moreover, the regularizer is prevented from immoderately favoring an arbitrary selection of validation data, provided by a selection of $\mathcal{V}$ according to (10).

The standard benchmark databases usually are provided with a disjoint test set $\mathcal{G}$ in addition to the training data $\mathcal{X}$. This dataset represents *unknown data* of the application and is applied to compare the performance of different approaches. This data is never applied to optimize any of the model weights or regularization parameters.

$$\mathcal{G} \cup \mathcal{T} = \emptyset, \quad \mathcal{G} \cup \mathcal{V} = \emptyset.$$

$\mathcal{X}$	available training data	$\mathcal{T} \subseteq \mathcal{X}$	optimization of weights
		$\mathcal{V} \subseteq \mathcal{X}$	estimation of generalization
$\mathcal{G}$ test database to benchmark the performance			

Table 1: Databases of a benchmark problem. $\mathcal{X}$ is available for optimization and $\mathcal{G}$ is used to test the performance.

REGULARIZING NEURAL CLASSIFIERS

Consider a K-class classification problem with features $\boldsymbol{x} \in \mathrm{I\!R}^D$ to be mapped on a corresponding binary target vector $\boldsymbol{t} \in \{0;1\}^K$, indicating feature $\boldsymbol{x}$ to be of class ω_i, if $t_i = 1$ and $t_j = 0 \ \forall j \neq i$. The database $\mathcal{X} = \{\boldsymbol{x};\boldsymbol{t}\}^P$ consists of P training tuples and is taken for validation $\mathcal{V} = \mathcal{X}$. The training set $\mathcal{T} \subset \mathcal{X}$ is composed of $P/2$ patterns from $\mathcal{X}$ with prior probabilities $\mathrm{p}(\omega_i|\mathcal{T}) = \mathrm{p}(\omega_i|\mathcal{X})$.

Let a given classifier model perform a differentiable decision function $\boldsymbol{f}(\boldsymbol{x},\boldsymbol{w}) : \mathrm{I\!R}^D \mapsto (0,1)^K$, parameterized by weight vector $\boldsymbol{w}$. Denoting f_i^j as the i-th output of the classifier for an input pattern $\boldsymbol{x}^j$, the classifier is trained in order to provide an estimation of the posterior probabilities $f_i^j \stackrel{!}{=} \mathrm{p}(\omega_i|\boldsymbol{x}^j)$ by minimizing the normalized negative log-likelihood cost function [1]

$$E(\boldsymbol{w},\mathcal{T}) = -\frac{1}{P}\sum_{j=1}^{P}\sum_{i=1}^{K}\left(t_i^j \ln(f_i^j) + (1-t_i^j)\ln(1-f_i^j)\right). \tag{12}$$

Let $\boldsymbol{f}(\boldsymbol{w}, \boldsymbol{x})$ be a L-layer feedforward neural network with layers $i = 1 \ldots L$ performing the nonlinear mapping

$$\boldsymbol{y}_i(\boldsymbol{x}) = \tanh(\boldsymbol{W}_i \, \boldsymbol{y}_{i-1}(\boldsymbol{x}) - \boldsymbol{s}_i), \quad \boldsymbol{y}_0(\boldsymbol{x}) := \boldsymbol{x}$$

The j-th row of weight matrix $\boldsymbol{W}_i$ contains the weights to neuron j of the i-th layer and the corresponding bias is the j-th component of vector $\boldsymbol{s}_i$. With the affine transformation $f_j(\boldsymbol{x}) = 0.5(1 + y_{L_j}(\boldsymbol{x}))$, the output values are normalized to the interval $(0, 1)$.

An appropriate regularizer for neural networks is given with the sum-of-squares penalty term (weight decay [5]) $p(\boldsymbol{w}) = \sum_i w_i^2/2$. Following [1], to ensure consistency of the weight decay with certain scaling properties of network mapping, we apply one penalty term for each layer and exclude the bias values $\boldsymbol{s}$ from regularization. Thus, the regularizer is

$$R(\boldsymbol{w}, \boldsymbol{r}) = \sum_{i=1}^{L} r_i \frac{\boldsymbol{w}_i^\mathsf{T} \boldsymbol{w}_i}{2} \tag{13}$$

with vector $\boldsymbol{w}_i$ containing all components of weight matrix $\boldsymbol{W}_i$. Let

$$\boldsymbol{g}_{V,i} := \frac{\partial}{\partial \boldsymbol{w}_i} E(\boldsymbol{w}, \mathcal{V}), \quad \boldsymbol{g}_{T,i} := \frac{\partial}{\partial \boldsymbol{w}_i} E(\boldsymbol{w}, \mathcal{T})$$

be the gradient w.r.t. weights $\boldsymbol{w}_i$, the least squares problem (7) with regularizer (13) has the solution

$$r_i = \frac{\boldsymbol{w}_i^\mathsf{T}(\boldsymbol{g}_{V,i} - \boldsymbol{g}_{T,i})}{\boldsymbol{w}_i^\mathsf{T} \boldsymbol{w}_i}, \quad i = 1 \ldots L. \tag{14}$$

In the first few iterations of the adaptive regularization training, (14) may provide invalid negative solutions for r_i. Then, instead of solving the quadratic programming problem (7) subject to inequality constraints $r_i \geq 0$, we suggest using the absolute value of the solution (14) for the regularization parameters.

SIMULATIONS

We used the Peterson Barney vowel benchmark database [8, 11] for verification of the presented adaptive regularization method. The database consists of $K = 10$ different vowels pronounced twice by 76 speakers (33 male, 28 female and 15 children), represented by the $D = 4$ first formant frequencies. The total number of 1520 input-output pairs were split into two disjoint data sets, i.e. a data set $\mathcal{X}$ with 760 patterns, and a test set $\mathcal{G}$ with 760 patterns to report the performance of the proposed adaptive regularization algorithm. Set $\mathcal{X}$ contains data from 16 male, 14 female and 8 child speakers, set $\mathcal{G}$ contains the remaining 17 male, 14 female and 7 child speakers. According to section 2, a subset $\mathcal{T} \subset \mathcal{X}$ of 8 male, 7 female and 4 child speakers was applied for training and the set $\mathcal{V} = \mathcal{X}$ was used for validation.

For the first simulation we applied a 2-layer neural network with 4 inputs, 10 hidden neurons and 10 output neurons, thus having a total number of 160 weights (including bias) and 2 regularization parameters. The reported results are based on 40 training runs and show the mean error with its standard deviation on the test set $\mathcal{G}$.

In the second simulation we used 100 hidden neurons (i.e. 1510 weights) to report the effect of adaptive regularization in a highly oversized network.

The results of the presented adaptive regularization (ADREG) method are compared with the early-stopping (ES) technique [1], a common applied heuristic to prevent the network from overtraining. Here, no regularization term limits the complexity. Instead, training is stopped at a minimum of the validation error. For the ES method, the same sets of training patterns $\mathcal{T}$, validation patterns $\mathcal{V}$ and test patterns $\mathcal{G}$ have been applied.

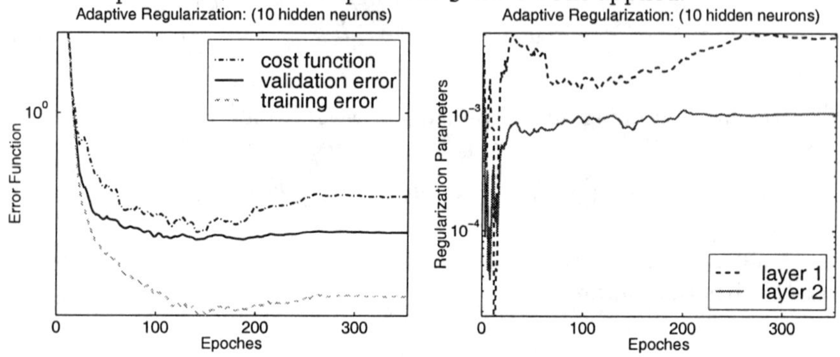

Figure 2: Typical run of the adaptive regularization

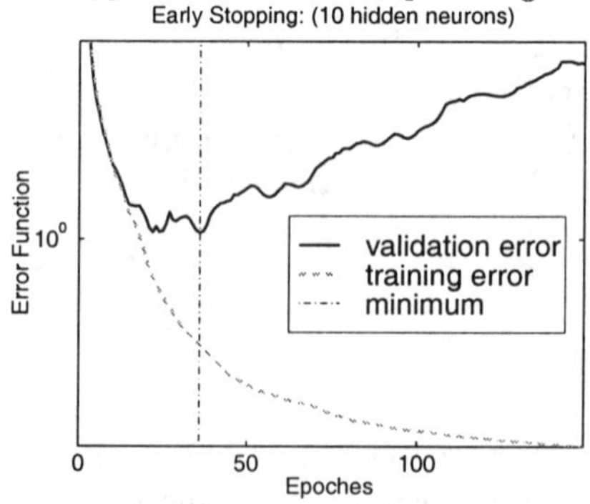

Figure 3: Typical run of the early stopping training

NN	ADREG	ES
4-10-10	12.3 ($\pm$0.74)	14.1 ($\pm$1.17)
4-100-10	11.7 ($\pm$0.52)	13.8 ($\pm$0.80)

Table 2: Misclassifications for the Peterson Barney test set $\mathcal{G}$ (in %) based on 40 runs

Lots of results for the applied database have been published recently, based on different classification strategies. Referring to [7], a different adaptive regularization method for neural classifiers combined with a pruning strategy was reported, misclassifying 12.2% of the test data without pruning and 11.9% after pruning. This method requires the computation of the inverse Hessian matrix to adapt the regularization parameters but has about the same performance as our conceptually simple gradient based update method. As reported in [7], the k-nearest neighbor (KNN) classification rule performs a higher error of 15.3% on this benchmark problem.

CONCLUSIONS

This paper presented a method to adapt regularization parameters for a given supervised training problem. The benefit of this approach to other concepts is based on the conceptually simple update rule for the regularization parameters, where only the gradient of the error function is required. Compared to other techniques working with a validation database (like early stopping), the presented method allows the use of all available data for training, after optimal regularization parameters are determined. Therefore, the model approximates the underlying data structure more exactly, resulting in an improvement of generalization. Is was shown, that with this method, even highly oversized models do not tend to overfit the given training data.

REFERENCES

[1] C. Bishop, **Neural Networks for Pattern Recognition**, Clarendon Press, 1995.

[2] R. Eigenmann and J. Nossek, "Constructive and Robust Combination of Perceptrons," in **Proc. of the International Conference on Pattern Recognition**, 1996, vol. IV, pp. 195–198.

[3] S. Fahlman, "The cascade-correlation learning architecture," in D. Touretzky, ed., **Advances in Neural Information Processing Systems 2**, San Mateo, CA: Morgan Kaufmann Publishers, 1990, pp. 524 532.

[4] S. Geman, E. Bienenstock and R. Dourstat, "Neural networks and the bias/variance dilemma," **Neural Computation**, vol. 4, no. 1, pp. 1–58, 1992.

[5] A. Krogh and J. Hertz, "A simple weight decay can improve generalization," in J. Moody, S. Hanson and R. Lippmann, eds., **Advances in Neural Information Processing Systems 4**, San Mateo, CA: Morgan Kaufmann Publishers, 1992, pp. 450–957.

[6] Y. Le Cun, J. Denker and S. Solla, "Optimal brain damage," in D. Touretzky, ed., **Advances in Neural Information Processing Systems 2**, San Mateo, CA: Morgan Kaufmann Publishers, 1990, pp. 598–603.

[7] L. Nonboe Andersen, J. Larsen, L. Hansen and M. Hintz-Madsen, "Adaptive Regularization of Neural Classifiers," in **Proc. of the IEEE Workshop on Neural Networks for Signal Processing**, 1997, pp. 24–33.

[8] G. Peterson and H. Barney, "Control Methods Used in a Study of the Vowels," **JASA**, vol. 24, pp. 175–184, 1952.

[9] R. Reed, "Pruning algorithms – a survey," **IEEE Trans. on Neural Networks**, vol. 4, no. 5, pp. 740–747, 1993.

[10] J. Sietsma and R. Dow, "Creating Artificial Neural Networks that Generalize," **Neural Networks**, vol. 4, pp. 67–79, 1991.

[11] R. Watrous, "Current Status of PetersonBarney Vowel Formant Data," **JASA**, vol. 89, pp. 2459–2460, 1991.

GLOBAL STABILITY IN DELAYED CELLULAR NEURAL NETWORKS

Jinde Cao
Adult Education College
Yunnan University
Kunming 650091
People's Republic of China
Phone:+86-871-5032762
Fax:+86-871-5153832
E-mail:jdcao@ynu.edu.cn

Abstract: In this paper, the problem of global asymptotic stability of a class of delayed cellular neural networks (DCNN) is further addressed by means of Lyapunov functional method and the inequality $a^3+b^3+c^3 \geq 3abc(a,b,c \geq 0)$ techniques, a set of criteria are found ensuring the global asymptotic stability of DCNN with more general output functions. These criteria have important significance in signal processing, especially in moving image treatment.

1. INTRODUCTION

Cellular neural networks(CNN) were introduced by L. O. Chua and L.Yang in Refs.[1,2] in 1988; they have found important applications in signal processing, especially in static image treatment. The stability of such networks are of prime importance in many applications and has been investigated in many papers, we refer to Refs.[1-7,9-14,16]. The basic circuit unit of a cellular neural network is called a cell. It contains linear and nonliear circuit elements, which typically are linear capacitors, linear resistors, linear and nonlinear controlled sources, and independent sources. The structure of the CNN is similar to that found in cellular automata, namely, any cell in a cellular neural network is connected only to its neighbor cells. The circuit diagram and connection pattern implementing for the CNN can be referred to Ref.[1]. However, processing of moving image require the introduction of a delay in the signals transmitted among the cells; CNN with delays (DCNN) were introduced in Ref.[3] and their dynamic behavior was considered in Refs.[3-7,9-14,16] and the references cited therein. In this paper, by introducing ingeniously real parameters $\alpha_j, \beta_j, \gamma_j, \alpha_j^*, \beta_j^*, \gamma_j^* \in R$ with $\alpha_j + \beta_j + \gamma_j = 1, \alpha_j^* + \beta_j^* + \gamma_j^* = 1 (j = 1, 2, \cdots, n)$, constructing different

0-7803-5673-X/99/$10.00 © 1999 IEEE

Lyapunov functionals and using inequalities $a^3+b^3+c^3 \geq 3abc(a,b,c \geq 0)$ analysis techniques, we derive a set of sufficient criteria of global asymptotic stability of the equilibrium of the DCNN, which are independent of delays. These possess important leading significance in the design and applications of globally stable DCNN, and are of great interest in many applications. Moreover, these results also extend and improve further results related in Refs.[3-7,9-14,16] and the references cited therein. In addition, the methods of this paper may extended to discuss more complicated systems such as Hopfield neural networks(HNN) in Ref.[8] and bidirectional associative memory (BAM) networks, and so on.
In the paper, we consider further the DCNN model described by the delayed differential equations:

$$\begin{aligned} x_i'(t) &= -c_i x_i(t) + \sum_{j=1}^{n} a_{ij} f_j(x_j(t)) \\ &+ \sum_{j=1}^{n} b_{ij} f_j(x_j(t-\tau_j)) + I_i,\ c_i > 0, i = 1,2,\cdots,n \end{aligned} \tag{1}$$

in which n corresponds a neural network with n units; $x_i(t)$ corresponds to the state of the i-th unit at time t; $f_j(x_j(t))$ denotes the output of the j-th unit at time t; a_{ij}, b_{ij}, I_i, c_i are constant, a_{ij} denotes the strength of the j-th unit on the i-th unit at time t, b_{ij} denotes the strength of the j-th unit on the i-th unit at time $t-\tau_j$, I_i denotes the external bias on the i-th unit, τ_j corresponds to the transmission delay along the axon of the j-th unit and is not negative constant, c_i represents the rate with which the i-th unit will reset its potential to the resting state in isolation when disconnected from the network and external inputs. In the following , we assume that each of the relation between the output of the cell $f_i(i=1,2,\cdots,n)$ and the state of the cell have following properties:

(H_1) $f_i(i=1,2,\cdots,n)$ is bounded on R;

(H_2) There is a number $\mu_i > 0$ such that $|f_i(u)-f_i(v)| \leq \mu_i|u-v|$ $(i = 1,2,\cdots,n)$ for any $u,v \in R$.

It is easy to find from (H_2) that f_i is a continuous function on R. In particular, if the relation between the output of the cell and the state of the cell is described by a piecewise-linear function $f_i(x) = \frac{1}{2}(|x+1|-|x-1|)$, then it is easy to see that the function f_i $(i=1,2,\ldots,n)$ clearly satisfy the hypotheses (H_1) and (H_2) above, and $\mu_i \equiv 1(i=1,2,\cdots,n)$.

2. STABILITY ANALYSIS OF DCNN

Lemma 1 Assume that the output of the cell function $f_i(i=1,2,\cdots,n)$ satisfy the hypotheses (H_1) and (H_2) above. Then there exists an equilibrium for the DCNN (1).

Proof If $x^* = (x_1^*, x_2^*, \cdots, x_n^*)^T$ denotes an equilibrium of the DCNN(1), then x^* satisfies the nonlinear algebraic equations:

$$\begin{aligned} x_i^*(t) &= c_i^{-1}[\sum_{j=1}^{n} a_{ij} f_j(x_j^*) + \sum_{j=1}^{n} b_{ij} f_j(x_j^*) + I_i] \\ &= \sum_{j=1}^{n} [c_i^{-1}(a_{ij} + b_{ij})] f_j(x_j^*) + c_i^{-1} I_i \end{aligned} \tag{2}$$

Let $B = [c_i^{-1}(a_{ij} + b_{ij})]_{n\times n}, I = (c_1^{-1}I_1, c_2^{-1}I_2, \cdots, c_n^{-1}I_n)^T, f(x^*) = (f_1(x_1^*), f_2(x_2^*), \cdots, f_n(x_n^*))^T$, Then the system (2) can be written in vector-matrix notation in the form

$$x^* = F(x^*) = Bf(x^*) + I \tag{3}$$

thus x^* is a fixed point of the map $F : R^n \rightarrow R^n$. The existence of a fixed point of the map F can be shown by the well known Brouwer's fixed point theorem. In fact, the i-th component of $F(x)$ satisfy the following form

$$\begin{aligned} |(F(x))_i| &= |\sum_{j=1}^{n} [c_i^{-1}(a_{ij} + b_{ij})] f_j(x_j) + c_i^{-1} I_i| \\ &\leq \sum_{j=1}^{n} |c_i^{-1}(a_{ij} + b_{ij})| |f_j(x_j)| + c_i^{-1} |I_i| \\ &\leq \sum_{j=1}^{n} |c_i^{-1}(a_{ij} + b_{ij})| M + c_i^{-1} |I_i| \end{aligned}$$

where $x = (x_1, x_2, \cdots, x_n)$, $M = \max_{1\leq i\leq n} \sup_s |f_i(s)|$. Let $K = \max_{1\leq i\leq n} (\sum_{j=1}^{n} |c_i^{-1}(a_{ij} + b_{ij})| M + c_i^{-1}|I_i|)$, then $F(R^n) \subset \Omega = \{(x_1, x_2, \cdots, x_n) \in R^n |\ |x_i| \leq K, i = 1, 2, \cdots, n\}$. It is easy to see that the map F is continuous. Let the map F be restricted on Ω, namely $F|_\Omega : \Omega \rightarrow \Omega$. Obviously $F|_\Omega$ maps the bounded closed and convex set Ω on R^n into itself; hence by the well known Brouwer's theorem, the map F has at least one fixed point say $x^* = (x_1^*, x_2^*, \cdots, x_n^*)^T$. This completes the proof.

Lemma 2 For the DCNN (1), suppose that the output of the cell $f_i(i = 1, 2, \cdots, n)$ satisfy the hypotheses (H_1) and (H_2) above. Then all solutions of the DCNN (1) remain bounded for $[0, +\infty)$.

Proof It is easy to observe that all solutions of the DCNN (1) satisfy differential inequalities of the form

$$-c_i x_i(t) - \lambda_i \leq x_i'(t) \leq -c_i x_i(t) + \lambda_i \tag{4}$$

where $\lambda_i = \sum_{j=1}^{n} (|a_{ij}| + |b_{ij}|) \sup_{s\in R} |f_j(s)| + |I_i|$.

Using (4) above, one can easily prove that solutions of the DCNN (1) remain bounded on $[0, +\infty)$.This completes the proof.

Remark 1: We note that Brouwer's theorem does not guarantee the uniqueness of the fixed point. However, in this paper, we derive a set of sufficient criteria on the DCNN (1) which will guarantee not only the uniqueness of the equilibrium but also its recall by its global asymptotic stability. The uniqueness of the equilibrium can easily follow from the global asymptotic stability to be established below.

Theorem For the DCNN (1), suppose that the output of the cell $f_i(i = 1, 2, \cdots, n)$ satisfies the hypotheses (H_1) and (H_2) above. Assume furthermore that the system parameters $a_{ij}, b_{ij}(i, j = 1, 2, \cdots, n)$ satisfy the following conditions:

$$\sum_{j=1}^{n}[(|a_{ij}|\mu_j^{3\alpha_i^*} + |a_{ij}|\mu_j^{3\beta_i^*} + |a_{ji}|\mu_i^{3\gamma_j^*})$$

$$+(|b_{ij}|\mu_j^{3\alpha_i} + |b_{ij}|\mu_j^{3\beta_i} + |b_{ji}|\mu_i^{3\gamma_j}]) < 3c_i, \ i = 1, 2, \cdots, n$$

in which $\mu_j(j = 1, 2, \cdots, n)$ is constant numbers of the hypotheses (H_2) above, and $\alpha_j, \beta_j, \gamma_j, \alpha_j^*, \beta_j^*, \gamma_j^*$ are any real numbers with $\alpha_j + \beta_j + \gamma_j = 1, \alpha_j^* + \beta_j^* + \gamma_j^* = 1 (j = 1, 2, \cdots, n)$. Then the equilibrium x^* of the DCNN (1) is globally asymptotically stable independent of delays.

Proof If $x^* = (x_1^*, x_2^*, \cdots, x_n^*)^T$ is an equilibrium of the DCNN (1), one can derive from (1) that the deviations $y_i(t) = x_i(t) - x_i^*(i = 1, 2, \cdots, n)$ satisfy

$$\begin{aligned} y_i'(t) &= -c_i y_i(t) + \sum_{j=1}^{n} a_{ij}[f_j(x_j^* + y_j(t)) - f_j(x_j^*)] \\ &+ \sum_{j=1}^{n} b_{ij}[f_j(x_j^* + y_j(t - \tau_j)) - f_j(x_j^*)] \end{aligned} \quad (5)$$

Obviously, $(0, 0, \cdots, 0)^T$ is an equilibrium of (5). To prove the global asymptotic stability of x^* of the DCNN (1), it is sufficient to prove the global asymptotic stability of the trivial solution of (5). The local existence of solutions of (1) and (5) follows by the method of steps, the existence on $[0, +\infty)$ will be a consequence of our analysis below. Now we consider the Lyapunov functional defined by

$$V(t) = V(y)(t) = \sum_{i=1}^{n}[\frac{1}{3}|y_i(t)|^3 + \frac{1}{3}\sum_{j=1}^{n}|b_{ij}|\mu_j^{3\gamma_i}\int_{t-\tau_j}^{t}|y_j(s)|^3 ds] \quad (6)$$

where $\gamma_j(j = 1, 2, \cdots, n)$ are any real numbers. The upper right derivative of $V(y)$ is defined by

$$D^+V(y)(t) = \limsup_{h \to 0+} \frac{V(y)(t + h) - V(y)(t)}{h}.$$

Calculating the upper right derivative D^+V of V along the solution of (5), we get

$$D^+V|_{(5)} \leq \sum_{i=1}^{n}[|y_i(t)|^2(-c_i|y_i(t)| + \sum_{j=1}^{n}|a_{ij}|\mu_j|y_j(t)|$$

$$
\begin{aligned}
&+ \sum_{j=1}^{n} |b_{ij}|\mu_j|y_j(t-\tau_j)|) \\
&+ \frac{1}{3}\sum_{j=1}^{n} |b_{ij}|\mu_j^{3\gamma_i}(|y_j(t)|^3 - |y_j(t-\tau_j)|^3)] \\
&= \sum_{i=1}^{n}[-c_i|y_i(t)|^3 + \sum_{j=1}^{n} |a_{ij}|\mu_j|y_i(t)|^2|y_j(t)| \\
&+ \sum_{j=1}^{n} |b_{ij}|\mu_j|y_i(t)|^2|y_j(t-\tau_j)|) \\
&+ \frac{1}{3}\sum_{j=1}^{n} |b_{ij}|\mu_j^{3\gamma_i}(|y_j(t)|^3 - |y_j(t-\tau_j)|^3)] \\
&= \sum_{i=1}^{n}[-c_i|y_i(t)|^3 + \sum_{j=1}^{n} |a_{ij}|(\mu_j^{\alpha_i^*}|y_i(t)|)(\mu_j^{\beta_i^*}|y_i(t)|)(\mu_j^{\gamma_i^*}|y_j(t)|) \\
&+ \sum_{j=1}^{n} |b_{ij}|(\mu_j^{\alpha_i}|y_i(t)|)(\mu_j^{\beta_i}|y_i(t)|)(\mu_j^{\gamma_i}|y_j(t-\tau_j)|) \\
&+ \frac{1}{3}\sum_{j=1}^{n} |b_{ij}|\mu_j^{3\gamma_i}(|y_j(t)|^3 - |y_j(t-\tau_j)|^3)] \qquad (7)
\end{aligned}
$$

where $\alpha_j, \beta_j, \gamma_j, \alpha_j^*, \beta_j^*, \gamma_j^*$ are any real numbers with $\alpha_j + \beta_j + \gamma_j = 1, \alpha_j^* + \beta_j^* + \gamma_j^* = 1 (j = 1, 2, \cdots, n)$. Now estimating the right side of (7) by using the inequality $3abc \leq a^3 + b^3 + c^3 (a, b, c \geq 0)$, we have

$$
\begin{aligned}
D^+V|_{(5)} &\leq \sum_{i=1}^{n}\{-c_i|y_i(t)|^3 + \sum_{j=1}^{n} |a_{ij}|\frac{1}{3}[(\mu_j^{\alpha_i^*}|y_i(t)|)^3 \\
&+ (\mu_j^{\beta_i^*}|y_i(t)|)^3 + (\mu_j^{\gamma_i^*}|y_j(t)|)^3] \\
&+ \sum_{j=1}^{n} |b_{ij}|\frac{1}{3}[(\mu_j^{\alpha_i}|y_i(t)|)^3 + (\mu_j^{\beta_i}|y_i(t)|)^3 + (\mu_j^{\gamma_i}|y_j(t-\tau_j)|)^3) \\
&+ \frac{1}{3}\sum_{j=1}^{n} |b_{ij}|\mu_j^{3\gamma_i}(|y_j(t)|^3 - |y_j(t-\tau_j)|^3]\} \\
&= \sum_{i=1}^{n}\{-c_i|y_i(t)|^3 + \sum_{j=1}^{n} |a_{ij}|\frac{1}{3}[(\mu_j^{\alpha_i^*}|y_i(t)|)^3 \\
&+ (\mu_j^{\beta_i^*}|y_i(t)|)^3 + (\mu_j^{\gamma_i^*}|y_j(t)|)^3] \\
&+ \sum_{j=1}^{n} |b_{ij}|\frac{1}{3}[(\mu_j^{\alpha_i}|y_i(t)|)^3 + (\mu_j^{\beta_i}|y_i(t)|)^3]
\end{aligned}
$$

$$
\begin{aligned}
&+ \frac{1}{3}\sum_{j=1}^{n}|b_{ij}|{\mu_j}^{3\gamma_i}|y_j(t)|^3\} \\
&= \sum_{i=1}^{n}[-c_i|y_i(t)|^3 + \sum_{j=1}^{n}\frac{1}{3}|a_{ij}|(\mu_j^{3\alpha_i^*} + \mu_j^{3\beta_i^*})|y_i(t)|^3 \\
&+ \sum_{j=1}^{n}\frac{1}{3}|a_{ji}|\mu_i^{3\gamma_j^*}|y_i(t)|^3 + \sum_{j=1}^{n}\frac{1}{3}|b_{ij}|(\mu_j^{3\alpha_i} + \mu_j^{3\beta_i})|y_i(t)|^3 \\
&+ \sum_{j=1}^{n}\frac{1}{3}|b_{ji}|\mu_i^{3\gamma_j}|y_i(t)|^3] \\
&= \sum_{i=1}^{n}\{-[c_i - \frac{1}{3}\sum_{j=1}^{n}((|a_{ij}|\mu_j^{3\alpha_i^*} + |a_{ij}|\mu_j^{3\beta_i^*} + |a_{ji}|\mu_i^{3\gamma_j^*}) \\
&+ (|b_{ij}|\mu_j^{3\alpha_i} + |b_{ij}|\mu_j^{3\beta_i} + |b_{ji}|\mu_i^{3\gamma_j}))]\}|y_i(t)|^3 \\
&\leq -r\sum_{i=1}^{n}|y_i(t)|^3 \qquad (8)
\end{aligned}
$$

where
$r = \min_{1\leq i\leq n}\{-[c_i - \frac{1}{3}\sum_{j=1}^{n}((|a_{ij}|\mu_j^{3\alpha_i^*} + |a_{ij}|\mu_j^{3\beta_i^*} + |a_{ji}|\mu_i^{3\gamma_j^*}) + (|b_{ij}|\mu_j^{3\alpha_i} + |b_{ij}|\mu_j^{3\beta_i} + |b_{ji}|\mu_i^{3\gamma_j}))]\} > 0.$
A consequence of (8) is that

$$V(y)(t) + r\int_0^t\sum_{i=1}^{n}|y_i(s)|^3 ds \leq V(y)(0) \qquad (9)$$

It follows from (9) that

$$\int_0^{+\infty}\sum_{i=1}^{n}|y_i(t)|^3 dt < +\infty \qquad (10)$$

According to Lemma 2, $x_i(t)$ is bounded on $(0, +\infty)$. This implies boundedness of $y_i(t), y_i'(t)$ and on $(0, +\infty)$; hence $|y_i(t)|, |y_i(t)|^3$ is uniformly continuous on $(0, +\infty)$. By Barbalat's Lemma in Ref.[15] , it easily follows that

$$\lim_{t\to+\infty}\sum_{i=1}^{n}|y_i(t)|^3 = 0 \qquad (11)$$

It follows from (11) that zero solution of the equation (5) is globally asymptotically stable for any delays, thus the equilibrium of the DCNN (1) is also globally asymptotically stable independent of delays. This completes the proof.
Remark 2: By comparing with these theorems in Refs.[7,9,10,16], it is easily found that the Theorem in this paper and these Theorems in Refs.[7,9,10,16]

do not contain each other, i.e. for any one of them there exists a network which satisfies it but does not the others.

3. CONCLUSION

By introducing ingeniously real parameters, constructing suitable Lyapunov functionals and applying the new inequality $3abc \leq a^3 + b^3 + c^3 (a, b, c \geq 0)$ techniques approach new results about the global asymptotic stability of DCNN with the hypotheses (H_1) and (H_2) above have been found. These criteria can be applied to design globally stable delayed cellular neural networks (DCNN). In addition, the new results in this paper are easily verifiable, the each of the output of the cell $f_j (j = 1, 2, \cdots, n)$ is quite general and our criteria possess many adjustable real parameters $\mu_j > 0, \alpha_j, \beta_j, \gamma_j, \alpha_j^*, \beta_j^*, \gamma_j^* \in R$ with $\alpha_j + \beta_j + \gamma_j = 1, \alpha_j^* + \beta_j^* + \gamma_j^* = 1 (j = 1, 2, \cdots, n)$, these are of important significance in the design and applications of DCNN and signal processing, especially in moving image treatment. Moreover, the methods of this paper may also extended to discuss some more complicated systems such as Hopfield neural networks(HNN) with delays and bidirectional associative memory (BAM) networks with delays.

ACKNOWLEDGMENT

This work is supported in part by the Natural Science Foundation of Yunnan Province, China. In addition, author acknowledges the valuable suggestions and comments of the reviewers and Dr. Jan Larsen(Program Chair) for improving this paper and would like to thank them very much for their comments.

References

[1] L. O. Chua and L. Yang, Cellular neural networks: Theory, **IEEE Transactions on Circuits and Systems**, vol. 35, no 10, pp. 1257-1272, 1988.

[2] L. O. Chua and L. Yang, Cellular neural networks: Applications, **IEEE Transactions on Circuits and Systems**, vol. 35, no 10, pp.1273-1290, 1988.

[3] T. Roska and L. O. Chua, Cellular neural networks with nonlinear and delay-type template, **International Journal of Circuit Theory and Applications**, vol. 20, pp. 469-481, 1992.

[4] Xiaoxin Liao, Mathematical theory (II) of cellular neural networks, **Science in China (Series A)**, vol. 24, no 10, pp. 1037-1046, 1994(in Chinese).

[5] P. P. Civalleri, M. Gilli and Pandolfi, On stability of cellular neural networks with delay, **IEEE Transactions on Circuits and Systems-I**, vol. 40, no 3, pp. 157-165, 1993.

[6] Hongtao Lu and Zhenya He, Unconditioned stability of cellular neural networks with delay, **Acta Electronica Sinica**, vol. 25, no 1, pp. 1-4, 1997(in Chinese).

[7] Jinde Cao, Global stability analysis in delayed cellular neural networks, **Physical Review E**, vol. 59, no 5, 1999.

[8] Jinde Cao and Shidong Wan, The global asymptotic stability of Hopfield neural network with delays, **Journal of Biomathematics**, vol. 12, no 1, pp. 60-63, 1997(in Chinese).

[9] Jinde Cao and Dongming Zhou, Stability analysis of delayed cellular neural networks, **Neural Networks**, vol. 11, no 9, pp. 1601-1605, 1998

[10] Jinde Cao, Stability in cellular neural networks with delays, **Proceedings of 14th IFAC'99 World Congress (E)**, Elsevier Science Ltd, 1999.

[11] M. Gilli, Stability of cellular neural networks and delayed cellular neural networks with nonpositive templates and nonmonotonic output functions, **IEEE Transactions on Circuits and Systems-I**, vol. 41, no 8, pp. 518-528, 1994.

[12] T. Roska, C. W. Wu and L. O. Chua, Stability of cellular neural networks with dominant nonlinear and delay-type template, **IEEE Transactions on Circuits and Systems-I**, vol. 40, no 4, pp. 270-272, 1993.

[13] T. Roska, C. W. Wu, M. Balsi and L. O. Chua, Stability and dynamics of delay-type general and cellular neural networks, **IEEE Transactions on Circuits and Systems-I**, vol. 40, no 4, pp. 487-490, 1993.

[14] S. Arik and V. Tavsanoglu, Global stability analysis of CNNs with delay. **IEEE Fourth International Workshop on CNNs and Their Applications**, pp.187-190, Seville, 1996.

[15] K. Gopalsamy, **Stability and Oscillation in Delay Differential Equations of Population Dynamics**, Kluwer Academic Dordrecht, 1992.

[16] Jinde Cao and Dongming Zhou, Global stability analysis of delayed cellular neural networks, **Journal of Biomathematics**, vol. 14, no 1, pp. 65-71, 1999(in Chinese).

WEIGHT ESTIMATION FOR THE LEARNING OF MODULAR PERCEPTRON NETWORKS

Y.P. Lee, Hsin-Chia Fu*
Department of Computer Science and Information Engineering
National Chiao Tung University,
Hsinchu, Taiwan, ROC

Abstract. In this paper, we propose a weight estimation method for the feed-forward neural networks. The proposed method includes two steps: (1) weight vector orientation estimation and (2) weight vector length estimation, such that the iteration learning of Modular Perceptron Networks (MPN) can be reduced. We have applied the proposed method to the Divide-and-Conquer learning (DCL) for Two-Spiral Problems (TPS) on an MPN. The experimental results show that the number of pattern presentation can be reduced 79.05% (from 190,980 to 40,012), and the number of subnets in MPN can also be reduced 20.38% (from 26.5 to 21.1).

INTRODUCTION

The supervised learning artificialneural networks, such as Multilayer Perceptrons (MLP), can perform pattern recognition, classification and approximation through training processes. However, when the problems or applications related to complicated data distribution, most neural networks were hindered from slow or failure learning. Therefore, investigations of the faster algorithms continue to be pursued and many iterative learning techniques have been developed. A. Van Ooyen proposed a modified back-propagation algorithm [1] to improve the rate of convergence on feedforward neural networks. By combining a bottom-up unsupervised learning process and a top-down supervised learning process, Karayiannis developed a hybrid learning schemes for fast training of feed-forward neural networks [2].

In [3], Chiang et al., proposed a Divide-and-Conquer techniques that partition data set first, and thus generate a new subnetworks to tackle complicated problems, such as Two-Sprial Problems.

In this paper, we propose a weight estimation method to provide a better initial weight to speed up the learning process of feed-forward neural networks. The proposed method includes two steps: one is to decide the weight

*The contact author (Prof. Hsin-Chia Fu) can be reached by Phone: 886-3-573-1930, Email: hcfu@csie.nctu.edu.tw, Fax: 886-3-572-4176.

0-7803-5673-X/99/$10.00 © 1999 IEEE

vector orientation that directs the initial weight vector towards the desired weight vector, and the other is to extend the initial weight vector to be even closer to the destination values. We applied the proposed weight estimation method on the Divide-and-Conquer learning (DCL) for some Modular Perceptron Networks. The experimental results show that the number of pattern presentations can be reduced 79.05% (from 190,980 to 40,012), and the number of subnets in MPN can also be reduced 20.38% (from 26.5 to 21.1).

The rest of this paper are arranged as follows: Section 2 present the weight estimation method that can be used to speed up the learning process. In Section 3, experimental results of various neural architecture are presented and compared. Finally, concluding remarks are given.

WEIGHT ESTIMATION METHOD

This section introduces the weight estimation method that can reduce the learning time on MLP. Without loss generality, we use an MLP containing one hidden layer with two hidden nodes (h_1, h_2) to present the proposed weight estimation method.

Weight vector orientation estimation The weight vector in an MLP acts like a data partition hyperplane in the data space. In [4], Diamantaras and Kung proposed the Oriented Principal Components Analysis (OPCA), which can optimally separate two classes along the direction of oriented principal component. In this paper, we applied the OPCA method to initiate the orientation of the weight vector for data partition hyperplane. According to the OPCA, the direction vector $\mathbf{w}$ that maximizes the energy ratio of two class z and v, both belong to x and $x \in R^n$. That is

$$J_{opc} = \frac{E\{(\mathbf{w}^T z)^2\}}{E\{(\mathbf{w}^T v)^2\}} = \frac{\mathbf{w}^T R_z \mathbf{w}}{\mathbf{w}^T R_v \mathbf{w}}, \tag{1}$$

where $R_z = \{zz^T\}$ and $R_v = \{vv^T\}$. The optimal solution $\mathbf{w}^*$ to (1) is called the principal oriented component of the two classes. (1) shows the principal component of z is steered by the distribution of v, it will be oriented toward the directions where v has minimum energy while trying to maximize the projection energy of z. The principal oriented component is the principal generalized eigenvector of the symmetric generalized eigenvalue problem.

$$R_z \mathbf{w} = \lambda R_v \mathbf{w} \tag{2}$$

where $\mathbf{w} = \{w_1, w_2, \ldots, w_n\}$, $\lambda = \{\lambda_1, \lambda_2, \ldots, \lambda_n\}$. And all the components w_i that maximize (1) should satisfy the following constraints:

$$w_i{}^T R_v w_j = 0; \quad w_i{}^T R_z w_j = 0, \quad \forall j < i. \tag{3}$$

For the initialization of weight vector, the largest generalized eigenvector w are assign to all the hidden nodes in the MLP.

From oriented principal component analysis, one can achieve only the direction of hyperplane for the separation of two classes in the data space. However the proper location along the OPC is still needed for the hyperplane. If we know the data z and v, distribution on the OPC axis, then we can find the intercept that is the optimal solution of discrimination for these data by Bayesian decision rule. As shown in Figure 1, the OPC axis of the two classes has been found, then we can calculate the distribution on the axis. The minimum error position for classification is b_k, so the decision boundary will be constructed.

Let us project the two classes of the data z and v onto the OPC axis, i.e.,

$$z' = w^T z$$
$$v' = w^T v.$$

Assuming the projected data to be D-dimension Gaussian distribution, with the means and variances as $\overline{z'}$, $\overline{v'}$, ${\sigma_{z'}}^2$, ${\sigma_{v'}}^2$, respectively. The conditional probability function $P(z' \mid C_z)$, $P(v' \mid C_v)$ for z' and v' are

$$P(x \mid C_i) = \frac{1}{(2\pi)^{\frac{D}{2}} |\Sigma_i|^{\frac{1}{2}}} \cdot \exp[-\frac{1}{2}(x - \overline{x_i}){\Sigma_i}^{-1}(x - \overline{x_i})] \tag{4}$$

where

$$\begin{aligned} \overline{x_i} &= [\overline{x}_{i1}, \overline{x}_{i2}, \cdots, \overline{x}_{iD}] \quad for \quad i = z, v \\ \Sigma_i &= diag[\sigma_{i1}^2, \sigma_{i2}^2, \cdots, \sigma_{iD}^2] \quad for \quad i = z, v \end{aligned}$$

and the priori probabilities of z and v are $P_z \ \ and \ \ P_v$.

According Bayesian decision rule, if

$$\log P(x \mid C_z) + \log P_z > \log P(x \mid C_v) + \log P_v \ \ for \ \ z \neq v, \tag{5}$$

then the input x belongs to class C_z.

Let us define a discriminate function ϕ as

$$\phi(x \mid C_z) = \frac{1}{2}\sum_{d=1}^{D}[\log \beta_{id} - \beta_{id}(x - \overline{x_i})^2] - \frac{D}{2}\log 2\pi + \log P_z \tag{6}$$

where $\beta_{id} = \frac{1}{\sigma_{id}^2}$.

The decision boundary b_k along OPC for class z and v can be achieved by solving the following equation.

$$\phi(b_k \mid C_z) - \phi(b_k \mid C_v) = 0 \tag{7}$$

Thus, the initial weight and bias for the partition hyperplane can be

$$w^T x + b_k = 0. \tag{8}$$

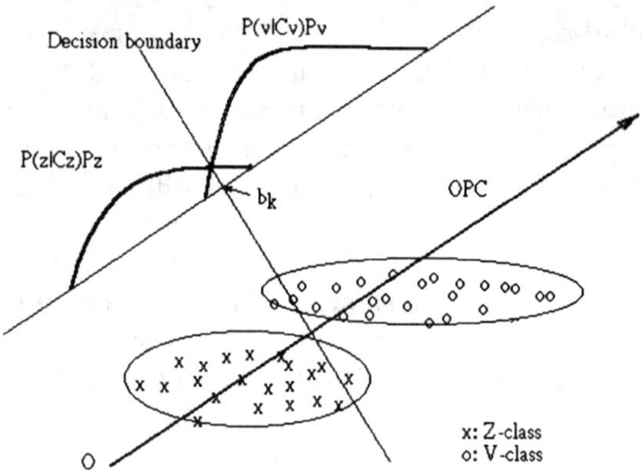

Figure 1: Data projection and decision boundary of two z and v along OPC.

As shown in Figure 1, we can see that the partition hyperplane $w^T x + b_k = 0$ decided by the OPCA and Bayesian decision rule to provide a good partition on the two classes of data z and v.

In the following, we would like to extend the estimated weight w and bias b_k to further closest to the desired weight vectors.

Weight vector length estimation Figure 2 shows an exemplar of error surface contour [5] on a weight space for a hidden node in a MLP. The label w_A is the estimated weight vector; w_B is a linear extension of w_A ($w_B = \gamma \cdot \mathrm{w}_A$); w_C is the weight vector of minimum error (Target weight vector). Suppose we can estimate the γ such that w_B is much closer to w_C, then the following neural learning can be greatly reduced. We define that the activation function of hidden node, f_h, is a hyperbolic function and the linear range is α. The activation function output can be

$$y_{ki} = f_h(w^T x_i + b_k)$$

where x_i is the input pattern i, y_{ki} is the output pattern i of the hidden node k and b_k is the bias of the hidden node k.

Let us look at the hidden node outputs distribution, as shown in Figure 3, the ranges of each hidden is from -1 to 1. The output values are generated from the estimated weight vector very condensed at the central area. It is easy to see that this kind of distribution is quite difficult to be separated by a linear hyperplane at the output node. Since the central region of the hyperbolic activation function is rather linear, thus we suggest to multiply a small constant γ, such that all these output data can be linearly spread in this area. As shown in Figure 4, one can chose a linear range $(-\alpha, +\alpha)$ from $(-0.6, +0.6)$ up to $(-0.9, +0.9)$.

Let y_k^* be the largest $\mid y_{ki} \mid$ for all the output patterns i of the hidden node k, then the linear extension factor can be estimated as that spread the

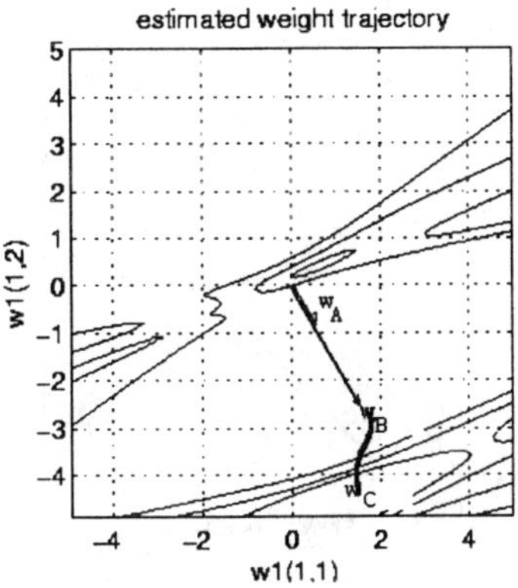

Figure 2: Learning weight trajectory in error surface.

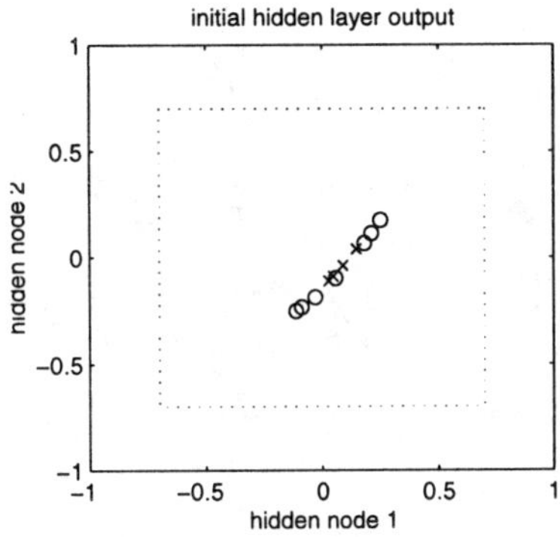

Figure 3: Hidden node output distribution computed according to the estimated weight vector from OPCA.

output

$$\gamma_k = \frac{\alpha}{y_k^*}$$

$$\begin{aligned} y_k' &= \gamma_k \cdot y_k \\ &= ({w'}^T x + b_k'), \end{aligned}$$

where y_k is the set for all the output patterns of the hidden node k.

The new estimated weights w' and b_k' are linear extension of w and b_k. Figure 5 shows the spread hidden node output with $\gamma_1 = 2.8\ and\ \gamma_2 = 3.1$. It is clear to see that the output node vector partitions this spread class more easier.

EXPERIMENTAL RESULTS

In order to demonstrate the learning capability of the weight estimation method, we choose Two-Spiral Problems to benchmark the weight estimation method. In the previous researches, the Cascade Correlation learning algorithm [6] uses 1,500 epochs or 1,500*194=291,000 pattern presentations and the Hybrid learning schemes for fast training method [2] need 28,847 epochs or 28,847*150=4,327,050 pattern presentations for 150 patterns. Both of their number of the pattern presentations are from 7 to 108 times more than the results of the proposed method with DCL. In [3], Chiang et al., provide DCL technique that can generate $25 \sim 28$ subnetworks, $1,800 \sim 2,500$ epochs and $151,810 \sim 230,130$ presentations to learn this problem. We apply the weight estimation method to DCL that has a significant improvement in learning time. In general, one epoch is defined as a single presentation of

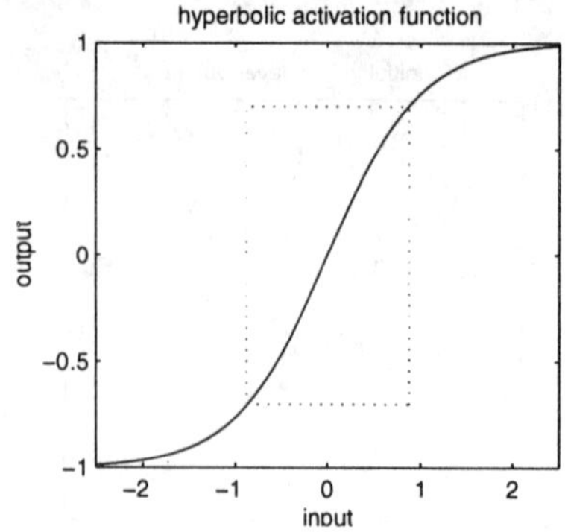

Figure 4: This is activation function of the hidden node. The dot line enclose linear regoin

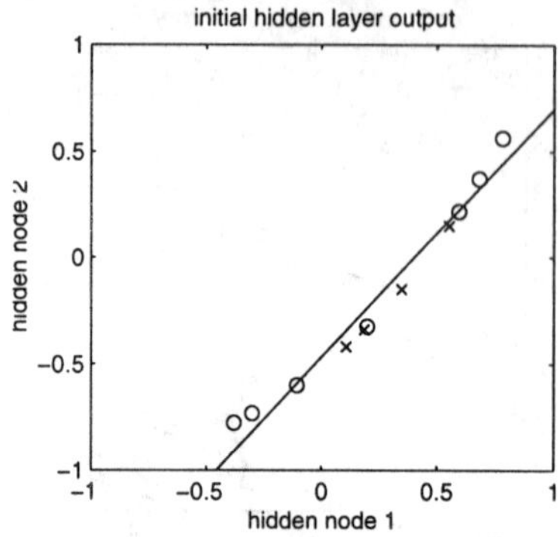

Figure 5: The output data of hidden layer with estimated weight and linear extension with $\gamma_1 = 2.8$, $\gamma_2 = 3.1$.

each of the I/O patterns in the training set. It should be noticed that it is appropriate to compare both methods in terms of number of presentations. This is because that the number of pattern presentations in each epoch will be lesser and lesser as the DCL iteration continues, so the presentation is a more natural measure to use in problems that do not cycle through the entire training set between weight-update. In this experiment, we use two different MLPs as the kernel model of subnetworks; one is a conventional MLP that contains one hidden layer with two hidden nodes, the other is a modified MLP called dynamic-threshold quadratic sigmoidal neural networks (QSNN) [7]. The QSNN has the same architecture as the conventional MLP, except it uses a different activation function called QSNN for each hidden neuron. We also applied the DCL methodology [3] on these two types of MLP.

The basic idea of DCL is described briefly as follows. For each training process, if the mean squared training error cannot be improved by at least 0.001 with 15 training epochs, then current data set is partitioned according to the Error Correlation Partitioning method[3] and the MLP is also split into two subsets to conquer each data subset. We also use $0.1225 (= 0.35^2)$ as the squared error tolerance of each training pattern.

Table 1: Experimental result of various MLP type neural networks by DCL and Weight Estimation for Two-Spiral Problems. *Experimental results were performed by [3].

Type of NN	No. of epochs	No. of presentations	No. of subnets
MLP	11266 ±1013	88525 ±7703	52.8 ±5.26
MLP w/est. *wt.*	7494 ±744	84847 ±4162	35.3 ±3.6
QSNN*	2150 ±350	190980 ±39060	26.5 ±1.5
QSNN w/est. *wt.*	2082 ±478	40012 ±6905	21.1 ±2.3

In Table 1, the learning performance according to the number of epochs, number of presentations, and number of subnetworks are presented. Using the weight estimation method on the MLP, the weight estimation can reduce the number of the epochs and the subnetworks about 33.48% and 33.14% respectively. Also, the number of presentation are reduced 4.15%. For the QSNN, the weight estimation can significantly reduce the number of the presentations and the subnetworks about 79.05% and 20.38% respectively, and the number of epochs is reduced about 3.16%.

We would like to comment on the poor improvement on the number of presentations for the MLP with weight estimation. As we can see in Figure 6, the two classes of data patterns in the TSP are mixed each other. Thus at the learning stages of DCL, weight estimation would not be very helpful. However, after a few iterations of data partition, the weight estimation becomes useful thus the number of epochs and number of subnetworks reduced significantly.

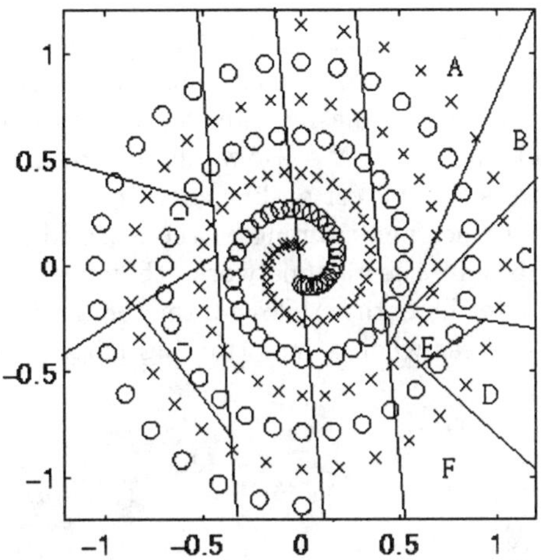

Figure 6: A part of learning process of the Two-Spiral problems

The generalization performance is an important index to evaluate the proposed method. We define a factor l that is a nearest distance of the two sampling points that belong to two classes of TSP. l is also multiplied by a scale (e.g. $0.1, 0.2, 0.3, 0.4$) as a normal distribution testing range corresponding to each sampling point of TSP. In these ranges, 1940 normal distribution testing data are generated and repeat 10 times to evaluate generalization performance. The test results can be observed from Table 2. The average generalization of the DCL method without weight estimation is 96.85%. The average generalization of the DCL with weight estimation using QSNN can reach 97.9%. The generalization is well for the TSP which is divided into some small simple subproblems, so the overall performance can keep a fairly good level through DCL methodology.

Table 2: Generalization of various MLP type neural networks by DCL and Weight Estimation for Two-Spiral Problems. We define a factor l that is a nearest distance of the two sampling points that belong to two classes of TSP. l is also multiplied by a scale (e.g. $0.1, 0.2, 0.3, 0.4$) as a normal distribution testing range corresponding to each sampling point of TSP. In these ranges, 1940 normal distribution testing data are generated and repeat 10 times to evaluate generalization performance. *Experimental results were not available in [3].

	performance in different testing distribution range			
Type of NN	0.1 l	0.2 l	0.3 l	0.4 l
MLP	98.92 ±0.37	97.80 ±0.49	96.30 ±0.61	94.36 ±0.69
MLP w/est. *wt.*	98.92 ±0.58	98.30 ±0.50	97.35 ±0.62	95.81 ±0.59
QSNN*	NA	NA	NA	NA
QSNN w/est. *wt.*	99.41 ±0.42	98.54 ±0.70	97.41 ±0.87	96.23 ±0.68

CONCLUDING AND REMARKS

In this paper, we propose the weight estimation method for MLP type neural networks. Although, we present the concept on a MLP containing one hidden layer with two hidden nodes. However, it can be easily extended to multiple hidden layers and multiple hidden nodes per layer. We also would like to claim that the MPN with weight estimation is capable for on-line and incremental learning problems. It is easy to see that when new training patterns are added to the original training set, there are only a few subnetworks that need to be trained. It is need not to recalculate the entire MPN parameters and to learn all training patterns learning process when a new training pattern is made available.

REFERENCES

[1] A.V. Ooyen, B. Nienhuis, " Improving the convergence of the back-propagation algorithm," **Neural Networks**, vol. 5, pp. 465-471, 1992.

[2] N.B. Karayiannis, " Hybrid learning schemes for fast training of feed-forward neural networks," **Mathematics and Computers in Simulation**, vol. 41, pp. 13-28, 1996.

[3] C.C. Chiang, H.C. Fu, " A divide-and-conquer methodology for modular supervised neural network design," **International Conference on Neural Networks,** pp.119-124, 1994.

[4] K.I. Diamantaras, S.Y. Kung, **Principal component neural networks**, John Wiley and Sons Inc., 1997.

[5] Christopher M. Bishop, **Neural Networks for Pattern Recognition** Clarendon press Oxford, 1995.

[6] S.E. Fahlman, C. Lebiere, " The cascade-correlation learning architecture," In David S. Touretzky, editor, **Advances in Neural Information Processing Systems,** vol. II, pp. 524-532, Morgan Kaufmann, San Mateo, CA, 1990.

[7] C.C. Chiang, H.C. Fu, "On the classification capability of dynamic threshold neural networks," **Pattern Recognition Letters**, vol. 15, no. 1, 1994.

General Statistical Inference by an Approximate Application of the Maximum Entropy Principle *

Lian Yan and David J. Miller
Department of Electrical Engineering
The Pennsylvania State University
University Park, PA 16802-2701
Phone: (814) 865-6510 E-mail: *miller@bilbo.ee.psu.edu*

Abstract

We propose a new learning method for building a general statistical inference engine, operating on discrete feature spaces. Such a model allows inference on *any* feature given values for the other features (or for a feature subset). *Bayesian networks* (BNs) are versatile tools that possess this inference capability. However, while the BN's explicit representation of conditional independencies is informative, this structure is not so easily learned. Typcially, learning methods for BNs use (sub-optimal) greedy search techniques. There is also a difficult issue of overfitting in these models. Alternatively, in 1983 Cheeseman proposed finding the *maximum entropy* (ME) joint pmf consistent with *arbitrary* lower order probability constraints. This approach has some potential advantages over BNs. However, the huge complexity required for learning the joint pmf has severely limited the use of this approach until now. Here we propose an *approximate* ME method which also allows incorporation of arbitrary lower order constraints, but while retaining quite tractable learning complexity. The new method approximates the joint feature pmf (during learning) on a *subgrid* of the full feature space grid. Experimental results on the UC-Irvine repository reveal significant performance gains over two BN approaches: Chow and Liu's dependence trees and Herskovits and Cooper's *Kutato*. Several extensions of our approach are indicated.

1 Introduction

We tackle the difficult problem of learning the joint probability mass function (pmf) for a discrete (possibly high-dimensional) feature space. The joint pmf is a rich

*This work was supported in part by National Science Foundation Career Award IIS-9624870.

0-7803-5673-X/99/$10.00 © 1999 IEEE

model that, once learned, allows one to address important inference tasks that appear in a variety of applications: computing *a posteriori* probabilities 1) for one feature given knowledge of the remaining feature values (the basis for pseudo-Bayes discrete feature space classification); 2) with some features *missing*; 3) for *any* (e.g. user-specified) feature dimensions given values for the other features (or given values for a feature subset). Capabilities 1) and 2) allow one to tackle classification/recognition. The more general inference objective 3) is a *generalized* classification task where, in principle, during model use, *any* subset of the features can be chosen as "class" features, to be inferred based on values for other features. Thus, the model must be learned consistent with its potential application to *any* of the (combinatoric !) inference tasks a user may select. The ability to address this problem is one of the main features of *probabilistic expert systems*, found useful in medical diagnosis [7], fault detection/troubleshooting [6], information retrieval [4], as well as other domains. As one medical example, symptoms may be consistent with simultaneous presence of *multiple* diseases. Here, we may wish to infer the presence of any one of several diseases, possibly with *or* without knowledge of the presence of other diseases. Each disease is thus a known feature (model input), a missing feature we do not need to infer, or a feature to be predicted (model output), depending on the inference problem a doctor needs to solve. Fault detection with multiple candidate faulty components is a similar example. Imputation of (arbitrary patterns of) missing fields in databases and questionnaires is a third application.

While there are many basic statistical classification approaches, e.g. neural networks, statistical methods, decision trees, there are far fewer techniques for tackling general inference (3)). To apply multilayer perceptrons, we would need to train one network for each feature that may need to be inferred. Moreover, if multiple features need to be inferred at the same time, each such network will have *multiple missing inputs* (the other features that need to be inferred) – the resulting system, based on some heuristic imputation strategy, would likely perform poorly. One model more realistic for task 3) is referred to as *probabilistic/Bayesian/belief networks* (BNs) [12], [8], [5],[1]. BNs represent the joint pmf for a feature space explicitly as a product of conditional probabilities over the constituent scalar features – they thus explicitly model conditional independence relations between features. While this is informative to a model user, the conditional independence structure is not so easily learned given (finite) training data[1] First, conditional independence is not so easily assessed given finite data. Second, optimizing over the set of possible structures is typically accomplished by sequential, greedy methods which are likely suboptimal. Third, associated with sequential learning of the model, there is the difficult issue of where to stop, so as to avoid overfitting. Finally, consider the case where, among a subset of the features (e.g. $\{F_1, F_2, \ldots, F_s\}$), there are *no* conditional independencies. In this case, the learning method may either i) incorrectly impose some or ii) express no conditional independencies, but then require estimation of high order probabilities, e.g. $P[F_1|F_2, F_3, \ldots, F_s]$ given finite data.

In this work, we avoid some of these problems by using an alternative joint pmf representation based on a set of model parameters rather than a set of conditional probabilities. Our representation does not require making explicit conditional

[1] In fact, originally these models were not learned from an existing database, but rather were formed by eliciting conditional independencies and probabilities from domain experts.

independence choices. Moreover, our model is amenable to an effective *joint* optimization learning technique which we propose here. While there is a potential overfitting issue for our model just as for BNs, we propose a strategy for model order selection that is simple and yet well-supported by our experimental results. One disadvantage of our model is that, since it does not make conditional independencies explicit, it is not informative to a user about which features are important for making inferences. In this sense, our model is closer to (black box) neural networks. However, we are willing to sacrifice informativeness for improved accuracy of inferences.

Our work builds on earlier work of Cheeseman [2] and Ku and Kullback [10]. Cheeseman proposed to find the *maximum entropy* (ME) joint pmf consistent with *arbitrary* lower order probability constraints. The ME principle has been championed as a general tool for statistical inference by Jaynes [9], and has been justified on information-theoretic grounds [13]. Successful application of this principle rests with the amount and accuracy of domain knowledge that is encoded in the joint pmf via the constraints. Cheeseman's suggestion to incorporate arbitrary lower order constraints is in principle quite powerful, allowing the joint pmf to express general dependencies between features. Such knowledge cannot be fully encoded in BNs.

Unfortunately, a powerful model often entails a difficult learning problem for estimating the model's *parameters*. Ku and Kullback [10] proposed a "simple" iterative algorithm for ME joint pmf estimation which satisfies one constraint at a time. However, at each step, their method requires marginalizations over the joint pmf for constraint satisfation with complexity exponential in the feature dimensionality. This computation is intractable for even moderately dimensioned feature spaces. In [10], results were only presented for dimension four, with two discrete values per feature. Moreover, the method in [10] computes the joint pmf values *directly* which, given that the computation is performed off-line (the only even remotely feasible way), implies enormous storage requirements.

Cheeseman recognized the potential significance of the ME approach for machine learning and showed how to use the model to do general inference. However, no practical learning method was proposed and experimental results were not presented. As a practical alternative to learning the true ME joint pmf, Herskovits and Cooper [8] proposed an algorithm called *Kutato* which applies the ME principle in a heuristic way for the construction of BNs. They recognized the work in [10] and [2], but noted that the intractable learning complexity for the general ME approach is one of the main reasons for seeking other models. Pearl [12] also cites complexity as a main barrier to using Cheeseman's approach. In summary, while interest in BNs has increased over the past 15 years, the general ME approach appears to have been largely forgotten since Cheeseman's 1983 paper, presumably due to the learning complexity.

In recent work [11], we reconsidered the ME problem with the specific aim of designing discrete space classifiers (models for inferring a single, fixed feature). We introduced an *approximate* ME solution whose key aspect was the restriction (during learning) of joint pmf support to a *subgrid* of the full feature space grid. What this "subgrid approximation" entails will be explained shortly. Here, we build on [11], extending this approach in a significant way so as to learn a (task 3)) general inference model. Like earlier ME work, our approach encodes *arbitrary* probability constraints, but unlike previous methods, while also retaining quite tractable learning complexity. We thus suggest a practical approach for approximately realizing ME inference. In the next section, we consider the ME problem in more detail to highlight inherent difficulties. We then develop our formulation for learning a general inference engine. Experimental results will confirm the effectiveness of our technique. We then identify several promising extensions.

2 The Maximum Entropy Joint PMF

Consider the general problem of obtaining the joint pmf for a feature vector $\underline{F} = (F_1, F_2, \ldots, F_N)$, with $F_i \in \mathcal{A}_i$, $\mathcal{A}_i$ a finite set $\{1, 2, 3, \ldots, |\mathcal{A}_i|\}$. The problem posed by [10] and [2] is that of choosing $P[\underline{F}]$ to maximize joint entropy while satisfying general equality constraints on lower order probabilities. To illustrate the difficulties, it suffices to consider second order (pairwise) pmf constraints. Thus, suppose we are given $\{P[F_m, F_n], \forall m, \forall n > m\}$. The ME joint pmf consistent with these $M = \frac{N(N-1)}{2}$ pairwise pmfs has the form:

$$P[\underline{F} = \underline{f}] = \frac{e^{\sum\limits_{l=1}^{M} \gamma_l^{(f_{i_1(l)}, f_{i_2(l)})}}}{\sum\limits_{\text{all } \underline{f}'} e^{\sum\limits_{l=1}^{M} \gamma_l^{(f'_{i_1(l)}, f'_{i_2(l)})}}}, \quad \underline{f} \equiv (f_1, \ldots, f_N) \in \mathcal{A}_1 \times \mathcal{A}_2 \times \cdots \times \mathcal{A}_N. \qquad (1)$$

Here, the γ variables are Lagrange multipliers, each corresponding to a pairwise constraint. Assuming the M constraints are ordered in some way, the index mappings $i_1(l)$ and $i_2(l)$ identify the lth feature pair. Note further that for each pair (F_m, F_n), there are $|\mathcal{A}_m||\mathcal{A}_n|$ possible Lagrange multipliers, one for each value taken on by the pair. Thus, l indicates the pair, with the remaining indices (superscripts) on γ identifying the 'active' pair value given a particular $\underline{f}$. While the form of the joint pmf is known, direct computation of (1) is generally intractable since calculation of the partition function in the denominator requires computing and summing $\prod\limits_{i=1}^{N} |\mathcal{A}_i|$ terms. This may not be problematic during the model's use phase, since one is generally not interested in $P[\underline{f}]$ directly, but rather in *a posteriori* probabilities of single features or feature groups. Such quantities, which depend on the joint pmf's Lagrange multipliers, but not directly on $P[\underline{f}]$, can often be tractably computed.

However, all of this assumes the model parameters $\Gamma = \{\gamma_l^{(j,k)}, l = 1, \ldots, M, (j,k) \in \mathcal{A}_{i_1(l)} \times \mathcal{A}_{i_2(l)}\}$ are known. The real difficulties lie in the learning problem for acquiring the γ's. We have already described the difficulties with the methods of [10], [2]. Since the form of the joint pmf is known, one is tempted to try a direct learning approach: i) write down the ME constrained optimization problem; ii) use the Lagrangian method to convert the constrained problem into an unconstrained one. The Lagrangian cost will have the form $D - TH$, wth H the joint entropy, D a cost function encapsulating the constraints, and T a Lagrange multiplier; iii) express H and D *explicitly* as functions of the joint pmf with optimal form (1); iv) minimize $D - TH$ over Γ. Some search for the appropriate value of T, satisfying constraints with maximum entropy, will be required.

There are two major difficulties with this approach. First, the optimization requires explicitly calculating $P[\underline{f}]$, intractable as already discussed. Second, the constraint cost D will require marginalizations over the joint pmf similar to those of Ku and Kullback's method – these are also generally intractable. While the method just outlined is clearly infeasible, the approximate method next proposed is somewhat inspired by the principle of this direct approach.

3 General Statistical Inference Formulation

Our objective is to learn the Lagrangian model Γ consistent with maximizing joint entropy while satisfying given lower order constraints (for concreteness, we assume pairwise pmf constraints). In order to achieve a learning method with tractable complexity we must avoid the normalization required for (1) and the intractable marginalizations required by [10],[2]. In recent work [11], we developed a method for designing a *statistical classifier* consistent with the ME principle. This approach introduced several key ideas for avoiding sources of intractability while (approximately) learning the ME model. Here, we build on this work in constructing a more general-purpose inference engine.

The intractability of (1) can be avoided by using an alternative representation for the joint pmf. Consider the feature vector $\underline{F} = (F_1, F_2, \ldots, F_N)$, where now we may wish to infer probabilities for *any* of the features F_i. For inference on F_i, the *a posteriori* probabilities have the (tractable) ME form:

$$P[F_i = f_i | \underline{F}^{(-i)} = \underline{f}^{(-i)}] = \frac{e^{\sum_{j=1, j\neq i}^{N} \gamma_{i,j}^{(f_i, f_j)}}}{\sum_{f_i'=1}^{A_i} e^{\sum_{i=1, j\neq i}^{N} \gamma_{i,j}^{(f_i', f_j)}}}, \quad f_i = 1, \ldots, A_i. \tag{2}$$

Here, $\gamma_{i,j}^{(f_i, f_j)}$ denotes the Lagrange multiplier associated with the pair (F_i, F_j), taking on pair value (f_i, f_j), with $\underline{F}^{(-i)}$ the feature vector $\underline{F}$ excluding F_i. Given the *a posteriori* probabilities, we can decompose the joint pmf with respect to each feature F_i as

$$P[\underline{F} = \underline{f}] = P[F_i = f_i | \underline{F}^{(-i)} = \underline{f}^{(-i)}] P[\underline{F}^{(-i)} = \underline{f}^{(-i)}], \ \forall \underline{f}, \tag{3}$$

with $P[f_i / \underline{f}^{(-i)}]$ expressed in its ME form (2), but, crucially, with $\{P[\underline{F}^{(-i)}], i = 1, \ldots, N\}$ instead specified by an exhaustive table of joint pmf values. By decomposing with respect to *each* feature, we obtain alternative representations for the joint pmf which, *collectively*, involve the full complement of Lagrange multipliers $\Gamma = \{\{\gamma_{i,j}^{(f_i, f_j)}\}, i = 1, \ldots, N\}$. So far, each such joint pmf representation introduces no suboptimality in principle, as there are choices for the "parameter sets" $\{\{\gamma_{i,j}^{(f_i,f_j)}\}, P[\underline{F}^{(-i)}]\}$ which achieve the ME joint pmf consistent with the given constraints. However, it does not appear that this alternate representation provides a practical advantage during learning. In fact, this representation appears to complicate learning, by adding the reduced pmfs $\{P[\underline{F}^{(-i)}], i = 1, \ldots, N\}$ to the set of parameters that require learning. However, while optimizing over $\{\{\gamma_{i,j}^{(f_i,f_j)}\}, P[\underline{F}^{(-i)}]\}$ is not viable, a related *approximation* of this approach will indeed yield tractable, effective (approximate) learning of the desired model Γ. The key idea is to view the reduced pmfs $\{P[\underline{F}^{(-i)}]\}$ as *auxiliary* variables which support the learning process, but which play *no* role during model use/inference. Thus, we will represent these pmfs only approximately if this is helpful in achieving tractable learning. The approximation may affect the accuracy of the learned model Γ, but will not sacrifice the *capability* for doing inference. So in particular, let us assume during learning that the joint pmf $P[\underline{F}]$, from which each of the reduced pmfs is formed, has non-zero probability support *only* on a *subgrid* $\mathcal{G}_s$ of the full feature space grid $\mathcal{G} \equiv \mathcal{A}_1 \times \mathcal{A}_2 \times \cdots \times \mathcal{A}_N$. How to choose this subgrid will be addressed shortly. For now we focus on developing our method and identifying what makes it computationally tractable.

The support restriction on the full joint pmf will induce a support restriction on each of the reduced joint pmfs. In particular, consider the reduced feature vectors $\underline{f_m}^{(-i)}, m = 1, \ldots, |\mathcal{G}_s|$, obtained by removing the ith component from each $\underline{f_m} \in \mathcal{G}_s$. If each full feature vector $\underline{f_m}$ yields a *distinct* reduced feature vector $\underline{f_m}^{(-i)}$, then the reduced joint pmf for $\underline{F}^{(-i)}$ is simply:

$$P[\underline{f_m}^{(-i)}] = P[\underline{f_m}], \;\; m = 1, \ldots, |\mathcal{G}_s|. \tag{4}$$

Alternatively, if more than one full feature vector yields the same reduced vector $\underline{f}^{(-i)}$, *i.e.* if there is a set $S(\underline{f}^{(-i)}) \equiv \{\underline{f_m} \in \mathcal{G}_s : \underline{f_m}^{(-i)} = \underline{f}^{(-i)}\}$ of size greater than one, then we have

$$P[\underline{f}^{(-i)}] = \sum_{m:\underline{f_m} \in S(\underline{f}^{(-i)})} P[\underline{f_m}]. \tag{5}$$

Using either (4) or the more general expression (5), we can derive each reduced joint pmf from the *same* (restricted support) full joint pmf $\{P[\underline{f_m}], \;\; m = 1, \ldots |\mathcal{G}_s|\}$. However, for convenience of expression we will assume in the sequel that reduction of all subgrid vectors $\underline{f_m}$ along the same feature dimension i produces *distinct* reduced vectors[2]. Thus, $|\mathcal{G}_s^{(-i)}| = |\mathcal{G}_s|$ and (4) is satisfied, for all i.

We can thus pose our general ME learning problem as choosing $\{\{\gamma_{i,j}^{(f_i,f_j)}\}, \{P[\underline{f}], \underline{f} \in \mathcal{G}_s\}\}$ to maximize joint entropy while satisfying the given constraints. Following the 'direct approach' to attacking the ME problem, we form the quantities H and D. The joint entropy can be written as

$$H \;\; = \;\; -\frac{1}{N} \sum_{k=1}^{N} (\sum_{m=1}^{|\mathcal{G}_s|} P[\underline{f_m}^{(-k)}] \log P[\underline{f_m}^{(-k)}]$$

[2]The subgrid $\mathcal{G}_s$ can be chosen to ensure this condition.

$$
\begin{aligned}
&+ \sum_{m=1}^{|\mathcal{G}_s|} P[\underline{f_m}^{(-k)}] \sum_{i=1}^{|\mathcal{A}_k|} P[F_k = i|\underline{f_m}^{(-k)}] \log P[F_k = i|\underline{f_m}^{(-k)}]) \\
&= -\sum_{m=1}^{|\mathcal{G}_s|} P[\underline{f_m}] \log P[\underline{f_m}] \\
&- \frac{1}{N} \sum_{k=1}^{N} \sum_{m=1}^{|\mathcal{G}_s|} P[\underline{f_m}] \sum_{i=1}^{|\mathcal{A}_k|} P[F_k = i|\underline{f_m}^{(-k)}] \log P[F_k = i|\underline{f_m}^{(-k)}], \qquad (6)
\end{aligned}
$$

with $\underline{f_m} \in \mathcal{G}_s, m = 1, \ldots, |\mathcal{G}_s|$. Suppose we wish to satisfy constraints on all pairwise pmfs $\{P[F_k, F_l]\}$. This can be accomplished by minimizing, i.e. *zeroing* a cross entropy criterion between the measured pmfs and those computed by our model. The *model's* pairwise pmf $P_M[F_k, F_l]$ can be calculated both as

$$
P_M[F_k = i, F_l = j] = \sum_{\underline{f_m} \in \mathcal{G}_s | f_k^{(m)} = i} P[F_l = j|\underline{f_m}^{(-l)}] P[\underline{f_m}], \qquad (7)
$$

and as

$$
P_M[F_k = i, F_l = j] = \sum_{\underline{f_m} \in \mathcal{G}_s | f_l^{(m)} = j} P[F_k = i|\underline{f_m}^{(-k)}] P[\underline{f_m}],
$$
$$
i = 1, \ldots, |\mathcal{A}_k|, \quad j = 1, \ldots, |\mathcal{A}_l|. \qquad (8)
$$

We must constrain both representations to agree with the measured pmf $P[F_k, F_l]$. Accordingly, our *overall constraint cost* D is formed as a sum of cross entropies:

$$
D = \sum_{k=1}^{N} \sum_{l=1, l \neq k}^{N} \sum_{i=1}^{|\mathcal{A}_k|} \sum_{j=1}^{|\mathcal{A}_l|} (P[F_k = i, F_l = j] \log \frac{P[F_k = i, F_l = j]}{P_M[F_k = i, F_l = j]}). \qquad (9)
$$

Given the quantities D and H, we form the Lagrangian cost $L \equiv D - TH$. Our ultimate goal is to minimize L such that we find the solution with maximum H under the constraint $D = 0$. We have found that this can be achieved by a *continuation* method, minimizing L for a sequence of decreasing T values, tracking parameter values from one T to the next, and stopping when D drops below a suitable threshold or stops changing, with the minimization at each T achieved by gradient descent. While this optimization approach is reminiscent of a *deterministic annealing* strategy, with $D - TH$ the Helmholtz free energy, D the 'energy' and T the 'temperature', we do not emphasize this here, especially since we have found in practice that little variation of T (annealing) is required – we can start from fairly small T (e.g. $T = 0.1$) and simply run the continuation approach until our stopping condition is reached. Starting from higher T does not typically yield better results. We have found that this algorithm, though somewhat complex, is *quite* tractable – we are easily able to obtain results (better than or competitive with alternative techniques) for high-dimensional problems from the UC-Irvine ML repository. For example, for *Zoo* (17 dimensions, 2-6 values per feature), with $|\mathcal{G}_s| = 70$, the learning time on a Sparc10 is $\sim$ 1.5 hours; for *Congress* (17 dimensions, 2-3 values per feature), with $|\mathcal{G}_s| = 300$, the learning time is 12 hours. However, this can be reduced substantially by decreasing $|\mathcal{G}_s|$, with little effect on performance. One important consideration untouched until now is the *choice* of $\mathcal{G}_s$. While we suspect it is quite difficult to directly optimize over the possible $\mathcal{G}_s$ of a given size, in practice this is not necessary – typically, there is a natural choice *posited by the available training data set itself*! Thus, in practice, if the training set size is not too great, we simply choose the training feature vectors as the subgrid locations. Otherwise, we randomly sample from the training set to form the subgrid. Given the resulting model $\{\gamma_{i,j}^{(f_i,f_j)}\}$, we can calculate the *a posteriori* probabilities of any single feature or collection of features, and do exact inference with missing values (this can be practically achieved if only a few values are missing).

4 Experimental Results

Experiments were performed on several sets from the UC-Irvine repository. The (training/test) split for each set was as follows: *Mushroom* (4000/4124), *Nursery* (6480/6480), *Congressional voting records* (300/135), *Hepatitis*[3] (100/55), and *Zoo*[4] (70/31). We compared our general inference ME method with *Kutato* and with the dependence trees of [3]. *Kutato* constructs a BN from a data set by applying the ME principle in a heuristic fashion. Chow and Liu's dependence tree is a simple BN with very stringent conditional independence constraints, but low learning complexity. We implemented two schemes based on dependence trees: 1) we used a single dependence tree to represent the joint pmf and to solve all inference problems; 2) we built *multiple* dependence trees, one for each possible value of each feature upon which inference may be performed. For each such tree, all probabilities are conditioned on this associated feature value. Inference on a particular feature is then performed using the subset of trees that condition on different values for this feature of interest. We refer to this approach as *multiple perspective* (MP) dependence trees. Because the data sets were (presumably) constructed for inference solely on the class label, some features have very poor predictablity based on other features. Thus, we only selected a *subset* of features as candidate class features. The selected subset, which we call the *inference subset*, consisted of the features for which prediction accuracy (averaged over all four inference methods) was above a threshold. We used the following performance criteria: 1) misclassification rate on the test set for the data set's class label; 2) same as 1) but with a single feature missing randomly during use; 3) average misclassification rate on the test set, based on predicting each feature in the inference subset; 4) misclassification rate on the test set, based on predicting a pair of randomly chosen features from the inference subset. In this case, an error is counted unless both features are correctly classified. Our results are shown in Table 1. The MP dependence trees encode a *subset* of *third* order conditional probabilities and *Kutato* encodes a subset of even higher order conditional probabilities. By contrast, we encoded only (second order) pairwise

[3]For *Mushroom* and *Hepatitis*, the indication of a missing feature was treated as another possible value ('missing'), during both training and use.

[4]For *Zoo*, we combined classes with small population, reducing the class cardinality to four.

pmfs, *but all* $\frac{N(N-1)}{2}$ *such.* The results confirm the consistent advantage of our ME approach – the method is significantly better than all the BN approaches except for a few cases[5]. We have also done some initial experiments on the inclusion of *higher order* constraints, using *Kutato* to (heuristically) seek them out – for each conditional pmf used by *Kutato* we added constraints on the corresponding joint pmf to our (second-order) model. However this approach was not found to yield performance gains over our fixed (all second-order constraint) method. Moreover, we found that imposing *all* third order constraints led to very poor generalization, further vindicating the choice of second order constraints.

Approaches		Mushroom	Congress	Nursery	Zoo	Hepatitis
ME	1	0.1918	0.0593	0.2369	0.0968	0.2545
	2	0.1901	0.0667	0.3029	0.0645	0.2364
	3	0.1529	0.1901	0.2635	0.1290	0.2400
	4	0.2733	0.2889	0.7006	0.2903	0.3455
Kutato	1	0.3172	0.0667	0.5265	0.0968	0.3091
	2	0.2956	0.0741	0.5316	0.0968	0.3091
	3	0.2211	0.2006	0.5373	0.1270	0.2727
	4	0.3773	0.3556	0.9182	0.3226	0.4000
C & L	1	0.5061	0.0889	0.6463	0.1613	0.4727
	3	0.5326	0.2340	0.6454	0.2177	0.4455
	4	0.6753	0.3630	0.9778	0.3226	0.6545
MP C & L	1	0.4258	0.1407	0.6451	0.3548	0.3637
	2	0.4554	0.1704	0.6483	0.4516	0.3636
	3	0.1903	0.1031	0.5892	0.1250	0.3491

Table 1: Test set misclassification rates of several methods on data sets from the UC Irvine repository.

5 Conclusions

Both BNs and general ME inference have been around for quite some time. Yet, while BNs have emerged as a widely used and much studied inference tool, the general ME model, in spite of its potential advantages, has received little attention beyond the work of Cheeseman in 1983, apparently due to the intractability of learning the model. Here, we reconsidered the ME learning problem, proposing an approximate method with quite tractable complexity and excellent performance, as demonstrated in comparison with several BN techniques. We thus provide a way of approximately achieving general ME inference. Future work will consider several promising extensions for: i) addressing mixed discrete/continuous feature space

[5]For criterion 3), over three data sets, our model was not only better than the single model *Kutato* produced using an entropy-based stopping criterion, but also better than the *best* model (not determinable in practice) from the sequence of *Kutato* models.

[14]; ii) tackling regression/time series prediction, possibly with missing features; and iii) *sequential model building*, i.e. adding constraints on *individual* probabilities of *any* order, with the chosen constraint on the probability least satisfied by the current model. We also plan to explore compelling inference applications such as diagnosis/fault detection, information retrieval, and imputation in databases.

References

[1] W. Buntine. A guide to the literature on learning probabilistic networks from data. *IEEE Transactions on Knowledge and Data Engineering*, 8:195–210, 1996.

[2] P. Cheeseman. A method of computing generalized Bayesian probability values for expert systems. In *Proceedings of the Eighth International Joint Conference on Artificial Intelligence*, volume 1, pages 198–202, 1983.

[3] C. Chow and C. Liu. Approximating discrete probability distributions with dependence trees. *IEEE Transactions on Information Theory*, 14:462–467, 1968.

[4] R. Fung and B. D. Favero. Applying Bayesian networks to information retrieval. *Communications of the ACM*, 38(3):42–48, 1995.

[5] D. Heckerman, A. Geiger, and D. M. Chickering. Learning Bayesian networks: the combination of knowledge and statistical data. *Machine Learning*, 20:197–243, 1995.

[6] D. E. Heckerman, J. S. Breese, and K. Rommelse. Decision-theoretic troubleshooting. *Communications of the ACM*, 38:49–57, 1995.

[7] D. E. Heckerman, E. J. Horvitz, and B. N. Nathwani. Toward normative expert systems: Part I. The Pathfinder project. *Methods of Inform. Med.*, 31:90–105, 1992.

[8] E. Herskovits and G. Cooper. Kutato: an entropy-driven system for construction of probabilistic expert systems from databases. In *Uncertainty in Artificial Intelligence 6*, pages 117–125, 1991.

[9] E. T. Jaynes. *Papers on Probability, Statistics and Statistical Physics.* Reidel, Dordrecht, 1982.

[10] H. H. Ku and S. Kullback. Approximating discrete probability distributions. *IEEE Transactions on Information Theory*, 15(4):444–447, 1969.

[11] D. J. Miller and L. Yan. Approximate maximum entropy joint feature inference for discrete space classification. To appear in *Proceedings of International Joint Conference on Neural Networks*, 1999.

[12] J. Pearl. *Probabilistic reasoning in intelligent systems: networks for plausible inference.* Morgan Kauffman, San Mateo, Ca., 1988.

[13] J. E. Shore and R. W. Johnson. Axiomatic derivation of the principle of maximum entropy and the principle of minimum cross-entropy. *IEEE Transactions on Information Theory*, 26:26–37, 1980.

[14] L. Yan and D. J. Miller. General statistical inference for discrete and mixed spaces by an approximate application of the maximum entropy principle. Submitted to *IEEE Transactions on Neural Networks*, 1999.

ON TRAINING PIECEWISE LINEAR NETWORKS

Amir Atiya
Department of Electrical Engineering
Caltech 136-93
Pasadena, CA 91125
amir@work.caltech.edu

Emad Gad
Department of Electrical Engineering
Carleton University
Ottawa, Canada

Samir Shaheen
Department of Computer Engineering
Cairo University
Giza, Egypt

Ayman El-Dessouky
Informatics Research Institute, MCSRTA
Alexandria, Egypt

Abstract. Piecewise-linear (PWL) neural networks are networks with piecewise-linear node functions. They have attractive features, such as speed of training and amenability to digital VLSI implementation. This paper presents a new algorithm for training PWL networks. The algorithm is general in that it can be used for the optimization of general PWL functions. It is based on moving from one linear region to the next. This is achieved by exploring $2N$ specific directions along the boundaries between the linear regions (N is the dimension), and moving along that direction that achieves a descent in the objective function. Convergence to the local minimum is proved, and simulation results confirm the computational efficiency of the algorithm.

INTRODUCTION:

Piecewise-linear neural networks (PWL NN) (Lin et al [4]) are networks where the node functions are piecewise linear. They have been a successful alternative to sigmoidal-type networks, and exhibit a number of useful features, such as amenability to digital VLSI implementation, speed and discrete nature of

0-7803-5673-X/99/$10.00 © 1999 IEEE

training. A number of training algorithms have been developed for PWL NN (see [1], [4], [5], and [7]). The drawback of these algorithms is that they are of the constructive-type. This means, one starts with a minimal network configuration, and then adds units in a prescribed manner to minimize the error. One problem with the constructive approach is that it is not committed to an upper bound in the number of network parameters (weights), and that makes it questionable as far as generalization performance is concerned. The initimate relationship between the number of weights and the generalization performance has been widely demonstrated.

In this paper, we propose a new algorithm for PWL networks. The proposed algorithm approaches the task by constructing a PWL objective function for the total error, and then seeks to minimize it, and it is thus not a constructive-type algorithm. The algorithm developed is a new algorithm for optimizing PWL functions, and is therefore a contribution to the general optimization field. There have been very few algorithms in the literature for the optimization of general PWL functions. We could locate only one such method ([3], it uses the branch and bound procedure). The proposed algorithm exhibits a number of advantages, including significant acceleration over the backpropagation method and some of its variants, convergence in a finite number of steps, the absence of a tuning parameter (there is no learning rate like in the backpropagation algorithm, and hence no time-consuming parameter tuning step).

OVERVIEW OF THE ALGORITHM:

By using the L_1 error function, the problem is converted into the minimization of a PWL function. Let N be dimension of the space (i.e. number of free parameters or number of weights in the network). Then a PWL function consists generally of regions of polyhedron shape where the function is linear. These regions are separated by hyperplane boundaries, that create discontinuity (in the linear slope) from region to next.

We first prove that the minimum is at the intersection of al least N hyperplane boundaries. This has effectively discretized the problem, and the goal now is to determine the N hyperplanes, at which intersection the minimum occurs. The idea of the algorithm is to locate a point at the intersection of N hyperplane boundaries, as a starting point. Starting from this point, we search along the one-dimensional boundary directions emanating from this point (there are typically $2N$ directions to search), until we find a descent direction w.r.t. the error function. We move along this direction until we hit another hyperplane boundary, hence arriving at a new point at the intersection of N hyperplanes. Starting from this point, we perform another similar search, and continue in this manner until we arrive at a point, where all one-dimensional boundary directions lead to an increase in the error function, hence a local minimum. We prove that, except for certain degenerate situations, the algorithm converges to a local minimum.

One can see that the algorithm is an "exchange"-type algorithm, and thus has similar flavor to the simplex algorithm for linear programming. The difference is that in the simplex algorithm we move from one vertex to another within a polyhedron region. In the proposed algorithm we move from one polyhedron region to the next.

THE OPTIMIZATION PROBLEM:

We consider a one-hidden-layer network, where the hidden node activation functions are the piecewise linear saturation nonlinearity (see Figures 1 and 2), i.e.

$$f(x) \equiv \mathrm{sat}_r(x) = \begin{cases} x & |x| \leq r \\ r\mathrm{sign}(x) & |x| > r \end{cases} \tag{1}$$

WLOG consider $r = 1$ for all nodes. For simplicity consider one output node, and assume the output node to be linear. Let $y(m)$ and $T(m)$ be the network output and desired output for example m respectively. We consider the L_1 error as the objective function to be minimized. Let $\mathbf{w}_j$ represent the weight vector of hidden node j (augmented with the bias), and $\mathbf{u}(m)$ be the input example m (augmented with 1 to account for the bias). Furthermore let $\mathbf{v} = (v_1, ..., v_J)^T$ and v_0 represent the weight vector and bias for the output node.

Using a simple variable transformation, the L_1 error becomes

$$E = \sum_{m=1}^{M} \left| v_0 + \sum_{j=1}^{J} \mathrm{sat}_{v_j}(\mathbf{w'}_j^T \mathbf{u}(m)) - T(m) \right| \tag{2}$$

where the $\mathbf{w'}_j$'s are the new transformed weight vectors.

We can see now that the problem of training the network is equivalent to the minimization of a piecewise linear function (Eq. 2).

THE PROPOSED METHOD:

As mentioned in Section 2, PWL functions (say $f(\mathbf{x})$, where $\mathbf{x}$ is an N-dimensional vector), consist generally of regions of a polyhedron shape where the function is linear, and these regions are separated by hyperplane boundaries. (The hyperplanes represent equality conditions, for example in the case of Eq. 2, $v_0 + \sum_{j=1}^{J} \mathrm{sat}_{v_j}(\mathbf{w'}_j^T \mathbf{u}(m)) - T(m) = 0$ or $\mathbf{w'}_j^T \mathbf{u}(m) = v_j$ represent two hyperplanes.) Defining a local minimum as a connected set L where $f(\mathbf{x}) = f_{min}$ for $x \in L$, and $f(\mathbf{x}) > f_{min}$ for $x \in B - L$, where B is some open neighborhood of L, we obtain the following theorem that characterizes a local minimum:

Theorem 1: If L is a connected set of local minima (as defined above), and

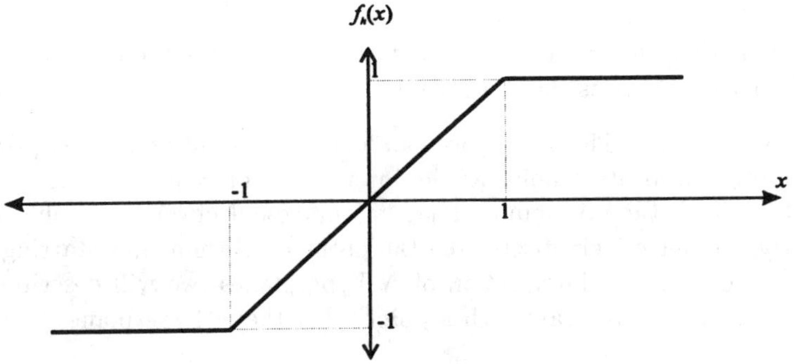

FIGURE 1. The activity function of the hidden neurons.

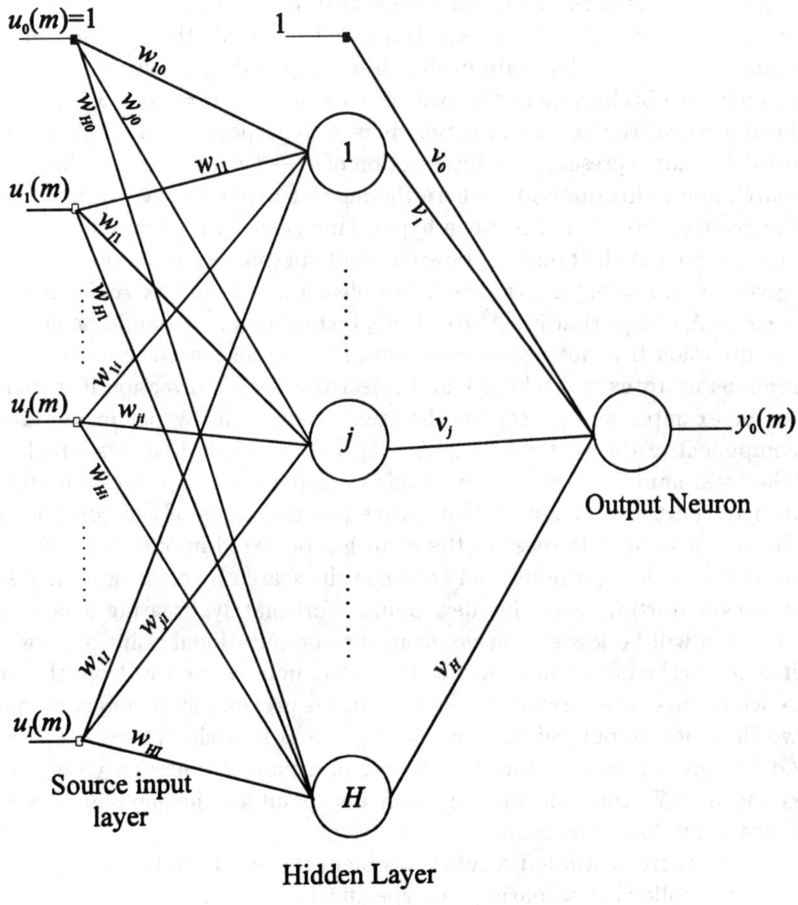

FIGURE 2 A schematic layout for the neural network model.

L is bounded, then there is $\mathbf{x}_0 \in L$ such that $\mathbf{x}_0$ is the intersection of at least N hyperplanes (N is the dimension of $\mathbf{x}$).

As a result of Theorem 1, we can focus our attention in the optimization procedure to locate a point at the intersection of N hyperplanes (or more), that achieves the minimum. That has narrowed down the problem significantly, and effectively discretized the problem. Assume our starting point is also a point at the intersection of N hyperplanes (we will describe later in the section how to locate such a point). Let the N hyperplanes be given as

$$\mathbf{A}\mathbf{x} = \mathbf{b} \tag{3}$$

For simplicity we assume in this paper that there are no degeneracies (for example $\mathbf{A}$ being singular, or the solution is at the intersection of more than N hyperplanes). The degenerate case needs a special treatment, and that is beyond the space constraints of this paper, and will be described in a future paper. If we perturb one of the elements in $\mathbf{b}$ (i.e. $b_k \equiv b_k + \nu$, hence $\mathbf{b}_\nu^T = [b_1, ..., (b_k + \nu), ..., b_N]$), where ν is a small constant, then we obtain a new point $\mathbf{x}_\nu = \mathbf{A}^{-1}\mathbf{b}_\nu$ The gain in the above step is that a new search direction can now be obtained using the two points $\mathbf{x}$ and $\mathbf{x}_\nu$: $\mathbf{h} = (\mathbf{x}_\nu - \mathbf{x})/\|(\mathbf{x}_\nu - \mathbf{x})\|$. Furthermore, the search direction $\mathbf{h}$ is a hyperplane of a single degree of freedom that represents the intersection of $N - 1$ hyperplanes. Thus, we can search along this direction, where the nearest hyperplane encountered along the positive direction of $\mathbf{h}$ (say a hyperplane given by $\mathbf{a'}_k\mathbf{x} = b'_k$) becomes the new member of the family of intersecting hyperplanes, while the old member (given by $\mathbf{a}_k\mathbf{x} = b_k$) is dismissed. We obtain a new matrix $\mathbf{A}_k$, which is the same as $\mathbf{A}$ except that its kth row is $\mathbf{a'}_k$ instead of $\mathbf{a}_k$. It should be noted that the direction $\mathbf{h}$ is not necessarily committed to decrease the error E, but it remains up to us to backtrack and reject this search direction if it increases it; for example we can try another search direction by perturbing another component of the vector $\mathbf{b}$ (e.g. $k + 1$), and proceeding as described above. The total number of search directions we explore is $2N$, corresponding to N positive perturbations and N negative perturbations of the components of the vector $\mathbf{b}$. If a decrease in the error has occurred in one search direction, we move to the new point and continue the search by creating a new search direction starting from this new point. Fortunately, creating a new search direction will be less expensive, from the computational point of view, since it requires the inversion of $\mathbf{A}_k$, which differs in only one row from the matrix $\mathbf{A}$ whose inverse is already available from the previous step, and consequently we shall not be obliged to compute $\mathbf{A}_k^{-1}$ from scratch. There is a recursive $O(N^2)$ procedure to obtain the inverse of a matrix where only one row has changed. We continue moving from one point to the next until we have reached the local minimum.

Now, there is still left a subtle problem that we have to face up to. Consider the following scenario: The specification of some of the vectors $\mathbf{a}_j$ is dependent on the configuration of linear regions we are in. Hence, any change to the configuration due to a crossing from one region to another necessitates a change in that row corresponding to the new configuration. Even the $2N$ search directions might correspond to different $\mathbf{A}$ matrices. The search directions become broken lines (see Figure 3 for an illustration). To remedy this problem, we think of the $2N$ directions as "transformed coordinate axes" (the positive and the negative sides of the axes). We think of the each region flanked by N axes like a "quadrant". Assume the matrix $\mathbf{A}$ and the gradient (say vector $\mathbf{g}$) are readily available for say the region flanked by the positive coordinate axes (corresponding to positive ν perturbations of the components of $\mathbf{b}$). Then, the N directions are:

$$\mathbf{h}_i = k\mathbf{A}^{-1}\mathbf{e}_i \tag{4}$$

where k is a positive constant and $\mathbf{e}_i$ is the unity vector. We can therefore explore the N directions in one shot by calculating $\mathbf{g}^T\mathbf{A}^{-1}$. We pick one of the negative components of the resulting row vector, since this indicates a negative dot product between the gradient and the corresponding $\mathbf{h}_i$ vector, hence indicating a descent direction.

As for the remaining N cordinate axes (search directions), we calculate the matrix $\mathbf{A}'$ and the gradient vector $\mathbf{g}'$ corresponding to the region flanked by these coordinate axes. By moving from the positive quadrant to the negative quadrant, we would "officially" be operating on different sides of all the N constraints, thus changing the configuration of the linear regions. The sign of the elements of the row vector $\mathbf{g}'^T\mathbf{A}'^{-1}$ determines the possible descent directions. We note that for our particular case of minimizing Eq. 2, the inverse of the matrix $\mathbf{A}'$ need not be calculated from scratch. We have shown that it can be obtained in terms of $\mathbf{A}^{-1}$ by performing some row addition and subtraction operations.

Exploring the $2N$ directions, by observing the signs of row vectors $\mathbf{g}^T\mathbf{A}^{-1}$ and $\mathbf{g}'^T\mathbf{A}'^{-1}$, determines a descent direction. In fact, exploring the $2N$ directions is sufficient to determine whether we have reached a local minimum. In other words, if the error increases along the $2N$ directions $\mathbf{h}_1, ..., \mathbf{h}_{2N}$, then we have reached a local minimum. This is given by the following theorem:

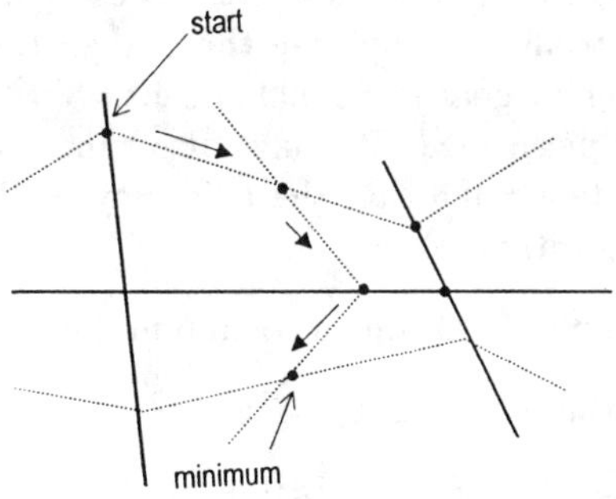

Figure 3: An illustration of the iterations of the algorithm

Theorem 2: Assume the matrix $\mathbf{A}$ is nonsingular. If $E(\mathbf{x}_0+\epsilon\mathbf{h}_j) > E(\mathbf{x}_0)$ (E as given by Eq. 2), for all $j = 1, ..., 2N$ and ϵ is a sufficiently small constant, then $\mathbf{x}_0$ is a local minimum of E, in the sense that $E(\mathbf{x}_0 + \epsilon\mathbf{h}) > E(\mathbf{x}_0)$ for any vector $\mathbf{h}$.

So far in this section, we have assumed that we could locate an initial point that is the intersection of N hyperplanes. Now we will describe a way how to attain that. We use the concept of *hypothetical hyperplanes.* These are dummy hyperplanes that do not correspond to real region-region boundaries. We start with any randomly chosen initial point, and assume that this point is at the intersection of N hyperplanes. We can take $\mathbf{A} = \mathbf{I}$ (the identity matrix) and $\mathbf{b}$ to equal the initial point. In this case the hypothetical hyperplanes are given by hyperplanes parallel to the coordinate axes. We then perform the optimization procedure described earlier, by searching along the intersections of the hypothetical hyperplanes. Gradually, the hypothetical hyperplanes will give way to real boundary-hyperplanes, and we proceed as usual.

SIMULATION RESULTS:

We have tested the proposed algorithm on a number of problems. Several other algorithms applied to sigmoidal-type networks have also been tested on the same problems to obtain a comparative idea about the speed of the developed algorithm. The algorithms used in the comparison were the back-propagation algorithm BP (using both of its update modes, the sequential or the individual update mode and the batch update mode) and its accelerated version, the Bold-Driver (BD) (see [2] and [8], it is based on adaptively adjusting the learning rate). In the test comparisons we used the abbreviation PWLO (piecewise-linear optimization) to stand for the new algorithm. In each test, the PWLO is run first (starting from five different points randomly chosen), until convergence to a local minimum. The final error level SSE, representing the sum of square error, is then recorded for each of the five trials. This error level is then set as an error goal for the other algorithms. The first run is used to obtain the best tuned values of the learning rate (η) and momentum (α) parameters. These parameters are then fixed for the four other trials. In any case these algorithms are stopped if they do not attain the error goal in 50,000 epochs. The following are a sample of the experiments performed. We have also run several other experiments, and they all consistently indicate the efficiency of PWLO, as compared to the other tested algorithms.

Example 1: A simple multi-dimensional function:

We considered the mapping function:

$$y = \frac{x_1x_2 + x_3x_4 + x_5x_6 + x_7x_8}{400} \tag{5}$$

where x_i is drawn at random from [0,10]. Table 1 shows the results of applying the new method, as well as the three other competing methods. One can see

Table 1. Learning the mapping of Eq. (5) using 8-4-1 network. (Summary of five trials)

Type of Algorithm	SSE error			Average No. of iterations	Average CPU Time in Sec.	Parameters	
	Best Case	Worst Case	Average			α	η
BP(ind.)	0.0569	0.2667	0.1063	50000	5500	0.0	0.001
BP(batch)	0.0372	0.0706	0.0558	44828	4931	0.0	0.001
BD	0.0358	0.153	0.0607	36996	4399	_	_
PWLO	0.0103	0.0706	0.0378	2295	730	_	_

Table 2. Bankruptcy prediction using a 5-4-1 network.

Type of Algorithm	SSE error			Average No. of iterations	Average CPU Time in Sec.	Parameters	
	Best Case	Worst Case	Average			α	η
BP(ind.)	9.3881	10.4892	9.9062	50000	5500	0.3	0.1
BP(batch)	10.837	18.536	16.324	50000	5434	0.5	0.005
BD	9.5147	10.027	9.7084	50000	5449	_	_
PWLO	0.0	3.8005	0.7601	1883	873	_	_

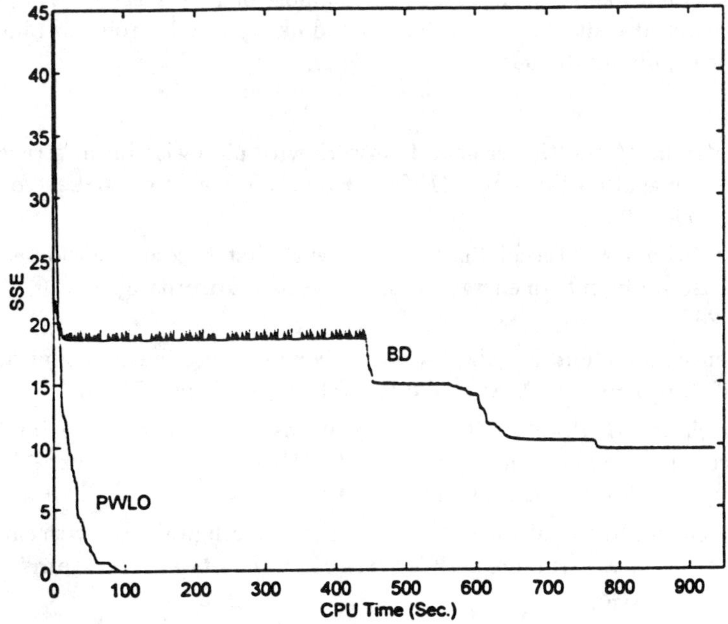

Figure 4: Learning curves for the bankruptcy prediction problem

that the proposed method achieves significantly faster convergence, in terms of both the number of iterations and the CPU time.

Example 2: The bankruptcy prediction problem:

We have used a data set that was supplied by Odom and Sharda [6]. They have used this data set in their work to apply neural networks to predict bankruptcies among firms. In this test, the inputs reflect a set of indicators defining the financial status of a given firm, while the desired output assumes a binary value to indicate the potential bankruptcy. The PWLO has been modified slighly by PWL saturation-type function at the output node. Table 2 shows the comparison results. One can see that the average number of iterations for PWLO was 1883, corresponding to an average CPU time of 873. On the other hand none of the other three methods was successful in converging to the target error within the specified 50,000 epochs. Figure 4 shows the learning curve of the PWLO and BD methods, for comparison. For the test set the results were similar, and the testing error against iterations exhibits a very similar behavior as in Figure 4. One can see from the Figure that the BD method (similarly for BP also) never arrives to an acceptable solution. The proposed method has demonstrated not only speed, but the ability to converge to better solutions (lower errors).

ACKNOWLEDGEMENTS:

A. Atiya would like to acknowledge the support of NSF's Engineering Research Center at Caltech. The authors would like to thank Professor Marcus Odom for supplying the data.

References:

[1] R. Batruni, "A multilayer neural network with piecewise-linear structure and backpropagation learning", **IEEE Trans. Neural Networks**, Vol. 2, 1991, pp. 395-403.

[2] R. Battiti and G. Tecchiolli, "Learning with first, second and no derivatives: a case study in high energy physics", **Neurocomputing**, Vol. 6, 1991, pp. 181-206.

[3] B. Benchekroun and J. Falk, "A nonconvex piecewise-linear optimization problem", **Computers, Math. Appl.**, Vol. 21, 1991, pp. 77-85.

[4] D. Hush and B. Horne, "Efficient algorithms for function approximation with piecewise linear sigmoidal networks", **IEEE Trans. Neural Networks**, Vol. 9, No. 6, 1998, pp. 1129-1141.

[5] J.-N. Lin and R. Unbehauen, "Adaptive nonlinear digital filter with canonical piecewise linear structure", **IEEE Trans. Circuits and Systems**, Vol. 37, 1990, pp. 347-353.

[6] M. Odom and R. Sharda, "A neural network model for bankruptcy prediction", **Proc. IEEE Int. Conf. Neural Networks**, 1990, pp. 163-168.

[7] M. Staley, "Learning with piecewise-linear neural networks", **Int. J. Neural Systems**, Vol. 6, 1995, pp. 43-59.

[8] T. Vogl, J. Mangis, A. Rigler, W. Zink, and D. Alkon, "Accelerating the convergence of the backpropagation method", **Biological Cybernetics**, Vol. 59, 1988, pp. 257-263.

A CONTROL THEORY FORMULATION FOR RANDOM VARIATE GENERATION

Malik Magdon-Ismail
malik@work.caltech.edu
Caltech Learning Systems Group
Department of Electrical Engineering
California Institute of Technology
136-93 Pasadena, CA, 91125

Amir F. Atiya
amir@work.caltech.edu
Caltech Learning Systems Group
Department of Electrical Engineering
California Institute of Technology
136-93 Pasadena, CA, 91125

Abstract. The need to simulate complex systems in a Monte Carlo manner necessitates efficient methods for generating random variates. In this paper we propose a new method for random variate generation. The method is based on a contol-theory formulation. We use a cascade structure consisting of a neural network "controller" and a density estimator ("plant"). The neural network "controller" acts as a density shaper, and is trained until the density of its output (as measured by the density estimator) is as close as possible to the given density. Once training is complete in the design phase, the generation of random numbers can be performed in a very fast manner.

INTRODUCTION

In many scientific problems in areas such as physics, biology, engineering and economics, it is now more the rule than the exception to be confronted with problems where analytic solutions are not possible and simulating the given problem would be the only feasible approach. Problems which possess some random element will necessitate an accurate and fast random variable generation procedure for the Monte Carlo simulation to be efficiently performed. Examples of applications include the simulation of a biological system, such as how nerve cells interact together, the simulation of a communication network where the packets are assumed to arrive according to a particular density function, the simulation of a control system where the plant disturbance is a random variable, and the simulation of the stock market in an attempt to estimate a valuation for some derivative instruments. In the past, the assumption of standard density functions such as Gaussian or exponential densities for these random variables often allowed analytical solutions for these problems. But now, with more accuracy needed, the next step has been to assume more realistic density functions for the random variables (possibly estimated from the data). Unfortunately, there are very few techniques that can generate a random variable possessing an arbitrary density function (see [6] for a review of the approaches). For example the transformation method [2], [3] cannot generate from an arbitrary density (it can generate only those densities where the inverse of the distribution function can be obtained analytically). The other well-known method, the rejection method [6], is more general. The problem, however, is that it can be very slow for large-tailed densities and for densities given by a sample drawn from it rather than an explicit formula.

In this paper we propose a method that formulates the problem using a "control theory" framework. The proposed idea can be implemented using multilayer neural networks. We propose a cascade structure that consists of a density shaping network (acting as the "controller"), and a density estimation network (acting as the "plant"). The controller is trained until the output of the "plant" (i.e. the resulting density) is as close as possible to the given density. For the method that we propose, the bulk of the time is spent in "learning" to generate the random variates. Once this learning is done, the actual generation of the random numbers is fast and can thus be used for Monte Carlo simulations.

CONTROL FORMULATION OF RANDOM VARIATE GENERATION

Suppose that we wish to generate random variates from a specified density $g(x)$. We propose here a method that is based on transforming a known density to the given density. Figure 1 illustrates the basic idea of the method. The structure consists of two cascaded units. A random variable x is generated according to any standard density function, for example a Gaussian or a uniform density. In principle, an unlimited number of x's can be generated. This random variable is the input to the composite structure. The first network acts as a nonlinear transformation that shapes the density of x. Let the distribution function of the transformed variable y be $Z(y)$. The second unit serves to estimate this distribution function. For example, one could use a standard kernel density estimator , [7], for this unit, i.e.

$$z(y) = \frac{1}{Nh^d}\sum_{n=1}^{N} K\left(\frac{y - y_n}{h}\right) \tag{1}$$

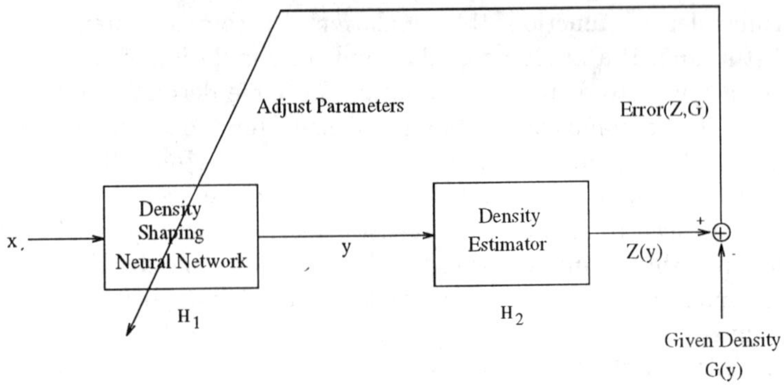

The first network shapes the density of x. The transformed density is estimated, and then compared with the true density. The parameters of the first network are altered until the transformed density matches the required density.

Figure 1: Control formulation for random variate generation.

where K is the kernel, h is the kernel width, d is the dimension of y and y_i are the N samples of y. Alternatively, we can use the neural network density estimator developed in [4] as the second unit (see Appendix for a short description of this density estimation method).

Thus, the output of the second unit is the estimate of the distribution function $Z(y)$ of network 1's output. Let the desired distribution function be $G(y)$. An error signal, $\mathcal{E}(G(y), Z_{w_1}(y))$ can now be generated where it can be seen that the distribution Z is a function of w_1, the weights of the first network. This error signal can be backpropagated to adjust the weights of network 1 so as to minimize $\mathcal{E}(G(y), Z_{w_1}(y))$. Such a training process would change network 1's mapping in such a way that would shape the distribution function of y to get it as close as possible to the desired distribution function. Once training is complete, we can generate numbers according to the distribution function $G(x)$, by generating x according to the standard density, and then passing it through network 1.

Note 1: This algorithm is not restricted to the case of scalar random variable, and so can be used to generate multi-dimensional variables.

Note 2: To generate from a density represented only by a set of data points, we use the same method above but replace $G(y)$ by $\hat{G}(y)$, an estimate of $G(y)$ using the data points. This estimate can be obtained using one of the standard density estimation methods, [4] [7].

Note 3: The entire learning process can be an involved process. However, once the learning has been done, the generation of random variates is fast, since one needs only to generate a uniform or a Gaussian density, and then pass it through a neural network.

The analogy with the control problem is as follows. The second unit represents the "plant," and network 1 is the "controller." The controller is trained, such that the plant produces the desired output[1]. The details of the algorithm are as follows:

I: Initialization:

1. Let the desired density be $g(y)$ and the corresponding distribution function $G(y)$. Generate N samples $x_1, \cdots, x_N \in \mathbf{R}^d$ according to any standard density such as a Gaussian density. The larger N is, the more accurate the final result will be.

2. Let $H_1(x, w)$ and $H_2(x, v)$ be the functions implemented by network 1 and network 2, where w and v represent parameters[2]. Network 1 has d outputs, and network 2 has one output. Initialize the weights of the networks.

3. Calculate network 1's output: $y_n(w) = H_1(x_n, w), \; n = 1, \cdots, N$.

II: Training Iterations:

4. Train network 2 (or construct the kernel estimator) to estimate the distribution of the y_n's. Thus, $H_2(y, v(w))$ is implementing the distribution function of y, where we have explicitly shown the w dependence.

5. Suppose the error function we wish to minimize is $\mathcal{E}(H_2(w), G)$. Possible error functions are an L_1 or L_2 norm, or a cross entropy (Kulback-Leibler distance). For the case of squared error, let $\{z_n\}_{n=1}^{M}$ be any set of points for calculating the error function ($\mathcal{E}$),

$$\mathcal{E} = \sum_{n=1}^{M} \Big[H_2(z_n, v(w)) - G(z_n) \Big]^2 \tag{2}$$

6. One alters the weights (w) of the first network in the direction of the negative gradient:

$$w(t+1) = w(t) - \eta \frac{\partial \mathcal{E}}{\partial w} \tag{3}$$

---

[1]This technique has a similar form to a neural network control structure (see [5]) where, typically, a neural network is trained to identify the plant (estimate the plant output), analogous in our set up to the density estimation unit. Then, a neural network controller is trained to control the plant – i.e., the identification network estimates the plant output and the controller network is altered until this output is as desired.

[2]In the case where network 2 is a kernel density estimator, the parameters v are the outputs of network 1 on the samples $x_1, \ldots, x_2$

For example in the case of squared error we have,

$$\frac{\partial \mathcal{E}}{\partial w} = 2\sum_{n=1}^{M}\Big[H_2(z_n, v(w)) - G(z_n)\Big]\frac{\partial H_2(z_n, v(w))}{\partial w} \tag{4}$$

The weights in the first network can now be updated using a standard standard backpropagation scheme [1, pg 115], where one backpropagates the δ's first through network 2, then through network 1 to adjust the weights of network 1.

7. Go to step 4 to perform another iteration of training, and continue doing so until the error becomes small enough.

8. After training is complete, the random numbers can be generated by generating x according to the standard density that was innitially used (e.g. Gaussian), and computing $y = H(x, w^*)$ (where w^* represents the final weights after the entire learning process). Now, y should be the distributed according to $g(y)$.

APPENDIX: DESCRIPTION OF A DENSITY ESTIMATION NETWORK

We mentioned in Section 2 that the second unit in the proposed structure can be implemented using the neural network density estimation procedure proposed in [4]. In this appendix we will briefly describe this method.

Let $x_n \in \mathbf{R}$, $n = 1, ..., N$ be the data points. Let the underlying density be $g(x)$ and its distribution function $G(x) = \int_{-\infty}^{x} g(t)dt$. Let the neural network output be $H(x, w)$, where w represents the set of weights of the network. Ideally, after training the neural network, we would like to have $H(x, w) = G(x)$. It can easily be shown that the density of the random variable $G(x)$ (x being generated according to $g(x)$) is uniform in $[0, 1]$. Thus, if $H(x, w)$ is to be as close as possible to $G(x)$, then the network output should have a density that is close to uniform in $[0, 1]$. This is what our goal will be. We will attempt to train the network such that its output density is uniform, then the network mapping should represent the distribution function $G(x)$. In [4] we have developed two algorithms to achieve that:

SLC (Stochastic Learning of the Cumulative)

In this method we use the N data points as inputs to the network. For every training cycle, we generate a different set of N network targets randomly from a uniform distribution in $[0, 1]$, and adjust the weights to map the data points (sorted in ascending order) to these generated targets (also sorted in ascending order). Thus we are training the network to map the data to a uniform distribution. This method applies only to the one-dimensional case.

It can be proved that this stochastic algorithm converges to a minimum of the sum of square error between the given distribution function $G(x)$ and the network output $H(x, w)$.

SIC (Smooth Interpolation of the Cumulative)

This method applies to the multi-dimensional case. Again, we have a multilayer network, to which we input the point $\mathbf{x}$, and the network outputs the estimate of the distribution function. Let $g(\mathbf{x})$ be the true density function, and let $G(\mathbf{x})$ be the corresponding distribution function. Let $\mathbf{x} = (x^1, ..., x^d)^T$. The distribution function is given by

$$G(\mathbf{x}) = \int_{-\infty}^{x^1} \cdots \int_{-\infty}^{x^d} g(\mathbf{x}) dx^1 \cdots x^d, \tag{5}$$

a straightforward estimate of $G(\mathbf{x})$ could be the fraction of data points falling in the area of integration:

$$\hat{G}(\mathbf{x}) = \frac{1}{N} \sum_{n=1}^{N} \Theta(\mathbf{x} - \mathbf{x}_n), \tag{6}$$

where Θ is defined as

$$\Theta(\mathbf{x}) = \begin{cases} 1 & \text{if } x^i \geq 0 \text{ for all } i = 1, \ldots, d, \\ 0 & \text{otherwise.} \end{cases}$$

The method we propose uses such an estimate for the target outputs of the neural network. The estimate given by (6) is discontinuous. The neural network method developed here provides a smooth, and hence more realistic estimate of the distribution function. The density can be obtained by differentiating the output of the network with respect to its inputs.

ACKNOWLEDGMENTS

We would like to acknowledge Yaser Abu-Mostafa and the Caltech Learning Systems Group for their useful input. We would like to acknowledge the support of NSF's Engineering Research Center at Caltech.

REFERENCES

[1] J. Hertz, A. Krogh, and R. G. Palmer. *Introduction to the Theory of Neural Computation.* Addison-Wesley, Redwood City, CA, 1991.

[2] G. Johnson. Construction of particular random processes. In *Proceedings IEEE*, volume 82(2), pages 270–281, February 1994.

[3] D. Knuth. *The Art of Computer Programming*, volume 1. Addison-Wesley, 1981.

[4] M. Magdon-Ismail and A. Atiya. Neural networks for density estimation. *Advances in Neural Information Processing Systems (NIPS 98), to appear*, 1998.

[5] K. Narendra and K. Parthasarathy. Identification and control of dynamical systems using neural networks. *IEEE Transactions on Neural Networks*, 1(1):4–27, 1990.

[6] W. Press, B. Flannery, S. Teukolsky, and W. Vetterling. *Numerical Recipes.* Cambridge University Press, Cambridge, UK, 1988.

[7] B. Silverman. *Density Estimation for Statistics and Data Analysis.* Chapman and Hall, London, UK, 1993.

A SPIKING NEURAL NETWORK ARCHITECTURE FOR NONLINEAR FUNCTION APPROXIMATION

Nicolangelo Iannella and Andrew Back
Brain Science Institute, RIKEN
2-1 Hirosawa, Wako–shi, Saitama 351-0198 Japan.
Phone: +81 48 462 1111 ex 7154,7173
Fax: +81 48 467 9684,+81 48 467 9694
Email: angelo@postman.riken.go.jp,back@brain.riken.go.jp

Abstract. Multilayer perceptrons have received much attention in recent years due to their universal approximation capabilities. Normally such models use real valued signals, although they are loosely based on biological neuronal networks which encode signals using spike trains. Spiking neural networks are of interest from both a biological point of view, but also from a method of robust signalling in particularly noisy or difficult environments. From a signal processing perspective, it is important to consider networks based on spike trains. A basic question that needs to be considered, is what type of architecture can be used to provide universal function approximation capabilities in spiking networks? In this paper, we propose a spiking neural network architecture using both integrate and fire units as well as delays which is capable of approximating a real valued function mapping to within a finite degree of accuracy.

INTRODUCTION

Multilayer perceptrons (MLPs) have received significant attention in recent years. The main reason for this is the fact that MLPs are capable of approximating any continuous function $\mathcal{F} \in \mathcal{R}^m$ arbitrarily well by means of spatially-local basis functions. This property is termed *universal approximation* [3, 6, 7]. This means that given a function $\mathcal{F}(\mathbf{x})$, a model can be constructed as

$$G(x) = \mathbf{B}\sigma\left(\mathbf{A}x + \varphi\right) \tag{1}$$

where $\mathbf{x} \in \mathcal{R}^m$ is the input, $\mathbf{A} \in \mathcal{R}^{m \times n}$ are the input layer weights, $\mathbf{B} \in \mathcal{R}^n$ are the output layer weights, $\varphi \in \mathcal{R}^n$ are the bias inputs and σ are some

0-7803-5673-X/99/$10.00 © 1999 IEEE

appropriately chosen basis functions, which can be shaped, offset and scaled [1], to approximate $\mathcal{F}(\mathbf{x})$ to an arbitrary degree of accuracy.

The results for universal approximation are normally made using real valued inputs, outputs and weights. However, in real biological systems, signals are encoded using spike trains, not real values. This raises the question of how function approximation (whether "universal" or not) can be achieved in spiking neural networks (SNNs). This issue is of interest for the development of spiking neural network models for practical applications as well as obtaining an understanding of possible biological computational mechanisms.

In contrast to real valued models, an important issue in spiking networks the *neural coding* scheme which by which we attach meaning to the spiking signals. Neural coding has been the subject of much debate and various spike train information encoding methods have been proposed in the literature [1, 2, 4, 5].

The earliest neural coding method proposed is the *mean or temporally averaged firing rate* for encoding information. More recently, there has been growing recognition that the traditional view of mean firing rate encoding is inadequate. Experiments from the fly visual system [1, 2] and studies of the middle temporal (MT) area of the monkey have indicated that the precise timing of spikes can be used to encode information, such a scheme is called *temporal coding. Population coding* is another scheme in which information is encoded in the activity of a given population of neurons firing within a small temporal window. Within this framework a notion of *rate coding* can be extracted as a (weighted) instantaneous average measure of firing activity in a given neuron population.

Function approximation methods have been demonstrated for several encoding schemes. Sanger [9] demonstrated this for a population of spiking neurons based on a rate coding scheme. He showed that a network of spiking neurons can be constructed such that the output neurons' firing probability is a smooth function of the input neurons' firing probability. Maass et. al. [8] showed that from a computational complexity theory, SNNs can in principle, emulate sigmoidal networks.

In this paper, we propose a SNN approach to function approximation based on temporal coding. In Section 2, we propose a spiking network model and show how it is able to approximate arbitrary nonlinear function maps. In Section 3, we give some examples of the network and conclusions are given in Section 4.

[1] These operations correspond to the transformations made by the hidden layer weights, bias weights and output layer weights respectively.

A SPIKING NETWORK ARCHITECTURE

Neural Encoding Method

We require a neural network capable of forming arbitrary nonlinear function mappings (not necessarily universal) between an input and output. To solve this problem we firstly define the neural encoding method used and then determine a neural architecture capable of forming the function mapping.

One must be aware that biological neurons actually receives a number of spikes, which forms a representation for the information received by the neuron. In our work we have chosen to use a temporal encoding scheme which is defined such that the *inter spike interval* (ISI), the temporal distance between two successive spikes, represents a real-valued quantity. The relationship between real valued inputs x and the ISI is given by

$$\begin{aligned} x_i(t) &= T_i(1) - T_i(0) \\ &= \Delta_i(0,1) \end{aligned}$$

where $x_i(t)$ is the effective real value of the ith input at time t, $T_i(0)$ and $T_i(1)$ correspond to the arrival times of the first and second spikes respectively, and Δ is used to denote the ISI.

A Network Architecture

The network architecture we propose is based on the idea of massive parallelism in biological nervous systems. The methodology here is to build a large network which temporally and spatially separates incoming spikes into different regions of the network, which in turn generate corresponding new spike pairs. These new output spikes may have any desired ISI. Hence the piecewise mapping can approximate any real valued function.

Our proposed model is given by

$$v(t) = \begin{cases} s(t) + s(t-\phi) & y_a > \theta \\ 0 & \text{otherwise} \end{cases} \tag{2}$$

$$y_a(t) = r(t) + r(t - \tau_d) \tag{3}$$

$$r(t) = \begin{cases} s(t) & t = T(j), \qquad j = 0, \ldots \\ 0 & \text{otherwise} \end{cases} \tag{4}$$

$$s(t) = C_i(\delta(t_0)) \tag{5}$$

$$= L \exp\left(-\frac{t - t_0}{\tau}\right). \tag{6}$$

where $r(t)$ is the input spike signal, $y_a(t)$ is the unit activation[2], found as the

[2]In a biologically-oriented model, $y_a(t)$ represents the total somatic membrane potential, given by the summation of all postsynaptic potentials. In biological neurons, this summation is spatio-temporally nonlinear. Currently, we only consider linear summation, since the addition of nonlinearities would simply increase the difficulty in achieving function mappings.

sum of $r(t)$ and a delayed[3] version of itself where the delay is τ_d, $s(t)$ is the time function of an individual spike which is modelled as a filtered impulse response of maximum amplitude L, $t_0 = T(j)$ $j = 0, ...,$ the initial firing time of each spike. If the activation y_a exceeds the threshold denoted by θ, then the unit 'fires' and generates a new pulse pair having an ISI of ϕ seconds at the output $v(t)$.

Conceptually the model is easy to understand. The equations above describe one input and one output, so to construct a network capable of function approximation, it is necessary to use many such units, each with their own associated τ_d and ϕ values. Each processing unit corresponds to each segment[4] as depicted in Figure (1).

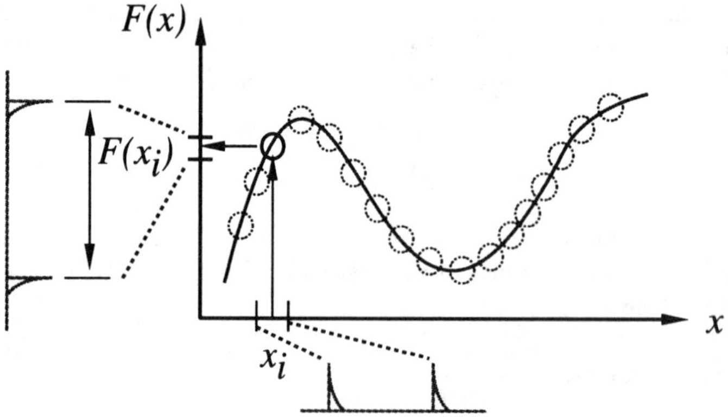

Figure 1: The input space is subdivided into regions, with signals corresponding to each region being passed into a particular processing unit.

Next we have to set the threshold value θ. This could be set simply at $\theta = 2L$, corresponding to $\tau_d = T(k) - T(k+1)$, for the kth set of spikes in a spike train. However this method would not be robust enough and would easily fail if the true maximum spike values were less than L or the spikes were anything less than exactly coincident. A solution is therefore to set θ to some lower value, but not too low that firing occurs for any spikes.

We consider next how to set θ in an appropriate way to allow to give some latitude in the input ISI for each unit.

[3]This corresponds well with biological spiking neurons where it is well known that the dendritic and axonal branches have transmission delays which primarily depend on their respective length and diameter. Furthermore, each spike arriving at different points on the dendritic tree evokes a current, which is transmitted up the dendrite to the soma and delayed by some amount. For simplicity, it is assumed that spikes occur at the same loci on the dendritic tree.

[4]This need not be the case in practice. Recently there has been some suggestion that biological networks may actually perform a type of 'multiplexing'.

Threshold Calculation

Unlike a coincidence detector, the network model we propose here performs robustly by allowing for some tolerance in the range of input ISIs that will evoke an output.

We specify a tolerance value ρ and derive θ such that a given unit produces an output spike pair if

$$|(T(k) - T(k+1)) - \tau_d| \leq \rho \qquad (7)$$

For the purpose of this paper, we specify the time decay characteristics τ of the input spikes. Choosing a threshold value of $1.5L$ gives

$$1.5L = L + Le^{-\rho/\tau} \qquad (8)$$

and hence

$$\tau = \frac{\rho}{ln(2)}. \qquad (9)$$

Thus ensuring that the neuron will fire when the time difference between spikes is ρ or less.

AN INTEGRATE-AND-FIRE NETWORK

If we are designing a spiking network, then the approach of specifying the spike decay times is reasonable. However for pedagogical reasons it is also of interest to consider networks comprised of continuous time integrate-and-fire (IAF) neural processing units.

An IAF model is given by

$$\frac{d}{dt}V(t) = -\frac{1}{\tau_m}V(t) + \mathcal{I}(t), \qquad (10)$$

where $V(t)$ represents the membrane potential, τ_m is the membrane time constant and $\mathcal{I}(t)$ is the total synaptic input given by

$$\begin{aligned} \mathcal{I}(t) &= U_1(t) + U_2(t) \\ &= \exp\left(-\frac{t-t_1}{\tau_s}\right)\mathcal{H}(t-t_0) + \exp\left(-\frac{t-t_0-\tau_d}{\tau_s}\right)\mathcal{H}(t-t_0-\tau_d) \end{aligned}$$

where t_i represent spike arrival times, τ_s is the (synaptic) time constant, τ_d is the (dendritic) delay and $\mathcal{H}(t)$ is the Heaviside step function.

In this case, since we cannot prespecify the spike decay time as before, a different approach is required to determine the threshold parameter θ. The method used to do this is as follows.

We select ρ and τ_d as before, and initially we set $\tau_s = \rho/ln(2)$. Noting that

$$\rho = |\tau_d - \Delta|, \qquad (11)$$

and using $t_1 = t_0 + \tau_d$, an analytical expression for the total membrane potential can be easily obtained by solving (10). The maximum total membrane potential and thus the required threshold, can be obtained as a function of time constants and the tolerance range, and is given by

$$\begin{aligned}\theta(\rho) &= \frac{1}{(\tau_m/\tau_s - 1)}\left[X_r^{\frac{\tau_s}{(\tau_m - \tau_s)}}\left(e^{\rho/\tau_m} + 1\right) - X_r^{\frac{\tau_m}{(\tau_m - \tau_s)}}\left(e^{\rho/\tau_s} + 1\right)\right] \\ X_r &= \frac{\tau_s}{\tau_m}\left(\frac{1 + e^{\rho/\tau_m}}{1 + e^{\rho/\tau_s}}\right). \qquad (12)\end{aligned}$$

This method of threshold determination is used in the next section to show the performance of the network architecture.

EXPERIMENTS

In this section, we present some simulation results to show the behaviour of the proposed model. In the experiments reported here, we approximated three nonlinear functions:

$$\begin{aligned} g_1(x) &= 1 + \sin(4\pi x) \\ g_2(x) &= (x - 1.6)^3 - x + 4 \\ g_3(x) &= (2x - 1.6)^3 - 2x + 4 \end{aligned}$$

Fig. (2) shows the approximation results for segment sizes is g_1,g_2: $\Delta x = 0.1$ units, and g_3: $\Delta x = 0.05$ units. It can be observed that the model approximates the functions as required. We have shown large segment sizes for clarity, however with a larger number of units (and smaller segment sizes), the approximation would become more accurate.

DISCUSSION

In this paper we have considered the issue of function approximation in spiking networks. Such networks, although widely considered in the biological literature, are not as well represented in the signal processing and control community. Yet for some practical applications spiking networks have particular merit. This paper is a contribution to understanding how it may be possible to obtain the same type of powerful approximation characteristics that we expect in real valued neural network models, to a spiking network structure. We have proposed an architecture which is simple, but is capable of nonlinear function approximation. Experimental results have been used to demonstrate the model.

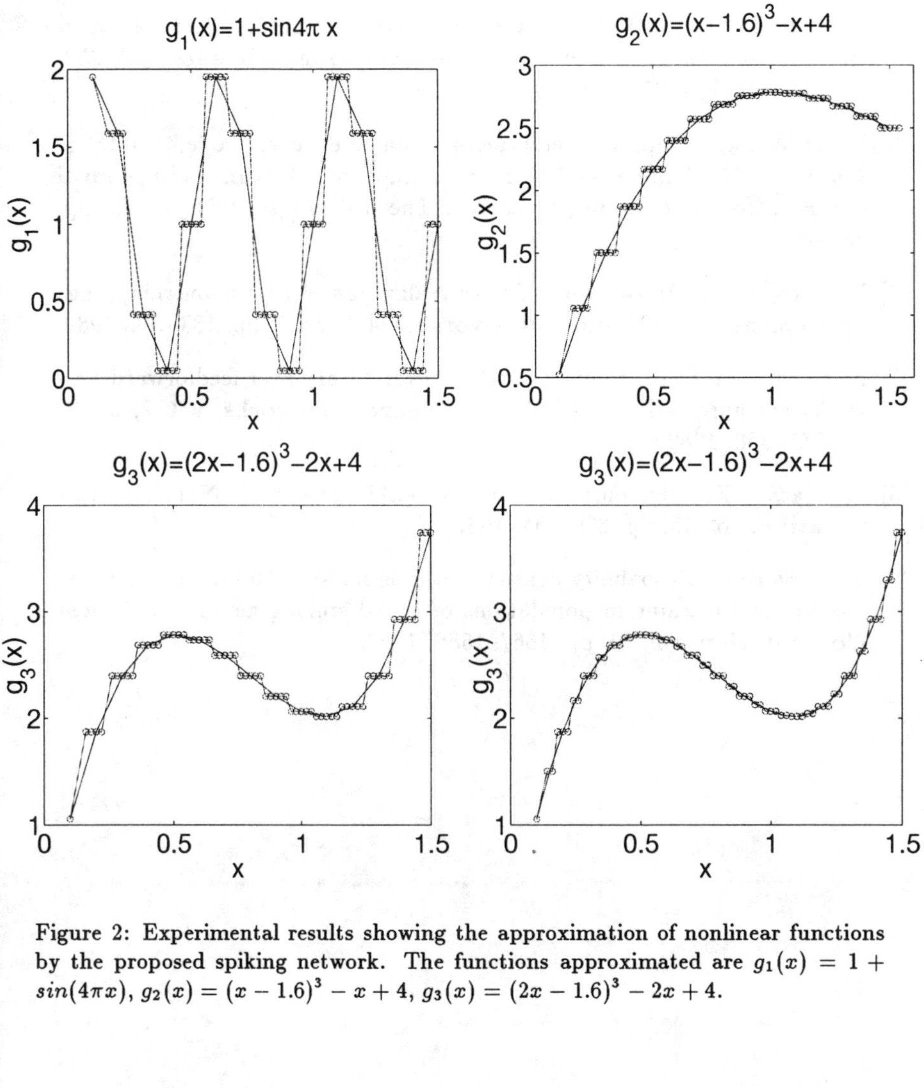

Figure 2: Experimental results showing the approximation of nonlinear functions by the proposed spiking network. The functions approximated are $g_1(x) = 1 + sin(4\pi x)$, $g_2(x) = (x - 1.6)^3 - x + 4$, $g_3(x) = (2x - 1.6)^3 - 2x + 4$.

REFERENCES

[1] W. Bialek, F. Rieke, R. R. de Ruyter van Stevenick and D. Warland, "Reading a neural code," **Science**, vol. 252, pp. 1854–1857, 1991.

[2] W. Bialek, F. Rieke, R. R. de Ruyter van Stevenick and D. Warland, **Spikes: Exploring the Neural Code**, MIT Press (Cambridge), 1997.

[3] G. Cybenko, "Approximation by superpositions of a sigmoidal function," **Mathematics of Control, Signals, and Systems**, vol. 2, no. 4, pp. 303–314, 1989.

[4] R. R. de Ruyter van Stevenick William Bialek, “Reliability and statistical efficiency of a blow fly movement-sensistive neuron,” **Science**, vol. 252, pp. 1854–1857, 1991.

[5] M. DeWeese, “Optimization principles for the neural code,” in D. S. Touretzky, M. C. Mozer and M. E. Hasselmo, eds., **Advances in Neural Information Processing Systems**, The MIT Press, 1996, vol. 8, pp. 281–287.

[6] K. Funahashi, “On the approximate realization of continuous mappings by neural networks,” **Neural Networks**, vol. 2, no. 3, pp. 183–192, 1989.

[7] K. Hornik, M. Stinchcombe and H. White, “Multilayer feedforward networks are universal approximators,” **Neural Networks**, vol. 2, no. 5, pp. 359–366, 1989.

[8] W. Maass, “Fast sigmoidal networks via spiking neurons,” **Neural Computation**, vol. 15, pp. 279–304, 1997.

[9] T. D. Sanger, “Probability density methods for smooth function approximation and learning in populations of tuned spiking neurons,” **Neural Computation**, vol. 10, pp. 1567–1586, 1998.

PARTIAL LIKELIHOOD METHODS FOR PROBABILITY DENSITY ESTIMATION *

Hongmei Ni, Tülay Adalı, and Bo Wang
Department of Computer Science and Electrical Engineering
University of Maryland, Baltimore County, Baltimore, MD 21250
{honi,adali,bwang1}@engr.umbc.edu

Abstract. Partial likelihood (PL) establishes a sufficiently general framework to develop and study statistical properties of nonlinear techniques in signal processing. In [1], we present the theorem by which the fundamental information-theoretic relationship for learning on the PL cost, the equivalence of likelihood maximization and relative entropy minimization, is established. In this paper, we reformulate the theorem to incorporate both the continuous and discrete probability modeling. We further show that, in both cases, the two conditions of the theorem are satisfied for the basic class of probability models, the exponential family, which includes many important network structures that can be effectively used as probability models. Hence we provide the prospect of using the PL cost in a wide class of applications with different models. We also propose several algorithms for learning/estimating the optimal model parameters by PL maximization. We give examples to illustrate the application of our general formulation and the learning algorithms and demonstrate the advantages of learning on the PL cost by simulation results.

PARTIAL LIKELIHOOD FORMULATION

The *partial likelihood* (PL) theory [5] is a recent extension of maximum likelihood (ML), and generalizes the ideas of conditional and marginal likelihood. Since it allows for inclusion of dependent observations, missing data and sequential processing, PL provides us with a general probabilistic framework which is very suitable for application to problems in which time-ordering is essential (e.g. time-series problems), or can be conveniently defined [1].

Partial likelihood can be defined as follows: Consider a time series $\{x_n\}$, $n = 1, 2, \cdots$, and its time-dependent covariates $\{\mathbf{y}_n\}$, and define $\mathcal{F}_{n-1} = \sigma\{1, [x_{n-1}, \cdots, x_1], [\mathbf{y}_n, \cdots, \mathbf{y}_1]\}$ as a collection of all relevant events upto

*RESEARCH SUPPORTED IN PART BY THE NATIONAL SCIENCE FOUNDATION CAREER AWARD, NSF NCR–9703161.

discrete (time) instant n. Since our goal is to estimate the probability distribution of x_i given the history $\mathcal{F}_{i-1}$, i.e., $p(x_i|\mathcal{F}_{i-1})$, we can use a nonlinear network/function with parameter $\boldsymbol{\theta}$ such that

$$p_{\boldsymbol{\theta}}(x_i|\mathcal{F}_{i-1}) = f_i(x_i, \boldsymbol{\theta}). \tag{1}$$

Note that, when x_n takes discrete values, we can model the probability mass function (pmf), and when x_n is continuous valued, we can choose probability models with continuous outputs, i.e., can model the probability density function (pdf). In both cases, the partial likelihood function relative to $\boldsymbol{\theta}$, $\mathcal{F}_i$ and the data $x_1, x_2, \cdots, x_N$ is written as the product

$$\mathcal{L}_N^p(\boldsymbol{\theta}) = \prod_{i=1}^{N} p_{\boldsymbol{\theta}}(x_i|\mathcal{F}_{i-1}) = \prod_{i=1}^{N} f_i(x_i, \boldsymbol{\theta}). \tag{2}$$

The relative entropy (RE), or the Kullback-Leibler distance, on the other hand, is a fundamental information-theoretic measure of how accurate the estimated probability distribution $p_{\boldsymbol{\theta}}$ is an approximation to the true probability distribution p and is given by $D(p\|p_{\boldsymbol{\theta}}) = E\left\{\log \frac{p}{p_{\boldsymbol{\theta}}}\right\}$ where the expectation is with respect to the true distribution p. Assume that for $\boldsymbol{\theta}_0$ $f_i(\cdot)$ defined in (1) achieves the true probability distribution and define $r_j(\boldsymbol{\theta}) \equiv \log \frac{f_j(x_j, \boldsymbol{\theta}_0)}{f_j(x_j, \boldsymbol{\theta})}$. We can write the *accumulated* relative entropy (ARE) as $\mathcal{I}_N(\boldsymbol{\theta}) = \sum_{j=1}^{N} i_j(\boldsymbol{\theta})$, where $i_j(\boldsymbol{\theta}) = E_{\boldsymbol{\theta}_0}\{r_j(\boldsymbol{\theta})\}$, and define $\mathcal{J}_N(\boldsymbol{\theta}) = \sum_{k=1}^{N} j_k(\boldsymbol{\theta})$, where $j_k(\boldsymbol{\theta}) = Var_{\boldsymbol{\theta}_0}\{r_k(\boldsymbol{\theta})\}$. Based on these definitions, we establish the relationship between PL maximization and ARE minimization for the cases of both dependent and independent observations by the following theorem:

Theorem: If, for each $\boldsymbol{\theta} \neq \boldsymbol{\theta}_0$, there exists a constant $\delta > 0$ such that, as $N \to \infty$,

$$P(\mathcal{I}_N(\boldsymbol{\theta})/N > \delta) \to 1 \tag{3}$$

and

$$\mathcal{J}_N(\boldsymbol{\theta})/N^2 \to 0 \text{ in probability} \tag{4}$$

then at least one arg $\min_{\boldsymbol{\theta}} \mathcal{I}_N(\boldsymbol{\theta})$ tends to one arg $\max_{\boldsymbol{\theta}} \overline{\mathcal{L}}_N^p(\boldsymbol{\theta})$ almost surely on $\Omega = \{\boldsymbol{\theta} \mid \mathcal{I}_N(\boldsymbol{\theta}) \uparrow \infty, \sum_{i=1}^{N} j_i(\boldsymbol{\theta})/\mathcal{I}_i^2(\boldsymbol{\theta}) < \infty\}$ where $\overline{\mathcal{L}}_N^p(\boldsymbol{\theta}) \equiv \ln \mathcal{L}_N^p(\boldsymbol{\theta})$.

This theorem is a generalization of the result in [1] as here we remove the restrictions on the probability model structure. Its proof however directly follows the proof in [1].

It is important to mention that, in the formulation, we do not make any assumptions about the underlying probability model and hence the theorem can be applied to a very wide class of probability models. Also, note that absence of any particular assumption on the dependence structure of the data makes the result applicable to the cases of both dependent and independent observations. As discussed in [2], for the simpler i.i.d. case, the widely referenced result on the equivalence of ML and RE, makes the assumption that the posterior probabilities are equal to the class labels, i.e., are either 0 or 1. In the theorem given above, we show the equivalence under two general conditions which do not require this equality assumption. Hence, our result is an alternate proof for the i.i.d. case as well and is one that bypasses the common but unrealistic equality assumption.

The theorem above establishes the equivalence of PL maximization and ARE minimization under two regularity conditions. The first condition, (3), represents the rate by which the Kullback-Leibler information accumulates with N, and guarantees that for each $\boldsymbol{\theta} \neq \boldsymbol{\theta}_0$, $\mathcal{I}_N(\boldsymbol{\theta}) \to \infty$ as $N \to \infty$, i.e., the information continues to accumulate. The second condition, (4), on the other hand implies asymptotical stability of variance. In what follows, we show that the two conditions are satisfied by the basic class of stochastic model, the exponential family.

INFORMATION-THEORETIC VIEW FOR THE EXPONENTIAL FAMILY OF DISTRIBUTIONS

Based on the definition of the exponential family in [4], we prove the following: *Proposition:* For the probability model defined in (1), if the family of the conditional distributions of x_n is an exponential family, i.e., if $f_n(x_n, \boldsymbol{\theta})$ can be written in the form:

$$f_n(x_n, \boldsymbol{\theta}) = \exp\{\boldsymbol{\alpha}_n^T(\boldsymbol{\theta})\boldsymbol{\gamma}(x_n) - \psi(\boldsymbol{\alpha}_n(\boldsymbol{\theta}))\}, \tag{5}$$

where all the functions $(\boldsymbol{\alpha}_n(\cdot), \boldsymbol{\gamma}(\cdot), \psi(\cdot))$ are bounded for bounded inputs, then under the assumption of compact parameter space Φ such that $\boldsymbol{\theta} \in \Phi$, the conditions of the theorem, (3) and (4), are satisfied.

Proof: If $f_n(x_n, \boldsymbol{\theta})$ belongs to an exponential family, we have

$$\ln f_n(x_n, \boldsymbol{\theta}) = \boldsymbol{\alpha}_n^T(\boldsymbol{\theta})\boldsymbol{\gamma}(x_n) - \psi(\boldsymbol{\alpha}_n(\boldsymbol{\theta})) \tag{6}$$

Using the definitions for the theorem, we can write

$$\begin{aligned} r_n(\boldsymbol{\theta}) &= \ln \frac{f_n(x_n, \boldsymbol{\theta}_0)}{f_n(x_n, \boldsymbol{\theta})} \\ &= (\boldsymbol{\alpha}_n(\boldsymbol{\theta}_0) - \boldsymbol{\alpha}_n(\boldsymbol{\theta}))^T \boldsymbol{\gamma}(x_n) - (\psi(\boldsymbol{\alpha}_n(\boldsymbol{\theta}_0)) - \psi(\boldsymbol{\alpha}_n(\boldsymbol{\theta}))) \end{aligned} \tag{7}$$

We assume all the functions $(\boldsymbol{\alpha}_n(\cdot), \boldsymbol{\gamma}(\cdot), \psi(\cdot))$ in (5) are bounded for bounded inputs, and that the parameter space is compact. Then, all the terms in (7) are bounded. Hence $r_n(\boldsymbol{\theta})$ and $i_n(\boldsymbol{\theta}) = E_{\boldsymbol{\theta}_0}\{r_n(\boldsymbol{\theta})\}$ are bounded. Also, from the definition of $i_n(\boldsymbol{\theta})$, we know $i_n(\boldsymbol{\theta}) > 0$, for $\boldsymbol{\theta} \neq \boldsymbol{\theta}_0$.

We can write

$$j_n(\boldsymbol{\theta}) = Var_{\boldsymbol{\theta}_0}\{r_n(\boldsymbol{\theta})\} = E_{\boldsymbol{\theta}_0}\{r_n^2(\boldsymbol{\theta})\} - i_n^2(\boldsymbol{\theta}) \tag{8}$$

Similarly, since $r_n(\boldsymbol{\theta})$ is always bounded, the first term in (8) $E_{\boldsymbol{\theta}_0}\{r_n^2(\boldsymbol{\theta})\}$ is bounded. We have also shown that $i_n(\boldsymbol{\theta})$ is bounded. Hence $j_n(\boldsymbol{\theta})$ is also bounded. From the definition of $j_n(\boldsymbol{\theta})$, we have $j_n(\boldsymbol{\theta}) \geq 0$.

When x_n is independent of $\mathcal{F}_{n-1}$, both the conditional information $i_n(\boldsymbol{\theta})$ and $j_n(\boldsymbol{\theta})$ are independent of n and are constants for each $\boldsymbol{\theta}$. In the general case where $\boldsymbol{\alpha}_n(\boldsymbol{\theta}) = \boldsymbol{\alpha}(\mathbf{y}_n, \boldsymbol{\theta})$, both $i_n(\boldsymbol{\theta})$ and $j_n(\boldsymbol{\theta})$ are random variables, i.e., functions of $\mathbf{y}_n$. For each $\boldsymbol{\theta} \neq \boldsymbol{\theta}_0$, we know that $i_n(\boldsymbol{\theta})$ is always positive and bounded, thus the expectation $E\{i_n(\boldsymbol{\theta})\}$ is positive and bounded. For stationary and ergodic $\{\mathbf{y}_n\}$, by Birkhoff-Khinchin theorem [6], as $N \to \infty$,

$$\sum_{n=1}^{N} i_n(\boldsymbol{\theta})/N \to E\{i_n(\boldsymbol{\theta})\} \quad \text{almost everywhere.} \tag{9}$$

We already have $E\{i_n(\boldsymbol{\theta})\} > 0$, so there must exist a constant $\delta > 0$, such that $P(\mathcal{I}_N(\boldsymbol{\theta})/N > \delta) \to 1$ as $N \to \infty$, i.e., the first condition of the theorem is satisfied. We have shown $j_n(\boldsymbol{\theta})$ is always bounded, thus $\mathcal{J}_N(\boldsymbol{\theta})/N^2 \to 0$ as $N \to \infty$. Hence both conditions in the theorem are satisfied for the exponential family defined in (5).

Note that we don't make any assumption regarding the process x_n, whether it takes discrete or continuous values, during the development. The theorem and the proposition hence are applicable to both the discrete and continuous cases. Thus we can use the PL cost in a wide range of applications in a number of ways.

THE LEARNING ALGORITHMS

In this section, we propose the general learning algorithms for maximum PL estimation of the parameters of the exponential family. Our goal is to minimize the negative log PL cost function

$$\mathcal{J}_N(\boldsymbol{\theta}) \equiv -\ln \mathcal{L}_N^p(\boldsymbol{\theta}) = \sum_{n=1}^{N} J_n(\boldsymbol{\theta}) \tag{10}$$

where $J_n(\boldsymbol{\theta}) = -\ln p_\theta(x_n|\mathcal{F}_{n-1})$. For the exponential family class, we have

$$J_n(\boldsymbol{\theta}) = -(\boldsymbol{\alpha}_n^T(\boldsymbol{\theta})\boldsymbol{\gamma}(x_n) - \psi(\boldsymbol{\alpha}_n(\boldsymbol{\theta}))) \tag{11}$$

Instantaneous gradient descent minimization of $J_n(\boldsymbol{\theta})$ results in the following parameter update:

$$\boldsymbol{\theta}_{n+1} = \boldsymbol{\theta}_n + \mu \mathbf{B}(\boldsymbol{\gamma}(x_n) - \frac{\partial \psi}{\partial \boldsymbol{\alpha}_n}) \tag{12}$$

where μ is the learning rate, $\mathbf{B}$ is a matrix with its jth column vector $B_j = \frac{\partial \alpha_{nj}}{\partial \boldsymbol{\theta}}$, and α_{nj} is the jth element of $\boldsymbol{\alpha}_n$.

For partial observed data (e.g., the hidden component index problem in the normal mixture model), we have $\boldsymbol{\gamma}(x_n) = (\boldsymbol{\gamma}^v, \boldsymbol{\gamma}^h)$ where $\boldsymbol{\gamma}^v$ is observed and $\boldsymbol{\gamma}^h$ is hidden. We can use the conditonal expectation $E\{\boldsymbol{\gamma}(x_n)|\boldsymbol{\gamma}^v\}$ instead of $\boldsymbol{\gamma}(x_n)$ for the parameter update, i.e., at time n, we perform

$$\eta_n = E_{\theta_n}\{\boldsymbol{\gamma}(x_n)|\boldsymbol{\gamma}^v\} \tag{13}$$

$$\boldsymbol{\theta}_{n+1} = \boldsymbol{\theta}_n + \mu\mathbf{B}(\eta_n - \frac{\partial\psi}{\partial\boldsymbol{\alpha}_n}) \tag{14}$$

This method is very similar to the popular EM algorithm in dealing with the partial observation, with (13) corresponding to the "Expectation" step and (14) to the "Maximization" step in EM algorithm. The EM algorithm maximizes the marginal likelihood and assumes that all the observations are independent, while our method maximizes the partial likelihood and doesn't make the unrealistic independence assumption.

When $\boldsymbol{\alpha}_n(\boldsymbol{\theta})$ is not a function of n, the negative log PL cost function can be written as

$$\begin{aligned} \mathcal{J}_N(\boldsymbol{\theta}) &= -(\boldsymbol{\alpha}^T(\boldsymbol{\theta})\sum_{n=1}^{N}\gamma(x_n) - N\psi(\boldsymbol{\alpha}(\boldsymbol{\theta}))) \\ &= -N(\boldsymbol{\alpha}^T(\boldsymbol{\theta})\overline{\gamma}_N - \psi(\boldsymbol{\alpha}(\boldsymbol{\theta}))) \end{aligned} \tag{15}$$

where $\overline{\gamma}_N = \frac{1}{N}\sum_{n=1}^{N}\gamma(x_n)$.

Its gradient becomes

$$\nabla\mathcal{J} \equiv \frac{\partial\mathcal{J}_N(\boldsymbol{\theta})}{\partial\boldsymbol{\theta}} = -N\mathbf{B}(\overline{\gamma}_N - \frac{\partial\psi}{\partial\boldsymbol{\alpha}}) \tag{16}$$

Then, a batch procedure for learning $\boldsymbol{\theta}$ can be obtained by solving the equation $\frac{\partial\mathcal{J}_N(\boldsymbol{\theta})}{\partial\boldsymbol{\theta}}\,|_{\boldsymbol{\theta}^*} = 0$.

For on-line learning, gradient descent minimization of $\mathcal{J}_n(\boldsymbol{\theta}) = \sum_{i=1}^{n} J_i(\boldsymbol{\theta})$ results in

$$\boldsymbol{\theta}_{n+1} = \boldsymbol{\theta}_n + \mu\mathbf{B}(\overline{\gamma}_n - \frac{\partial\psi}{\partial\boldsymbol{\alpha}}) \tag{17}$$

where $\overline{\gamma}_n = \frac{1}{n}\sum_{i=1}^{n}\gamma(x_i)$ is the time average for $\boldsymbol{\gamma}(x_n)$.

Another alternative is to estimate $E\{\boldsymbol{\gamma}(x_n)\}$ and use it instead of $\overline{\gamma}_n$ for parameter updates. We define $\overline{\eta} = E\{\boldsymbol{\gamma}(x_n)\}$ and update $\overline{\eta}$ sequentially by varying the contribution of the new sample. We can then perform the following two steps iteratively:

$$\overline{\eta}_n = (1-\epsilon_n)\overline{\eta}_{n-1} + \epsilon_n\gamma(x_n); \tag{18}$$

$$\boldsymbol{\theta}_{n+1} = \boldsymbol{\theta}_n + \epsilon_n\mathbf{B}(\overline{\eta}_n - \frac{\partial\psi}{\partial\boldsymbol{\alpha}}) \tag{19}$$

where ϵ_n is a learning rate selected as a decreasing sequence. If we set $\epsilon_n = 1/n$, we have $\overline{\eta}_n = \frac{1}{n}\sum_{i=1}^{n}\gamma(x_i) = \overline{\gamma}_n$ and thus we are implementing (17). The updates given in (18) and (19) allow for flexible choice of weighting for computing the average for $\boldsymbol{\gamma}(x_n)$.

Hence, our formulation not only provides the prospect of using PL in a very wide range of probability models, it also provides various and effective ways to learn/estimate the optimal model parameters. In the following section, we give application examples to illustrate this point.

APPLICATION EXAMPLES

In this section, we give two examples to show the application of our formulation and the learning algorithms to probability distribution estimation of continuous and discrete random variables.

Discrete Case

Assume that we have a training set $\mathcal{T}$ of N related input and output pairs $\mathcal{T} = \{x_n, \mathbf{y}_n\}_{n=1}^{N}$ and the problem is to train the classifier such that for a given observation $\mathbf{y}_n$, x_n takes a value from the finite alphabet $\mathcal{S} = \{a_0, a_1, \cdots, a_{M-1}\}$. Define the class labels C_k, $k = 0, 1, \cdots, M-1$, such that $\mathbf{y}_n \in C_k$ if $x_n = a_k$. Bayes classifier will assign $\mathbf{y}_n$ to class C_k if $P(C_k|\mathbf{y}_n) > P(C_j|\mathbf{y}_n) \quad \forall j \neq k$, a choice which minimizes the classification error probability. To estimate the posterior probabilities $P(C_k|\mathbf{y}_n)$ for $k = 0, 1, \cdots, M-1$, we can use a feedforward neural network probability model with parameter $\boldsymbol{\theta}$ such that

$$P_\theta(C_k|\mathbf{y}_n) = g_k(\boldsymbol{\theta}, \mathbf{y}_n). \tag{20}$$

For the general case of multiple classes, in order to ensure that all the network outputs are valid probabilities (i.e., they sum up to one), we also need a normalization stage. We can introduce softmax normalization for a single hidden layer feedforward neural classifier with logistic activation function $h(\cdot)$ as

$$g_k(\boldsymbol{\theta}, \mathbf{y}_n) = \frac{\exp(\phi_{k_n})}{\sum_{j=0}^{M-1} \exp(\phi_{j_n})} \quad k = 0, 1, \cdots, M-1 \tag{21}$$

where $\phi_{j_n} = \sum_{i=1}^{q} h\left(\mathbf{y}_n{}^T \mathbf{w}^i\right) v^{ij}$ and $\boldsymbol{\theta} = (\mathbf{w}, \mathbf{v})$ is the network weight vector.

By defining

$$\alpha_{ni} = \phi_{i_n}/\phi_{0_n}, \quad \gamma_i = \delta_i(x_n), \quad i = 1, \cdots, M-1 \tag{22}$$

and

$$\psi(\boldsymbol{\alpha}_n) = \ln(1 + \sum_{i=1}^{M-1} \exp(\alpha_{ni})) \tag{23}$$

we have

$$p_\theta(x_n|\mathcal{F}_{n-1}) \equiv f_n(x_n, \boldsymbol{\theta}) = \exp(\boldsymbol{\alpha}_n{}^T \boldsymbol{\gamma} - \psi(\boldsymbol{\alpha}_n)). \tag{24}$$

Since the softmax perceptron model (21) belongs to the exponential family, by our proposition, it satisfies the regularity conditions of the theorem. We can learn the weights of a softmax network by instantaneous gradient descent, i.e., using Equation (12). In [3], we apply the softmax perceptron model to perform multi-class classification. We note the inefficiency of training a softmax network when $M > 2$ and propose an efficient multi-class classifier structure based on binary coding of of the classes.

By eliminating the inherent redundancy in (21), we can define a modified softmax model with $M-1$ outputs instead of M outputs as follows:

$$g_k(\boldsymbol{\theta}, \mathbf{y}_n) = \frac{\exp(\phi_{k_n})}{1+\sum_{j=1}^{M-1}\exp(\phi_{j_n})} \quad k = 1, \cdots, M-1 \tag{25}$$

Similarly, using (22) and (23), we can show that the modified softmax model (25) belongs to the exponential family.

Note that, for the binary case, we only need one network output to perform the classification, and hence, can write

$$g(\boldsymbol{\theta}, \mathbf{y}_n) = \frac{\exp(\phi_n)}{1+\exp(\phi_n)} = \frac{1}{1+\exp(-\phi_n)}. \tag{26}$$

where $\phi_n = \sum_{i=1}^{q} h\left(\mathbf{y}_n{}^T \mathbf{w}^i\right) v^i$. This is a binary MLP network with logistic sigmoid output function.

Continuous Case

For the aforementioned training set $\mathcal{T} = \{x_n, \mathbf{y}_n\}_{n=1}^N$, when the objective is to model the probability distribution of $\mathbf{y}_n$, we can obtain a PL formulation similar to (2) by a factorization of the likelihood as given in [8]:

$$\mathcal{L}_N^p(\boldsymbol{\theta}) = \prod_{n=1}^{N} p_\theta(\mathbf{y}_n|\mathcal{F}_{n-1}). \tag{27}$$

where $\mathcal{F}_{n-1} = \sigma\{1, [x_n, \cdots, x_1][y_{n-1}, \cdots, y_1]\}$.

Usually $\mathbf{y}_n$ is completely determined by $\{x_i\}_{i=1}^n$, i.e., $p_\theta(\mathbf{y}_n|\mathcal{F}_{n-1}) = p_\theta(\mathbf{y}_n|x_n, \cdots, x_1)$. We can choose a model such that $p_\theta(\mathbf{y}_n|\mathcal{F}_{n-1}) = p_\theta(\mathbf{y}_n|\mathbf{x}_n)$, where $\mathbf{x}_n = [x_n, x_{n-1}, \cdots, x_{n-L+1}]^T$. Then the PL function becomes $\mathcal{L}_N^p(\boldsymbol{\theta}) = \prod_{n=1}^N p_\theta(\mathbf{y}_n|\mathbf{x}_n)$.

Because x_n takes a value from a finite alphabet of size M, we can map $\mathbf{x}_n$ to a discrete variable z which takes values $\{0, 1, \cdots, K\}$ with $K = M^L - 1$. Assume that $\mathbf{y}_n$ is normally distributed $N(\mu_i, \Sigma_i)$ when $z = i$, we have

$$p_\theta(\mathbf{y}_n|z) = \sum_{i=0}^{K} \frac{\delta_i(z)}{(\sqrt{2\pi})^d|\Sigma_i|^{1/2}} \exp\{-\frac{1}{2}(\mathbf{y}_n - \mu_i)^T\Sigma_i{}^{-1}(\mathbf{y}_n - \mu_i)\}, \tag{28}$$

where $\delta_i(z) = 1$ when $z = i$ and 0 otherwise, d is the dimension of $\mathbf{y}_n$, $\mathbf{y}_n = [y_1, \cdots, y_d]^T$.

We assume that all the components of $\mathbf{y}_n$ are conditionally independent given $\mathbf{x}_n$. In channel equalization, for example, the output vector $\mathbf{y}_n$ given the transmitted symbols $x_n, x_{n-1}, \cdots$, is independent. So the covariance matrix Σ_i is diagonal, $\Sigma_i = diag[\sigma_{i1}^2, \ldots, \sigma_{id}^2]$, $i = 0, 1, \cdots, K$. Let $\zeta_i = [\sigma_{i1}^{-2}, \ldots, \sigma_{id}^{-2}]^T$, $i = 0, 1, \cdots, K$, we can use $\boldsymbol{\theta} = [\mu_0^T, \cdots, \mu_K^T, \zeta_0^T, \cdots, \zeta_K^T]^T$

to denote the vector of model parameters to be estimated. By using the condition $\sum_{i=0}^{K} \delta_i(z) = 1$ and defining

$$
\begin{aligned}
\gamma_j &= y_j, \\
\alpha_j &= \mu_{0j}\sigma_{0j}^{-2}, \\
\gamma_{d+j} &= y_j^2, \\
\alpha_{d+j} &= -\frac{1}{2}\sigma_{0j}^{-2}, \\
\gamma_{2d+i} &= \delta_i(z), \\
\alpha_{2d+i} &= \ln \frac{|\Sigma_i|^{-1/2}}{|\Sigma_0|^{-1/2}} - \frac{1}{2}\mu_i^T \Sigma_i^{-1} \mu_i + \frac{1}{2}\mu_0^T \Sigma_0^{-1} \mu_0 \\
\gamma_{2d+K+(i-1)d+j} &= y_j \delta_i(z), \\
\alpha_{2d+K+(i-1)d+j} &= \mu_{ij}\sigma_{ij}^{-2} - \mu_{0j}\sigma_{0j}^{-2}, \\
\gamma_{2d+K+Kd+(i-1)d+j} &= y_j^2 \delta_i(z), \\
\alpha_{2d+K+Kd+(i-1)d+j} &= \frac{1}{2}\sigma_{ij}^{-2} - \frac{1}{2}\sigma_{0j}^{-2},
\end{aligned}
\tag{29}
$$

where $i = 1, \cdots, K$, $j = 1, \cdots, d$, we can show that this model belongs to the exponential family,

$$p_\theta(\mathbf{y}_n|z) = \exp\{\boldsymbol{\alpha}^T(\boldsymbol{\theta})\boldsymbol{\gamma} - \psi(\boldsymbol{\alpha}(\boldsymbol{\theta}))\} \tag{30}$$

where

$$\psi(\boldsymbol{\alpha}) = -\ln p_0|\Sigma_0|^{-1/2} + \frac{1}{2}\mu_0^T \Sigma_0^{-1} \mu_0 + \ln(\sqrt{2\pi})^d. \tag{31}$$

Since $\boldsymbol{\alpha}(\boldsymbol{\theta})$ is a function independent of n, we can use Equation (18) and (19) iteratively to learn the optimal parameter $\boldsymbol{\theta}^*$ of this finite normal mixture (FNM) model.

Given $p_\theta(\mathbf{y}_n|z)$, i.e., $p_\theta(\mathbf{y}_n|\mathbf{x}_n)$, we can write the class conditional probabilities $p_\theta(\mathbf{y}_n|C_k), k = 0, 1, \cdots, M-1$, as

$$p_\theta(\mathbf{y}_n|C_k) = \sum_{j=1}^{Q} \pi_j p_\theta(\mathbf{y}_n|\mathbf{x}_n^{(j)}) \tag{32}$$

where the superscript $(\cdot)^j$ indicates all the $Q = M^{L-1}$ possible combinations of $\mathbf{x}_n$ when $x_n = a_k$. In supervised learning, the priors π_j, $j = 1, \cdots, Q$ and the class priors $P(C_k), k = 0, 1, \cdots, M-1$ are easy to obtain. Given $p_\theta(\mathbf{y}_n|C_k)$ and $P(C_k)$, we can use Bayes rule to perform the classification.

Simulation Results

We apply our formulation and the learning algorithms to solve adaptive channel equalization problem. We consider transmission of binary data $x_n \in \{-1, 1\}$ through a nonlinear multipath channel such that the received signal is given by $o(n)$. At time n, the d-dimensional observation vector is given by $\mathbf{y}_n = [o(n), o(n-1), \cdots, o(n-d+1)]^T$. In Figure 1, we show the two-dimensional observation space for the channel characterized by $o(n) = o_l(n) - 0.2o_l^2(n) + \eta(n)$ where $o_l(n) = 0.3482x_n + 0.8704x_{n-1} + 0.3482x_{n-2}$ and $\eta(n)$ is the Gaussian noise. We can observe that, the FNM model provides a very good match to the distribution of the output of a multipath channel plus additive noise, a typical output observed in communications applications.

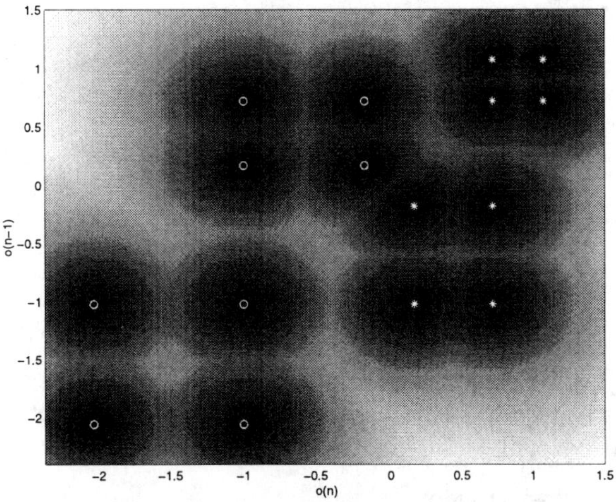

Figure 1: Observation space with "o" denoting centers for class C_1 and "*" for C_2

We use the FNM model and MLP classifiers to perform the channel equalization. As discussed before, the binary MLP classifier with logistic activation function is a special case of the modified softmax model. Hence it belongs to the exponential family [1] and satisfies the regularity conditions of the theorem. For performance comparison, two MLP equalizers are used, one trained with the mean square error (MSE) cost, the other with the PL (or RE) cost function. We call them the MLP-MSE and the MLP-LRE equalizers respectively. The structure of the MLP equalizers is chosen such that they have similar complexity with the FNM model.

We update the parameters of the MLP-LRE equalizer using Equation (12) and those of the MLP-MSE equalizer by instantaneous gradient descent minimization of the MSE cost. The parameters of the FNM model are updated using Equations (18) and (19). Number of training samples is chosen such that the algorithm converges to a predetermined threshold τ, that is, 120 samples for the FNM equalizer, 1000 samples for the MLP equalizers. Their bit error rate (BER) curves are given in Figure 2. We can see that, the FNM and the MLP-LRE equalizers trained with the PL (or RE) cost perform better than the MLP-MSE equalizer. Especially, the FNM equalizer is very efficient compared to the MLP-MSE and the MLP-LRE equalizers such that it needs much smaller number of training samples to converge while it still achieves better BERs. This efficiency is a result of the good model match observed in Figure 1. However, it is also important to note that, the cor-

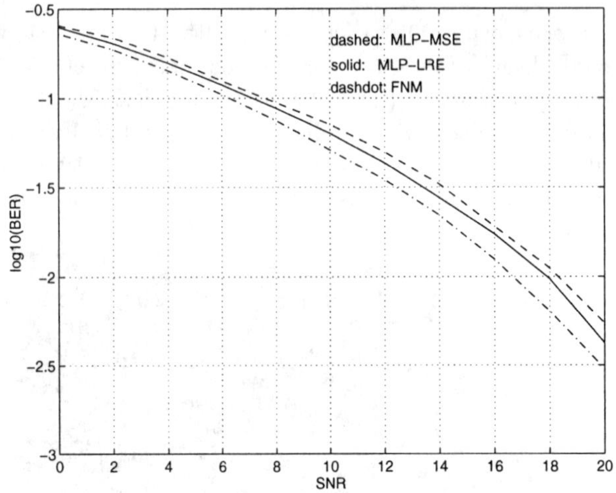

Figure 2: BER curves for the FNM equalizer with 16 mixtures, the MLP-MSE (2-18-1 MLP), and the MLP-LRE (2-18-1 MLP) equalizers

rect order selection is important [7] and when the channel memory increases, the complexity (order) of the FNM model and the computational burden increases substantially.

ACKNOWLEDGMENT

The authors would like to thank Yi Cai for his help with the plots.

REFERENCES

[1] T. Adalı, X. Liu, and M. K. Sönmez, "Conditional distribution learning with neural networks and its application to channel equalization," *IEEE Trans. Signal Processing*, vol. 45, no. 4, pp. 1051-1064, Apr. 1997.

[2] T. Adalı, "A general probabilistic formulation for neural classifiers," in *Proc. IEEE Workshop on Neural Networks for Signal Processing (NNSP)*, Cambridge, England, Sep. 1998, pp. 145-154.

[3] T. Adalı, H. Ni, and B. Wang, "Partial likelihood for estimation of multi-class posterior probabilities," in *Proc. IEEE Int. Conf. Acoust., Speech, Signal Processing (ICASSP)*, Phoenix, AZ, March 1999, vol. 2, pp. 1053-1056.

[4] S. Amari, "Information Geometry of the EM and em algorithms for Neural Networks," *Neural Networks*, vol. 8, No. 9, pp. 1379-1408, 1995.

[5] D.R. Cox, "Partial likelihood," *Biometrika*, vol. 62, pp. 69-72, 1975.

[6] R. M. Gray and L. D. Davisson, *Random Processes: A Mathematical Approach for Engineers*, Englewood Cliffs, NJ: Prentice-Hall, 1986.

[7] B. Wang, T. Adali, X. Liu, and J. Xuan, "Partial likelihood for real-time signal processing with finite normal mixtures," in *Proc. IEEE Workshop on Neural Networks for Signal Proc. VIII*, Cambridge, England, Sep. 1998.

[8] W. H. Wong,"Theory of partial likelihood," *Ann. Statist.*, 14, pp. 88-123, 1986.

WHY A NONLINEAR SOLUTION FOR A LINEAR PROBLEM?*

Tülay Adalı
Department of Computer Science and Electrical Engineering
University of Maryland Baltimore County
Baltimore, MD 21250, USA
adali@umbc.edu

Abstract. We emphasize a key point that when there is noise in the system, even if the system is linear, a nonlinear solution is more desirable. We derive a simple expression that shows that for a linear regression model, the logistic nonlinearity will be the natural match for modeling posterior class probabilities, and that the steepness of this logistic function is inversely proportional to the level of noise in the system. We note a problem that matches this data generation mechanism, equalization of an infinite impulse response channel, and show that for this example, the logistic type equalizer not only achieves lower bit error rate than its linear counterpart but is very efficient as well.

INTRODUCTION

The logistic function has played an important role in the neural network literature as the nice smooth, monotonic nonlinearity, or the "squashing" function that maps the real line into a bounded interval. Recently, with the growing interest in the probabilistic view of a neural network classifier, its connections with the well-established method of logistic regression have been pointed out in the neural network literature [1, 4, 8] with emphasis on why it does arise as the "natural" choice for the nonlinearity to be used in a probabilistic setting. In this paper, we point out another one of these observations and show that for a general linear regressive type data generation process, the logistic nonlinearity is again the natural choice. We select an application which matches this data generation mechanism, the inverse system identification, or the *equalization* problem that arises in different forms in a wide variety of signal processing problems. Using the equalization of a *linear* channel with binary inputs as an example, we emphasize a key point, that we would actually prefer to use a nonlinear equalizer for the equalization

*RESEARCH SUPPORTED IN PART BY THE NATIONAL SCIENCE FOUNDATION CAREER AWARD, NSF NCR–9703161.

of a linear channel *when there is noise in the system.* We show this advantage by a simple expression that directly links the steepness of the logistic nonlinearity to the noise level in the system.

It has been noted before that the perceptron type equalizers using logistic nonlinearity could produce better bit error rates (BERs) than their linear counterparts for equalization of a linear channel [5, 6]. However, the perceptron equalizer has also been recognized as very slow in convergence compared to a linear equalizer [5, 6, 9] and its use in a real-time application has been considered impractical [9]. We show that if the type of the nonlinearity, the sigmoidal nonlinearity in this case, matches the data generation process, then the nonlinear equalizer can be more efficient than its linear counterpart. This advantage is more pronounced when there is significant noise in the system, i.e., when the signal to noise ratio (SNR) is poor.

BINARY LOGISTIC REGRESSION CLASSIFIER

In a supervised time-series classification problem, given a training set $\mathcal{T}$ of N ordered input and output pairs $\mathcal{T} = \{x_n, \mathbf{y}_n\}_{n=1}^N$, the problem is to train the classifier using $\mathcal{T}$, such that, for a given observation $\mathbf{y}_n$, the input x_k will be assigned to one of m classes $C_1, C_2, \cdots, C_m$. Given the posterior class probabilities $P(C_i|\mathbf{y}_n)$ for $i = 1, 2, \cdots, m$, Bayes classifier will assign $\mathbf{y}_n$ to class C_i if $P(C_i|\mathbf{y}_n) > P(C_j|\mathbf{y}_n) \;\; \forall j \neq i$, a choice which minimizes the classification error probability.

To design the classifier, we can estimate the posterior probabilities directly by choosing an appropriate probability model $P_\theta(C_i|\mathbf{y}_n)$ parameterized by $\boldsymbol{\theta}$, such as the logistic/perceptron type model. Alternatively, we can estimate the class conditional density $p_\theta(\mathbf{y}_n|C_i)$ and can perform classification using Bayesian relationship along with a specification of the marginal probabilities $P(C_i)$. For the conditional probabilities, we can use models such as the finite normal mixtures (FNM) or the generalized FNM, which usually requires the estimation of a much larger number of parameters than the posterior probability models. However, if the model is a good match to the data generation process, as in the example of equalization of a finite impulse response (FIR) channel with a FNM equalizer [2], then the estimation will be highly efficient. The logistic function model for the posterior probabilities, on the other hand, requires estimation of a smaller number of parameters in general and is also more robust to model specification mismatches. Most often, though, we do have to trade in the efficiency in learning/estimation when using logistic models.

Consider the binary case where x_n takes a value from the binary alphabet $\mathcal{S} = \{0, 1\}$. Note that for this case we only need to estimate $P_\theta(x_n = 1|\mathbf{y}_n)$ as $P_\theta(x_n = 0|\mathbf{y}_n) = 1 - P_\theta(x_n = 1|\mathbf{y}_n)$. It is easy to recognize that for the binary classification problem where the conditional class densities belong to the exponential family with equal dispersion parameters, the posterior probabilities will be logistic function of the components of $\mathbf{y}_n$ [4, 8]. This result follows from the following form which uses Bayes relationship to write the binary posterior probabilities in terms of the class conditional properties as

$$P_\theta(x_n = 1|\mathbf{y}_n) = \frac{1}{1 + \exp\{-\log \frac{P_\theta(\mathbf{y}_n|x_n=1)}{P_\theta(\mathbf{y}_n|x_n=0)} - \log \frac{P(x_n=1)}{P(x_n=0)}\}}. \quad (1)$$

Since the normal distribution assumption arises naturally in many problems, we can provide connections for the data generation mechanisms for the FNM model quite easily. For the equalization example, in the FNM model, the normal component means can be shown to be functions of channel coefficients and the normal component variances directly proportional to the additive noise in the system (discussed in more detail in the next section). We can relate these parameters to the posterior probability model parameters as well, for example by using the exponential family representation for the FNM model with hidden variables given in [2] and the relationship above (1). This, however, will result in a complex expression and the way different system parameter effects are reflected in the model parameters will be hard to identify. Next, we derive a simple relationship that links the model parameters to the sigmoidal posterior probability (classifier) parameters directly and show that the logistic function is not only robust but can be efficient as well for an important class of linear problems when there is noise in the system. We then show that this model specification exactly matches the problem of equalization of an infinite impulse response channel.

Assume the regressive type relationship between two random processes z_n and y_n such that

$$z_n = a_0 + a_1 y_n + a_2 y_{n-1} + \cdots a_{k+1} y_{n-k} + \beta v_n \quad (2)$$

where β is a constant and the v_n are independent and identically distributed (i.i.d.) random variables that are logistically distributed $v_n \sim f(v) = \frac{e^v}{(1+e^v)^2}$. Define the binary random process x_n as

$$x_n = \begin{cases} 1, & \text{if } z_n \geq \tau \\ 0, & \text{if } z_n < \tau \end{cases} \quad (3)$$

We can then show that the binary posterior probability is given by:

$$P_{\theta_*}(x_n = 1|\mathbf{y}_n) = \frac{1}{1 + \exp(-\boldsymbol{\theta}_*{}^T \mathbf{y}_n)} \quad (4)$$

where

$$\boldsymbol{\theta}_* = \frac{1}{\beta}\left[a_0 - \tau,\ a_1,\ ,\cdots,\ a_{k+1}\right] \quad (5)$$

and

$$\mathbf{y}_n = \left[1,\ y_n,\ y_{n-1},\ \cdots,\ y_{n-k}\right]. \quad (6)$$

Hence the regression model of (2) with logistic errors implies the logistic posterior probability. Notice that the linear relationship given in (2) is nothing

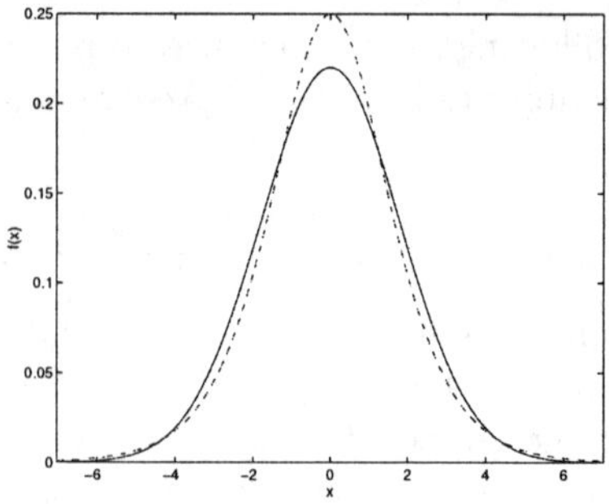

Figure 1: Logistic (solid) vs Normal (dashed) density

but the moving average (MA) model, or if we let $z_n \equiv y_{n+1}$, gives us the auto-regressive (AR) model. Since the derivation of (4) is based on the i.i.d. assumption of the logistic error term v_n, the form given in (4)-(6) will hold for both the MA and the AR cases. Note that the model in (2) with different definitions for z_n is a fairly general model, and that this type of data generation mechanisms can arise in a number of time series problems. Before we introduce the linear equalization example which matches this specification, note the closeness of the logistic distribution to the normal distribution. Figure 1 shows the comparison of the logistic density $f(x) = \frac{e^x}{(1+e^x)^2}$ which has 0 mean and variance $\pi^2/3$ with the density $\mathcal{N}(0, \pi^2/3)$. Hence, in our example, we do not need to assume logistic errors and can stick to the standard description of the equalization problem using our favorite, the normal distribution.

EQUALIZATION EXAMPLE

The standard difference equation shown in (2) arises in a number linear systems applications. One good example is the equalization (deconvolution or the inverse system identification) problem. Consider the channel equalization setting shown in Figure 2 where the aim is to recover the original transmitted sequence x_n at the output of the equalizer. If we let the channel response be

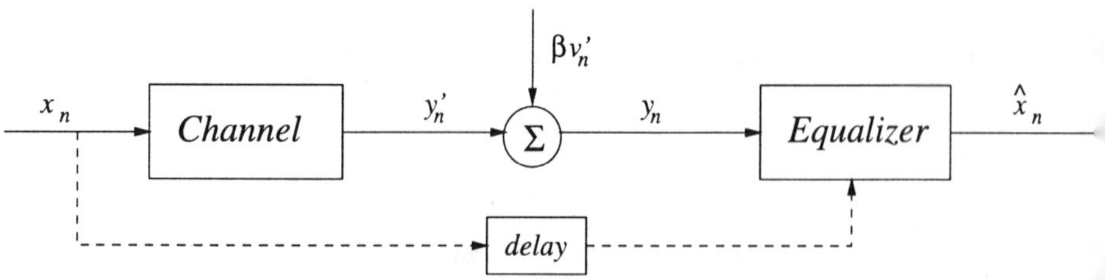

Figure 2: Equalization Problem

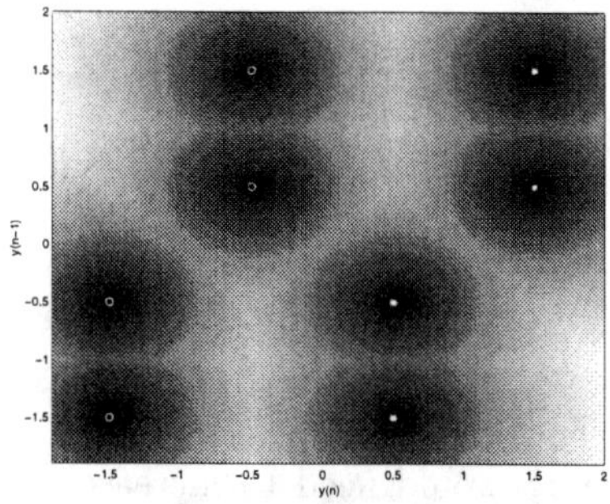

Figure 3: Distribution of the observation $\mathbf{y}_n$ for multipath channel of $1 + 0.5z^{-1}$ plus additive noise for $x_n \in \{-1, 1\}$ ("o" denotes cluster centers for $x_n = -1$ and "*" for $x_n = 1$)

infinite impulse response (IIR) given as

$$H(z) = \frac{1}{a_1 + a_2 z^{-1} + \cdots + a_{k+1} z^{-k}}$$

and assume that the channel output is corrupted by additive white Gaussian noise (which is the standard assumption in these problems), for binary inputs $x_n \in \{0, 1\}$ the output posterior probabilities will be very close to the form given in (4) and exactly equal if the output is corrupted by logistic distributed noise.

To observe this, let channel output be written as y'_n. We can then express the input to the channel, say z_n (or x_n for this example) as

$$z_n = a_1 y_n + a_2 y_{n-1} + \cdots + a_{k+1} y_{n-k} + \beta v_n \tag{7}$$

where $v_n = -\sum_{i=0}^{k} a_{i+1} v'_{n-i}$. Then the posterior probabilities $P(z_n > 0.5|\mathbf{y}_n) = P(x_n = 1|\mathbf{y}_n)$ if $x_n \in \{0, 1\}$ and are given by (4)-(6).

When there is no noise in the system, i.e., $\beta = 0$, there is no uncertainty in the system and $P(x_n = 1|\mathbf{y}_n) = 1$, as the noise level increases, $\beta \longrightarrow \infty$, we have $P(x_n = 1|\mathbf{y}_n) \longrightarrow 0.5$. Obviously the nonlinearity plays an important role as the level of noise in the system increases and is reflected in the expression given by (4)-(6) as the decreasing steepness in the logistic posterior probability with increasing noise, β.

It is easy to visualize the effect of increasing noise levels in terms of increasing normal component variances when modeling the distribution of the observations $\mathbf{y}_n$, the class conditional density $p_\theta(\mathbf{y}_n|x_n = 1)$. To observe this, consider using the FNM model for the conditional density, i.e., let $p_\theta(\mathbf{y}_n|x_n = 1) = \sum_{k=1}^{K} \pi_k g(\mathbf{y}_n, \mu_k, \mathbf{\Sigma}_k|\mathbf{x}_n^{(i)})$ where K is the number of normal clusters $g(\cdot)$, π_k is the prior, μ_k are the means and $\mathbf{\Sigma}_k$ is the covariance matrix given by $\mathbf{\Sigma}_k = \sigma_k^2 \mathbf{I}$ if past labeled data are available, i.e., $\mathbf{x}_n^{(k)}$, all K combinations of the sequence $\mathbf{x}_n = [x_n x_{n-1} \cdots x_{n-L+1}]$ are available. In this case, for increasing noise levels, to match to the observed data the

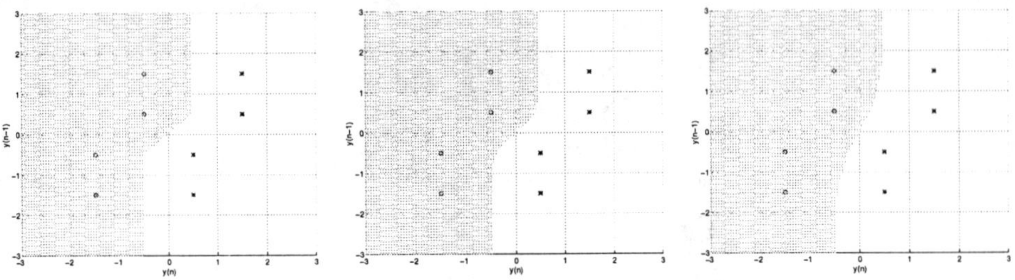

Figure 4: Decision boundaries with increasing noise levels

model parameter variance σ_k^2 would need to increase. Figure 3 shows the distribution of observations for $\mathbf{y}_n = [y_n \;\; y_{n-1}]$, for transmission of binary data $x_n \in \{-1, 1\}$ though the multipath channel $1 + 0.5z^{-1}$. The noise is assumed to be additive at the channel output as shown in Figure 2. The change of the shape of the optimal decision boundary for this FIR equalization example with increasing noise levels is given in Figure 4. When modeling the posterior probability $P_\theta(x_n = 1|\mathbf{y}_n)$, however, the increasing noise effect is reflected in the change of the steepness of the logistic function, steepness decreasing with the level of expected noise in the system. This is what we have observed in the expression given in (4)-(6) for the posterior probability as a direct relationship between the steepness of the optimal regression model and the noise level reflected in β.

If we use a simple nonlinear equalizer whose structure matches that given in (4), then the optimal coefficients $\boldsymbol{\theta}_*$ of the equalizer are given by (5), and we do expect convergence to this optimum when we use an appropriate cost function, either one that *approximates* the posterior probabilities, such as the mean square error (MSE) cost, or one that directly aims at estimating the probabilities, such as the likelihood or the relative entropy cost. This is also observed in our numerical experiments together with faster convergence for the nonlinear equalizer of (4) compared to the linear equalizer, especially as the noise level increases. Note that this is different than the use of a "soft-decision" at the output of the linear equalizer and indicates a very good match to the problem characteristics in the case of inverse identification of an IIR channel.

The discussion above also highlights a point made earlier [3] on the assumption, $P(C_i|\mathbf{y}_n) = 1$ when $\mathbf{y}_n \in C_i$, which is needed for training with the relative entropy cost. To observe this, consider the relative entropy function written as

$$D_j(p||p_\theta) = \sum_{i=1}^{m} P(C_i|\mathbf{y}_j) \log \frac{P(C_i|\mathbf{y}_j)}{P_\theta(C_i|\mathbf{y}_j)} \tag{8}$$

where $P(C_i|\mathbf{y}_j)$ are the true posteriors. The *accumulated* relative entropy (ARE) is defined as the total Kullback-Leibler discriminatory information contained in the training set $\mathcal{T}$ as

$$\mathcal{I}_N(\theta) = \sum_{j=1}^{N} D_j(p\|p_\theta). \tag{9}$$

Since the true posterior probabilities are not available in practice for training, we make the assumption $P(C_i|\mathbf{y}_n) = 1$ when $\mathbf{y}_n \in C_i$ in a practical implementation. The true posterior probabilities will not be unity for the true class label when there is uncertainty present in the system given the finite length observation $\mathbf{y}_n$ which can either be an unpredictable component, noise, as seen in (4) or the fact that sufficient covariate information is not supplied for classification. In [1], we show that equivalence of training with the ARE and the partial likelihood (PL) objectives, and show desirable large sample properties of the PL estimator such as consistency and asymptotic normality. Since the PL estimator does not require the specification of the true posteriors, the equivalence result also justifies the use of this common assumption, i.e., letting $P(C_i|\mathbf{y}_n) = 1$ when $\mathbf{y}_n \in C_i$ for training on the ARE cost. In the simulations presented in the next section, we use the least relative entropy algorithm derived by simple steepest descent minimization of the ARE cost (9) for training along with steepest descent minimization of the MSE cost.

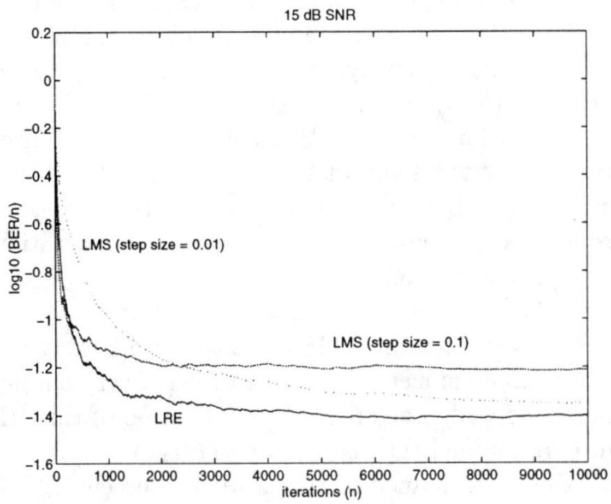

Figure 5: Convergence of Logistic and Linear Equalizers (LRE, LMS with step-size 0.1 and with step size 0.01) at 15 dB SNR

SIMULATION RESULTS

Consider the equalization problem shown in Figure 2 where the binary sequence $x_n \in \{0, 1\}$ is being transmitted through the channel $H(z) = 1/(1+$

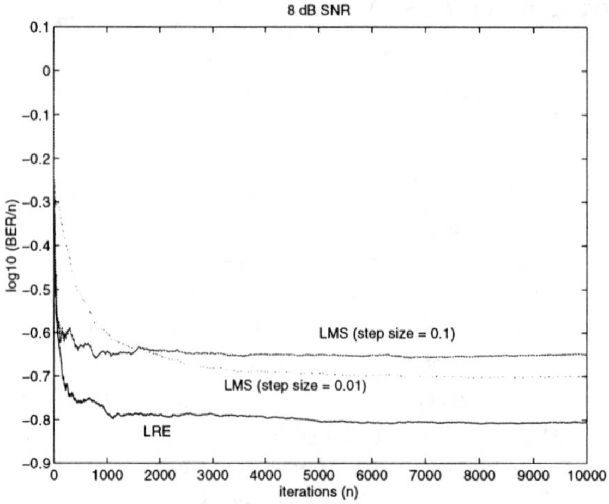

Figure 6: Convergence of Logistic and Linear Equalizers (LRE, LMS with step-size 0.1 and with step size 0.01) at 8 dB SNR

$0.75z^{-1} + 0.25z^{-2} - 0.25z^{-3}$) and is corrupted by the Gaussian noise βv_n at the channel output. We implement a linear equalizer and a sigmoidal (logistic, single layer perceptron) equalizer and use gradient descent learning on the MSE cost for the linear equalizer, i.e., use the least mean squares (LMS) algorithm for training and use the gradient optimization of the relative entropic cost for the sigmoidal equalizer, the least relative entropy (LRE) algorithm [1]. The sigmoidal equalizer requires an additional bias term when the transmitted sequence x_n is not symmetric (for the $a_0 - \tau$ term in (5)), hence the order of the LRE equalizer is 5 and that of the LMS equalizer is 4 (i.e., the optimal values for the given channel for both). Figures 5 and 6 compare the convergence of the sigmoidal equalizer with the linear equalizer trained with the normalized LMS at two noise levels. Because the sigmoidal LRE equalizer provides a very good match to the data generation process, the estimation is very efficient. Smaller step size for the LMS leads to a better BER that approaches that of the LRE equalizer but the convergence is very slow and larger step size for the LMS leads to considerably higher BER at the output of the LMS equalizer. Note that this is a very simple problem for the linear equalizer as well, since perfect equalization of this IIR channel is possible with an FIR channel that is equal to $1/H(z)$.

We have shown the performance of the gradient descent on relative entropy (partial likelihood) cost for training in the figures. The MSE cost also results in performance similar to that of LRE. Obviously the difference is in the equalizer structure, the use of a simple perceptron equalizer which provides a nice match to this problem when there is noise in the system, resulting in not only fast convergence but also better optimization of the selected cost, i.e., resulting in smaller information-theoretic distance between the true and estimated distributions for the ARE cost and lower residual error for the

MSE cost for the sigmoidal equqlizer.

Also, we show two cases for the LMS as discussed, one with a smaller step size resulting in lower BER floor but with slow convergence, the other approaching the LRE performance in terms of convergence but resulting in much higher BER. In the simulation, for this case, LMS convergence is observed to be highly affected by the step size choice, while that of LRE was quite robust to the choice of the step size resulting in better performance both in terms of the final BER and the convergence characteristics. Again, it is important to note that the performance difference between the two equalizers is more noticeable at lower SNR.

ACKNOWLEDGMENT

The author would like to thank Yi Cai, Hongmei Ni, and Levent Demirekler for helpful discussions.

REFERENCES

[1] T. Adalı, X. Liu, and M. K. Sönmez, "Conditional distribution learning with neural networks and its application to channel equalization," *IEEE Trans. Signal Processing*, vol. 45, no. 4, pp. 1051-1064, Apr. 1997.

[2] H. Ni, T. Adalı, and B. Wang, "Partial Likelihood Methods for Probability Density Estimation" to appear *Proc. IEEE Workshop on Neural Networks for Signal Processing (NNSP)*, Wisconsin, Madison, Aug. 1999.

[3] T. Adalı, "A general probabilistic formulation for neural classifiers," in *Proc. IEEE Workshop on Neural Networks for Signal Processing (NNSP)*, Cambridge, England, Sep. 1998, pp. 145-154.

[4] C. Bishop, "Neural Networks for Pattern Recognition," Oxford University Press Inc. New York, 1995.

[5] S. Chen, G. J. Gibson, C. F. N. Cowan, and P. M. Grant, "Adaptive equalization of finite non-linear channels using multilayer perceptrons," *Signal Processing*, 10, pp. 107-119, 1990.

[6] G. J. Gibson, S. Siu, and C. F. N. Cowan, "The application of nonlinear structures to the reconstruction of binary signals," *IEEE Trans. Signal Processing*, vol. 39, no. 8, pp. 1877-1884, Aug. 1991.

[7] D.R. Cox, "Partial likelihood," *Biometrika*, vol. 62, pp. 69-72, 1975.

[8] M. I. Jordan, "Why the logistic function? A tutorial discussion on probabilities and neural networks," *Computational Cognitive Science tech. Report*, 9503, 1995.

[9] S. Haykin, "Neural Networks: A comprehensive foundation," Macmillan College Publishing Co., NY, 1998.

[10] E. Slud and B. Kedem, "Partial likelihood analysis of logistic regression and autoregression," *Statistica Sinica*, vol. 4, no. 1, pp. 89-106, 1994.

[11] B. Mulgrew, "Applying radial basis functions," *IEEE Signal Proc. magazine*, pp. 50-65, March 1996.

APPROXIMATION BY RANDOM NETWORKS WITH BOUNDED NUMBER OF LAYERS

Erol Gelenbe *
School of Computer Science
University of Central Florida
Orlando, FL 32816
erol@cs.ucf.edu
Fax: 407 823 5419

Zhi-Hong Mao and Yan-Da Li †
Department of Automation
Tsinghua University
Beijing 100084
P.R. China
maozh@jerry.au.tsinghua.edu.cn

Abstract. This paper discusses the function approximation properties of the "Gelenbe" random neural network (GNN) [4, 5, 6]. We use an extension of the basic model: the bipolar GNN (BGNN) [13]. We limit the networks to being feedforward and consider the case where the number of hidden layers does not exceed the number of input layers. We show that the feedforward BGNN with s hidden layers (total of $s+2$ layers) can uniformly approximate continuous functions of s variables.

INTRODUCTION

A novel neural network model – the GNN or "Gelenbe's Random Neural Network" [4, 5, 6] – has had significant applications in various engineering areas [13, 7, 11, 14, 8, 1, 9, 10], using the network's learning algorithm [6] or its ability to act as an optimizing network. These models differ significantly from standard connexionist models in that information travels between neurons in this model travels in the form of random spike trains similar to biological neural networks. They have a mathematical structure which is significantly different from that of the sigmoidal connexionist model, the Hopfield model, or the Boltzman machine [15]. Thus the approximation capability of these networks also needs to be established in a manner distinct from that of previous models [3]. In particular, the "Gelenbe" random neural network model [4, 5, 6] does not use sigmoid functions which are basic to the standard models' approximation capabilities. In recent work [12] we have studied the approximation of arbitrary continuous functions on $[0,1]^s$ using the GNN. We have shown that the clamped GNN and the bipolar GNN [13] have the universal

*This author's research was supported by the Office of Naval Research under Grant No. N00014-97-1-0112.

†These authors' work was supported by the National Science Foundation of the P.R. of China under Grant No. 69682010.

approximation property, using a constructive method which actually exhibits networks constructed from a polynomial approximation of the function to be approximated. However the constructions in [12] place no restrictions on the structure of the networks except for limiting them to being feedforward. In this paper we discuss the design of GNN approximators with a bounded number of layers. In Section 2, a brief introduction to the GNN and the bipolar GNN (BGNN) is given. In Section 3, we establish the technical premises for our main results. Then in Section 4 we prove the universal approximation capability of the feedforward BGNN.

THE GNN AND ITS EXTENSIONS

Consider a GNN [4, 5, 6] with n neurons in which "positive" and "negative" signals circulate. The $i-th$ neuron's state is represented at any time t by its "potential" $k_i(t)$, which is a non-negative integer. In the RNN (Gelenbe (1989,90) [4, 5]) signals in the form of spikes of unit amplitude circulate among the neurons. Positive signals represent excitation and negative signals represent inhibition. Each neuron's state is a non-negative integer called its potential, which increases when an excitation signal arrives to it, and decreases when an inhibition signal arrives. An excitatory spike is interpreted as a "+1" signal at a receiving neuron, while an inhibitory spike is interpreted as a "−1" signal. Neural potential also decreases when the neuron fires. Thus a neuron i emitting a spike, whether it be an excitation or an inhibition, will lose potential of one unit, going from some state whose value is k_i to the state of value $k_i - 1$. In general, this is a "recurrent network" model, *i.e.* a network which is allowed to have feedback loops of arbitrary topology.

The state of the n-neuron network at time t, is represented by the vector of non-negative integers $k(t) = (k_1(t), \ldots, k_n(t))$, where $k_i(t)$ is the potential or integer state of neuron i. We will denote by k and k_i arbitrary values of the state vector and of the i-th neuron's state. Neuron i will "fire" (i.e. become excited and send out spikes) if its potential is *positive*. The spikes will then be sent out at a rate $r(i) \geq 0$, with independent, identically and exponentially distributed inter-spike intervals. Spikes will go out to some neuron j with probability $p^+(i,j)$ as excitatory signals, or with probability $p^-(i,j)$ as inhibitory signals. A neuron may also send signals out of the network with probability $d(i)$, and $d(i) + \sum_{j=1}^{n}[p^+(i,j) + p^-(i,j)] = 1$.

Exogenous excitatory signals arrive to neuron i in a Poisson stream of rate $\Lambda(i)$. Similarly exogenous inhibitory signals arrive to neuron i in a Poisson stream of rate $\lambda(i)$. These different Poisson streams for $i = 1, ..., n$ are independent of each other. To simplify the notation, in the sequel we will write:

$$\omega^+(i,j) = r(i)p^+(i,j), \quad (1)$$

$$\omega^-(i,j) = r(i)p^-(i,j). \quad (2)$$

The state transitions of the network are represented by Chapman-Kolmogorov equations [2] for the probability distribution:

$$p(k,t) = Prob[k(t) = k], \tag{3}$$

where $k = (k_1, \ldots, k_n)$ denotes a particular value of the state vector.

Let $\lambda^+(i)$ and $\lambda^-(i)$ denote the average arrival rates of positive and negative signals to each neuron i. The key results about the GNN developed in [4, 5, 6] are summarized below.

Theorem 1. (Proposition 1 in the Appendix of [6]) *A non-negative solution* $(\lambda^+(i) \geq 0,\ \lambda^-(i) \geq 0)$ *always exists to the equations:*

$$\lambda^+(i) = \Lambda(i) + \sum_{j=1}^{n} q_j \omega^+(j,i), \tag{4}$$

$$\lambda^-(i) = \lambda(i) + \sum_{j=1}^{n} q_j \omega^-(j,i), \tag{5}$$

for $i = 1, ..., n$ where

$$q_i = \frac{\lambda^+(i)}{r(i) + \lambda^-(i)}. \tag{6}$$

The next important result concerns the stationary joint probability distribution of network state:

$$p(k) = \lim_{t \to \infty} p(k,t). \tag{7}$$

Theorem 2. (Theorem 1 of [4]) *For an n neuron GNN, let the vector of neuron potentials at time t be $k(t) = (k_1(t), k_2(t), ..., k_n(t))$, and let $k = (k_1, k_2, ..., k_n)$ be an n-vector of non-negative integers. Then if the q_i in (6) satisfy $0 \leq q_i < 1$, the stationary joint probability of network state is given by:*

$$p(k) = \prod_{i=1}^{n} (1 - q_i) q_i^{k_i}. \tag{8}$$

Note that if the conditions of Theorem 2 are satisfied then the stationary probability distribution of the state of neuron i denoted by $p(k_i) = \lim_{t \to \infty} p(k_i(t) = k_i)$, is given by:

$$p(k_i) = (1 - q_i) q_i^{k_i}, \tag{9}$$

and

$$q_i = \lim_{t \to \infty} Prob\{k_i(t) > 0\}. \tag{10}$$

The BGNN Model

In order to represent bipolar patterns taking values such as $\{+1, -1\}$, and to strengthen the associative memory capabilities of the GNN, in some early work Gelenbe, Stafylopatis and Likas [13] extended the original model by introducing the artifact of "positive and negative" neurons. The resulting Bipolar GNN (BGNN) can also be viewed as the coupling of two complementary standard GNN models.

In the BGNN the two types of neurons have opposite roles. A positive neuron behaves exactly as a neuron in the original GNN. A negative neuron has a completely symmetrical behavior, namely only negative signals can accumulate at this neuron, and the role of positive signals arriving to a negative neuron is to eliminate negative signals which have accumulated in a negative neuron's potential. A positive signal arriving to a negative neuron i cancels a negative signal (adds $+1$ to the neuron's negative potential), and has no effect if $k_i = 0$.

This extension is in fact mathematically equivalent to the original GNN described above, with respect to the specific form taken by the stationary solution (Theorems 1 and 2). However the use of both positive and negative neurons allows the BGNN to become convenient universal approximator for continuous functions because of the possibility of using both positive and negative valued functions of the input variables. Let P and N denote, respectively, the indeces of the positive and negative neurons in the network. In the BGNN the state of the network is represented by the vector $k(t) = (k_1(t), \ldots, k_n(t))$ so that $k_i(t) \geq 0$ if $i \in P$ and $k_i(t) \leq 0$ if $i \in N$.

In the BGNN, the emission of signals from a positive neuron is the same as in the original GNN. Similarly, a negative neuron may emit negative signals. A signal leaving negative neuron i arrives to neuron j as a negative signal with probability $p^+(i,j)$ and as a positive signal with probability $p^-(i,j)$. Also, a signal departs from the network upon leaving neuron i with probability $d(i)$. Other assumptions and denotations retain as in the original model. Let us consider a BGNN with n nodes. Since negative signals account for the potential of negative neurons, we will use negative values for k_i if neuron i is negative. If we take into account the distinction between positive and negative neurons, Theorems 1 and 2 can be summarized as follows for the BGNN. The flow of signals in the network is described by the following equations:

$$\lambda^+(i) = \Lambda(i) + \sum_{j \in P} q_j \omega^+(j,i) + \sum_{j \in N} q_j \omega^-(j,i),$$

$$\lambda^-(i) = \lambda(i) + \sum_{j \in P} q_j \omega^-(j,i) + \sum_{j \in N} q_j \omega^+(j,i),$$

and

$$q_i = \frac{\lambda^+(i)}{r(i) + \lambda^-(i)}, \quad i \in P, \tag{11}$$

$$q_i = \frac{\lambda^-(i)}{r(i) + \lambda^+(i)}, \quad i \in N. \tag{12}$$

Using a direct extension of the results for the conventional GNN, it can be shown that a non-negative solution $\{\lambda^+(i), \lambda^-(i), i = 1, ..., n\}$ exists to the above equations. If the $q_i < 1$, $i = 1, ..., n$, then the steady-state joint probability distribution of network state is given by [13]:

$$p(k) = \prod_{i=1}^{n} (1 - q_i) q_i^{|k_i|}, \tag{13}$$

where the quantity q_i is the steady-state probability that node i is "excited". Note the $|k_i|$ exponent in the above product form, since the $k_i' s$ can be positive or negative, depending on the polarity of the $i - th$ neuron. In the sequel we will consider how the BGNN, as well as a simpler extension of the feedforward (i.e. non-recurrent) GNN, can be used to approximate arbitrary continuous functions.

APPROXIMATION OF FUNCTIONS OF ONE VARIABLE BY THE GNN WITH A BOUNDED NUMBER OF LAYERS

All feedforward models considered in this section are guaranteed to have an unique solution for the q_i,$i = 1, ..., n$ as a result of Theorems 2 and 3 of [5, 6]. Thus from now on we do not revisit this issue.

Consider a continuous function $f : [0,1]^s \mapsto R$ of an input vector $X = (x_1, ..., x_s)$. Since an $[0,1]^s \mapsto R^w$ function can always be separated into a group of w distinct functions $[0,1]^s \mapsto R$, we will only consider outputs in one dimension. The sequel of this paper is therefore devoted to how a continuous function $f : [0,1]^s \mapsto R$ can be approximated by neural networks derived from the GNN model. To approximate f, we will construct s-input, 1-output, L-layer feedforward GNN's. We will use the index (l, i) for the $i - th$ neuron at the $l - th$ layer. Furthermore, when we need to specify this, we will denote by M_l the number of neurons in the $l - th$ layer.

The network under consideration is organized as follows: in the first layer, i.e. the input layer, we set $\Lambda(1, i) = x_i$, $\lambda(1, i) = 0$, $r(1, i) = 1$, so that $q_{1,i} = x_i$, for $i = 1, ..., s$.

In the l-th layer ($l = 2, ..., L$), $\Lambda(l, i)$, $\lambda(l, i)$, and $r(l, i)$ are adjustable parameters, and $q_{l,i}$ is given by

$$q_{l,i} = \frac{\Lambda(l,i) + \sum_{1 \le h < l} \sum_{1 \le j \le M_h} q_{h,j} \omega^+((h,j),(l,i))}{\lambda(l,i) + r(l,i) + \sum_{1 \le h < l} \sum_{1 \le j \le M_h} q_{h,j} \omega^-((h,j),(l,i))} \tag{14}$$

where the connection "weights" $\omega^+(\cdot, \cdot)$ and $\omega^-(\cdot, \cdot)$ are also adjustable parameters. In the $L-th$ or output layer there is only one neuron. As suggested in [4] we can use the output function

$$A_{L,1} = \frac{q_{L,1}}{1 - q_{L,1}} \tag{15}$$

whose physical meaning is that it is the average potential of the output neuron as the output of the network. In this manner, we will have $A_{L,1} \in [0, +\infty)$, rather than just $q_{L,1} \in [0, 1]$.

Technical Premises

Before we proceed with the developments concerning GNN approximations we need some technical results. They are similar to some technical results used in [12] concerning continuous functions $f : [0,1] \mapsto R$ for a single input variable x. The generalization to $f : [0,1]^s \mapsto R$ is direct and will be examined in Section 4. The proofs are given in the Appendix.

Lemma 1. *For any continuous function $f : [0,1] \mapsto R$ and for any $\epsilon > 0$, there exists a polynomial*

$$P(x) = c_0 + c_1(\frac{1}{1+x}) + .. + c_m(\frac{1}{1+x})^m, x \in [0,1], \tag{16}$$

such that $\sup_{x\in[0,1]} |f(x) - P(x)| < \epsilon$ is satisfied.

The second technical result is concerning the relationship between polynomials of the form (16) and the GNN.

Lemma 2. *Consider a term of the form*

$$\frac{1}{(1+x)^v},$$

for $0 \leq x \leq 1$, and any $v = 1, 2, \cdots$. There exists a feedforward GNN with a single output neuron $(v+1, 1)$ and input $x \in [0,1]$ such that

$$q_{v+1,1} = (\frac{1}{1+x})^v. \tag{17}$$

The following Lemma shows how an arbitrary polynomial of the form (16) with non-negative coefficients can be realized by a feedforward GNN.

Lemma 3. *Let $P^+(x)$ be a polynomial of the form (16) with the restriction that $c_v \geq 0$, $v = 1, ..., m$. Then there exists a feedforward GNN with a single output neuron (O) such that:*

$$q_O = \frac{P^+(x)}{1 + P^+(x)}, \tag{18}$$

so that the average potential of the output neuron is $A_O = P^+(x)$.

The fourth technical result will be of use in proving the approximating power of the "clamped GNN' discussed below.

Lemma 4. *Consider a term of the form*

$\frac{x}{(1+x)^v}$,

for $0 \leq x \leq 1$, *and any* $v = 1, ..., m$. *There exists a feedforward GNN with a single output neuron* $(v+1, 1)$ *and input* $x \in [0, 1]$ *such that*

$$q_{v+1,1} = (\frac{x}{1+x})^v. \tag{19}$$

We state without proof another lemma, very similar to Lemma 3, but which uses terms of both forms of $1/(1+x)^v$ and $x/(1+x)^v$ to construct polynomials. It's proof uses Lemma 3 and 4, and follows exactly the same lines as Lemma 3.

Lemma 5. *Let* $P^o(x)$ *be a polynomial of the form*

$$P^o(x) = c_0 + \sum_{v=1}^{m}[c_v \frac{1}{(1+x)^v} + d_v \frac{x}{(1+x)^v}], \tag{20}$$

for $0 \leq x \leq 1$, *with non-negative coefficients, i.e.* $c_v, d_v \geq 0$, $v = 1, ..., m$. *Then there exists a feedforward GNN with a single output neuron* (O) *such that:*

$$q_O = \frac{P^o(x)}{1 + P^o(x)}, \tag{21}$$

so that the average potential of the output neuron is $A_O = P^o(x)$.

The next lemma is a technical premise of Lemma 7.

Lemma 6. *For any* $(\frac{1}{1+x})^i$ $(0 \leq x \leq 1, i = 1, 2, \cdots)$ *and for any* $\epsilon > 0$, *there exists a function*

$$P_1(x) = b_0 + \frac{b_1}{x + a_1} + \frac{b_2}{x + a_2} + .. + \frac{b_r}{x + a_r}, x \in [0, 1] \tag{22}$$

where $a_k > 0$, $k = 1, ..., r$, *such that*
$\sup_{x \in [0,1]} |(\frac{1}{1+x})^i - P_1(x)| < \epsilon$ *is satisfied.*

The following lemma is the preparation for the construction of a single-hidden-layered BGNN for the approximation of one dimensional continuous function.

Lemma 7. *For any continuous function* $f : [0, 1] \mapsto R$ *and for any* $\epsilon > 0$, *there exists a function* $P_1(x)$ *in the form of (22) such that* $\sup_{x \in [0,1]} |f(x) - P_1(x)| < \epsilon$ *is satisfied.*

Proof: This is a direct consequence of Lemma 1 and Lemma 6. **Q.E.D.**

BGNN Approximation of Continuous Functions of One Variable

The technical results given above now pave the way for the use of the Bipolar GNN (BGNN) with a bounded number of layers. Specifically in Theorem 4, we show that a BGNN with a single hidden layer can uniformly approximate functions of one variable. The multivariable case is discussed in Section 4.

Let us first recall a result from [12] concerning the case when the number of layers is not bounded.

Theorem 3. *For any continuous function* $f : [0,1] \mapsto R$ *and any* $\epsilon > 0$, *there exists a BGNN with one positive output neuron* $(O,+)$, *one negative output neuron* $(O,-)$, *the input variable* x, *and the output variable* $y(x)$ *such that:*

$$y(x) = A_{O,+} + A_{O,-}, \tag{23}$$

$$A_{O,+} = \frac{q_{O,+}}{1 - q_{O,+}}, \tag{24}$$

$$A_{O,-} = \frac{-q_{O,-}}{1 - q_{O,-}}, \tag{25}$$

and $\sup_{x\in[0,1]} |f(x) - y(x)| < \epsilon$. *We will say that the BGNN's output uniformly approximates* $f(x)$.

The next theorem shows the approximation capability of a BGNN with a single hidden layer.

Theorem 4. *For any continuous function* $f : [0,1] \mapsto R$ *and any* $\epsilon > 0$, *there exists a BGNN of three layers (only one hidden layer), one positive output neuron* $(O,+)$, *one negative output neuron* $(O,-)$, *the input variable* x, *and the output variable* $y(x)$ *determined by (23) such that* $\sup_{x\in[0,1]} |f(x)-y(x)| < \epsilon$.

APPROXIMATION OF CONTINUOUS FUNCTIONS OF S VARIABLES

Now that the process for approximating a one-dimensional continuous functions with the BGNN having a single hidden layer is well understood, consider the case of continuous functions of s variables, i.e. $f : [0,1]^s \mapsto R$. As a starting point, consider the straightforward extension of Lemma 1 to the case of $s-$inputs such that there is a polynomial:

$$P(x) = \sum_{m_1\geq 0,..,m_s\geq 0,\sum_{v=1}^{s} m_v = m} c(m_1,...,m_s). \; \Pi_{v=1}^{s} \frac{1}{(1+x_v)^{m_v}}, \tag{26}$$

with coefficients $c(m_1,...,m_s)$ which approximates f uniformly. We now extend Lemma 2 to Lemma 8 and Theorem 5 which are given below.

Lemma 8. *Consider a term of the form*

$$\frac{1}{(1\;+\;x_{z1})^{m_{z1}}}\cdot\cdot\frac{1}{(1\;+\;x_{zK})^{m_{zK}}}$$

for $0 \leq x_{zj} \leq 1$, *positive integers* $m_{zj} > 0$ *and* $j = 1,...,K$. *There exists a feedforward GNN with a single output neuron* $(\mu+1,1)$ *and input* $x \in [0,1]$ *such that*

$$q_{\mu+1,1} = \frac{1}{(1\;+\;x_{z1})^{m_{z1}}}\cdot\cdot\frac{1}{(1\;+\;x_{zK})^{m_{zK}}}. \tag{27}$$

Theorem 5. *For any continuous function* $f : [0,1]^s \mapsto R$ *and any* $\epsilon > 0$, *there exists a BGNN with one positive output neuron* $(O,+)$, *one negative output neuron* $(O,-)$, s *input variables* $X = (x_1, ..., x_s)$, *and the output variable* $y(X)$ *such that:*

$$y(X) = A_{O,+} + A_{O,-}, \tag{28}$$

$$A_{O,+} = \frac{q_{O,+}}{1 - q_{O,+}}, \tag{29}$$

$$A_{O,-} = \frac{-q_{O,-}}{1 - q_{O,-}}, \tag{30}$$

and $\sup_{x\in[0,1]} |f(X) - y(X)| < \epsilon$. *We will say that the BGNN's output uniformly approximates* $f(X)$.

Proof: The proof follows the proof of Theorem 3, using the polynomial of (26). Lemma 7 establishes that the terms of such a polynomial can be realized by a GNN. We then construct two polynomials, one with non-negative coefficients only, and the other with negative coefficients, and show how they are realized with the BGNN. We will not go through the steps of the proof since it is a step by step duplicate of the proof of Theorem 3. **Q.E.D.**

We now extend Lemma 7 to the case of s-inputs.

Lemma 9. *For any continuous function* $f : [0,1]^s \mapsto R$ *and for any* $\epsilon > 0$, *there exists a function of the form*

$$P_s(x) = \sum_{i=1}^{r} \sum_{0\le m_1 \le 1, ..., 0 \le m_s \le 1} b(m_1, ..., m_s, i).$$

$$\Pi_{v=1}^{s} \frac{1}{(a_{v,i} + x_v)^{m_v}}, \tag{31}$$

where $a_{v,i} > 0$, $v = 1, ..., s$, $i = 1, 2, ..$ *such that* $\sup_{x\in[0,1]} |f(x) - P_s(x)| < \epsilon$ *is satisfied.*

Proof: This is simply an extension of Lemma 7. **Q.E.D.**

As a consequence we can now establish the following general result.

Theorem 6. *For any continuous function* $f : [0,1]^s \mapsto R$ *and any* $\epsilon > 0$, *there exists a BGNN of no more than* $s + 2$ *layers (*s *hidden layers), one positive output neuron* $(O,+)$, *one negative output neuron* $(O,-)$, s *input variables* $X = (x_1, ..., x_s)$, *and the output variable* $y(X)$ *determined by (28) such that*

$\sup_{x\in[0,1]} |f(X) - y(X)| < \epsilon$.

REFERENCES

[1] C. Cramer, E. Gelenbe and H. Bakircioglu, "Low bit rate video compression with neural networks and temporal subsampling," **Proceedings of the IEEE**, vol. 84, no. 10, pp. 1529–1543, 1996.

[2] W. E. Feller, **An Introduction to Probability Theory and Its Applications**, New York: J. Wiley, 1968.

[3] K. Funahashi, "On the approximate realization of continuous mapping by neural network," **Neural Networks**, vol. 2, pp. 183–192, 1989.

[4] E. Gelenbe, "Random neural networks with negative and positive signals and product form solution," **Neural Computation**, vol. 1, no. 4, pp. 502–511, 1989.

[5] E. Gelenbe, "Stability of the random neural network model," **Neural Computation**, vol. 2, no. 2, pp. 239–247, 1990.

[6] E. Gelenbe, "Learning in the recurrent random neural network," **Neural Computation**, vol. 5, no. 1, pp. 154–164, 1993.

[7] E. Gelenbe and F. Batty, "Minimum cost graph covering with the random neural network," in O. Balci, ed., **Computer Science and Operations Research**, New York: Pergamon, 1992, pp. 139–147.

[8] E. Gelenbe, C. Cramer, M. Sungur and P. Gelenbe, "Traffic and video quality in adaptive neural compression," **Multimedia Systems**, vol. 4, pp. 357–369, 1996.

[9] E. Gelenbe, T. Feng and K. Krishnan, "Neural network methods for volumetric magnetic resonance imaging of the human brain," **Proceedings of the IEEE**, vol. 84, no. 10, pp. 1488–1496, 1996.

[10] E. Gelenbe, A. Ghanwani and V. Srinivasan, "Improved neural heuristics for multicast routing," **IEEE J. Selected Areas in Communications**, vol. 15, no. 2, pp. 147–155, 1997.

[11] E. Gelenbe, V. Koubi and F. Pekergin, "Dynamical random neural network approach to the traveling salesman problem," **Proc. IEEE Symp. Syst. Man. Cybernetics**, pp. 630–635, 1993.

[12] E. Gelenbe, Z. H. Mao, and Y. D. Li, "Function approximation with the random neural network," **IEEE Trans. Neural Networks**, vol. 10, no. 1, pp. 3–9, 1999.

[13] E. Gelenbe, A. Stafylopatis and A. Likas, "Associative memory operation of the random network model," **Proc, Int, Conf, Artificial Neural Networks**, pp. 307–312, 1991.

[14] A. Ghanwani, "A qualitative comparison of neural network models applied to the vertex covering problem," **Elektrik**, vol. 2, no. 1, pp. 11–18, 1994.

[15] J. McClelland, D. Rumelhart and et al, **Parallel Distributed Processing**, MIT Press, 1986.

NATURAL POWER METHOD FOR FAST SUBSPACE TRACKING

Yingbo Hua, Yong Xiang, Tiangping Chen*, Karim Abed-Meraim**, Yongfeng Miao

Department of Electrical and Electronic Engineering
University of Melbourne
Parkville, Victoria 3052, Australia
email: yhua@ee.mu.oz.au

*Fudan University, Shanghai, China.
**Telecom Paris, France.

ABSTRACT

This paper elaborates on a natural version of the power method for fast estimation and tracking of principal subspace or/and principal components of a vector sequence. The natural power method has the fastest convergence rate among a class of power-based methods such as the Oja method, the projection approximation subspace tracking (PAST) method, and the novel information criterion (NIC) method. Like the above three methods, the natural power method can also be implemented with only O(*np*) flops of computation at each iteration but maintain the fastest convergence rate, where *n* is the dimension of the vector sequence and *p* is the dimension of the principal subspace. Also like other power-based methods, the natural power method can be easily adopted for principal components tracking, constrained subspace tracking, and detection of the dimension of the principal subspace. In great contrast to non-power-based methods such as MALASE and OPERA, the natural power method is globally convergent.

1. INTRODUCTION

Fast estimation and tracking of the principal subspace (or/and principal components) of a sequence of random vectors is a classic problem which is encountered widely in such areas as data compression, data filtering, parameter estimation, pattern recognition, and neural analysis. For this classic problem, numerous solutions or methods are available as shown in the 1990 survey paper

0-7803-5673-X/99/$10.00 © 1999 IEEE

[1] by Comon and Golub. The marriage of signal processing and neural networks in the early 1990's [2] and the apparent popularity of subspace methods in the signal processing community in the past 15 years have sparked a renewed interest in this classic problem. In particular, the Oja method originated by a mathematical biologist [3] has received a special attention and been applied in practical applications [4]. Thorough analyses [5-6] of the Oja method have shown that the Oja method with a very small step size is (almost always) globally convergent. The convergence speed of its digital implementation is, however, very slow in comparison to some other methods.

It is established in [7-8] that the Oja method can be viewed as an approximation of gradient descent of a mean square error (MSE) function. Following the gradient descent idea, a number of much faster subspace tracking methods have been developed by researchers. Among them are the projection approximation subspace tracking (PAST) method [8] and the novel information criterion (NIC) method [9]. The PAST method was developed from the same MSE function as for the Oja method except that a heuristic "projection approximation" was applied. The PAST method converges much faster than the Oja method. The NIC method is another gradient flow based method. The criterion used by the NIC method has little resemblance to the MSE function. But the NIC has all the desired properties needed for principal subspace estimation, and the NIC method turns out to be a robust generalization of the PAST method. Recently, we have shown [14] that the three methods: Oja, PAST and NIC, can all be viewed as some heuristic variations of the classic method for subspace computation known as power method [10].

In this paper, we focus on a natural version of the power method and discuss its implementations with varied computational complexities. Like Oja, PAST and NIC, the natural power method can also be implemented with linear complexity of computation, i.e., O(*np*), but maintain the fastest convergence rate. In contrast to other fast methods such as MALASE [12] and OPERA [13], the natural power method is not only computationally superior (or at least in an equivalent order) but also has a global convergence property.

2. THE NATURAL POWER METHOD

The data under consideration are a sequence of $n \times 1$ random vectors: $\{\mathbf{x}(k)\}$. The covariance matrix of the sequence is $\mathbf{C} = E\left[\mathbf{x}(k)\mathbf{x}(k)^H\right]$. The principal subspace spanned by the sequence, of dimension $p<n$, is known to be the span of the p principal eigenvectors of the covariance matrix. It is known that the first principal eigenvector of the covariance matrix can be obtained by the power method as follows:

$$\mathbf{w}(i+1) = \mathbf{C}\mathbf{w}(i) \qquad (1)$$

where the $n \times 1$ weight vector $\mathbf{w}(i)$ is the estimate of the first principal eigenvector at the ith iteration. The norm of the weight vector needs to be scaled during the iterations to prevent it from becoming too large or too small. A natural choice of the scaling can be done as follows:

$$\mathbf{w}(i+1) = \mathbf{C}\mathbf{w}(i)\left(\mathbf{w}^H(i)\mathbf{C}^2\mathbf{w}(i)\right)^{-1/2} \tag{2}$$

where the scaling $\left(\mathbf{w}^H(i)\mathbf{C}^2\mathbf{w}(i)\right)^{-1/2}$ guarantees that $\mathbf{w}(i+1)$ has norm one. It is known that the weight vector of (1) converges exponentially to the desired vector (up to a scalar). The convergence rate is governed by the ratio of the first and second largest eigenvalues of the covariance matrix. In tracking applications, one can simply replace the covariance matrix in (2) by its recursive version:

$$\mathbf{C}(i+1) = \alpha\mathbf{C}(i) + \mathbf{x}(i+1)\mathbf{x}(i+1)^H \tag{3}$$

at the (i+1)th iteration, where $\mathbf{C}(i)$ is the sample covariance matrix using the data available up to time i, and α is a forgetting factor chosen between (0, 1].

For computing a principal subspace of dimension p, one can use a natural generalization of (2) as follows:

$$\mathbf{W}(i+1) = \mathbf{C}(i+1)\mathbf{W}(i)\left(\mathbf{W}^H(i)\mathbf{C}^2(i+1)\mathbf{W}(i)\right)^{-1/2} \tag{4}$$

where $\mathbf{W}(i)$ is a $n \times p$ matrix, and the square root inverse of the matrix $\mathbf{W}^H(i)\mathbf{C}^2\mathbf{W}(i)$ is defined by the eigenvalue decomposition of the matrix with its eigenvalues replaced by their square root inverses. It is easy to show that if the initial weight matrix $\mathbf{W}(0)$ satisfies $\mathbf{W}^H(0)\mathbf{C}^2\mathbf{W}(0) > 0$ (positive definite) or equivalently $\mathbf{W}^H(0)\mathbf{C}\mathbf{W}(0) > 0$, then $\mathbf{W}(i)$ remains orthonormal, i.e.,

$$\mathbf{W}^H(i)\mathbf{W}(i) = \mathbf{I} \tag{5}$$

where i>0. The algorithm represented by (4) is referred to as the basic (abstract) form of the natural power method. It can be shown [14] that this algorithm converges to an orthogonal basis of the principal subspace at an exponential rate governed by the ratio of the pth and (p+1)th eigenvalues of $\mathbf{C}$.

2.1 O($n^2 p$) implementation – NP1

A direct implementation of (4) requires $n^2 p$ flops to compute $\mathbf{C}(i+1)\mathbf{W}(i)$, about $\frac{1}{2}np^2$ additional flops to compute (conjugate symmetric) $\mathbf{W}^H(i)\mathbf{C}^2(i+1)\mathbf{W}(i)$, O($p^3$) additional flops to compute the inverse square root (by computing the eigendecomposition and taking the inverse square roots of the eigenvalues), and a further np^2 flops to complete (4). The above approach needs O(n^2p)+O(np^2)+O(p^3) flops. Alternatively, one can apply the QR factorization of the product **C**(i+1)**W**(i):

$$\mathbf{C}(i+1)\mathbf{W}(i) = \mathbf{Q}(i+1)\mathbf{R}(i+1) \tag{6}$$

where **Q**(i+1) is orthonormal and **R**(i+1) is upper (or lower) triangular. Using (6) in (4) yields

$$\mathbf{W}(i+1) = \mathbf{Q}(i+1)\mathbf{H}(i+1) \tag{7}$$

where

$$\mathbf{H}(i+1) = \mathbf{R}(i+1)\left(\mathbf{R}^H(i+1)\mathbf{R}(i+1)\right)^{-1/2} \tag{8}$$

It is easy to verify that **H**(i+1) is a unitary matrix, and hence **W**(i+1) and **Q**(i+1) are equivalent for subspace estimation. Therefore, the direct implementation of the natural power method (4) is equivalent (up to a unitary rotation) to the well known QR power method [10]:

$$\mathbf{C}(i+1)\mathbf{Q}(i) = \mathbf{Q}(i+1)\mathbf{R}(i+1) \tag{9}$$

This algorithm requires n^2p to compute the product on the left side of (10) and O(np^2) additional flops [1] to compute the QR factorization on the right side. Either (4) or (9) guarantees the orthogonality of the weight matrix at each iteration.

2.2 O(np^2) implementation – NP2

To further reduce the complexity of the natural power method, we need to consider the following (using (3)):

$$\mathbf{C}(i+1)\mathbf{W}(i) = \alpha\mathbf{C}(i)\mathbf{W}(i) + \mathbf{x}(i+1)\mathbf{x}^H(i+1)\mathbf{W}(i) \tag{10}$$

The first term in (10) does not contain any information carried in the new vector **x**(i+1) while the second term does. The first term also needs much more flops than the second term. This motivates the idea that the first term should be approximated by some historical information to save the computational complexity. A simple approximation is

$$\mathbf{C}(i)\mathbf{W}(i) \approx \mathbf{C}(i)\mathbf{W}(i-1) \tag{11}$$

which is clearly valid if the weight matrix **W**(i) is slowly varying with i. With the approximation (11), the product **C**(i+1)**W**(i) requires only $2np$ flops to update at each iteration according to (10). The approximation (11) also follows from the "projection approximation" $\mathbf{W}^H(i)\mathbf{x}(k) \approx \mathbf{W}^H(k-1)\mathbf{x}(k)$ for all $i \geq k$-1 as proposed in [8] although the latter expression is intuitively less acceptable than (11).

It is important to note that the approximation (11) is justifiable by the natural power method (4) when $\mathbf{W}(i)$ is close to a desired matrix $\mathbf{U}_1\mathbf{T}(i)$ where $\mathbf{U}_1$ is the (orthonormal) principal-eigenvector matrix of $\mathbf{C}(i+1)$ and $\mathbf{T}(i)$ is a unitary matrix, but the approximation (11) is *not* justifiable by the QR power method (9). The reasons are as follows.

"projection approximation" justifiable by the natural power method:

Let $\mathbf{C}(i+1) = \mathbf{U}_1\Sigma_1(i+1)\mathbf{U}_1{}^H + \mathbf{U}_2\Sigma_2(i+1)\mathbf{U}_2{}^H$ where the first term consists of the principal component. If $\mathbf{W}(i) \approx \mathbf{U}_1\mathbf{T}(i)$, then

$$\mathbf{C}(i+1)\mathbf{W}(i) \approx \mathbf{U}_1\Sigma_1(i+1)\mathbf{T}(i)$$

$$\left(\mathbf{W}^H(i)\mathbf{C}^2\mathbf{W}(i)\right)^{-1/2} \approx \mathbf{T}(i)^H\Sigma_1(i+1)^{-1}\mathbf{T}(i)$$

and hence, from (4), $\mathbf{W}(i+1) \approx \mathbf{W}(i)$. The notation $\approx$ means that both sides of the equation differ only by an additive small matrix.

"projection approximation" not justifiable by the QR power method:

Consider the desired situation $\mathbf{Q}(i) \approx \mathbf{U}_1\mathbf{S}(i)$ where $\mathbf{S}(i)$ is a unitary matrix. Then $\mathbf{C}(i+1)\mathbf{Q}(i) = \mathbf{U}_1\Sigma_1(i+1)\mathbf{S}(i) + \mathbf{P}(i+1)$ where $\mathbf{P}(i+1)$ is small perturbation matrix. By applying the QR factorization (9), we have $\mathbf{Q}(i+1) \approx \mathbf{U}_1\mathbf{S}(i+1)$ where $\mathbf{S}(i+1)$ is a unitary matrix that is not only rotationally perturbed by the small matrix $\mathbf{P}(i+1)$ but also significantly affected by $\Sigma_1(i+1)$ unless the latter is proportional to the identity matrix. Therefore, (11) is in general not satisfied by $\mathbf{Q}(i)$ of the QR power method.

To continue the discussion of the natural power method, we now define $\mathbf{Y}(i+1) \triangleq \mathbf{C}(i+1)\mathbf{W}(i)$. Then, with the approximation (11), (10) becomes:

$$\mathbf{Y}(i+1) = \alpha\mathbf{Y}(i) + \mathbf{x}(i+1)\mathbf{y}^H(i+1) \qquad (12)$$

where

$$\mathbf{y}(i+1) = \mathbf{W}^H(i)\mathbf{x}(i+1) \qquad (13)$$

Let $\mathbf{Z}(i+1) \triangleq \mathbf{Y}^H(i+1)\mathbf{Y}(i+1)$. It then follows that

$$\mathbf{Z}(i+1) = \alpha^2\mathbf{Z}(i) + \mathbf{z}(i+1)\mathbf{y}^H(i+1) + \mathbf{y}(i+1)\mathbf{z}^H(i+1) + \gamma(i+1)\mathbf{y}(i+1)\mathbf{y}^H(i+1) \tag{14}$$

with

$$\mathbf{z}(i+1) = \alpha\mathbf{Y}^H(i)\mathbf{x}(i+1) \tag{15}$$

$$\gamma(i+1) = \mathbf{x}^H(i+1)\mathbf{x}(i+1) \tag{16}$$

Note that like **Y**(i+1), the matrix **Z**(i+1) also requires only O(np) flops to update at each iteration according to (14). The natural power method (4) now becomes

$$\mathbf{W}(i+1) = \mathbf{Y}(i+1)\mathbf{Z}^{-1/2}(i+1) \tag{17}$$

The above implementation has reduced the complexity to O(np^2)+O(p^3) flops, which is linear in n. Like (4) and (9), (17) also guarantees the orthogonality of the weight matrix at each iteration.

2.3 O(np) implementation – NP3

To further reduce the complexity of the natural power method, we need to accommodate asymmetric square root inverse to be applied to (17). We now write

$$\mathbf{Z}^{-1/2}(i+1) = \frac{1}{\alpha}\left(\mathbf{Z}^{1/2}(i)\left(\mathbf{I} + \mathbf{b}\mathbf{a}^H + \mathbf{a}\mathbf{b}^H + \gamma\mathbf{a}\mathbf{a}^H\right)\mathbf{Z}^{1/2^H}(i)\right)^{-1/2} \tag{18}$$

where the dependence of the vectors on the index i+1 is dropped for convenience, and

$$\mathbf{a} = \frac{1}{\alpha}\mathbf{Z}^{-1/2}(i)\mathbf{y}(i+1)$$
$$\mathbf{b} = \frac{1}{\alpha}\mathbf{Z}^{-1/2}(i)\mathbf{z}(i+1) \tag{19}$$

Since $\mathbf{b}\mathbf{a}^H + \mathbf{a}\mathbf{b}^H + \gamma\mathbf{a}\mathbf{a}^H$ is conjugate symmetric and of rank 2, we can write

$$\mathbf{I} + \mathbf{b}\mathbf{a}^H + \mathbf{a}\mathbf{b}^H + \gamma\mathbf{a}\mathbf{a}^H = \mathbf{I} + \lambda_1\mathbf{e}_1\mathbf{e}_1^H + \lambda_2\mathbf{e}_2\mathbf{e}_2^H \tag{20}$$

where $\mathbf{e}_i$ (orthonormal) and λ_i (real) can be easily computed from the eigendecomposition of a 2×2 matrix as follows. Let $[\mathbf{e}_1 \ \ \mathbf{e}_2] = [\mathbf{a} \ \ \mathbf{b}]\mathbf{T}$ where **T** is a 2×2 matrix. Using this in (20) leads to $\mathbf{RT} = \mathbf{T}diag(\lambda_1 \ \ \lambda_2)$ where **R** is a 2×2 matrix defined by

$$\mathbf{R} = \left\{[\mathbf{a} \quad \mathbf{b}]^H [\mathbf{a} \quad \mathbf{b}]\right\}^{-1} [\mathbf{a} \quad \mathbf{b}]^H \left(\mathbf{b}\mathbf{a}^H + \mathbf{a}\mathbf{b}^H + \gamma\mathbf{a}\mathbf{a}\right)[\mathbf{a} \quad \mathbf{b}]$$

and **T** is the eigenvector matrix of **R**. It is easy to verify that the computational complexity of (20) is negligible in comparison to O(np) when n and p are large.

With (20), an asymmetric square root inverse of **Z**(i+1) is

$$\mathbf{Z}^{-1/2}(i+1) = \frac{1}{\alpha}\left[\mathbf{I} - \tau_1 \mathbf{e}_1 \mathbf{e}_1{}^H - \tau_2 \mathbf{e}_2 \mathbf{e}_2{}^H\right]\mathbf{Z}^{-1/2}(i) \tag{21}$$

where $\tau_i = 1 - \frac{1}{\sqrt{\lambda_i + 1}}$. The asymmetric square root inverse is valid in the sense that

$$\mathbf{Z}^{-1/2}(i+1)\mathbf{Z}(i+1)\mathbf{Z}^{-1/2^H}(i+1) = \mathbf{I} \tag{22}$$

The asymmetric square root inverse of **Z**(i+1) differs from the symmetric square root inverse of **Z**(i+1) by a unitary rotation, i.e., the symmetric square root inverse pre-multiplied by any unitary matrix still satisfies (22). This unitary rotation tends to weaken the assumption (11). However, unlike the case of the QR power method, this unitary rotation is a slightly rotated version of the identity matrix when $\mathbf{x}(i+1)\mathbf{x}^H(i+1)$ is much smaller than α**C**(i), which holds in normal situations.

For analytical convenience, we now use a variation of (17) as follows:

$$\mathbf{W}(i+1) = \mathbf{Y}(i+1)\mathbf{Z}^{-1/2^H}(i+1) \tag{23}$$

which is equivalent to (17) if the symmetric square-root-inverse is used. When an asymmetric square-root-inverse is used, (23) differs from (17) by a post-multiplied unitary rotation.

With the expression (21), we can now write (23) as:

$$\begin{aligned} &\mathbf{W}(i+1) \\ &= \left(\alpha\mathbf{Y}(i) + \mathbf{x}\mathbf{y}^H\right)\mathbf{Z}^{-1/2^H}(i)\left[\mathbf{I} - \tau_1 \mathbf{e}_1 \mathbf{e}_1{}^H - \tau_2 \mathbf{e}_2 \mathbf{e}_2{}^H\right]\frac{1}{\alpha} \\ &= \mathbf{W}(i)\left[\mathbf{I} - \tau_1 \mathbf{e}_1 \mathbf{e}_1{}^H - \tau_2 \mathbf{e}_2 \mathbf{e}_2{}^H\right] + \mathbf{x}\mathbf{y}'^H\left[\mathbf{I} - \tau_1 \mathbf{e}_1 \mathbf{e}_1{}^H - \tau_2 \mathbf{e}_2 \mathbf{e}_2{}^H\right]\frac{1}{\alpha} \end{aligned} \tag{24}$$

where the dependence of all vectors on i+1 is dropped for convenience, and

$$\mathbf{y}' = \mathbf{Z}^{-1/2}(i)\mathbf{y} \tag{25}$$

The expression (24) provides an algorithm which does not involve any matrix-matrix multiplication and in fact requires only O(np) flops of computation at each iteration. Furthermore, if an exact (symmetric or asymmetric) square root inverse of the initial matrix **Z**(0) of full rank is used, the algorithm (24) guarantees the orthogonality of the weight matrix at each iteration.

3. OTHER POWER-BASED METHODS

This section summarises a detailed discussion shown in [14].

3.1 Oja's Method

An abstract form of the Oja method is

$$\mathbf{W}(i+1) = \mathbf{W}(i) + \eta\left[\mathbf{C}\mathbf{W}(i) - \mathbf{W}(i)\mathbf{W}^H(i)\mathbf{C}\mathbf{W}(i)\right] \qquad (26).$$

where the power term $\mathbf{CW}(i)$ is the heart of the algorithm. It is shown in [5-6] that if the step size is very small, the weight matrix (without any orthogonality condition) of (26) is (almost) globally convergent to an orthonormal matrix whose column span is the principal subspace. For digital implementation, however, the Oja method converges very slowly due to the fact that only a very small step size is allowed. A larger step size would make the Oja method divergent. Using the approximation (11), (26) can be implemented as follows:

$$\mathbf{W}(i+1) = \mathbf{W}(i) + \eta\left[\mathbf{Y}(i+1) - \mathbf{K}(i+1)\right]$$

where (also see (12)-(13))

$$\mathbf{K}(i+1) = \alpha\mathbf{K}(i) + \mathbf{W}(i)\mathbf{y}(i+1)\mathbf{y}^H(i+1)\ .$$

The Oja method needs O(np) flops at each iteration and does not involve any inverse or square root as the natural power method does. The original Oja method [3] chose α=0 and p=1.

3.2 The PAST Method

An abstract form of the PAST method is

$$\mathbf{W}(i+1) = \mathbf{C}\mathbf{W}(i)\left(\mathbf{W}^H(i)\mathbf{C}\mathbf{W}(i)\right)^{-1} \tag{27}$$

where $\mathbf{CW}(i)$ is the power term and the matrix inverse serves as a normalizer. The original PAST method [8] follows directly from (27) when the approximation (11) is used, and the computational complexity is O(np). Unlike the Oja method, the PAST method involves (scalar) inverses. It can be shown [14] that the PAST method does not always converge to an orthonormal matrix.

3.3 The NIC Method

An abstract form of the NIC method is

$$\mathbf{W}(i+1) = (1-\beta)\mathbf{W}(i) + \beta\mathbf{CW}(i)\left(\mathbf{W}^H(i)\mathbf{CW}(i)\right)^{-1} \qquad (28)$$

which is a generalized form of (27) with a leakage factor β. The NIC method has been shown in [9] to be (almost) globally convergent when β is very small. Simulations have also suggested that the NIC method converges for all values of β chosen between (0,1) although a rigorous proof is yet to be found. Like Oja and PAST, the NIC method can be implemented [9] with only O(np) flops at each iteration when the approximation (11) is used.

4. OBSERVATIONS FROM SIMULATIONS

A large number of computer simulations have been conducted. The initial $\mathbf{C}(0)$ was chosen to be a scaled identity matrix and the initial $\mathbf{W}(0)$ was a random matrix. Our main observations are as follows.

- The three versions of the natural power method: NP1, NP2 and NP3 are all globally convergent with exponential convergence rates although the NP3 requires a relatively large $\mathbf{C}(0)$ which causes some initial delay of convergence.
- With respect to subspace errors, the NP3 has a convergence rate similar to those of the PAST and NIC (with β close to 1) methods and much faster than that of Oja's. In terms of orthogonality errors, the performances of NP3, NIC and PAST are in descending order. Note that if an exact (symmetric or asymmetric) square root inverse of the initial matrix $\mathbf{Z}(0)$ is used, the NP3 guarantees the orthogonality of the weight matrix at each iteration.
- In comparison to the two recent methods: MALASE [12] (O(np)) and OPERA [13] (O(np^2)), all of the power based methods are much better choices because of their global convergence property. The former two methods do not work unless the initial weight matrix is a very good estimate of the basis of the principal subspace which does not change rapidly at any time.

5. FINAL REMARKS

Undoubtedly, the natural power method is important for multi-channel data processing, such as target tracking with multiple sensors. Like the Oja method, the natural power method can also be used as an on-line (but faster) learning algorithm for a class of linear neural networks. The additional complexity (of the NP3) is mainly some scalar inverses and scalar square roots. The natural power method can also easily be adopted for principal component extraction, detection of principal components, and constrained subspace tracking [14].

6. REFERENCES

[1] P. Comon and G. H. Golub, "Tracking a few extreme singular values and vectors in signal processing," *Proc of the IEEE*, Vol. 78, No. 8, pp. 1327-1343, Aug 1990.

[2] B. H. Juang, S. Y. Kung, and C. A. Kamm, ed., *Proc of the 1991 IEEE Workshop on Neural Networks for Signal Processing*, Princeton, NJ, Sept 30 - Oct 2, 1991.

[3] E. Oja, "A simplified neuron model as a principal component analyzer", *J. Math. Biology*, Vol. 15, pp. 267-273, 1982.

[4] S. Y. Kung, K. I. Diamantaras, and J. S. Taur, "Adaptive principal component extraction (APEX) and applications," *IEEE Trans on Signal Processing*, Vol. 42, No. 5, pp. 1202-1217, May 1994.

[5] W. Y. Yan, U. Helmke, J. B. Moore, "Global analysis of Oja's flow for neural networks," *IEEE Trans on Neural Networks*, Vol. 5, No. 5, pp. 674-683, 1994.

[6] T. Chen, Y. Hua, and W. Yan, "Global convergence of Oja's subspace algorithm for principal component extraction," *IEEE Trans on Neural Networks*, Vol. 9, No. 1, pp. 58-67, Jan 1998.

[7] L. Xu, "Least mean square error reconstruction principle for self-organizing neural nets," *Neural Networks*, Vol. 6, pp. 627-648, 1993.

[8] B. Yang, "Projection approximation subspace tracking," *IEEE Trans on Signal Processing*, Vol. 43, No. 1, pp. 95-107, Jan 1995.

[9] Y. Miao and Y. Hua, "Fast subspace tracking and neural network learning by a novel information criterion," *IEEE Trans on Signal Processing*, Vol. 46, No. 7, pp. 1967-1979, July 1998.

[10] G. H. Golub, and C. F. Van Loan, *Matrix Computations*, Baltimore, MD: Johns Hopkins Univ Press, 1989.

[11] E. Oja, "Neural networks, principal components, and subspaces," *International Journal of Neural Systems*, Vol. 1, No. 1, pp. 61-68, 1989.

[12] C. Riou, and T. Chonavel, "Fast adaptive eigenvalue decomposition: a maximum likelihood approach," *Proc. of IEEE ICASSP'97*, pp. 3565-3568, 1997.

[13] C. S. MacInnes, "Fast, accurate subspace tracking using operator restriction analysis," *Proc. of IEEE ICASSP'98,* pp. 1357-1360, 1998.

[14] Y. Hua, Y. Xiang, T. Chen, K. Abed-Meraim, Y. Miao, "A new look at the power methods for fast subspace tracking," submitted to *IEEE Transactions on Signal Processing*.

ROBUST NEURAL NETWORK ONLINE-LEARNING IN TIME-VARIANT REGRESSION MODELS

Thomas Briegel and Volker Tresp
Siemens AG, Corporate Technology
Dept. Information and Communications
Otto-Hahn-Ring 6, 81730 Munich, Germany
{Thomas.Briegel, Volker.Tresp}@mchp.siemens.de

Abstract. We consider robust online learning in time-variant neural network regression models. Using a state space representation for the neural network's weight evolution in time we derive weight estimates by maximizing posterior modes via the Fisher scoring algorithm. By taking the family of t-densities as the output error cost function we get a robust error measure suitable for handling additive outliers. Fisher scoring was implemented using a forward-backward pass of fixed length through the data set for every time step resulting in so-called online smoothing algorithms. Furthermore, we derive an EM-type algorithm for approximate maximum likelihood estimation of unknown hyperparameters. Our experiments show that online posterior mode weight smoothing outperforms standard online methods like online backpropagation and extended Kalman filtering, both for Gaussian measurements and non-Gaussian measurements with additive outliers.

INTRODUCTION

Online learning algorithms are of great interest in many applications which exhibit non-stationary behavior or where data arrivial is inherently sequential. These include time series forecasting, tracking, signal and image processing, navigation and fault detection. This paper focuses on the special case that the systems are nonlinear and are modeled using neural networks. A sensible way to sequentially estimate neural network models is to use a state space formulation and the extended Kalman filter algorithm, see for instance [9]. Such state space models are typically based on the assumption that errors are distributed as multivariate Gaussians. Methods based on the Gaussian error assumption like the extended Kalman filter, however, are known to be nonrobust against observation outliers [7] which might result in highly fluctuating weights and possible instability when estimating the neural network weight vector online. Especially for online learning tasks, one is interested in avoiding inappropriate weight changes due to measurement outliers to maintain stable online learning capability. In this paper we replace the Gaussian assumption

0-7803-5673-X/99/$10.00 © 1999 IEEE

with a heavy-tailed observation error distribution to obtain robust online learning. Special emphasis is on the Student t-family. Based on related work by Ormoneit [8] we tackle online learning problems by considering so-called *online smoothing* algorithms which derive weight updates by pairs of sliding window forward-backward passes through the data set. Online smoothing can be carried out by the computational efficient *Fisher scoring* method which was introduced to the neural network community by the authors in the context of offline inference and learning in general Gaussian nonlinear state space models [2]. Here, we apply Fisher scoring to obtain an online smoothing method for time-variant regression models with t-distributed measurement noise. Mathematically, Fisher scoring performs posterior mode estimation of the neural network weights by optimizing a penalized likelihood criterion. As a by-product we develop a new neural network online learning algorithm for Gaussian observation densities based on posterior modes. Furthermore, we present an *online EM-type algorithm* for approximate maximum likelihood (ML) estimation of hyperparameters, e.g. for estimating the degrees of freedom of the t-distribution and the measurement noise variance. In our experiments we show that online smoothing outperforms the traditional filtering approach to online estimation. Furthermore, we show that posterior mode estimation leads to better models than training based on extended Kalman online smoothing and online backpropagation, both for the Gaussian case and the non-Gaussian case with additive outliers.

TIME-VARIANT REGRESSION VIA STATE SPACE MODELING

We employ a state space representation to model the neural network's evolution in time. The measurement equation models the nonlinear relation between unnoisy, known inputs $u_t \in \mathbf{R}^k$ and noisy outputs $y_t \in \mathbf{R}$ by assuming a time-variant nonlinear model, e.g.

$$y_t = g(u_t, w_t) + v_t. \tag{1}$$

Here, $g(., w_t)$ denotes a neural network model characterized by weight vector $w_t \in \mathbf{R}^n$, e.g. a multi-layer perceptron (MLP). The neural network weights are interpreted as time-varying hidden states in a state space model. Assuming w_t follows a first order random walk with normally distributed increments leads to the state transition density

$$p(w_t|w_{t-1}) = \mathcal{N}(w_t|w_{t-1}, Q_t) \tag{2}$$

where $\mathcal{N}(x|\mu, \Sigma)$ denotes a multivariate Gaussian with mean μ and covariance Σ evaluated at x. (1) implies the observation density

$$p(y_t|u_t, w_t) = p_{v_t}(y_t - g(u_t, w_t)) \tag{3}$$

where v_t is an uncorrelated noise term with density $p_{v_t}(.)$. We define a Gaussian prior on w_0, e.g. $w_0 \sim \mathcal{N}(a_0, \Sigma_0)$. Each time a new measurement y_t and input u_t arrive, our belief about the weight vector w_t at time t can be updated using the state prediction density

$$p(w_t|Y_{t-1}, U_{t-1}) = \int p(w_t|w_{t-1})\, p(w_{t-1}|Y_{t-1}, U_{t-1})\, dw_{t-1} \tag{4}$$

and the filtering density

$$p(w_t|Y_t, U_t) = \frac{p(y_t|u_t, w_t)\, p(w_t|Y_{t-1}, U_{t-1})}{\int p(y_t|u_t, w_t)\, p(w_t|Y_{t-1}, U_{t-1})\, dw_t} \tag{5}$$

with $Y_t = (y_1, \ldots, y_t)^\top$ and $U_t = (u_1^\top, \ldots, u_t^\top)^\top$. While for linear Gaussian observation densities the Kalman filter may be used to evaluate $p(w_t|Y_{t-1}, U_{t-1})$ and $p(w_t|Y_t, U_t)$ analytically, for the general nonlinear case $g(., w_t)$ involving either Gaussian or non-Gaussian observation densities $p_{v_t}(.)$, these integrals are analytically intractable. Tresp & Hofmann [11] have approximated (4) and (5) using Gibbs sampling. de Freitas *et. al.* [3] apply sequential Monte Carlo methods to this problem. While being feasible for small networks these methods become computationally intractable for large networks in real world tasks. A well known computationally tractable approximation for nonlinear Gaussian observation densities $p(y_t|u_t, w_t) = \mathcal{N}(y_t|g(u_t, w_t), \sigma_v^2)$ is the extended Kalman filter algorithm [10]. It propagates Gaussian approximations to $p(w_t|Y_{t-1}, U_{t-1})$ and $p(w_t|Y_t, U_t)$ forward at each time step. For robust online learning we drop the Gaussian assumption in the observation equation and replace it with a heavy-tailed density, e.g. a t-density. It is generally recognized that the extended Kalman filter is nonrobust against outliers and cannot be applied to this situation. In the next section we propose a method that produces *smoothed* online estimates of the weight vector sequence rather than just *filtered* estimates as the extended Kalman filter does. Smoothing means that current values of the time series are taken to improve estimates of earlier weight vectors resulting in a non-causal filter. The proposed method is based on optimization of a certain penalized likelihood criterion leading to online posterior mode estimators.

PENALIZED LIKELIHOOD WEIGHT ESTIMATION

We start by considering offline learning where data sets U_T and Y_T are given. An offline smoothed estimate for the weight sequence $W_T = (w_0^\top, \ldots, w_T^\top)^\top$ is obtained by maximizing the log posterior $\log p(W_T|Y_T, U_T)$ or, equivalently, $\log p(W_T) + \log p(Y_T|W_T, U_T)$ with respect to W_T. Inserting the model assumptions and substituting $\rho = -\log p_{v_t}$ leads to the penalized log-likelihood cost function

$$\begin{aligned} pll(W_T) &= \sum_{t=1}^{T} \rho(y_t - g(u_t, w_t)) + \frac{1}{2}(w_0 - a_0)^\top Q_0^{-1}(w_0 - a_0) \\ &\quad + \frac{1}{2}\sum_{t=1}^{T}(w_t - w_{t-1})^\top Q_t^{-1}(w_t - w_{t-1}) \end{aligned} \tag{6}$$

which has to be minimized to obtain $W_T^{\mathrm{MAP}} = (w_{0|T}^\top, \ldots, w_{T|T}^\top)^\top$, the posterior mode estimate. The first term in (6) is a robust measure for the distance between data and model fit when assuming a heavy-tailed density for p_{v_t}, e.g. the t-density. This density is given up to a normalizing constant by

$$p_{v_t}(z) \propto (1 + z^2/\nu_t\sigma_{v_t}^2)^{-(\nu_t+1)/2}, \quad \nu_t, \sigma_{v_t} > 0 \tag{7}$$

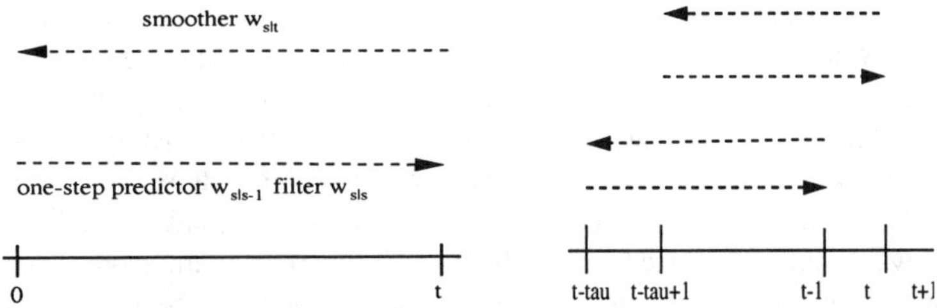

Figure 1: (left) Offline smoothing consisting of one forward step where step-one predictions $w_{s|s-1}$ and filter estimates $w_{s|s}$ are computed followed by one backward pass where smoothed estimates $w_{s|t}, s < t$ are derived. (right) Online smoothing consisting of a new pair of forward and backward steps each time a new measurement is available.

with scale factor σ_{v_t} and ν_t degrees of freedom. The second and third term in (6) serve as regularization terms penalizing roughness of the weight sequence [8]. A suitable way to determine a stationary point of $pll(.)$ is to apply *Fisher scoring*. With the current estimate W_T^{old} we get a better estimate $W_T^{\text{new}} = W_T^{\text{old}} + \eta\gamma$ for the unknown weight sequence W_T where γ is the solution of

$$\mathcal{S}(W_T^{\text{old}})\gamma = s(W_T^{\text{old}}) \tag{8}$$

with the score function $s(W_T) = \partial pll/\partial W_T$ and the expected information matrix $\mathcal{S}(W_T) = \mathrm{E}[-\partial^2 pll/\partial W_T \partial W_T^{\mathrm{T}}]$.[1] For computing these quantities one needs the score function and the expected information of the negative log t-density. The score function $\psi(.)$ and the expected information Ψ of the negative log t-density are given by

$$\psi(z) = (\nu_t + 1)(\nu_t + z^2/\sigma_{v_t}^2)^{-1}\sigma_{v_t}^{-2} z \tag{9}$$

and

$$\Psi = (\nu_t + 1)(\nu_t + 3)^{-1}\sigma_{v_t}^{-2}, \tag{10}$$

respectively [6]. By applying the ideas given in Fahrmeir & Kaufmann [4] to time-variant neural network regression models it turns out that solving (8), e.g. to compute the inverse of the expected information matrix, can be performed by Cholesky decomposition in one forward and backward pass through the set of data. Note that the expected information matrix is a positive definite block-tridiagonal matrix. The forward-backward steps have to be iterated to obtain the posterior mode estimate W_T^{MAP} for W_T.

The above algorithm provides estimates only after a block of T consecutive filter steps. For *online* posterior mode smoothing, it is of interest to smooth backwards after each filter step t. If Fisher scoring steps are applied sequentially for $t = 1, 2, \ldots$, then $w_{0|t-1}, \ldots, w_{t-1|t-1}$ (posterior mode smoother for time $t-1$) together with the step-one predictor $w_{t|t-1}$ are reasonable starting values for obtaining posterior mode smoothers $w_{s|t}$ at time steps $s < t$ when a new observation y_t becomes available. Fig. 1 compares the offline smoothing algorithm with online smoothing. On

[1]Note that the difference between Gauss-Newton and Fisher scoring is that in the former we take the expectation of the information matrix. This ensures positive definiteness.

the left part of the figure we see the described offline smoothing algorithm consisting of one forward and one backward pass through the whole data set. Smoothing backwards just takes place once after all measurements were recorded. On the right, we see that each time a new measurement is available the algorithm smoothes backwards to obtain posterior mode smoothers for earlier time steps $s < t$. These estimates are used to obtain improved step-one prediciton $w_{s|s-1}$ and filtered values $w_{s|s}$ in a new forward pass since these are based on evaluations of the score function of re-computed smoothed estimates $w_{s|t}, s < t$ (see Appendix). The improvement in step-one prediction and filtering, however, results in a greater computational load since smoothing backwards takes place after *each* filter step. One can reduce the compuational load by limiting the backward pass to a sliding time window, e.g. the last τ_t time steps, which is a reasonable assumption in non-stationary environments for online purposes. Hereby we use the underlying assumption that in most cases a new measurement y_t should not change estimates too drastically implying that a *single* Fisher scoring step often suffices to obtain the new posterior mode estimate at time t. The resulting single Fisher scoring step algorithm with lookback parameter τ_t can be found in the Appendix. As a by-product, we have also derived an algorithm for Gaussian measurement noise since the Gaussian density is a special case of the t-density in the limit of infinite degrees of fredom ($\nu_t \to \infty$).

ONLINE ESTIMATION OF HYPERPARAMETERS

So far, we have assumed that the measurement noise variance $\sigma_{v_t}^2$ and the degrees of freedom ν_t of the t-density are known. In practical applications one has to estimate these quantities from the data set. In the following, we apply results from Fahrmeir & Künstler [5] and from Lange et al. [6] to our nonlinear assumptions and develop an online EM-type algorithm for approximate ML estimation of these quantities. We assume the scale factors $\sigma_{v_t}^2$ and the degrees of freedom ν_t being fixed quantities in a certain time window of length τ_t, e.g. $\sigma_{v_t}^2 = \sigma_v^2, \nu_t = \nu, t \in \{t-\tau_t, t\}$. For deriving ML estimates we take advantage of the fact that a t-distributed random variable can be generated as a mixture $y/\sqrt{z/\nu}$ with independently distributed $y \sim \mathcal{N}(0, \sigma_v^2)$ and chi-squared mixture variables $z \sim \chi_\nu^2$ [1]. Therefore we can treat the weight sequence w_t together with the mixture variables z_t as missing. The E-step is given by

$$\mathrm{E}[\log p(Y_{t-\tau_t,t}, Z_{t-\tau_t,t}, W_{t-\tau_t,t}, U_{t-\tau_t,t}) | Y_{t-\tau_t,t}, U_{t-\tau_t,t}, \phi^{\mathrm{old}}] \qquad (11)$$

with $Z_{t-\tau_t,t} = (z_{t-\tau_t}, \ldots, z_t)^\top$ as the sequence of mixture variables (and the other vectors analogically defined) and $\phi^{\mathrm{old}} = (\sigma_v^2, \nu)^\top$ as the vector of unknown hyperparameters to be estimated. The expectation can be decomposed into

$$\begin{aligned} \mathrm{E}\big[&\log p(Y_{t-\tau_t,t} | Z_{t-\tau_t,t}, W_{t-\tau_t,t}, U_{t-\tau_t,t}) + \log p(Z_{t-\tau_t,t}) \\ &+ \log p(W_{t-\tau_t,t}) | Y_{t-\tau_t,t}, U_{t-\tau_t,t}, \phi^{\mathrm{old}}\big]. \end{aligned} \qquad (12)$$

We have to solve

$$\frac{\partial}{\partial \sigma_v} \sum_{s=t-\tau_t}^{t} \mathrm{E}[\log p(y_s | z_s, w_s, u_s) | Y_{t-\tau_t,t}, U_{t-\tau_t,t}, \phi^{\mathrm{old}}] = 0 \qquad (13)$$

for deriving the ML estimator for σ_v^2 and

$$\frac{\partial}{\partial \nu} \sum_{s=t-\tau_t}^{t} \mathrm{E}[\log p(z_s) | Y_{t-\tau_t,t}, U_{t-\tau_t,t}, \phi^{\mathrm{old}}] = 0 \tag{14}$$

for obtaining update rules for ν. By approximating posterior expectations with posterior modes and posterior variances with curvatures a somehow lengthy derivation results in (approximate) ML update rules for σ_v^2 and ν given in the Appendix.

EXPERIMENTS

Input-output data vectors were generated from the following nonlinear time-variant regression model

$$y_t = 30\,\theta_1 u_t^2 + 10\,\theta_2 \sin(6u_t) - 10 + v_t, \quad t = 1, \ldots, 600 \tag{15}$$

where u_t was uniformly sampled from the interval $[-1, 1]$ and $\theta = (\theta_1, \theta_2)^\top$ was chosen as

$$\theta = \begin{cases} (0.5, 0.2)^\top & , \quad t = 1, \ldots, 200 \\ (0.2, -0.5)^\top & , \quad t = 201, \ldots, 400 \\ (0.2, \frac{t-400}{200} - 0.75)^\top & , \quad t = 401, \ldots, 600 \end{cases} \tag{16}$$

to simulate abruptly and slowly changing dynamics, respectively. In the first experiment we take Gaussian output noise $v_t \sim \mathcal{N}(0, 1)$. We use a multi-layer perceptron (MLP) with five hidden units to implement the nonlinearity $g(., w_t)$. Q_t was chosen as $10^{-2} I_{n,n}$ and $\eta = 0.05$. Table 1 (left) shows the results for four different online learning methods: online backpropagation (BP) with the update rule

$$\hat{w}_t = \hat{w}_{t-1} + \eta \frac{\partial}{\partial w_{t-1}} \mathcal{N}(y_t, g(u_t, \hat{w}_{t-1}), 1), \tag{17}$$

extended Kalman filtering (EKF), extended Kalman smoothing with a lookback parameter of 10 (EKS-10) and the single-step Fisher scoring algorithm with the same lookback parameter (FS-10). Shown are the out of sample average online sum square error (AOSSE)[2] and its standard deviation averaged over 20 independent simulations of the experiment. Both smoothing algorithms (EKS-10 and FS-10) outperform the standard online learning methods (BP, EKF) by far: EKS-10 achieves a 20% reduction in AOSSE compared to EKF, FS-10 achieves 47% reduction, respectively. Furthermore, FS-10 can significantly improve the results obtained from EKS-10 (33% reduction in AOSSE). In the second experiment we want to compare the robustness of the different algorithms towards additive outliers. We use Gaussian noise terms as above with probability $1 - \pi$ and uniformly distriduted additive outliers from $[-10, 10]$ with probability π. Table 1 (right) shows the results for the four different methods when using the t-density with $\nu = 5$ degrees of freedom for $p_{v_t}(.)$ in the FS-10 algorithm. The performance improvements are significantly higher when increasing π, the probability of outliers in the model.

---

[2]AOSSE $= \frac{1}{20} \sum_{t=1}^{T} \left(y_t - v_t - g(u_t, w_t^\star)\right)^2$ where $w_t^\star = \hat{w}_{t-1}$ in the BP case and $w_t^\star = w_{t|t-1}$ for the EKF, EKS and FS approach.

Table 1: AOSSE for Gaussian output noise and additive outliers

	Gaussian		additive outliers			
			$\pi = 0.1$		$\pi = 0.3$	
MODEL	AOSSE	std	AOSSE	std	AOSSE	std
BP	5430	279	5843	302	7971	485
EKF	3282	201	5471	213	7227	318
EKS-10	2613	138	4878	163	6598	248
FS-10	1739	98	2488	105	2987	121

CONCLUSIONS

We considered robust online learning in time-variant neural network regression models. Using a state space representation for the neural network's weight evolution in time we maximize posterior modes by optimizing a penalized likelihood criterion for the weight sequence. By taking the family of t-densities we get a robust error measure against additive outliers. Posterior mode estimation was implemented using Fisher scoring with *one* forward-backward pass with time lag τ_t resulting in an efficient online *smoothing* algorithm. Furthermore, we derived an EM-type algorithm for approximate ML estimation of unknown hyperparameters. Our preliminary experimental results show that online smoothing is superior compared to the standard online filtering approach. Posterior mode online smoothing outperforms standard extended Kalman online smoothing. Using the t-density as a robust measure for additive outliers resulted in improved modeling performance of the Fisher scoring algorithm compared to all other benchmark models.

APPENDIX

We initialize the *online single-Fisher scoring step algorithm* with $w_{0|0} = a_0, \Sigma_{0|0} = Q_0$ and $\gamma_0 = s_0(w_{0|0})$ and choose a possibly time-variant lookback parameter τ_t. For $t = 1, 2, \ldots$ one applies the following two steps: First, one performs the set of forward recursions ($s = t - \tau_t, ..., t$)

$$\Sigma_{s|s-1} = \Sigma_{s-1|s-1} + Q_s \tag{18}$$

$$B_s = \Sigma_{s-1|s-1}\Sigma_{s|s-1}^{-1} \tag{19}$$

$$\Sigma_{s|s} = \left(\Sigma_{s|s-1}^{-1} + G_s^\top(w_{s|t-1})\Psi G_s(w_{s|t-1})\right)^{-1} \tag{20}$$

$$\gamma_s = s_s(w_{s|t-1}) + B_s^\top \gamma_{s-1} \tag{21}$$

with the initial steps $w_{t|t-1} = w_{t-1|t-1}$ and $\gamma_{t-\tau_t-1} = s_{t-\tau_t-1}(w_{t-\tau_t-1|t-1})$ and the final corrector step $w_{t|t} = w_{t|t-1} + \eta\Sigma_{t|t}\gamma_t$. Then, one performs the set of backward smoothing recursions ($s = t, ..., t - \tau_t$)

$$D_{s-1}^{-1} = \Sigma_{s-1|s-1} - B_s\Sigma_{s|s-1}B_s^\top \tag{22}$$

$$w_{s-1|t} = w_{s-1|t-1} + B_s(w_{s|t} - w_{s|t-1}) + \eta D_{s-1}^{-1}\gamma_{s-1} \tag{23}$$

$$\Sigma_{s-1|t} = D_{s-1}^{-1} + B_s\Sigma_{s|t}B_s^\top \tag{24}$$

where $G_s(z) = \frac{\partial g(u_s, w_s)}{\partial w_s}|_{w_s=z}$ and $s_s(z) = \frac{\partial pll(W_t)}{\partial w_s}|_{w_s=z}$ is the score function of $pll(.)$ evaluated at weight vector z.

The *online EM update equation* for the measurement variance σ_v^2 is given by

$$\sigma_v^{2,\text{new}} = \frac{1}{\tau_t} \sum_{s=t-\tau_t}^{t} \alpha_{s|t} \Big[\big(y_s - g(u_s, w_{s|t})\big)^2 + G_s(w_{s|t}) \Sigma_{s|t} G_s(w_{s|t})^\top \Big] \quad (25)$$

where $\alpha_{s|t} = (\nu^{\text{old}} + 1)/(\nu^{\text{old}} + \delta_{s|t})$ with $\delta_{s|t} = \big(y_s - g(u_s, w_{s|t})\big)^2 / \sigma_v^{2,\text{old}}$. A new value for ν is obtained by solving the one-dimensional nonlinear optimization problem for ν^{new}

$$\frac{\tau_t}{2} + \frac{\tau_t}{2} \log(\frac{\nu^{\text{new}}}{2}) - \frac{\tau_t}{2} \text{DG}(\frac{\nu^{\text{new}}}{2}) + \frac{1}{2} \sum_{s=t-\tau_t}^{t} \beta_{s|t} - \frac{1}{2} \sum_{s=t-\tau_t}^{t} \alpha_{s|t} = 0 \quad (26)$$

where $\beta_{s|t} = \text{DG}(\frac{\nu^{\text{old}}+1}{2}) - \log\big(\frac{1}{2}(\nu^{\text{old}} + \delta_{s|t})\big)$ with the Digamma function $\text{DG}(x) = \frac{\partial \Gamma(x)}{\partial x}$.

REFERENCES

[1] D. Andrews and C. Mallows, "Scale Mixtures of Normal Distributions," Journal of the Royal Statistical Society, B 36, pp. 99-102, 1974.

[2] T. Briegel and V. Tresp, "Fisher Scoring and a Mixture of Modes Approach for Approximate Inference and Learning in Nonlinear State Space Models," to appear in Advances in Neural Information Processing Systems 11, eds. M. Kearns, S. Solla, D. Cohn, MIT Press, Cambridge, MA, 1999.

[3] J. de Freitas, S. Johnson, M. Niranjan and A. Gee, "Global Optimisation of Neural Network Models via Sequential Sampling-Importance Resampling," ICSLP, Sidney, Australia, 1998.

[4] L. Fahrmeir and H. Kaufmann, "On Kalman Filtering, Posterior Mode Estimation and Fisher Scoring in Dynamic Exponential Family Regression," Metrika, 38, pp. 37-60, 1991.

[5] L. Fahrmeir and R. Künstler, "Penalized likelihood smoothing in robust state space models," Discussion paper 111, Seminar für Statistik, Universität München, 1998.

[6] K. Lange, L. Little and J. Taylor, "Robust Statistical Modeling Using the t-Distribution," Journal of the American Statistical Association, 84, pp. 881-896, 1989.

[7] R. Meinhold and N. Singpurwalla, "Robustification of Kalman Filter Models," Journal of the American Statistical Association, 84, pp. 470-496, 1989.

[8] D. Ormoneit, "Neural and Nonparametric Approaches to Learning in Nonstationary Systems," Continuous Learning Workshop, NIPS*11, Breckenridge, CO, 1998.

[9] G. Puskorius and L. Feldkamp, "Neurocontrol of Nonlinear Dynamical Systems with Kalman Filter Trained Recurrent Networks," IEEE Transactions on Neural Networks, 5:2, pp. 279-297, 1994.

[10] A. Sage and J. Melsa, "Estimation Theory with Applications to Communications and Control," McGraw-Hill, New York, 1971.

[11] V. Tresp and R. Hofmann, "Missing and noisy data in nonlinear time-series prediction," IEEE Int. Workshop on Neural Networks for Signal Processing 5, pp. 1-10, New York, 1995.

HIDDEN MARKOV MIXTURES OF EXPERTS FOR PREDICTION OF NON–STATIONARY DYNAMICS

Stefan Liehr, Klaus Pawelzik
University of Bremen, Institute of Theoretical Neurophysics,
Kufsteiner Str., D–28334 Bremen, Germany
email: {sliehr,klaus}@physik.uni-bremen.de

Jens Kohlmorgen, Steven Lemm, Klaus–Robert Müller
GMD FIRST, Rudower Chaussee 5, D–12489 Berlin, Germany
email: {jek,lemm,klaus}@first.gmd.de

Abstract. The prediction of non–stationary dynamical systems may be performed by identifying appropriate sub–dynamics and an early detection of mode changes. In this paper, we present a framework which unifies the mixtures of experts approach and a generalized hidden Markov model with an input–dependent transition matrix: the Hidden Markov Mixtures of Experts (HMME). The gating procedure incorporates state memory, information about the current location in phase space, and the previous prediction performance. The experts and the hidden Markov gating model are simultaneously trained by an EM algorithm that maximizes the likelihood during an annealing procedure. The HMME architecture allows for a fast on–line detection of mode changes: change points are detected as soon as the incoming input data stream contains sufficient information to indicate a change in the dynamics.

INTRODUCTION

Non–stationarity is a severe problem in classification and prediction of dynamical systems. A basic framework for dealing with non–stationarity is the mixtures of experts (ME) architecture, introduced by Jacobs et al. [3]. The mixtures of experts framework aims at separating the seemingly complex global behavior into a couple of lower dimensional sub–dynamics which can be modeled more easily.

For example, the Lorenz system [5] exhibits switching between two different oscillatory modes which are globally non–linear, while each single oscillation can be assumed to be approximately linear near the corresponding fix–point. In this case, two linear models would be the optimal choice in order to resolve the dynamical structure of the system. Then, the non–linearity can be

0-7803-5673-X/99/$10.00

incorporated into the gating procedure. A central problem of using a set of experts is therefore the calculation of the activities of each expert — called the gating problem.

Many solutions have been proposed for dealing with the gating problem [1, 2, 3, 4, 7, 9, 10]. In its original formulation [3], the mixtures of experts method can be applied to systems, where different regimes do not overlap in phase space (i.e. the input space). The expert activities are provided by a feed–forward gating network given the current location in phase space [3, 10]. The use of a recurrent gating network [2] allows to distinguish also between overlapping regimes.

An alternative, non–recurrent approach to distinguish between overlapping regimes is the annealed competition of experts (ACE) method [7]. It has its roots in statistical mechanics and is a purely performance–driven concept, which considers a moving average prediction error for estimating the activities instead of using a gating network.

In contrast to these approaches, we use the concept of hidden Markov models (HMM) and associate each prediction expert with a hidden state of the system. Moreover, we introduce a non–linear gating network that models the conditional probabilities of transitions between the predictors depending on the actual location in phase space and the previous prediction performance. Hence, this approach unifies previous approaches by integrating (1) input information, (2) performance information, (3) state information for modeling the gating probabilities, *and* (4) strict consistency between training and prediction task. It is therefore also substantially more general than related HMM based methods [1, 9], which either do not make use of performance information in the gating process [1] or do not use input information in the gating process [9].

Simulation results show that mode changes can be detected earlier compared to other methods, if all the three types of information are incorporated into the gating process. Likewise, the prediction performance can be improved significantly.

THE HMME ALGORITHM

In the following, we assume that the reader is already familiar with the basic principles of hidden Markov models (HMMs). For a thorough introduction, we refer to the tutorial of Rabiner [8].

The Hidden Markov Mixtures of Experts (HMME) architecture consists of three information processing components, whose interactions are illustrated in Figure 1.

1. Experts

Consider a set of K models $\{f^k\}_{k=1}^K$, which will also be called experts or predictors. At time step t, $1 \leq t \leq T$, each expert provides a prediction $y_t^k =$

$f^k(\vec{x}_t, \vec{\Theta}^k)$, which might be e.g. the estimate of a future value $y_t = x_{t+\tau}$ of a time series $\{x_t\}$ given a vector of past values $\vec{x}_t = (x_t, x_{t-\tau}, \ldots, x_{t-(d-1)\tau})$. The parameter d is called the embedding dimension and τ is called the delay parameter. Note that the extension to multivariate time series is straightforward.

The parameter vector of each model is denoted by $\vec{\Theta}^k$, the combined parameter vector of the experts is $\vec{\Theta} = (\vec{\Theta}^1, \ldots, \vec{\Theta}^K)$. Under a Gaussian assumption, the probability density that a particular predictor k might have produced an observation y_t is given by

$$r_d(y_t|k) \;=\; \sqrt{\frac{\beta}{\pi}} \exp\left(-\beta\epsilon_t^k\right) \quad \text{with} \quad \epsilon_t^k = \left(y_t - y_t^k\right)^2 \tag{1}$$

The parameter β can be interpreted as an inverse–temperature and is used for deterministic annealing during the training process.

By using Bayes' theorem, the probability that expert k has generated a given observation y_t, can be written as

$$r_t^k = r(k|y_t) \;=\; \frac{\exp\left(-\beta\epsilon_t^k\right)}{\sum_{l=1}^K \exp\left(-\beta\epsilon_t^l\right)}. \tag{2}$$

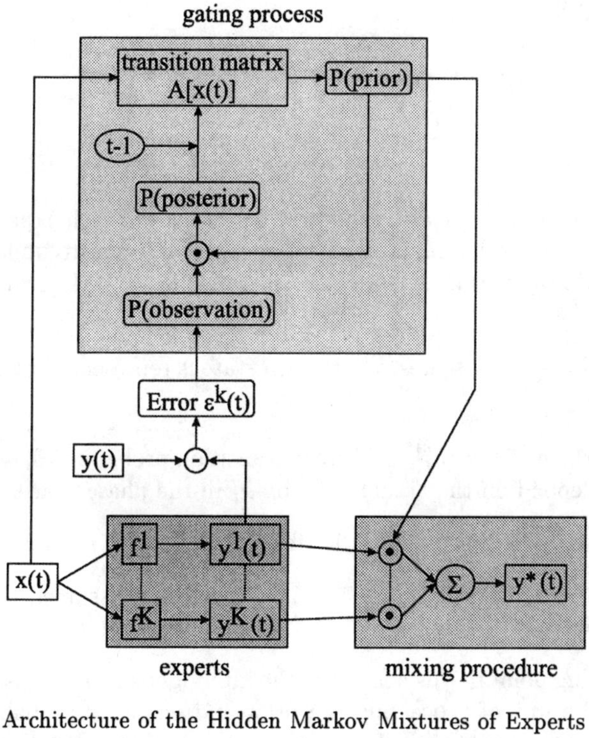

Figure 1: Architecture of the Hidden Markov Mixtures of Experts (HMME). The diagram shows the interactions of experts, hidden Markov gating, and mixing component. The different kinds of probabilities are denoted more intuitively by P(observation)$= r_t^k$, P(prior)$= q_t^k$ and P(posterior)$= p_t^k$.

In eq. (2), we anneal the "temperature" $1/\beta$ to zero during training, which promotes the initial diversification of the experts and leads to an exclusive assignment of training data points to experts (hard segmentation). However, in order to obtain the best performance of the prediction system, we finally use the set of parameters $\vec{\Theta}$ at the temperature with the lowest prediction error.

2. Mixing

The joint prediction of the ensemble is given by a weighted sum of the individual outputs

$$y_t^* = \sum_{k=1}^{K} q_t^k y_t^k. \tag{3}$$

In figure 1 this is highlighted as the mixing procedure. The calculation of the mixing factors q_t^k is performed causally (eq. 5), i. e. no future information is used for calculating the estimate y_t^*. This is important for a strict consistency between the training and the application of the algorithm.
The probability distribution for observing y_t, using the entire prediction system, is given by

$$p_d(y_t|\vec{x}_t, \vec{\Theta}) = \sqrt{\frac{\beta}{\pi}} \exp\left(-\beta\left(y_t - y_t^*\right)^2\right) \tag{4}$$

3. Gating

The mixing factors q_t^k are also called the activations of each expert. They are calculated in the hidden Markov gating process. In order to understand how this calculation is performed, we have to consider an HMM, which consists of

a) a set $S = \{s^k\}$ of states, where each state is represented by a prediction expert f^k,

b) a matrix $A(\vec{x}_t) = \{a_t^{k|k'}\}$ of state transition probabilities, which, in our case, depend on the actual location $\vec{x}_t$ in the phase space,

c) an observation probability distribution $p(y_t|s^k) = r_d(y_t|k)$,

d) an initial state distribution $\pi = \{\pi^k\}$, which is assumed to be equally distributed.

The interesting point in this context is the input–dependent transition matrix $A(\vec{x}_t)$ and the way of processing the prior and posterior probabilities. The activation q_t^k is given by the *a priori* probability of being in state k at time t, which depends on the input–dependent transition probabilities $a_t^{k|k'}$ and the *a posteriori* state probabilities p_{t-1}^k from the previous time step:

$$q_t^k = \sum_{k'=1}^{K} a_t^{k|k'} p_{t-1}^{k'} \quad \text{with} \quad p_t^k = \frac{r_t^k q_t^k}{\sum_{l=1}^{K} r_t^l q_t^l}. \tag{5}$$

In eq. (5), the posterior probabilities p_t^k are given, according to Bayes' theorem, by the prior probabilities q_t^k and the observation probabilities r_t^k, with an adequate normalization. Therefore, the posterior probabilities p_t^k contain information about the target value y_t, whereas the prior probabilities do not. The columns of the transition matrix and the probabilities sum to one, which is a requirement for the deduction of the learning rule:

$$\sum_k a_t^{k|k'} = \sum_k q_t^k = \sum_k p_t^k = \sum_k r_t^k \quad = \quad 1 \tag{6}$$

The transition matrix is represented by an input–dependent gating network[1] h with a $(K \times K)$–output matrix and a parameter vector $\vec{\Theta}^g$,

$$a_t^{k|k'} = h^{k|k'}(\vec{x}_t, \vec{\Theta}^g).$$

Training is performed by a gradient descent on the expected likelihood L. The log–likelihood can be decomposed into the free energy F, which provides a quality measure of the expert system, and the Kullback–Leibler "distance" KL, which yields to the same adaptation rule for the gating network as the second term of $\log L$:

$$\log L \quad = \quad \frac{1}{T} \sum_{t=1}^{T} \left(\log p_d(y_t|\vec{x}_t, \vec{\Theta}) + \sum_{k,k'} \xi_t^{kk'} \log a_t^{k|k'} \right) \tag{7}$$

$$F = -\frac{1}{\beta T} \sum_{t=1}^{T} \log p_d(y_t|\vec{x}_t, \vec{\Theta}) \quad \text{and} \quad KL = \frac{1}{T} \sum_{t=1}^{T} \sum_{k,k'} \xi_t^{kk'} \log \frac{\xi_t^{kk'}}{a_t^{k|k'}} \tag{8}$$

By minimizing KL the gating network learns to predict the joint probabilities $\xi_t^{kk'}$, which are calculated in accordance with the theory of hidden Markov models using the forward and backward probabilities α_t^k and β_t^k:

$$\xi_t^{kk'} \quad = \quad \frac{\beta_t^k r_t^k a_t^{kk'} \alpha_{t-1}^{k'}}{\sum_{ll'} \beta_t^l r_t^l a_t^{ll'} \alpha_{t-1}^{l'}} \tag{9}$$

In order to calculate the gradients of eq. (8), we use the method of Lagrange multipliers for incorporating the normalization conditions in eq. (6). The training can efficiently be performed by Expectation–Maximization (EM). The E–step consists of estimating the probabilities, the M–step adapts the models by minimizing the objective functions using gradient descent. Since in the M–step the probabilities are considered to be constant, the derivatives of the objective functions can be simplified drastically.

[1] We use a radial basis function (RBF) network of the Moody–Darken type [6] (look there for architecture and initialization) for the gating network.

EXPERIMENTAL RESULTS

We applied the HMME to different well-known systems, the deterministically switching logistic map and the Lorenz-System [5]. As shown in table 1, in both cases the HMME yields significantly better predictions than the ACE method.

Deterministically switching logistic map

The first example consists of a switching system of a noisy logistic map, $y_{t+1} = x_{t+1} = 4x_t(1 - x_t) + \eta_t$, and its "inverse", $y^I_{t+1} = x^I_{t+1} = 1 - 4x^I_t(1 - x^I_t) + \eta_t$, with uniform noise $\eta_t \in [-0.01, 0.01]$. The dynamics jumps from one mode to the other whenever $x_t \in [0.45, 0.55]$ holds. This is a system which exhibits non-linear behaviour, totally overlapping input spaces and a transition probability depending on the location in phase space. Additionally, the mean length of a mode is only about 16 time steps and therefore relatively short compared to previous applications of gated prediction systems. Modeling the jump processes is therefore very important for obtaining a high prediction quality. For the experts we use two radial basis function (RBF) networks of Moody-Darken type [6] with 6 RBFs each, and another network of the same type but with 10 RBFs for the gating network.
In table 1 the performance of our method is shown in comparison with the ACE-algorithm [7], which uses a lowpass filter instead of modeling transition probabilities. The advantage of the HMM-based method is first a fast detection of change points, which is possible because it does not depend on a fixed smoothing algorithm. Second, the method allows a recursive iteration of the prediction system with the possibility of predicting self-driven mode changes. Both properties are shown in Fig. 2.

Lorenz System

The second example is the Lorenz system [5], which is given by a set of three coupled differential equations. With the chosen parameters the Lorenz

system	algorithm	training	test 1	test 2
switching logistic map	HMME	0.00609	0.0125	0.0110
	ACE	0.00452	0.404	0.537
		$\tau_{ac} = 3$	$\tau_c = 2$	
Lorenz system	HMME	0.0270	0.0282	0.0279
	ACE	0.0287	0.0397	0.0381
		$\tau_{ac} = 9$	$\tau_c = 5$	

Table 1: Comparison of normalized mean squared errors (NMSE). Note that the filter used in the annealed competition of experts algorithm (ACE) is a-causal on the training data (τ_{ac}), while it is causal on the test data (τ_c). The Lorenz system is calculated with the parameters $\sigma = 16$, $b = 4$ and $r = 45.92$.

system exhibits a switching behaviour between oscillations around two fix-points. The system is globally non-linear, with the strongest non-linearity near the switching area from one oscillatory wing to the other, while each single oscillation can be assumed to be approximately linear near the corresponding fix-point. Therefore, we choose two linear experts and a non-linear gating network for modeling the Lorenz system. The non-linearity is thus incorporated into the gating procedure. The input and output of the experts are given by the state vector (X, Y, Z).

As shown in table 1, our algorithm yields significantly better predictions than ACE. In contrast to the logistic map example, however, the relative improvement is not that large. This can be explained first by deviations from a pure linear oscillation around the fix-points and second by the more difficult segmentation task due to the strong non-linear and non-instantaneous

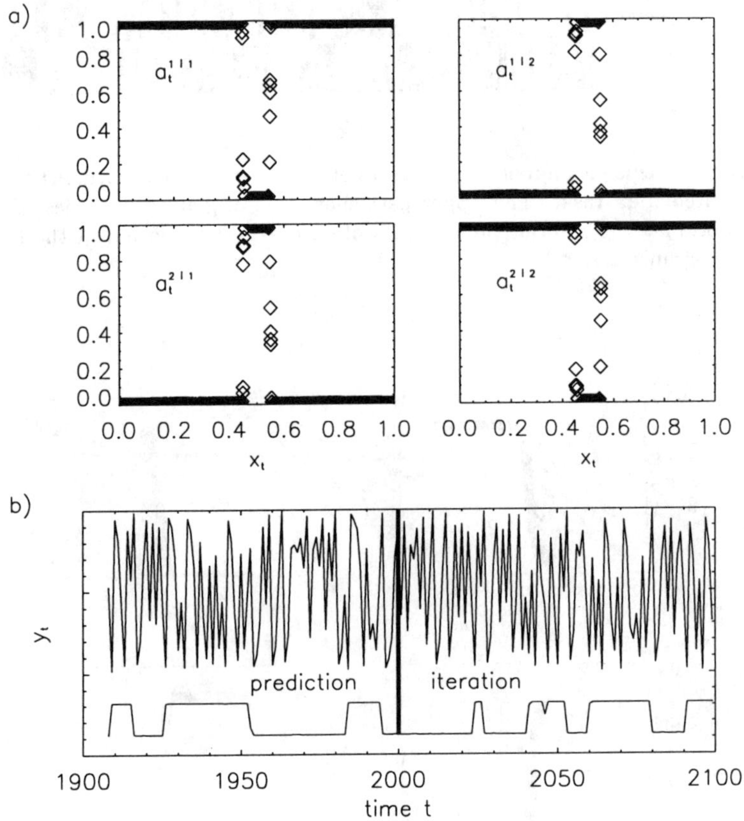

Figure 2: a) Input-dependent "elements" of the 2×2-transition matrix. The columns sum up to zero according to the normalization condition of eq. (6). b) Original time series of a validation dataset. The lower part shows the evolution of the activity of one expert during the one-step prediction task ($t < 2000$) and the iterated prediction task ($t > 2000$).

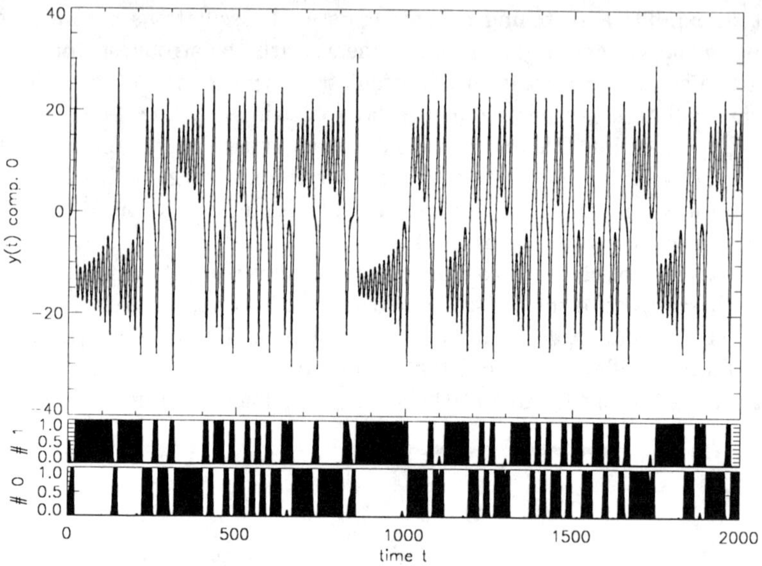

Figure 3: Segmentation of the dynamics of the Lorenz system during the one-step prediction task. The upper part shows the original time series of the X-coordinate. The prior probabilities of each expert is plotted in the lower two diagrams.

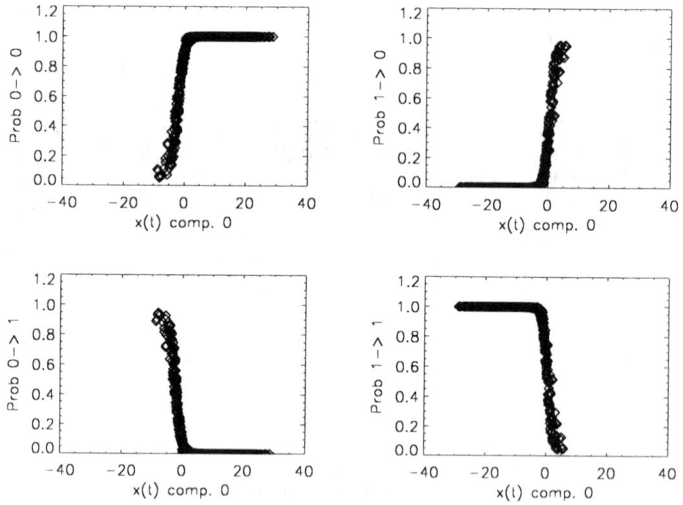

Figure 4: Representation of the input-dependent 2 × 2-output matrix of the gating network projected onto the plane (X/Prob). 'Prob' denotes the corrsponding transition probabilities.

switching process, as mentioned above.
Figure 3 demonstrates the segmentation of the dynamics. Obviously, each expert specializes on one oscillation. The algorithm follows even short–term mode changes. The time scale of the detection of change points can be seen in Figure 5. It clearly points out that the input–dependent gating procedure leads to an earlier detection of mode changes than the ACE algorithm. The final estimation of the input–dependent transition matrix projected onto the plane (X/Prob) between the X–coordinate of the state vector and the transistion probability is shown in Figure 4. It reflects the switching behavior of the Lorenz dynamics near $X = 0$.

CONCLUSION

We presented a generalized framework for unsupervised segmentation, identification, and prediction of switching dynamics. The architecture is composed of a mixture of experts and a hidden Markov model with an input–dependent state transition matrix. In contrast to existing methods, the system is trained in consistency with the prediction task and makes use of all available sources of information: input information from phase space, prediction error infor-

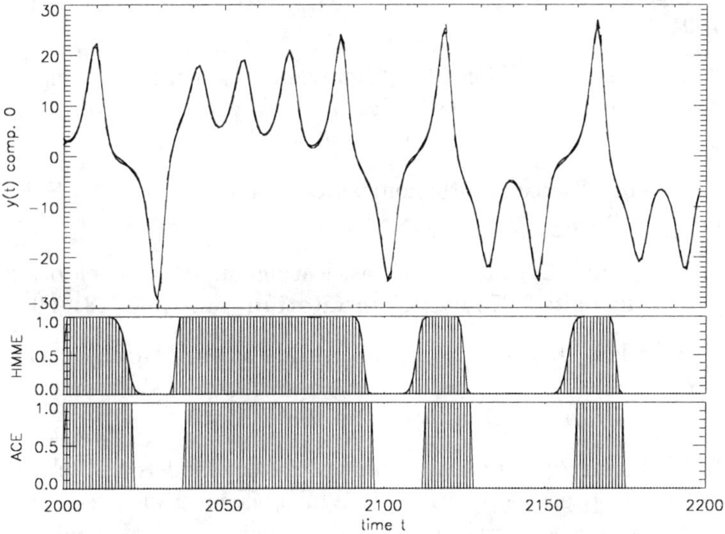

Figure 5: *Top:* Original dynamics and one–step prediction of the HMME–gated mixture of experts algorithm are almost identical. *Bottom:* The upper diagram shows the activation of one expert of the HMME–system, the lower diagram shows the activation of the corresponding expert of an expert system trained by the ACE–algorithm (without gating). The HMME detects the switching to the other wring much earlier and follows the drift process instantaneously. The non–gated expert system only performs a sudden switch, which is late compared to the real dynamical process.

mation, and HMM state information (memory). In particular, this allows for a fast detection of change points in on–line scenarios. Thereby, it can improve the prediction performance significantly. We expect that the method will be useful for the prediction of a wide range of natural signals, as e.g. climatologic or financial data.

Acknowledgement: We acknowledge support of the Deutsche Forschungsgemeinschaft (grants Pa569/2–1 and Ja379/51).

REFERENCES

[1] Y. Bengio and P. Frasconi, "An input output HMM architecture," in **Advances in Neural Information Processing Systems 7: NIPS 1994**, MIT Press, pp. 427–434, 1995.

[2] T. W. Cacciatore and S. J. Nowlan, "Mixtures of controllers for jump linear and non–linear plants," in **Advances in Neural Information Processing Systems 6: NIPS 1993**, MIT Press, pp. 719–726, 1994.

[3] R. A. Jacobs, M. I. Jordan, S. J. Nowlan and G. E. Hinton, "Adaptive mixtures of local experts," **Neural Computation**, vol. 3, pp. 79–87, 1991.

[4] A. Kehagias and V. Petridis, "Time series segmentation using predictive modular neural networks," **Neural Computation**, vol. 9, pp. 1691–1710, 1997.

[5] N. Lorenz, "Deterministic non–periodic flow," **J. Atmos. Sci.**, vol. 20, pp. 130, 1963.

[6] J. Moody and C. J. Darken, "Fast learning in networks of locally–tuned processing units," **Neural Computation**, vol. 1, pp. 281–294, 1989.

[7] K. Pawelzik, J. Kohlmorgen and K.-R. Müller, "Annealed competition of experts for a segmentation and classification of switching dynamics," **Neural Computation**, vol. 8, pp. 340–356, 1996.

[8] L. R. Rabiner, "A tutorial on hidden markov models and selected applications in speech recognition," in A. Waibel and K. Lee, eds., **Readings in Speech Recognition**, Morgan Kaufmann, pp. 267–296, 1988.

[9] S. Shi and A. S. Weigend, "Taking time seriously: Hidden markov experts applied to financial engineering," **Proceedings of the IEEE/IAFE Conference on Computational Intelligence for Financial Engineering**, pp. 244–252, 1997.

[10] A. S. Weigend, M. Mangeas and A. N. Srivastava, "Non–linear gated experts for time series: discovering regimes and avoid overfitting," **International Journal of Neural Systems**, vol. 6, pp. 373–399, 1995.

A COMPARISON AMONG FEATURE SELECTION METHODS BASED ON TRAINED NETWORKS

Mercedes Fernández Redondo, Carlos Hernández Espinosa.
Universidad Jaume-I. Departamento de Informática.
Campus Riu Sec. Edificio TI. Castellón. Spain.
Phone: 964728271 Fax: 964728435
E-mail: espinosa@inf.uji.es

Abstract. In this paper, we present a review of feature selection methods, based on the analysis of a trained multilayer feedforward network, which have been applied to neural networks. Furthermore, a methodology that allows evaluating and comparing feature selection methods is carefully described. This methodology is applied to the 19 reviewed methods in a total of 15 different real world classification problems. We present an ordination of methods according to its performance and it is clearly concluded which method performs better and should be used. We also discuss the applicability and computational complexity of the methods.

INTRODUCTION

Neural networks (NNs) are used in many different real world applications, in these problems one can usually measure a large number of variables that can be used as potential inputs. One clear example is the extraction of features for object recognition [1], many different types of features can be utilized, for example: geometric features, morphological, etc. However, usually not all variables that can be collected are equally informative: they may be noisy, irrelevant or redundant.

Feature selection is the problem of choosing a small subset of features ideally necessary and sufficient to perform the classification task, from a larger set of candidate features. Feature selection has long been one of the most important topics in pattern recognition and it is also an important issue in NNs. If one could select a subset of variables one could reduce the size of the NN, the amount of data to process, the training time, and possibly increase the generalization performance. This last result is known in the bibliography and ratified in our results.

Feature selection is also a complex problem, we need a criterion to measure the importance of a subset of variables and that criterion depends on the classifier. A subset of variables can be optimal for one system and very inefficient for another. In the bibliography there are several potential ways to determine the best subset of features: analyze all subsets, genetic algorithms, a heuristic stepwise analysis and direct estimations.

In the case of NNs direct estimation methods are preferred because of the computational complexity of training a NN. Inside this category we can perform another classification: methods based on the analysis of the training set, [2],

0-7803-5673-X/99/$10.00

methods based on the analysis of a trained multilayer feedforward network [1], [3-16], and methods based on the analysis of other specific architectures [18].
The purpose of this paper is to make a brief review of the methods based on the analysis of a trained multilayer feedforward network and present the first empirical comparison among them.
In the next section we will briefly review 19 different methods, in section 3 we present the comparison methodology, the experimental results and an ordination of the methods according to its performance and we finally conclude in section 4.

THEORY

Many methods based on the analysis of a trained multilayer feedforward network try to define what it is called the relevance of an input unit S_i, one input I_i is considered more important if its relevance S_i is larger. They also define the relevance s_{ij} of a weight w_{ij} connected between the input unit i and the hidden unit j. The relation between S_i and s_{ij} is in equation (1), where Nh is the number of hidden units.

$$S_i = \sum_{j=1}^{Nh} s_{ij} \tag{1}$$

The criteria for defining weight relevance are varied. Some of them are based on direct weight magnitude. For instance, the criterion proposed by Belue [3] (from here named BL2) is in equation (2).

$$s_{ij} = (w_{ij})^2 \tag{2}$$

And the one proposed by Tetko [9] (from here named TEKA) is in equation (3).
These criteria are based on the heuristic principle that, as a result of the learning process, the weights of an important input should have a larger magnitude than other weights connected to a useless and may be random input.

$$s_{ij} = \left| w_{ij} \right| \tag{3}$$

Other criteria of weight relevance are based on an estimation of the change in the m.s.e. (mean square error), E, when setting the weight to 0, this estimation is calculated by using the hessian matrix H, as a result of considering the Taylor expansion of the m.s.e., E, with respect to the weight. One example is the method proposed by Cibas [8] which is in equation (4) (from here CIB) where we denote by w_k the weight w_{ij}.

$$s_k = s_{ij} = \frac{1}{2} \cdot h_{kk} \cdot w_k^2 \tag{4}$$

In the above expression h_{kk} is a diagonal element of hessian matrix H.
And the criterion proposed by Tetko [9] (from here TEKE) in equation (5).

$$s_k = s_{ij} = \frac{w_k^2}{2 \cdot [H^{-1}]_{kk}} \tag{5}$$

The hessian matrix can be exactly calculated with the algorithm and expressions described in [17].

Another method of estimating weight relevance is based on an estimation of the change in E when setting w_{ij} equal to 0, but it does not use the hessian matrix It was proposed by Tetko [9] (named TEKC from here) and the value of weight relevance is in equation (6). Where the sum over t represents a sum for all iteration steps of the training process from the initialization t=0, until the iteration of convergence, w^i_{ij} is the initial value of weight w_{ij} and w^f_{ij} is the value of that weight at the iteration of convergence, $\Delta w_{ij}(t)$ is the change in the weight calculated by using the learning algorithm at the iteration t. In order to apply the method, we should keep a record of the appropriate information during the learning process for calculating the weight relevance.

$$s_{ij} = \sum_{t=0} \frac{\partial E}{\partial w_{ij}}(t) \cdot \Delta w_{ij}(t) \cdot \frac{w_{ij}(t)}{w^i_{ij} - w^f_{ij}} \qquad \textbf{(6)}$$

Other methods define the relevance S_i of input i by a calculation related to the variance of weights w_{ij} of input i, they are based on the heuristic that an input unit with small variance will behave as a threshold and therefore have little importance. One example is the criterion defined by Devena [1] (from here DEV) which is in equation (7).

$$S_i = \mathrm{var}_i = \sum_j \left(w_{ij} - \left(\frac{\sum_j w_{ij}}{Nh} \right) \right)^2 \qquad \textbf{(7)}$$

Another example is the criterion proposed by Deredy [10] (from here DER3) in equation (8).

$$S_i = \frac{Nh \cdot \mathrm{var}_i}{\sum_j w_{ij}} \qquad \textbf{(8)}$$

Another way to define the relevance of input i S_i is by using the sensitivity of outputs o_j with respect to the input I_i. It is based on the heuristic that a higher sensitivity means a larger variation in the output values with respect to a change in the input, and therefore we can suppose that the input is more important. For example, Belue [3] (from here BL1) uses the definition in equation (9). Where S if the training set, D is a set of points in the input space equally distributed along the range of variability of inputs, *No* is the number of outputs and N the sum of cardinals of S and D.

$$S_i = \frac{1}{N} \cdot \sum_{x \in S} \sum_{j=1}^{No} \sum_{x \in D} \left| \frac{\partial o_j}{\partial I_i}(x, w) \right| \qquad \textbf{(9)}$$

A similar method was proposed by Cloete [4] (from here named CLO) and it is in equations (10) and (11), where *Ns* is the number of training samples.

$$S_i = max(A_{ij})\ \forall j \tag{10}$$

$$A_{ij} = \sqrt{\frac{\sum_{x \in S}\left(\frac{\partial o_j}{\partial I_i}(x,w)\right)^2}{Ns}} \tag{11}$$

Priddy [5] also proposed a method (from here called PRI) based on sensitivities and described in equation (12).

$$S_i = \sum_{x \in S}\sum_{j=1}^{No}\sum_{x \in D}\left|\sum_{k \neq j}\frac{\partial o_k}{\partial I_i}(x,w)\right| \tag{12}$$

This method tries to estimate the variation of the probability of classification error with respect to the variation of the input I_i. See the reference for more details.
The method proposed by Sano [16] (from here named SAN) also uses sensitivities, see equation (13).

$$D(k,i) = max\left\{\left|\frac{\partial o_k}{\partial I_i}(x,w)\right| \quad \forall x \in S\right\} \tag{13}$$

This method gives a matrix of sensitivities D(k,i) with k=1, ..., Number of outputs, and i=1, ..., Number of inputs, and the relevance S_i of input i is considered to be larger than the relevance S_j of input j if D(k,i)>D(k,j) for a number of values of k greater than No/2.
And, finally Deredy [10] proposes the use of logarithmic sensitivities (from here named DER2), equations (14) and (15), where t_k is the target of output o_k. The purpose of this logarithmic sensitivity is to avoid the saturation term $o_k \cdot (1-o_k)$, where o_k is an output value, which appears in the calculation of the other sensitivities.

$$S_i = max(B_{ik})\ \forall k \tag{14}$$

$$B_{ik} = \frac{\partial \ln(|t_k - o_k|)}{\partial \ln I_i} \tag{15}$$

Some other methods try to evaluate the contribution of one input to the output taking into account the values of all weights in the NN structure. For example, equation (16) proposed by Tetko [9] (from here we call this method TEKB).

$$S_i^s = \sum_{j=1}^{M}\left(\frac{w_{ij}}{max|w_{aj}|\ \forall a}\right)^2 \cdot S_j^{s+1} \tag{16}$$

This equation tries to estimate the overall importance S_i^s of unit i in layer s over the units in the next layer s+1, w_{ij} is the weight between unit i in layer s and unit j in layer s+1, M is the number of units in layer s+1. The equation is recursive, we

set S_j equal to 1 for all outputs and calculate S_i for all hidden units, applying the equation again we calculate the input relevance.
An similar method also proposed by Tetko [7] (named TEK from here) is based on the equation (17), where $E[a_k^s]$ is the mean value of the output of unit k in layer s. We also set S_j equal to one for the outputs and recursively calculate S_i for the inputs.

$$S_i^s = \sum_{j=1}^{M} \frac{(w_{ij})^2 \cdot E\left[a_i^s\right]^2}{max\left\{(w_{kj})^2 \cdot E\left[a_k^s\right]^2\right\} \ \forall k} \cdot S_j^{s+1} \quad \textbf{(17)}$$

Another method proposed by Mao [12] (named MAO from here) calculates an estimation of the m.s.e. increase when deleting input i, the estimation is made again by a Taylor expansion of the m.s.e. with respect to the value of the input I_i. That value is used as the relevance S_i of the input. The equations are (18) and (19), where the sum of equation (18) is for all patterns in the training set N_S.

$$S_i = \sum_{k=1}^{Ns} \Delta E_k(I_i) \quad \textbf{(18)}$$

$$\Delta E_k(I_i) = \frac{\partial E_k}{\partial I_i} \cdot \Delta I_i + \frac{1}{2} \cdot \frac{\partial^2 E_k}{\partial I_i^2} \cdot (\Delta I_i)^2 \quad \textbf{(19)}$$

In equation (19) ΔI_i should be $0 - I_i$ (which is the effect of setting I_i equal to 0). The derivatives can be calculated recursively. For the output units we can use equation (20).

$$\frac{\partial E_k}{\partial o_i} = o_i - t_i \qquad \frac{\partial^2 E_k}{\partial o_i^2} = 1 \quad \textbf{(20)}$$

And the relationships between the derivatives of two layers l+1 and l are in equations (21) and (22). In these equations g denotes the sigmoid function and its first and second derivatives are in equation (23).

$$\frac{\partial E_k}{\partial y_i^l} = \sum_{j=1}^{N(l+1)} \frac{\partial E_k}{\partial y_j^{l+1}} \cdot g' \cdot w_{ij} \quad \textbf{(21)}$$

$$\frac{\partial^2 E_k}{\partial y_i^{l2}} = \sum_{j=1}^{N(l+1)} \frac{\partial^2 E_k}{\partial y_j^{(l+1)2}} \cdot (g' \cdot w_{ij})^2 + \sum_{j=1}^{N(l+1)} \frac{\partial E_k}{\partial y_j^{l+1}} \cdot g'' \cdot w_{ij}^2 \quad \textbf{(22)}$$

$$g' = y_j^{l+1} \cdot (1 - y_j^{l+1}) \qquad g'' = y_j^{l+1} \cdot (1 - y_j^{l+1}) \cdot (1 - 2 \cdot y_j^{l+1}) \quad \textbf{(23)}$$

There are two very simple methods that calculate the effect of substituting an input by its mean value in the training set. They are based on the heuristic that if this substitution has little effect in the performance the input nearly behaves as a threshold and has little importance.

The first one proposed by Lee [6] (called LEE in the paper), calculates the percentage correct in the test set, with one input substituted by its mean value. The input is considered more relevant if the value of the tested percentage is lower, because the performance decrease is larger.
The second one proposed by Utans [13] (called UTA in the paper), focus on the m.s.e., E, and calculates its increment when substituting an input for its mean value. One input is considered more relevant if the increment of E is higher.
Bowles [15] proposed another method (called BOW from here) that should also keep an information record of the training process. The relevance of one input S_i is defined as the sum of equation (24) over all iteration steps until the convergence point T. Where w_{ij} is the weight between input i and hidden unit j, and δ_j is the backpropagated error of hidden unit j.

$$S_i = \sum_{t=0}^{T} \left| \sum_{j=1}^{Nh} \delta_j \cdot w_{ij} \right| \tag{24}$$

Finally, Younes [14] proposed another method that we have used in the comparison (we call it YOU), but we will not describe it because it is rather complex, its computational cost is high and the applicability limited, we got division by zero errors in 6 of 15 problems.
It is very important to point out that every method reviewed allows obtaining an ordination of inputs or features according to its importance or relevance. Obviously, the ordination of two methods will not be, in general, the same and therefore its performance will also be different.
Furthermore, we can get an ordination of inputs and we will know which inputs should be first discarded from the training set (the least important ones), but there is not simple and efficient procedure to know the cardinal, k, of the final subset of inputs. We do not know the optimal number of inputs that should be kept in the training set.
As we saw before, every method is based on heuristic principles and the only way to compare them may be empirically because of the complexity of the problem. This will be described and accomplished in the following section.

EXPERIMENTAL RESULTS

In order to compare the 19 methods, we have applied them to 15 different classification problems, which are from the UCI repository of machine learning databases. They are: Abalone (AB), Balance Scale (BL), Cylinder Bands (BN), Liver Disorders (BU), Credit Approval (CR), Display 1 (DI), Glass identification (GL), Heart Disease (HE), Mushroom (LE), the Mok's Problems (M1, M2, M3), Pima Indians Diabetes (PI), Voting Records (VO) and Wisconsin Breast Cancer (WD). The complete data and a full description of the problems can be found in the repository.
In all problems, we have included a first useless input generated at random inside [0,1]. It is interesting to see how important this input is considered by every method.

Furthermore, we have normalized the range of variability of every input to the interval [0,1], this is important because the range of variability of an input influences the magnitude of the weights connected to the input and almost all the relevance measurements described in the theory section.
In order to apply the methods we should use al least one neural network trained for each problem. Well, we have trained 30 different neural networks (with different initialization weights) for every problem, we have applied the methods to the 30 neural networks to obtain 30 relevance measurements S_i for each input and problem. And we have obtained a final value of relevance S_i' by averaging the 30 values of S_i.
We have followed this procedure because we wanted to obtain what we can call a general performance of the method and avoid results of the relevance biased by a concrete neural network, which obviously depends on the initialization weights.
Furthermore, the number of hidden units of the neural network for each problem was carefully obtained before training the 30 neural networks by a trial and error procedure.
Following the above methodology, for each problem and method we have obtained an ordination of feature importance according to its relevance value. For example, the ordination of all methods for problem PI is in Table 1.

TABLE 1. INPUT IMPORTANCE FOR PROBLEM PI AND ALL METHODS.

Method	Least Important								Most
BL1	1	6	5	9	8	2	4	3	7
BL2	5	6	1	8	4	2	9	7	3
CLO	1	6	5	9	8	2	4	3	7
PRI	1	6	5	9	8	2	4	3	7
DEV	5	6	8	1	4	2	9	3	7
LEE	3	7	8	5	4	9	6	2	1
TEK	6	5	8	2	9	1	4	7	3
SAN	1	6	5	8	9	2	4	3	7
BOW	6	9	5	8	2	4	1	7	3
TEKC	6	8	2	9	7	4	3	5	1

Method	Least Important								Most
TEKA	5	6	1	8	4	2	9	7	3
TEKB	6	5	1	8	4	2	9	3	7
DER2	1	6	9	5	8	2	4	3	7
DER3	4	2	9	3	8	7	6	1	5
MAO	6	5	8	2	1	9	4	7	3
UTA	6	1	5	8	9	4	2	7	3
CIB	6	5	8	1	2	9	4	7	3
TEKE	3	4	7	2	8	9	5	6	1
YOU	6	5	8	9	2	1	4	7	3

After that, we obtained several inputs subsets by removing successively the least important input until a final subset of one input. For example, using the results of Table 1 for method UTA, the first subset is obtained by deleting input {6}, the following subset is obtained by deleting inputs {6,1} and the final subset of one input is {3}.
For every subset we wanted to obtain the performance of a classifier to see how good the subset is, the classifier of interest is Multilayer Feedforward. We trained several multilayer feedforward networks to get a mean performance independent of initial conditions (initial weights), and also an error for the mean by using standard error theory.
The performance criterion was the percentage correct in the test set. The number of trained neural networks for each subset was a minimum of ten, in many cases we trained much more than ten neural networks in order to diminish the error in the mean, the maximum permitted error in the measurement was 3%. This value is an appropriate one resulting of a tradeoff between precision and computational cost.

In table 2, we can see the results of all methods for problem PI.

TABLE 2. PERFORMANCE IN PROBLEM PI FOR ALL SUBSETS AND METHODS.

	Number of Omitted Inputs							
	1	2	3	4	5	6	7	8
BL1	75.5±0.3	75.7±0.2	76.8±0.2	77.0±0.2	76.6±0.1	75.5±0.1	75.8±0.1	63.3±0.3
BL2	74.8±1.1	76.1±0.2	76.8±0.2	76.2±0.2	77.2±0.2	77.0±0.3	75.8±0.1	74.4±0
CLO	75.5±0.3	75.7±0.3	76.8±0.2	77.0±0.2	76.6±0.1	75.5±0.1	75.8±0.1	63.3±0.3
PRI	75.5±0.3	75.7±0.4	76.8±0.2	77.0±0.2	76.6±0.1	75.5±0.1	75.8±0.1	63.3±0.3
DEV	74.8±1.0	76.1±0.2	76.2±0.2	76.2±0.2	77.2±0.2	77.0±0.3	75.8±0.1	63.3±0.3
LEE	66.3±0.6	69.7±0.7	66.9±0.5	67.9±0.7	67.8±0.5	66.9±0.3	64.4±0.1	65.2±0
TEK	75.2±0.2	76.1±0.2	76.2±0.2	73.6±1.4	75.1±0.3	75.5±0.1	75.8±0.1	74.4±0
SAN	75.5±0.3	75.7±0.2	76.8±0.2	76.2±0.2	76.7±0.1	75.5±0.1	75.8±0.1	63.3±0.3
BOW	75.2±0.2	76.0±0.1	73.7±1.4	76.5±0.1	75.1±0.3	73.4±1.3	75.8±0.1	74.4±0
TEKC	75.2±0.2	75.1±0.3	74.8±0.4	75.5±0.3	74.6±0.3	73.7±1.0	65.0±0.1	65.2±0
TEKA	74.8±1.1	76.1±0.2	76.8±0.2	76.2±0.2	77.2±0.2	77.0±0.3	75.8±0.1	74.4±0
TEKB	75.2±0.2	76.1±0.2	76.8±0.2	76.2±0.2	77.2±0.2	77.0±0.3	75.8±0.1	63.3±0.3
DER2	75.5±0.3	75.7±0.2	74.4±1.1	77.0±0.2	76.6±0.1	75.5±0.1	75.8±0.1	63.3±0.3
DER3	76.3±0.3	75.4±0.1	75.8±0.2	64.8±0.5	60.4±0.2	65.9±0.2	65.0±0.1	65.6±0.1
MAO	75.2±0.2	76.1±0.2	76.2±0.2	73.6±1.4	75.7±0.4	75.5±0.1	75.8±0.1	74.4±0
UTA	75.2±0.2	75.7±0.2	76.8±0.2	76.2±0.2	76.6±0.1	77.3±0.2	74.8±0.1	74.4±0
CIB	75.2±0.2	76.1±0.2	76.2±0.2	76.2±0.2	75.7±0.4	75.5±0.1	75.8±0.1	74.4±0
TEKE	66.3±0.6	65.3±0.7	70.8±0.4	70.8±0.9	69.1±0.8	65.9±0.2	66.1±0.1	65.2±0
YOU	75.2±0.2	76.1±0.2	76.2±0.2	76.5±0.1	75.1±0.3	75.5±0.1	75.7±0.1	74.4±0

Then, for each method and problem we can obtain what we call the optimal subset. This subset is the one which provides the best performance, and in the case of two subsets with indistinguishable performance, the one with a lower number of inputs, because it provides a simpler neural network model.

In order to see if the performances are distinguishable we have performed t-tests. The hypothesis tested was the null hypothesis $\mu_A=\mu_B$, assuming that the two mean performances are indistinguishable. In the case that this null hypothesis can be rejected we conclude that the difference between the two measurements is significant. The significance level of test, α, was 0.1.

For example, from the results of Table 2, the performance of method UTA for the subset with 6 omitted inputs is distinguishable of the rest and it is the best, so it is the optimal subset. Also for CIB there are three subsets with best performance and indistinguishable among them, subsets of 2, 3 and 4 omitted inputs. The optimal subset is the one of 4 omitted inputs because it has a lower number of inputs.

After obtaining the optimal subsets for each method and problem, we can compare the performance of different methods in the same problem by comparing the performance of their optimal subsets. Again, we performed t-tests to see significant differences.

For example, the results for method UTA and CIB in Table 2 are distinguishable and we can conclude that the performance of method UTA is better for the problem PI.

By comparing all the methods, two by two, we can obtain another table where we can find whether one method is better or worse than another for a concrete problem. An extract of this table is in Table 3.

TABLE 3. PERFORMANCE COMPARISON OF UTA AND CIB.

	UTA
CIB	Best UTA: AB,BN,BU,CR,GL,VO,LE,PI,WD
	Equal: BL,DI,M1,M2,M3
	Best CIB: HE

We can see that method UTA is better than CIB in a larger number of problems and conclude that method UTA performs better than CIB.
Following this methodology and this type of comparison with the full results, we can get the following ordination:

UTA > TEKA > BL2 > DEV > TEKB =DER2 = MAO > CLO > BL1 = = BOW = YOU > PRI = SAN > CIB =TEKE > TEK > TEKC >DER3 >LEE

The best method is UTA and the worst is LEE. We can further discuss the applicability of every method. The unique method with limited applicability was YOU as commented in the theory section.
Another important question is the computational complexity. The methods with a higher computational cost were CIB and TEKE because of the calculation of the hessian matrix and also YOU which requires an iteration over all samples of the training set and the calculation of two integrals for each iteration.

CONCLUSIONS

We have presented a review of the feature selection methods based on an analysis of a trained multilayer feedforward network, which have been applied to neural networks. We have also carefully presented a methodology that allows selecting an optimal subset, evaluating and comparing feature selection methods. This methodology was applied to the 19 reviewed methods in a total of 15 different real world classification problems. We presented an ordination of methods according to its performance and it was clearly concluded which method performs better and should be used. We have also discussed the applicability and computational complexity of the methods.

REFERENCES

[1] L. Devena, "Automatic selection of the most relevant features to recognize objects". **Proc. of the Int. Conf. on Artificial NNs**, vol.2, pp.1113-1116, 1994.

[2] R. Battiti, "Using mutual information for selecting features in supervised neural net learning". **IEEE Trans. on Neural Networks**, vol. 5, n. 4, pp. 537-550, 1994.

[3] L.M. Belue and K.W. Bauer, "Determining input features for multilayer perceptrons," **Neurocomputing**, vol. 7, n. 2, pp. 111-121, 1995.
[4] A.P. Engelbrecht and I. Cloete, "A sensitivity analysis algorithm for pruning feedforward neural networks", **Proceedings of the International Conference on Neural Networks**, vol. 2, pp. 1274-1277, 1996.
[5] K.L. Priddy, S.K. Rogers, D.W. Ruck, G.L. Tarr and M. Kabrisky, "Bayesian selection of important features for feedforward neural networks", **Neurocomputing**, vol. 5, n. 2&3, pp. 91-103, 1993.
[6] H. Lee, K. Mehrotra, C. Mohan and S. Ranka, "Selection procedures for redundant inputs in neural networks", **Proceedings of the World Congress on Neural Networks**, vol. 1, pp. 300-303, 1993.
[7] I.V. Tetko, V.Y. Tanchuk and A.I. Luik, "Simple heuristic methods for input parameter estimation in neural networks", **Proceedings of the IEEE International Conference on Neural Networks**, vol. 1, pp. 376-380, 1994.
[8] T. Cibas, F.F. Soulié, P. Gallinari and S. Raudys, "Variable selection with neural networks", **Neurocomputing**, vol. 12, pp. 223-248, 1996.
[9] I.V. Tetko, A.E.P. Villa and D.J. Livingstone, "Neural network studies. 2. Variable selection", **Journal of Chemical Information and Computer Sciences**, vol. 36, n. 4, pp. 794-803, 1996.
[10] W. El-Deredy and N.M. Branston, "Identification of relevant features in HMR tumor spectra using neural networks", **Proceedings of the 4th International Conference on Artificial Neural Networks**, pp. 454-458, 1995.
[11] J.M. Steppe and K.W. Bauer, "Improved feature screening in feedforward neural networks", **Neurocomputing**, vol. 13, pp. 47-58, 1996.
[12] J. Mao, K. Mohiuddin and A.K. Jain, "Parsimonious network design and feature selection through node pruning", **Proceedings of the 12th IAPR International Conference on Pattern Recognition**, vol. 2, pp. 622-624, 1994.
[13] J. Utans, J. Moody and S. Rehfuss, "Input variable selection for neural networks: Application to predicting the U.S. business cycle", **Proceedings of IEEE/IAFE 1995 Computational Intelligence for Financial Engineering.**, pp. 118-122, 1995.
[14] B. Younes and B. Fabrice, "A neural network based variable selector", **Proceedings of the Artificial Neural Network in Engineering,(ANNIE'95)**, pp. 425-430, 1995.
[15] A. Bowles, "Machine learns which features to select", **Proceedings of the 5th Australian Joint Conference on Artificial Intelligence**, pp. 127-132, 1992.
[16] H. Sano, A. Nada, Y. Iwahori and N. Ishii, "A method of analyzing information represented in neural networks", **Proceedings of 1993 International Joint Conference on Neural Networks**, pp. 2719-2722, 1993.
[17] C. Bishop, "Exact calculation of the hessian matrix for the multilayer perceptron", **Neural Computation**, vol. 4, pp. 494-501, 1992.
[18] R. Watzel, A. Meyer-Bäse, U. Meyer-Bäse and H. Hilberg, "Identification of irrelevant features in phoneme recognition with radial basis classifiers", **Proceedings of 1994 International Symposium on Artificial Neural Networks**, pp. 507-512, 1994.

SIZING OF THE MULTILAYER PERCEPTRON VIA MODULAR NETWORKS

Hema Chandrasekaran[1], Kyung K. Kim[2], and Michael T. Manry[3]
[1,3]Department of Electrical Engineering
The University of Texas at Arlington
Arlington, TX 76019
e-mail : manry@uta.edu
[2]Lockheed Martin Tactical Aircraft Systems, Mail Zone 1772
P.O. Box 748, Fort Worth, Texas 76101

Abstract

A fast method for sizing the multilayer perceptron is proposed. The principal assumption is that a modular network with the same theoretical pattern storage as the multilayer perceptron has the same training error. This assumption is analyzed for the case of random patterns. Using several benchmark datasets, the validity of the approach is demonstrated.

I. INTRODUCTION

Multilayer perceptrons (MLPs) with N_h nonlinear units in a single hidden layer have been established as universal function approximators [1,2]. Many investigators have proposed methods for choosing N_h, a process known as sizing. These methods include the leave-one-out technique, growing methods [3,4], pruning methods [5,6], and approaches based upon Akaike's information criterion [7,8]. Unfortunately, these approaches are very time consuming. Unlike in the leave-one-out and Akaike approaches, our goal is to predict MLP training errors for a wide range of N_h values, in less time than that required to train a single MLP.

It has been observed that different kinds of nonlinear networks, having the same theoretical pattern storage (or pattern memorization), produce very similar values of the training error E [9]. In order to verify the observation with networks having many free parameters, however, very efficient training methods are required. In our paper, we develop a fast sizing approach based upon the observation. In section II, we summarize a convergent technique for the fast design of piecewise linear (PWL) modular networks. In section III, we relate the pattern storage of the PWL and MLP networks, and describe the resulting sizing algorithm. In section IV, we analyze the bounds on the training error versus pattern storage for both PWL and MLP for the case of random training patterns. In section V, we demonstrate the algorithm on several well-known benchmark data sets.

0-7803-5673-X/99/$10.00 © 1999 IEEE

II. PIECEWISE LINEAR MODULAR NETWORKS

The PWL modular network for N inputs and M outputs contains N_{mod} modules, each of which has an N-dimensional cluster center vector $\boldsymbol{m}_n$ and a matrix $\mathbf{W}_{pwl}(n)$ of dimension M by (N+1). The network employs a weighted sum of squares distance measure d(·). If the input vector $\boldsymbol{x}$ belongs to the nth cluster such that $d(\boldsymbol{x}, \boldsymbol{m}_n) = \min_k d(\boldsymbol{x}, \boldsymbol{m}_k)$ then the output vector is $\boldsymbol{y} = \mathbf{W}_{pwl}(n) \cdot (\boldsymbol{x}^T : 1)^T$. In this section, we summarize a PWL modular network training algorithm and give the network's pattern storage.

A. Training Algorithm

A convergent training algorithm for the PWL modular network [9] is described in the following.

(1) Weights of the distance measure are found which de-emphasize less useful inputs, following the approach of [10].
(2) We initialize the network with one cluster centered at the mean vector of input vectors. Two new cluster centers are found by perturbing this mean and reassigning patterns. A linear mapping is designed for each module or cluster through regression.
(3) We add a new cluster by splitting the cluster contributing most to the mapping error. Again, a linear mapping is designed for each module or cluster through regression.
(4) In each iteration, the number of clusters increases at most by one, till the specified number of clusters have been generated.

B. Convergence

Note that each regression step above will tend to reduce the network training error or leave it the same. However, it is possible for a new cluster to attract patterns from nearby clusters, which are not well approximated by a linear mapping, resulting in an increase in the network training error. If this happens, we delete the new module and pick a new cluster to split. Whenever the mapping error decreases, we store the network configuration. If we fail to achieve a lower mean square mapping error after making exhaustive attempts to split clusters, we backtrack the network to the previous optimum configuration and stop. The design algorithm clearly converges. The pattern storage of the PWL modular network is N_{mod} multiplied by the storage per module,

$$S_{PWL} = N_{\text{mod}} \cdot (N + 1) \tag{1}$$

Although the PWL modular network trains well, the resulting mapping is discontinuous at the boundaries between adjacent modules. When a discontinuous mapping is unacceptable, one may use a global type network such as the MLP.

III. SIZING ALGORITHM

In this section we describe a sizing algorithm for a fully connected MLP, which includes bypass weights, thresholds in the hidden layer, and thresholds in the output layer.

A. MLP Pattern Storage

It is well known that the Volterra filter can memorize a number of training patterns equal to the number of coefficients associated with one output. The upper bound on the MLP's pattern storage is the same [11]. Its pattern storage, S_{MLP}, is bounded above by P_{MLP}/M, where P_{MLP} is the total number of free parameters in the network. It has been shown [11] that this bound is fairly tight. Therefore we assume

$$S_{MLP}(N_h) = \left(\frac{(N+1+M)}{M}\right) \cdot N_h + (N+1) \tag{2}$$

$$N_h(S_{MLP}) = \left(\frac{M}{N+1+M}\right) \cdot S_{MLP} - \frac{(N+1) \cdot M}{N+1+M} \tag{3}$$

B. Fundamental Assumption

The fundamental assumption we make is that the PWL modular network and MLP have the same error performance if their respective theoretical pattern storages are the same. This assumption is certainly true in the following three situations:

- If the networks' outputs are constant for all training patterns, then both the PWL modular network and MLP have the same error performance and they both memorize 1 pattern only.
- If the networks' outputs are linear functions of the inputs (as when the PWL modular network has one module and the MLP has only bypass weights and thresholds), both networks are linear, have the same error performance, and can memorize N+1 patterns.
- If the both the PWL modular network and MLP memorize all the training patterns, then they both have the same training error (zero).

Our observation is that at intermediate storage capacities, the assumption also usually holds. Note, however that mappings can be constructed which are counterexamples to our assumption.

C. Algorithm

Consider the PWL modular network trained for N_{it} iterations. Let $E_t(i)$ be the training error and $N_{mod}(i)$ be the number of modules in the ith iteration. The pattern storage $S_{PWL}(i)$ in the ith iteration is given by

$$S_{PWL}(i) = N_{\text{mod}}(i) \cdot (N+1) \tag{4}$$

Equating the pattern memorization of the MLP in equation (2) to that of the PWL modular network in equation (4), we have

$$N_h(i) \equiv N_h(S_{PWL}(i)) = \frac{M \cdot (N+1) \cdot (N_{\text{mod}}(i) - 1)}{N + M + 1} \tag{5}$$

This formula helps us to estimate the number of hidden units in an equivalent MLP for various $E_t(i)$, i = 1,2,...N_{it}.

Case $M = 1$: When the number of outputs $M = 1$, equation (5) approximately reduces to $N_h = (N_{mod} - 1)$.

Case $M > 1$: In this case, overly large MLP networks result when the desired outputs are correlated and many free parameters are redundant. Then, a singular value decomposition (SVD) technique is employed to detect whether outputs can be compressed without significantly degrading the MSE performance. Compressing the outputs allows the sizing algorithm to predict a smaller, less complex MLP. The resulting mean square error E_t' after we compress the M outputs down to M' is given by

$$E_t' = E_t + \sum_{i=M'+1}^{M} \lambda_i$$

where λ_i*'s* are the *i*th singular values of the desired outputs' covariance matrix. For each value of *i*, the sizing algorithm generates all possible MLP configurations and their predicted training MSE using M' values between 1 and M. The algorithm sifts through these potential configurations and saves those that have the lowest training MSE for a given number of hidden units. The final output of our algorithm is the sequence of ordered pairs, $\{E_t'(i), N_h(i)\}$.

IV. ERROR VERSUS PATTERN STORAGE

Our goal in this section is to show that a failure of our principal assumption, as depicted in figure 1, does not happen when the training patterns are chosen randomly (all elements of patterns are statistically independent). For the following discussion, assume there are N_v random training patterns having N-dimensional input vectors, M-dimensional desired output vectors t_p. We also assume that the outputs t_p and $-t_p$ are equally likely.

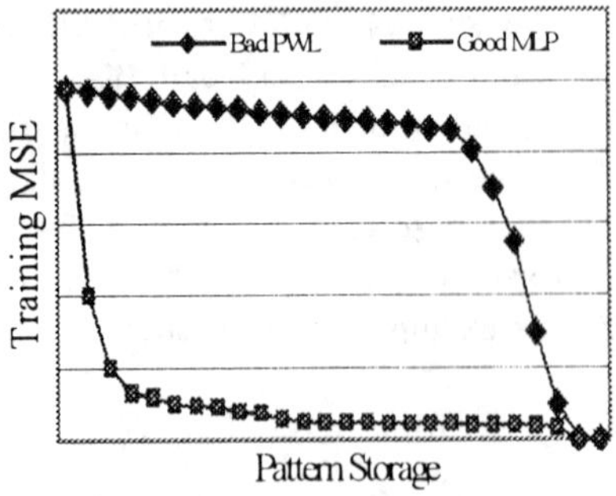

Figure 1 : Failure of Principal Assumption

A. Upper Bound on the MSE versus N_{mod} Curve

Each module in the PWL network memorizes (N+1) patterns perfectly. We assume (N_{mod}–1) modules memorize (N_{mod}–1)·(N+1) training patterns giving zero training error, leaving the rest of the training patterns to be handled by the last module. Let the output vector of the last module be $\boldsymbol{y} = E[\boldsymbol{t}_p]$ where $\boldsymbol{t}_p$ is the desired output vector. Here E[] denotes the expectation over all randomly chosen training sets, from a given joint probability density. Thus the total mean square training error E_t is given by

$$\begin{aligned} E_t &= \{N_v - (N_{\text{mod}} - 1)\cdot(N+1)\}\cdot M \cdot E\left[(t_p - E(t_p))^2\right] \\ &= \{N_v - (N_{\text{mod}} - 1)\cdot(N+1)\}\cdot M \cdot \sigma_t^2 \\ &= \{N_v + (N+1)\}\cdot M \cdot \sigma_t^2 - N_{\text{mod}}\cdot(N+1)\cdot M \cdot \sigma_t^2 \end{aligned} \qquad (6)$$

where $\sigma_t^2 = E\left[(t_p - E(t_p))^2\right]$ is the variance of each output and $N_{mod} \geq 1$. We see that equation (6) represents a straight line with a slope of $-(N+1)\cdot M\cdot\sigma_t^2$. We conclude that the linear case is an upper bound on the E_t versus N_{mod} and E_t versus S_{PWL} curves.

B. Upper Bound on the MSE versus N_h Curve

In this subsection we show that on the average the E versus N_h curve for the MLP has a linear upper bound, for the case of random patterns.

We assume that the MLP with N_h hidden units has completely memorized N_v training patterns and we have applied the Gram-Schmidt(GS) procedure to orthonormalize and order the output basis functions. Also, we assume that we have eliminated any linearly dependent basis functions (hidden units). Consider any two consecutive hidden units whose basis functions are numbered j and (j+1). Let the initial raw basis functions be $\sigma_1, \sigma_2, \sigma_3, \ldots \sigma_{Nu}$ where $N_u = N+1+N_h$, N is the input vector dimension, N_h is the number of hidden units, 1 for the threshold. Let the corresponding orthonormal basis functions after the GS procedure be $\varphi_1, \varphi_2, \varphi_3, \ldots \varphi_{Nu}$. Next, removing the effects of all the previous (j-1) basis functions from the rest of the basis functions we obtain

$x_k = \sigma_k - D_1\varphi_1 - D_2\varphi_2 - D_3\varphi_3 - D_4\varphi_4 \ldots\ldots\ldots - D_{j-1}\varphi_{j-1}$ where $j \leq k \leq N_u$,

$D_1 = \langle \varphi_1, \sigma_k \rangle$, $D_2 = \langle \varphi_2, \sigma_k \rangle$, ... $D_{j-1} = \langle \varphi_{j-1}, \sigma_k \rangle$. Here

$\langle \phi_l, \sigma_k \rangle = \frac{1}{N_v}\sum_{p=1}^{Nv} \phi_{lp} \cdot \sigma_{lp}$ following the definition of [15]. Now,

$t_i' = t_i - w_{i1}\varphi_1 - w_{i2}\varphi_2 - w_{i3}\varphi_3 - w_{i4}\varphi_4 \ldots\ldots\ldots - w_{i\,j-1}\varphi_{j-1}$ where $1 \leq i \leq M$

where $w_{i1}, w_{i2}, \ldots w_{i\,j-1}$ are the orthonormal weights connecting the basis functions to the ith output.

For the following derivation we consider only one output. Consider once again the basis functions j, (j+1). Without loss of generality, we refer to them as basis functions 1 and 2 from now on. Let the one output be t'. Let w_1, w_2 be the output weights for orthonormal basis functions 1 and 2. Define the following :

$$P_{11} = \langle x_1, x_1 \rangle = \frac{1}{N_v}\sum_{p=1}^{Nv} x_{1p} \cdot x_{1p}, \qquad P_{12} = \langle x_1, x_2 \rangle = \frac{1}{N_v}\sum_{p=1}^{Nv} x_{1p} \cdot x_{2p}$$
$$Q_1 = \langle x_1, t \rangle = \frac{1}{N_v}\sum_{p=1}^{Nv} x_{1p} \cdot t'_p, \qquad Q_2 = \langle x_2, t \rangle = \frac{1}{N_v}\sum_{p=1}^{Nv} x_{2p} \cdot t'_p \tag{7}$$

From modified Gram-Schmidt orthonormalization procedure we have the following :

$$x'_1 = x_1 / \sqrt{P_{11}}$$
$$x'_2 = \frac{x_2 - \langle x'_1, x_2 \rangle x'_1}{\left\| x_2 - \frac{P_{12}}{P_{11}} \cdot x_1 \right\|} = \frac{x_2 - \frac{P_{12}}{P_{11}} \cdot x_1}{\sqrt{P_{22} - \frac{P_{12}^2}{P_{11}}}} = \frac{P_{11}x_2 - P_{12}x_1}{\sqrt{P_{11}} \cdot \sqrt{P_{11}P_{22} - P_{12}^2}} \tag{8}$$

$$w_1 = \langle x'_1, t' \rangle = Q_1 / \sqrt{P_{11}}$$
$$w_2 = \langle x'_2, t' \rangle = \frac{P_{11}Q_2 - P_{12}Q_1}{\sqrt{P_{11}} \cdot \sqrt{P_{11}P_{22} - P_{12}^2}} \tag{9}$$

If we choose x_2 as the first basis function, then the corresponding orthonormal weight would be,

$$w'_1 = \langle x'_2, t' \rangle = Q_2 / \sqrt{P_{22}} \tag{10}$$

We state the following properties resulting from Schmidt procedure ordering and complete memorization:

1. All the basis functions we have are linearly independent; if not we can always eliminate those dependent hidden unit basis functions. This means that none of the orthonormal weights connecting the basis functions to the output are zero.
2. The training error E = 0 due to perfect memorization of the patterns.
3. $w_1^2 \geq w_1'^2$, a property of Schmidt procedure ordering. Therefore,
$$\frac{Q_1^2}{P_{11}} > \frac{Q_2^2}{P_{22}} \Rightarrow P_{11}P_{22}Q_1^2 > P_{11}^2Q_2^2$$
4. $P_{11}P_{22} > P_{12}^2$, since $(P_{11}P_{22} - P_{12}^2) = \| x_2 - \langle x_1', x_2 \rangle \|$, thus positive.

Consider $w_1^2 - w_2^2 = \left(P_{11}P_{22}Q_1^2 - P_{11}^2Q_2^2\right) + 2\left(P_{11}P_{12}Q_1Q_2 - P_{12}^2Q_1^2\right)$ (11)

The term $\left(P_{11}P_{22}Q_1^2 - P_{11}^2Q_1^2\right)$ is always positive because of property 3. However, we can not say whether the second term $\left(P_{11}P_{12}Q_1Q_2 - P_{12}^2Q_1^2\right)$ is positive or negative for a particular realization of the network. So, we consider the ensemble average of $\left(P_{11}P_{12}Q_1Q_2 - P_{12}^2Q_1^2\right)$, which may be written as

$$E\left[P_{11}P_{12}Q_1Q_2 - P_{12}^2Q_1^2\right]=$$

$$\sum_{j=1}^{Nv}\sum_{k=1}^{Nv}\sum_{m=1}^{Nv}\sum_{n=1}^{Nv} E\left[x_{1k}^2 x_{1m} x_{2m} x_{1n} t_n x_{2j} t_j\right] - \sum_{j=1}^{Nv}\sum_{k=1}^{Nv}\sum_{m=1}^{Nv}\sum_{n=1}^{Nv} E\left[x_{1k} x_{2k} x_{1j} x_{2j} x_{1m} t_m x_{1n} t_n\right] \quad (12)$$

Here the subscripts j,k,m, and n denote the pattern numbers in a given data set. The training patterns in a data set are assumed to be statistically independent. This is a valid assumption since we are considering completely independent random inputs and outputs.

We notice that the terms like $E[x_{1m}^i x_{2m}]$ are zero for every positive integer i and terms like $E[x_{1m}(x_{2m})^{2i} t_m^j]$ are zero for every nonnegative i and j for the following reason: The network performance is not affected if one or all of the basis functions and their corresponding output weights change sign simultaneously. Thus equation (12) now reduces to

$$E\left[P_{11}P_{12}Q_1Q_2 - P_{12}^2Q_1^2\right] = \sum_{j=1}^{Nv}\sum_{k=1}^{Nv} E\left[x_{1k}^2\right] E\left[x_{1j}^2 x_{2j}^2 t_j^2\right] \quad (13)$$

which is net positive. We have proved that in an ensemble sense $E\left[w_1^2 - w_2^2\right] > 0$. *This result is true even if we have multiple outputs.* Thus we have proved that the average E versus N_h curve for a multilayer perceptron and the E_t versus N_{mod} curve for PWL net lie below a linear curve.

Since the average E_t versus S_{PWL} curve and E versus S_{MLP} curves are below a linear curve, the worst case depicted in Figure 1 does not occur, on the average, for random patterns.

V. NUMERICAL RESULTS

A. Data Sets With One Output

X.tra : This training data set was created from a chaotic time series generated by the Mackey-Glass delay-difference equation[12] with $a = 0.2$, $b = 0.1$, and $\tau = 17$. The desired output is the sample at time T+6. The training data set consists of 4 inputs and 1 output and contains 2654 training patterns. The predicted and actual E versus N_h curves, illustrated in figure 2, show good performance by the sizing algorithm.

Hwang-t3 : The "hwang–t3" data set contains artificial data derived from a real valued function of two variables, used by J.N.Hwang and others to test non-parametric regression methods[13]. The function y is defined to be

$$y = 42.659\cdot\left(0.1 + x_1\cdot\left(0.05 + x_1^4 - 10\cdot x_1^2\cdot x_2^2 + 5\cdot x_2^4\right)\right)$$

The functions are all real-valued, and are defined on the two-dimensional domain [0,1] x [0,1]. There are 8170 training patterns with 2 inputs and 1 output. The predicted and actual E versus N_h curves, illustrated in figure 3, again show that the sizing algorithm is very effective.

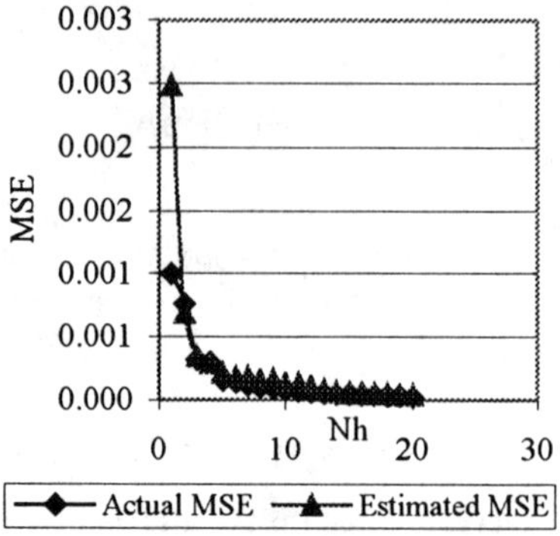

Figure 2 : MLP Sizing - x.tra

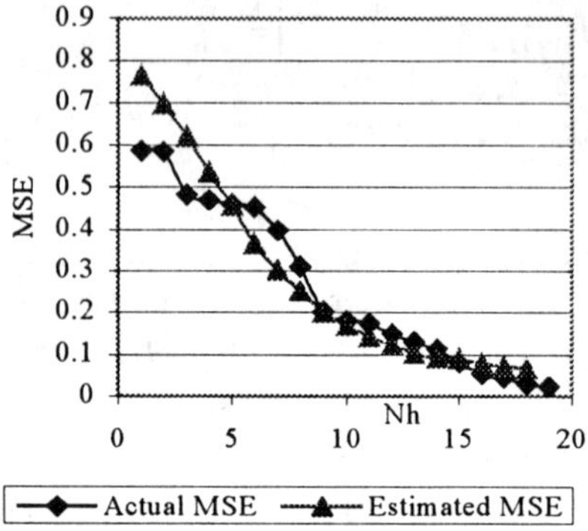

Figure 3 : MLP Sizing - hwang-t3.tra

Kin-8nm.tra : This data is part of the DELVE data set and contains 4918 training patterns. The data is artificial and describes forward kinematics of an 8 link robot arm . The data set is characterized by 8 inputs, 1 output, high non-linearity, and medium noise. For a given value of N_h in figure 4, the estimated training error is off by much less than an order of magnitude.

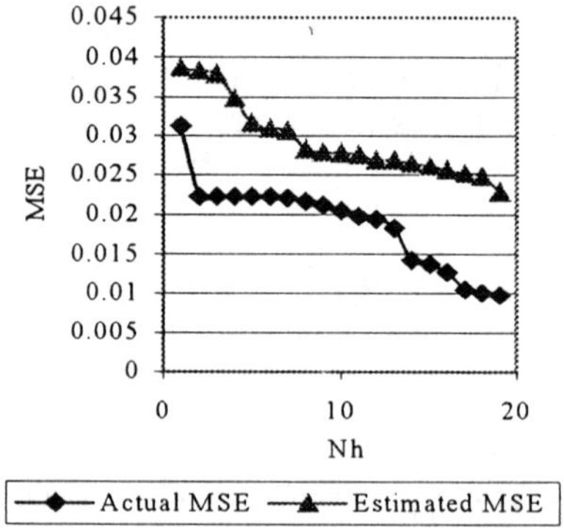

Figure 4 : MLP Sizing -kin-8nm.tra

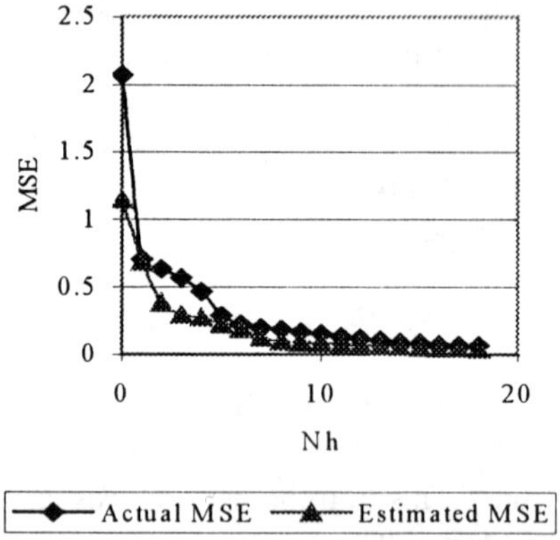

Figure 5 : MLP Sizing - single2.tra

B. Data Sets With Multiple Outputs

Single2.tra : The data set *Single2* has 16 inputs and 3 outputs, and represents the training set for inversion of surface permittivity ε, the normalized surface rms roughness kσ, and the surface correlation length kL found in backscattering models from randomly rough dielectric surfaces [14]. The training set contains 5992 patterns. The 3 outputs can be compressed down to 1 with less than 1% increase in training MSE of an equivalent MLP. The results, illustrated in figure 5, show good performance by the sizing algorithm.

Build3.tra : This file is part of the Proben1 benchmarking data sets. The data set represents the problem of predicting the hourly consumption of electrical energy, hot water, and cold water based on the date, time of day, outside temperature, outside air humidity, solar radiation, and wind speed. There are 8 inputs, and 3 outputs and 2104 training patterns in this data set. All the three outputs are relevant and no compression of the outputs is possible. The predicted and actual E versus N_h curves, illustrated in figure 6, coincide very well.

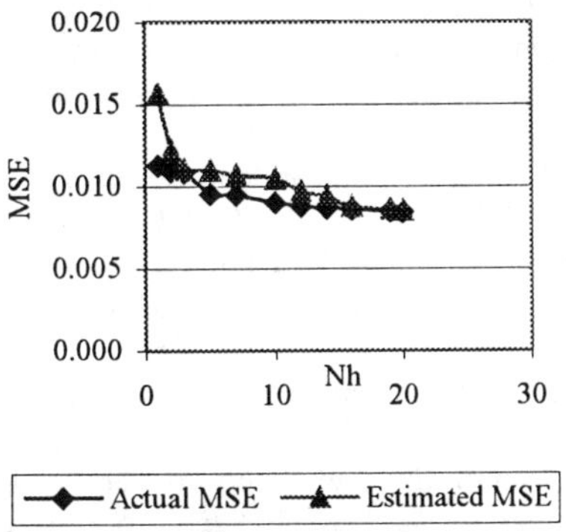

Figure 6 : MLP Sizing - build3.tra

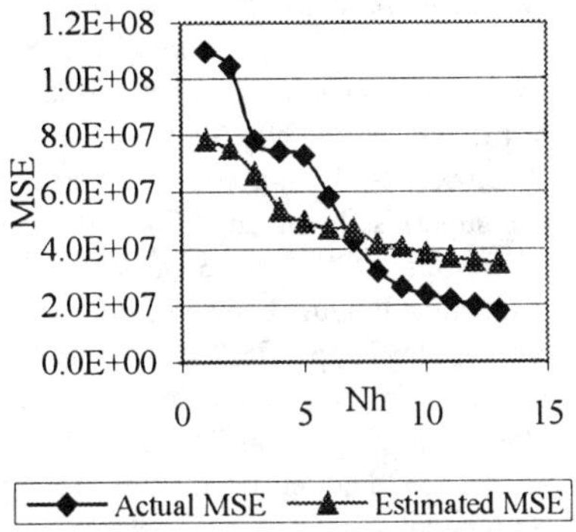

Figure 7 : MLP Sizing - f17.tra

F17.tra : The third multiple-output example is the data set *F17*, which contains 2823 training patterns for onboard *Flight Load Synthesis* (FLS) in helicopters. There are 17 inputs and 9 outputs for each pattern. This data was obtained from the M430 flight load level survey conducted in Mirabel Canada in early 1995 by Bell Helicopter Textron of Fort Worth. The 9 outputs can be compressed down to 2 with less than 1% increase in training MSE of an equivalent MLP. The predicted and actual E versus N_h curves, illustrated in figure 7, are not off by very much.

VI. CONCLUSIONS

In this paper we have described an efficient, accurate and fast method to size an MLP using PWL modular networks. We have shown theoretically that the principal assumption behind our algorithm is a good one, for the case of random patterns. Using several standard benchmark data sets, we have experimentally verified that the sizing algorithm works and is an order of magnitude faster than the brute force design of multiple MLPs. Much work remains to be done. Bounds on the error of our principal assumption need to be found for nonrandom datasets. We need to size networks, from training data, so that testing error is minimized. Finally, we need to determine when PWL modular networks are acceptable substitutes for the MLP.

Acknowledgement

This work was funded by the Advanced Technology Program of the state of Texas under grant 003656-063.

VI. REFERENCES

[1] K.Hornik, M.Stinchcombe, and H.White, "Multilayer feedforward networks are universal approximators", *Neural Networks,* Vol. 2, No. 5, pp. 359-366, 1989.

[2] M.Leshno, V.Lin, A.Pinkus, and S.Schocken, "Multilayer feedforward networks with a nonpolynomial activation function can approximate any function", *Neural Networks,* Vol. 6, No. 6, pp. 861-867, 1993.

[3] B.Fritzke, "Growing cell structures – A self-organizing network for unsupervised and supervised learning", *Neural Networks,* Vol. 7, No. 9, pp. 1441-1460, 1994.

[4] J.N.Hwang, S.S.You, S.R.Lay, and I.C.Jou, "Casacade Correlation Learning: A Projection Pursuit Learning Perspective", *IEEE Transactions on Neural Networks*, Vol. 7, No. 2, pp. 278-289, Mar 1996.

[5] R.Reed, "Pruning Algorithms – A Survey", *IEEE Transactions on Neural Networks*, Vol. 4, No. 5, pp. 740-747, Sep 1993.

[6] Z.J.Yang, "Hidden-layer size reducing for multilayer neural networks using the orthogonal least-squares method", *Proceedings of the 1997 36th IEEE Society of Instrument and Control Engineers (SICE) Annual Conference*, pp. 1089-1092, July 1997.

[7] H.Akaike, "A New Look at the Statistical Model Identification", *IEEE Transactions on Automatic Control,* Vol. AC-19, No. 6, pp. 716-723, Dec 1974.

[8] D.B.Fogel, "An Information Criterion for Optimal Neural Network Selection", *IEEE Transactions on Neural Networks*, Vol. 2, No. 5, pp. 490-497, Sep 1991.

[9] Hema Chandrasekaran and Michael T. Manry, "Convergent Design of a Piecewise Linear Neural Network", to appear in *International Joint Conference on Neural Networks (IJCNN'99)*, July 1999.

[10] S.Subbarayan, K.Kim, M.T.Manry, V.Devarajan and H.Chen, "Modular Neural Network Architecture using Piecewise Linear Mapping", *Thirtieth Asilomer Conference on Signals, Systems & Computers,* Vol. 2, pp 1171-1175, Nov 1996.

[11] A.Gopalakrishnan, X.Jiang, M.S.Chen, and M.T.Manry, "Constructive Proof of Efficient pattern Storage in the Multilayer Perceptron", *Twentyseventh Asilomar Conference on Signals, Systems & Computers,* Vol. 1, pp 386-390, Nov 1993.

[12] A.Lapedes and R.Farber, "Nonlinear signal processing using neural networks: prediction and system modelling", *Technical Report LA-UR-87-2662*, Los Alamos National Laboratory, Los Alamos, NM, 1987.

[13] Hwang, J.-N., Lay, S.-R., Maechler, M., Martin, R. D., and Schimert, J. "Regression modeling in back-propagation and projection pursuit learning", *IEEE Transactions on Neural Networks*, vol. 5, no. 3, pp. 342-353, 1994.

[14] A.K.Fung, Z.Li, and K.S.Chen, "Backscattering from a randomly rough dielectric surface", *IEEE Transactions on Geoscience and Remote Sensing,* Vol. 30, No. 2, pp. 356-369, 1992.

[15] A.Papoulis, *Probability, Random Variables, and Stochastic Processes*, McGraw-Hill, 1991.

ON OPTIMAL DATA SPLIT FOR GENERALIZATION ESTIMATION AND MODEL SELECTION

Jan Larsen and Cyril Goutte
Department of Mathematical Modeling, Building 321
Technical University of Denmark, DK-2800 Lyngby, Denmark
E-mail: jl,cg@imm.dtu.dk, Web: eivind.imm.dtu.dk

OVERVIEW

Modeling with flexible models, such as neural networks, requires careful control of the model complexity and generalization ability of the resulting model. Whereas general asymptotic estimators of generalization ability have been developed over recent years (e.g., [9]), it is widely acknowledged that in most modeling scenarios there isn't sufficient data available to reliably use these estimators for assessing generalization, or select/optimize models. As a consequence, one resorts to resampling techniques like cross-validation [3, 8, 14], jackknife or bootstrap [2]. In this paper, we address a crucial problem of cross-validation estimators: how to split the data into various sets.

The set $\mathcal{D}$ of all available data is usually split into two parts: the design set $\mathcal{E}$ and the test set $\mathcal{F}$. The test set is exclusively reserved to a final assessment of the model which has been designed on $\mathcal{E}$ (using e.g., optimization and model selection). This usually requires that the design set in turn is split in two parts: training set $\mathcal{T}$ and validation set $\mathcal{V}$. The objective of the *design/test* split is to both obtain a model with high generalization ability and to assess the generalization error reliably. The second split is the *training/validation* split of the design set. Model parameters are trained on the training data, while the validation set provides an estimator of generalization error used to e.g., choose between alternative models or optimize additional (hyper) parameters such as regularization or robustness parameters [10, 12]. The aim is to select the split so that the generalization ability of the resulting model is as high as possible.

This paper is concerned with studying the very different behavior of the two data splits using *hold-out cross-validation*, *K-fold cross-validation* [3, 14] and *randomized permutation cross-validation*[1] [1], [13, p. 309]. First we describe the theoretical basics of various cross-validation techniques with the purpose of reliably estimating the generalization error and optimizing the

[1] Also called monte-carlo cross-validation or repeated learning-testing methods.

0-7803-5673-X/99/$10.00 © 1999 IEEE

model structure. The next section deals with the simple problem of estimating a single location parameter. This problem is tractable as non-asymptotic theoretical analysis is possible, whereas mainly asymptotic analysis and simulation studies are viable for the more complex AR-models and neural networks discussed in the subsequent sections.

TRAINING AND GENERALIZATION

Suppose that our model $\mathcal{M}$ (e.g., neural network) is described by the function $\boldsymbol{f}(\boldsymbol{x};\boldsymbol{w})$ where $\boldsymbol{x}$ is the input vector and $\boldsymbol{w}$ is the vector of parameters (or weights) with dimensionality m. The objective is to use the model for approximating the true conditional input-output distribution $p(\boldsymbol{y}|\boldsymbol{x})$, or some moments thereof. For regression and signal processing problems we normally model the conditional expectation $E\{\boldsymbol{y}|\boldsymbol{x}\}$. Define the training set $\mathcal{T} = \{\boldsymbol{x}(k);\boldsymbol{y}(k)\}_{k=1}^{N_\mathcal{T}}$ of $N_\mathcal{T}$ input-output examples sampled from the unknown but fixed joint input-output probability density $p(\boldsymbol{x},\boldsymbol{y})$. The model is trained by minimizing a cost function $C_\mathcal{T}(\boldsymbol{w})$, usually the sum of a loss function (or training error), $S_\mathcal{T}(\boldsymbol{w})$, and a regularization term $R(\boldsymbol{w},\boldsymbol{\kappa})$ parameterized by a set of regularization parameters $\boldsymbol{\kappa}$:

$$C_\mathcal{T}(\boldsymbol{w}) = S_\mathcal{T}(\boldsymbol{w}) + R(\boldsymbol{w},\boldsymbol{\kappa}) = \frac{1}{N_\mathcal{T}}\sum_{k\in\mathcal{T}} \ell\left(\boldsymbol{y}(k),\widehat{\boldsymbol{y}}(k)\right) + R(\boldsymbol{w},\boldsymbol{\kappa}) \tag{1}$$

where $\ell(\cdot)$ measures the cost of estimating the output $\boldsymbol{y}(k)$ with the model prediction $\widehat{\boldsymbol{y}}(k) = \boldsymbol{f}(\boldsymbol{x}(k);\boldsymbol{w})$, e.g., log-likelihood loss or the simple squared error loss function $\ell(\boldsymbol{y},\widehat{\boldsymbol{y}}) = \|\boldsymbol{y}-\widehat{\boldsymbol{y}}\|^2$. Training provides the estimated weight vector $\widehat{\boldsymbol{w}} = \mathrm{argmin}_{\boldsymbol{w}}\, C_\mathcal{T}(\boldsymbol{w})$. *Generalization error* is defined as the expected loss on a future independent sample $(\boldsymbol{x},\boldsymbol{y})$,

$$G(\widehat{\boldsymbol{w}}) = E_{\boldsymbol{x},\boldsymbol{y}}\{\ell(\boldsymbol{y},\widehat{\boldsymbol{y}})\} = \int \ell(\boldsymbol{y},\widehat{\boldsymbol{y}})\, p(\boldsymbol{x},\boldsymbol{y})\, d\boldsymbol{x}d\boldsymbol{y} \tag{2}$$

The *average generalization error* Γ is defined by averaging $G(\widehat{\boldsymbol{w}})$ over all possible training sets (see also [11]):

$$\Gamma = E_\mathcal{T}\{G(\widehat{\boldsymbol{w}})\} = \int G(\widehat{\boldsymbol{w}})\, p(\mathcal{T})\, d\mathcal{T}. \tag{3}$$

Optimization of the model structure, including e.g., regularization parameters [12], is done by minimizing an empirical estimate of the generalization error based on the validation data. Finally, the test data provides an unbiased empirical estimate of the generalization error of the resulting model.

Generalization Assessment

Given a data set $\mathcal{D} = \{\boldsymbol{x}(k);\boldsymbol{y}(k)\}_{k=1}^{N}$ of N *independent* input-output examples, let us first consider the split of the data into the design and test sets, denoted by $\mathcal{E}$ and $\mathcal{F}$ respectively. The purpose is to design a model achieving maximum generalization performance and assess this performance as reliably as possible. We consider three methods:

Hold-Out Cross-Validation (HO). An empirical estimate of (2) is obtained by splitting the data once into design and test sets. Define γ as the split ratio leaving $N_{\mathcal{F}} = \gamma N$ for testing and $N_{\mathcal{E}} = (1-\gamma)N$ for design[2]. The HO estimate of generalization error for model $\widehat{y}$ (with weights $\widehat{w}$) designed on $\mathcal{E}$ is given by

$$\widehat{G}_{\mathrm{HO}}(\widehat{w}) = \frac{1}{N_{\mathcal{F}}} \sum_{k \in \mathcal{F}} \ell(y(k), \widehat{y}(k)). \tag{4}$$

The quality of the estimate is evaluated by considering the mean squared error:

$$\begin{aligned}\mathrm{MSE}_{\mathrm{HO}}(\gamma) &= E_{\mathcal{D}}\left\{\left(\widehat{G}_{\mathrm{HO}}(\widehat{w}) - G(w^*)\right)^2\right\} \\ &= \underbrace{E_{\mathcal{D}}\left\{\left(\widehat{G}_{\mathrm{HO}}(\widehat{w}) - G(\widehat{w})\right)^2\right\}}_{\text{variance}} + \underbrace{E_{\mathcal{D}}\left\{(G(\widehat{w}) - G(w^*))^2\right\}}_{\text{bias}}. \end{aligned} \tag{5}$$

where $G(w^*)$ is the minimum achievable error within the model, i.e., $w^* = \operatorname{argmin}_w G(w)$. The bias term is the excess generalization of our model, and decreases with γ. The variance term measures the reliability of the estimator and it increases when γ decreases. We therefore expect an optimal γ to solve the bias/variance trade-off: $\gamma_{\mathrm{opt}} = \operatorname{argmin}_\gamma \mathrm{MSE}_{\mathrm{HO}}(\gamma)$. This optimal choice has been studied asymptotically for non-linear models [11], using Vapnik-like bounds [7], and in the context of pattern recognition [6]. Surprisingly, $\gamma_{\mathrm{opt}} \to 1$ as $N \to \infty$, indicating that most data should be used for testing. For finite sample sizes, theoretical investigations are limited to simple models (see below).

K-Fold Cross-Validation (KCV). The average over all training sets in (3) is simulated by resampling the design and test set. In KCV, the data set is split into K disjoint subsets $\mathcal{F}_j$ of approximately equal sizes, $\bigcup_{j=1}^{K} \mathcal{F}_j = \mathcal{D}$. For $\gamma < 1/2$, the split ratio is the ratio of the size of the subsets to the total amount of data, i.e., $K = \lfloor 1/\gamma \rfloor$. We evaluate on each subset the model designed on the remaining data $\mathcal{E}_j = \mathcal{D} \setminus \mathcal{F}_j$.[3] The cross-validation estimator is obtained by averaging the K estimates of generalization error:

$$\widehat{\Gamma}_{\mathrm{KCV}} = \frac{1}{N} \sum_{j=1}^{K} \sum_{k \in \mathcal{F}_j} \ell(y(k), \widehat{y}^{-j}(k)) \tag{6}$$

where $\widehat{y}^{-j}$ is the model designed without subset $\mathcal{F}_j$. It is easy to show that $\widehat{\Gamma}_{\mathrm{KCV}}$ is an unbiased estimate of $\Gamma = E_{\mathcal{E}}\{G(\widehat{w})\}$, the average generalization

[2] For practical reasons γN is restricted to be an integer, i.e., $\gamma = i/N$ where $i = 1, 2, \cdots, N-1$.

[3] For $\gamma > 1/2$ the roles of the design and test set are inverted such that we design on each subset and test on the remaining data.

error based on $N_\mathcal{E}$ data. The quality of (6) is assessed by:

$$\begin{aligned}\mathrm{MSE}_{\mathrm{KCV}}(\gamma) &= E_\mathcal{D}\left\{\left(\widehat{\Gamma}_{\mathrm{KCV}} - G(\boldsymbol{w}^*)\right)^2\right\} \\ &= \underbrace{E_\mathcal{D}\left\{\left(\widehat{\Gamma}_{\mathrm{KCV}} - \Gamma\right)^2\right\}}_{\text{variance}} + \underbrace{E_\mathcal{D}\left\{(\Gamma - G(\boldsymbol{w}^*))^2\right\}}_{\text{bias}}. \qquad (7)\end{aligned}$$

Randomized permutation cross-validation (PCV). This involves resampling test sets by randomly selecting $N_\mathcal{F} = N\gamma$ samples for the test set, and the rest for the design set. This can be repeated at most $K \le \binom{N}{N_\mathcal{F}}$ times. For each permutation, a model $\widehat{\boldsymbol{y}}_j$ with weights $\widehat{\boldsymbol{w}}_j$ is designed, then averaging the K empirical estimates of generalization obtained on each test set yields $\widehat{\Gamma}_{\mathrm{PCV}} = K^{-1}\sum_{j=1}^{K} \widehat{G}(\widehat{\boldsymbol{w}}_j)$. The mean squared error $\mathrm{MSE}_{\mathrm{PCV}}(\gamma)$ is defined as in (7).

Model Selection/Optimization

The design of a model is usually done by estimating model parameters on the training data $\mathcal{T}$, and selecting among alternative models, doing early stopping or tuning various additional hyper parameters on the basis of the validation set $\mathcal{V}$. Either of the 3 methods described above (HO, KCV and PCV) can be used for that purpose. However, the relevant criterion for choosing the optimal split ratio γ should now be the performance of the resulting model. In non-parametric modeling, the ultimate goal is usually to obtain good generalization. The optimal split ratio will then be the value of γ for which the resulting model minimizes the "true" (average) generalization error. In the context of model selection:

$$\gamma_{\mathrm{opt}} = \operatorname*{argmin}_{\gamma} E_\mathcal{E}\left\{G_{\widehat{\mathcal{M}}}(\gamma)\right\} \qquad (8)$$

where $G_{\widehat{\mathcal{M}}}(\gamma)$ is the generalization of the model $\widehat{\mathcal{M}}$ which minimizes the cross-validation estimator with split ratio γ. On the other hand, in the context of feature or model selection, the optimal split ratio is one which maximizes the probability of selecting the "correct model". However, as mentioned in e.g., [13, sec. 7.4], selecting a model according to estimated generalization error typically does not result in a consistent selection, i.e., the probability of selecting the correct model does not tend to one. Typically, oversized models will be selected. We shall indeed illustrate in the following examples that those two goals, good generalization and consistent model selection, potentially lead to conflicting decision rules regarding γ.

LOCATION PARAMETER MODEL

In this simple setting, we consider a simple Gaussian variable $y \sim \mathcal{N}(w^\circ, \sigma^2)$ with known σ. This problem has been extensively studied, e.g., [4, 5, 12]. The true generalization of a candidate parameter $\widehat{w}$ is simply $G(\widehat{w}) = \sigma^2 + (w^\circ - \widehat{w})^2$, such that the minimum achievable generalization is σ^2. For the model selection we consider two models: $\mathcal{M}_1$ is a Gaussian with unknown mean estimated from the data, while $\mathcal{M}_0$ is a Gaussian variable with fixed, zero mean $\mathcal{N}(0, \sigma^2)$ (the pruned model).

Design/Test split. For HO cross-validation, tedious but straightforward calculations lead to:

$$\begin{aligned} \mathrm{MSE}_{\mathrm{HO}}(\gamma) &= \frac{2\sigma^4}{N\gamma}\left(1 + \frac{2}{(1-\gamma)N}\right) + \frac{3\sigma^4}{(1-\gamma)^2 N^2} \\ \gamma_{\mathrm{opt}} &= 1 - \frac{8}{A^{1/3}} + \frac{A^{1/3}}{6N} + \frac{2}{3NA^{1/3}} + \frac{1}{3N} \\ \text{with } A &= -324N^2 - 144N + 8 + 12N\sqrt{3(243N^2 + 472N - 28)} \end{aligned} \tag{9}$$

Accordingly, the optimal split ratio converges rather slowly towards 1, as $1 - \gamma_{\mathrm{opt}} = O(N^{-1/3})$, $N \to \infty$. This means, in order to obtain an accurate HO estimate of the generalization error, one should asymptotically reserve the bulk of the data for validation. This is confirmed by the experiments reported in figure 1 (left). All curves are averaged over 40000 replication of the data for each size. When N increases, the optimal γ increases towards 1. Note that the MSE curves flatten, indicating that a wide interval of possible split ratios are near optimal (see also [7]). For K-fold CV:

$$\mathrm{MSE}_{\mathrm{KCV}}(\gamma) = \begin{cases} \dfrac{\sigma^4(2\gamma^3 N - 2\gamma^2 - 6N\gamma^2 + 7\gamma + 6N\gamma - 7 - 2N)}{N^2(\gamma-1)^3} & \gamma \le 0.5 \\ \dfrac{\sigma^4(-4N\gamma^2 - 9\gamma + 8 + 2N\gamma + 2\gamma^2 + 2\gamma^3 N)}{N^2(\gamma-1)^2\gamma} & \gamma \ge 0.5 \end{cases} \tag{10}$$

It is easy to see that $\partial \mathrm{MSE}_{\mathrm{KCV}}(\gamma)/\partial\gamma > 0$ for all $0 \le \gamma \le 1$ and N. That is, $\mathrm{MSE}_{\mathrm{KCV}}$ is minimum for $\gamma_{\mathrm{opt}} = 1/N$, i.e., leave-one-out (LOO) – irrespective of the size of N. This is supported by the simulation results of figure 1 (right). An interesting feature of these curves is the discontinuity in slope for $\gamma = 1/2$, due to change from overlapping design sets to overlapping test sets. The use of PCV gives estimators with uniformly lower (or equal) MSE compared to KCV (for K sufficiently large). This, however, does not change the qualitative result, as the minimum MSE is always reached for leave-one-out.

Model selection. In the case of model selection, the results are only a function of the normalized variable $\theta \equiv w^\circ\sqrt{N}/\sigma$. We therefore use a single sample size $N = 25$, which gives a good compromise between resolution in

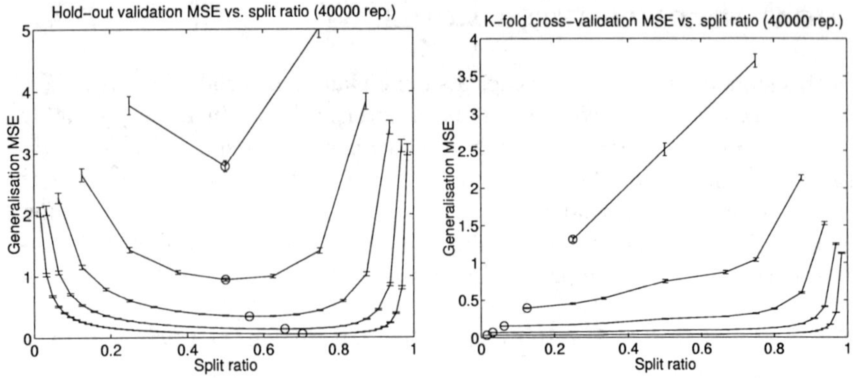

Figure 1: Location parameter example. Left: MSE curves for hold-out cross-validation – the optimal split ratio tends (slowly) to 1 as N increases. Right: K-fold cross-validation – the optimal split ratio is always LOO. In each plot, from top to bottom, $N = 4, 8, 16, 32, 64$ samples. Circle indicates the minimum of each curve, error bars indicate 2 standard deviations.

the possible split ratios and computing requirements, and vary θ through w°. All results are again averaged over 40000 replications of N observations sampled from $\mathcal{N}(w^\circ, 1)$. Figure 2 shows the generalization of the resulting model, for HO, for increasing values of w°. The largest split ratio is optimal for small values of w°. This is illustrated on the right plot by the fact that in that case, the pruned model is almost always selected, and produces a better estimate than the full model, in agreement with [13, sec. 7.4]. For $\theta = 1$, i.e., $w^\circ = \sigma/\sqrt{N} = 0.2$ there is a complete shift and $\gamma = 1/N$ (LOO) becomes optimal. On the left plot, the corresponding curve is almost flat, and at the precise value where the phase transition occurs, both small and large values of γ are within error bars of the optimum. This phase transition occurs because for moderate w° there are two ways of getting good performance: either by selecting the pruned model (for which the excess generalization error is $(w^\circ)^2$), or having enough data for the true model (with average excess generalization error σ^2/N) to perform well. As w° increases, the latter will outperform the former with probability 1. Note that when w° increases some more, the optimal split ratio grows again towards 1 with a very slow asymptotic rate. Figure 2 (right) shows that this is because the split ratio yielding most correct model (consistency) tends to 1. The curves becomes flatter and flatter as w° increases, indicating (as expected) that almost all choices of γ will tend to choose the correct model, and get near-optimal generalization.

The effect of K-fold cross-validation (figure 3) is slightly more subtle. As before, the largest split ratio is optimal for small values of w°. But there are now two transitions, the first one to $\gamma_{\text{opt}} = 1/2$, around $w^\circ = \sigma/\sqrt{N} = 0.2$ and the second one to leave-one-out for a slightly larger value. The first transition occurs as before between the default (pruned) model and the most consistent model. However, due to the overlap in training or validation sets, $\gamma = 1/2$, not LOO, provides the most consistent estimator for small values

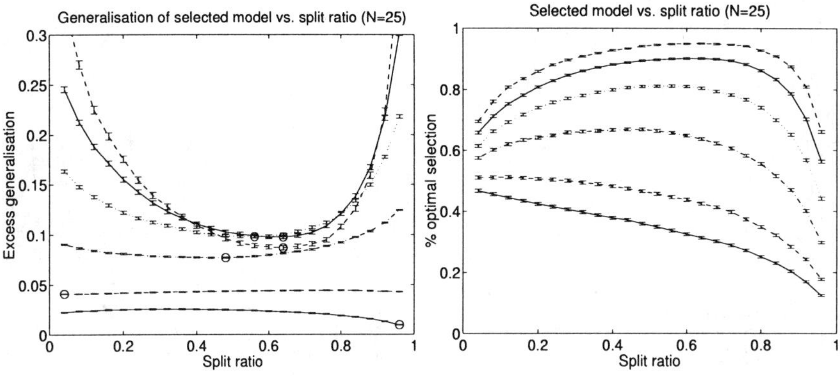

Figure 2: Location parameter example: model selection with hold-out cross-validation, for $w^\circ = 0$ (bottom) to 1 (top) in 0.2 increments. Left: MSE on the hold-out generalization estimator (circle indicates minimum). Right: % correct model selected.

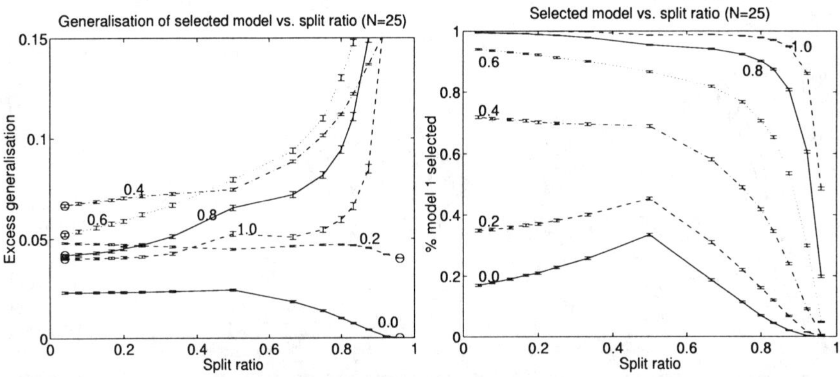

Figure 3: Location parameter example: Model selection with K-fold cross-validation, for $w^\circ = 0$ to 1 in 0.2 increments. Left: excess generalization error of the K-fold generalization estimator (circle indicates minimum). Right: % correct model selected. The curves are labeled with th value of w°.

of w°. The second transition occurs when LOO starts yielding more correct models than 2-fold (fig. 3, right). Additional differences between the K-fold and HO estimators are: 1) in the former, leave-one-out stays optimal as w° grows, 2) the minimum excess generalization error is lower, and 3) the proportion of correctly selected models grows faster towards 1. Note that as the phase transition thresholds are inversely proportional to N, for any $w^\circ \neq 0$, the asymptotically optimal split ratio is $1/N$.

The effect of permutation cross-validation is again similar, and the qualitative conclusion ($\gamma_{\text{opt}} = \text{LOO}$) identical. There is no discontinuity in $\gamma = 1/2$ thanks to the better averaging strategy for intermediate split ratios. There is therefore only one γ-transition, from one extreme value to the other.

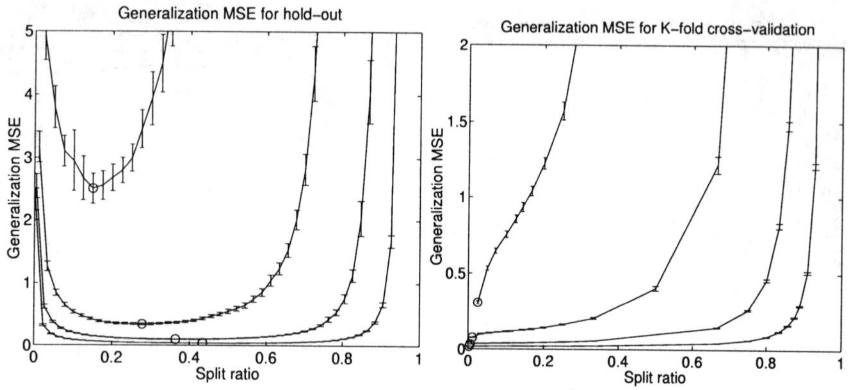

Figure 4: Generalization MSE for the AR filter using hold-out (left) and K-fold (right) cross-validation. From top to bottom, $N = 50, 100, 200, 400$. Circle indicates the minimum of the average MSE, error bars are 2 standard deviations.

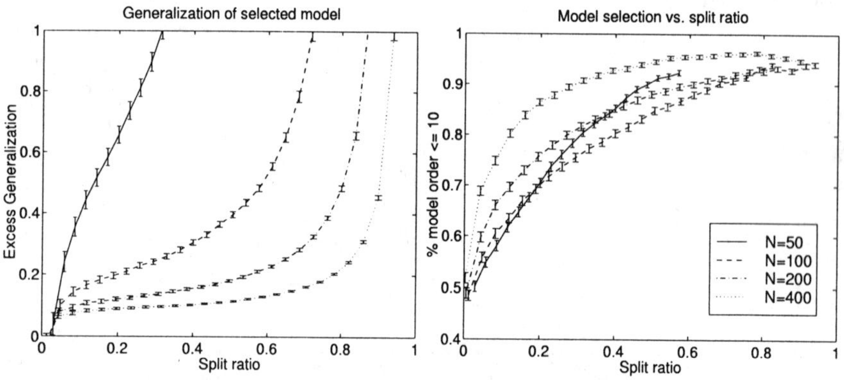

Figure 5: Model order selection. Left: generalization of the selected model. Right: % of non-overestimated models. The legend in the left plot also applies to the right plot.

AUTOREGRESSIVE MODEL

Let us now consider the estimation of a linear autoregressive (AR) filter. The target is a low-pass filter of order 10, with coefficients [−3.99, 8.09, −10.48, 9.42, −6.08, 2.84, −0.94, 0.21, −0.028, 0.0018]. Data are generated by filtering white noise $\epsilon \sim \mathcal{N}(0, \sigma^2)$, with $\sigma^2 = 1$. The minimum achievable generalization error $G(\boldsymbol{w}^*)$ is therefore 1.

Design/Test split. In order to assess the optimal split ratio with respect to generalization estimation, we study our cross-validation schemes using a 10th order linear filter. The coefficients are estimated using regularized least squares, $\widehat{\boldsymbol{w}} = \left(\boldsymbol{X}^\top \boldsymbol{X} + \kappa \boldsymbol{I}\right)^{-1} \boldsymbol{X}^\top \boldsymbol{Y}$, with $\boldsymbol{X}$ and $\boldsymbol{Y}$ the input and output matrices, respectively, $\boldsymbol{I}$ the unit matrix, and κ set to 10^{-6} times the largest eigenvalue of the covariance matrix. As shown on figure 4, all MSE curves behave in a manner qualitatively similar to the simple Gaussian variable above. Minimum MSE is obtained for increasing γ for HO, and for LOO in the case of K-fold. This suggests the same asymptotics as before: for hold-out $\gamma_{\text{opt}} \to 1$ with $(1-\gamma_{\text{opt}})N \to +\infty$ and $\gamma_{\text{opt}} = 1/N$ for KCV (and PCV). As before, the curves get flatter with increasing N, meaning that a wide range of split ratios become near-optimal.

Model selection. Hold-out cross-validation is used to select the order of the AR model, between 8 and 14. Experiments are reported by averaging over 10000 independent data sets of increasing sizes N = 50, 100, 200 and 400. Figure 5 shows that generalization and model selection lead to conflicting optimal decisions. Clearly, small split ratios give better generalization, but they tend to overestimate the model order. On the other hand, large split-ratios select more parsimonious models but yield poor generalization. Note that due to the small contribution from the last two filter parameters, models of order 8 and 9 are often selected. These results are consistent with [13, sec. 7.4]. This suggests that moderate values of γ might asymptotically provide a good trade-off between model consistency and generalization abilities, though the optimum would depend on a particular weighting of both effects.

NEURAL NETWORKS

We also considered non-linear modeling using neural networks. The target system is the Hénon map: $y(k) = 1-1.4y(k-1)+0.3y(k-2)+\epsilon$, $\epsilon \sim \mathcal{N}(0,\sigma^2)$ and σ^2 is tuned such that the signal-to-noise ratio is approximately 10%. Data sampled from this system are modeled using a standard feed-forward multi-layer perceptron with one hidden layer of 5 hidden units, and an input layer with time-delayed inputs and task-dependent size.

Experiments indicate that the behaviour of the hold-out estimator is similar to what has been described above. Due to space limitation, the detailed results will be presented at the workshop.

SUMMARY

We addressed the problem of choosing the optimal split ratio for cross-validation estimators. We showed that different cross-validation strategy (the design/test and the training/validation splits), and different objectives (reliable assessment of generalization, best generalization or model consistency) lead to different quality measures (MSE, resulting generalization, probability of correct selection), and potentially result in conflicting decision strategies. For hold-out cross-validation, $\gamma_{\text{opt}} \to 1$ as $N \to \infty$ seems to be well supported theoretically and experimentally. On the other hand, for K-fold and randomized permutation CV, the asymptotically optimal split-ratio is highly dependent on the task and on the model. In particular, we have illustrated that best generalization and model consistency lead to opposite optimal choices

($\gamma_{\text{opt}} \to 0$ and $\gamma_{\text{opt}} \to 1$ respectively).

Acknowledgments. Research supported by the Danish Research Councils through the Danish Computational Neural Network Center (CONNECT), the THOR Center for Neuroinformatics and the European Union, through BIOMED II grant number BMH4-CT97-2775. Lars Kai Hansen is acknowledged for valuable discussions.

REFERENCES

[1] P. Burman, "A Comparative Study of Ordinary Cross-Validation, v-fold Cross-validartion, and the Repeated Learning-Testing Methods," **Biometrika**, vol. 76, no. 3, pp. 503–514, 1989.

[2] B. Efron and R. Tibshirani, **An Introduction to the Bootstrap**, New York, NY: Chapman & Hall, 1993.

[3] S. Geisser, "The Predictive Sample Reuse Method with Applications," **Journal of the American Statistical Association**, vol. 50, pp. 320–328, 1975.

[4] C. Goutte and L. K. Hansen, "Regularization with a pruning prior," **Neural Networks**, vol. 10, no. 6, pp. 1053–1059, 1997.

[5] L. K. Hansen and C. E. Rasmussen, "Pruning from Adaptive Regularization," **Neural Computation**, vol. 6, pp. 1223–1232, 1994.

[6] W. H. Highleynman, "The Design and Analysis of Pattern Reocgnition Experiments," **The Bell Systems Technical Journal**, pp. 723–744, March 1962.

[7] M. Kearns, "A Bound on the Error of Cross Validation Using the Approximation and Estimation Rates, with Consequences for the Training-Test Split," **Neural Computation**, vol. 9, no. 5, pp. 1143–1161, 1997.

[8] M. Kearns, Y. Mansour, A. Y. Ng and D. Ron, "An Experimental and Theoretical Comparison of Model Selection Methods," **Machine Learning**, vol. 27, no. 1, pp. 7–50, April 1997.

[9] J. Larsen, **Design of Neural Network Filters**, Ph.D. thesis, Electronics Institute, The Technical University of Denmark, March 1993.

[10] J. Larsen, L. N. Andersen, M. Hintz-Madsen and L. K. Hansen, "Design of Robust Neural Network Classifiers," in **Proceedings of IEEE ICASSP'98**, IEEE, May 1998, vol. 2, pp. 1205–1208.

[11] J. Larsen and L. K. Hansen, "Empirical generalization assessment of neural network models," in F. Girosi, J. Makhoul, E. Manolakos and E. Wilson (eds.), **Neural Networks for Signal Processing V – Proceedings of the 1995 IEEE Workshop**, Piscataway, New Jersey: IEEE, 1995, pp. 42–51.

[12] J. Larsen, C. Svarer, L. N. Andersen and L. K. Hansen, "Adaptive Regularization in Neural Network Modeling," in **Neural Networks: Tricks of the Trade**, Germany: Springer-Verlag, 1998, no. 1524 in Lecture Notes in Computer Science.

[13] J. Shao and D. Tu, **The Jackknife and Bootstrap**, New York, NY: Springer-Verlag, 1995.

[14] M. Stone, "Cross-validatory choice and assessment of statistical predictions," **Journal of the Royal Statistical Society B**, vol. 36, no. 2, pp. 111–147, 1974, with discussion.

NEAREST-PROTOTYPE CLASSIFIER DESIGN BY DETERMINISTIC ANNEALING WITH RANDOM CLASS LABELS*

Ertem Tuncel and Kenneth Rose
Signal Compression Laboratory
Department of Electrical and Computer Engineering
University of California, Santa Barbara, CA 93106, USA
E-mail: ertem@laurel.ece.ucsb.edu, rose@ece.ucsb.edu

Abstract.

The design of nearest-prototype (NP) classifiers is a challenging problem because of the prevalence of poor local minima, and the piecewise constant nature of the cost function which is incompatible with gradient-based techniques. This paper extends the deterministic annealing (DA) method for NP-classifier design in two ways. First, the association between prototypes and class labels is also randomized, and the corresponding association probabilities are added to the set of parameters to be optimized. Second, the multiplicity (or the mass) of prototypes are optimized. During the design, all parameters are optimized so as to minimize the expected misclassification rate for a given level of randomness. The *joint entropy*, which measures the level of randomness, is gradually reduced while optimizing the cost Lagrangian. As the entropy approaches zero, the method seeks a deterministic classifier that minimizes the rate of misclassification.

INTRODUCTION

The nearest-prototype (NP) classifier is an efficient tool for classifying data, that has the capability of implementing *nonconvex* or even *disconnected* partitions which are determined by a small number of parameters, namely, the prototype locations. Moreover, although the NP-classifier decision regions are restricted to be unions of Voronoi cells, any classifier can be approximated by NP-classifiers, with a sufficient number of prototypes.

*THIS WORK WAS SUPPORTED IN PART BY THE NATIONAL SCIENCE FOUNDATION UNDER GRANT NO. NCR-9314335, THE UNIVERSITY OF CALIFORNIA MICRO PROGRAM, CISCO SYSTEMS, INC., DIALOGIC CORP., FUJITSU LABORATORIES OF AMERICA, INC., GENERAL ELECTRIC CO., HUGHES NETWORK SYSTEMS, INTEL CORP., LERNOUT & HAUSPIE SPEECH PRODUCTS, LOCKHEED MARTIN, LUCENT TECHNOLOGIES, INC., QUALCOMM, INC., ROCKWELL INTERNATIONAL CORP., PANASONIC TECHNOLOGIES, INC., AND TEXAS INSTRUMENTS, INC.

A known difficulty with the design of NP classifiers is that the optimal labeling, or allocation of prototypes to classes, is not known a priori. In practice, the allocation is made in a heuristic fashion. For example, prototype labels may be distributed among classes uniformly, or according to the class label distribution in the training set. Such heuristic initialization often results in suboptimal distribution of prototypes and compromises the ultimate performance. Known algorithms, such as the learning vector quantizer (LVQ) design [1], as well as the existing deterministic annealing (DA) approach [2], suffer from this weakness.

In this paper, this problem is tackled by a significant extension of the DA algorithm of [2], which itself builds on the DA approach to clustering [3] (see also [4] for a tutorial). In the DA algorithm proposed here, the classifier randomization is extended to include the assignment of labels to prototypes. In other words, each prototype is assigned to a class label *in probability*. The randomness is controlled by monitoring the *joint entropy* of the system.

NP CLASSIFIER FORMULATION AND DESIGN

In general, a classifier is defined [5] as a mapping $C : \mathcal{R}^m \to \mathcal{K}$, where $\mathcal{R}^m$ is the feature space and $\mathcal{K}$ is any index set of labels. It induces a partitioning of the feature space into regions $R_k \equiv \{x \in \mathcal{R}^m : C(x) = k\}$, where $\bigcup_k R_k \equiv \mathcal{R}^m$ and $\bigcap_k R_k \equiv \emptyset$.

NP Classification

A *nearest-prototype* classifier is a composite mapping $C = Q \circ V$ where $V : \mathcal{R}^m \to \mathcal{J}$ is the partitioning function and $Q : \mathcal{J} \to \mathcal{K}$ is the labeling function, and where $\mathcal{J}$ is an intermediate (prototype) index set. The partition function is

$$V(x) = \arg\min_{j \in \mathcal{J}} \{d(x, s_j)\}\,, \tag{1}$$

where, $S = \{s_j : j \in \mathcal{J}\}$ is the set of *prototypes* and $d(\cdot,\cdot)$ is any appropriately defined *distance or distortion measure* on $\mathcal{R}^m$. Each vector $x \in \mathcal{R}^m$ is first mapped to the nearest prototype $s_{V(x)}$, and then classified to the corresponding label by $Q(\cdot)$. Accordingly, each classifier region R_k is the union of Voronoi cells:

$$\begin{aligned} R_k &\equiv \bigcup_{j:Q(j)=k} V_j \text{ with} \\ V_j &\equiv \{x \in \mathcal{R}^m : V(x) = j\}\,. \end{aligned} \tag{2}$$

Let $\mathcal{T} = \{(x_n, c_n)\}$ be a training set of N pairs of vectors consisting of $x_n \in \mathcal{R}^m$ and corresponding class labels $c_n \in \mathcal{K}$. The performance of a classifier C is measured by the empirical error rate

$$P_e(C) = \frac{1}{N} \sum_{n=1}^{N} \rho(c_n, C(x_n)), \tag{3}$$

where $\rho(c_n, k) = 1$ if $c_n \neq k$ and 0 otherwise. For reasons that will soon become obvious, it is more convenient to rewrite $P_e(C)$ as

$$P_e(C) = \frac{1}{N} \sum_{n} \sum_{j \in \mathcal{J}} \sum_{k \in \mathcal{K}} v_{nj} q_{jk} \rho(c_n, k), \tag{4}$$

where

$$v_{nj} = \begin{cases} 1 & \text{if } V(x_n) = j, \\ 0 & \text{otherwise}. \end{cases}$$

$$q_{jk} = \begin{cases} 1 & \text{if } Q(j) = k, \\ 0 & \text{otherwise}. \end{cases} \tag{5}$$

Design with Deterministic Annealing

In the DA approach for NP-classifier design [2], $Q(\cdot)$ was fixed as a deterministic mapping, and the mapping $V(\cdot)$ was randomized. In other words, points in the feature space $\mathcal{R}^m$ were assigned to prototypes *in probability* during the design phase. More specifically,

$$P[x_n \in V_j] = \frac{e^{-\gamma d(x_n, s_j)}}{\sum_i e^{-\gamma d(x_n, s_i)}}, \tag{6}$$

where γ is a parameter governing the "peakiness" of the distribution. The classifier output was declared as the fixed-label of the chosen prototype. A "randomness level" of $V(\cdot)$, measured by the *Shannon entropy*

$$H = -\frac{1}{N} \sum_{n} \sum_{j \in \mathcal{J}} P[x_n \in V_j] \log P[x_n \in V_j], \tag{7}$$

was imposed on the solution, and this level was gradually decreased. At the limit of zero entropy, a "hard" classifier is obtained, in which case $\gamma \to \infty$, and hence $V(\cdot)$ goes to (1). At each intermediate level of randomness, the objective was to minimize the misclassification rate over the training set, which can still be expressed as in (4) with the modification $v_{nj} = P[x_n \in V_j]$, subject to the entropy constraint. From a different perspective, the same objective could be stated as "to maximize the entropy while gradually lowering the level of misclassification rate."

It was shown [2] with several synthetic and "real-world" examples that the fixed-label DA algorithm outperforms LVQ-based classification [1]. However, the open question of how $Q(\cdot)$ should be fixed, i.e., how many prototypes should be associated with each class, remained unresolved. In fact, there is no analytical way to determine the optimum $Q(\cdot)$ prior to the actual classifier design. There are several known *ad hoc* ways, such as distributing the prototype labels among classes uniformly, or according to the frequency of the class labels in the training set. Each of these heuristics has its limitations. We next propose a remedy for this shortcoming.

Random-label DA

We proceed to randomize $Q(\cdot)$ as well. We redefine q_{jk} as the *probability* that prototype j belongs to class k, so that the misclassification rate is still given by (4). However, if we reinterpret the meaning of $\mathcal{J}$ as the index set for *distinct* prototypes, and denote the number of *identical* prototypes located at a distinct position by λ_j, the overall misclassification rate becomes

$$P_e(C) = \frac{1}{N}\sum_n \sum_{j\in\mathcal{J}} \sum_{k\in\mathcal{K}} \lambda_j v_{nj} q_{jk} \rho(c_n, k)\,. \tag{8}$$

We also optimize λ_j as done for clustering in [6] and [4], to obtain the *correct* multiplicity at each distinct prototype. We do not require λ_j to be an integer, so it may be viewed as the *mass* of prototype s_j, where the total mass to be distributed among prototypes is M.

The randomized composite mapping $C = Q(V(\cdot))$ is in fact a two-stage Markov chain, whose overall entropy is

$$\begin{aligned} H &= -\frac{1}{N}\sum_n \sum_{j\in\mathcal{J}} \sum_{k\in\mathcal{K}} \lambda_j v_{nj} q_{jk} \log\left(v_{nj} q_{jk}\right) \\ &= -\frac{1}{N}\sum_n \sum_{j\in\mathcal{J}} \lambda_j v_{nj} \log v_{nj} - \frac{1}{N}\sum_n \sum_{j\in\mathcal{J}} \sum_{k\in\mathcal{K}} \lambda_j v_{nj} q_{jk} \log q_{jk}\,. \end{aligned} \tag{9}$$

With this modification we follow a derivation similar to that of [2], and consider the expected *structural cost* induced by this random partitioning of the feature space:

$$D = \frac{1}{N}\sum_n \sum_{j\in\mathcal{J}} \lambda_j v_{nj} d(x_n, s_j)\,. \tag{10}$$

Our objective is to minimize the misclassification rate (8) where $H = H^*$ is given as a constraint. So an equivalent objective is the following:

$$\begin{aligned} \text{Minimize} \quad & L = \beta P_e - H \\ \text{with respect to} \quad & \gamma, s_j, \lambda_j, \text{and } q_{jk} \\ \text{subject to} \quad & \sum_{k\in\mathcal{K}} q_{jk} = 1\,, \forall j \in \mathcal{J} \\ & \sum_{j\in\mathcal{J}} \lambda_j = M \\ & v_{nj} = \frac{e^{-\gamma d(x_n, s_j)}}{\sum_{i\in\mathcal{J}} \lambda_j e^{-\gamma d(x_n, s_i)}}\,, \end{aligned} \tag{11}$$

where β is the "reciprocal temperature" controlling the joint entropy level. Initially $\beta = 0$, in which case the sole objective becomes the maximization of the joint entropy, which is achieved by $\gamma = 0$ and an arbitrary set of identical prototypes. When $\beta \to \infty$, the objective becomes the minimization of the misclassification rate, which is, of course, our ultimate objective. This minimum is achieved only by $H = 0$, i.e., $\gamma \to \infty$ and all prototypes are distinct. It is easy to see from (11), that in this case the classifier is a hard NP-classifier. When β is gradually increased from 0 to ∞, the prototypes undergo a sequence of prototype splits, or *bifurcations*, which correspond to *phase transitions* in the statistical physics analogy. One significant problem with the above formulation as it stands is that it allows too much freedom in the location of prototypes until β reaches the critical value for the second phase transition. To be more specific,

- When $0 \leq \beta < \beta_1$, that is, before the first phase transition, any set of identical prototypes will minimize the cost function.
- When $\beta_1 \leq \beta < \beta_2$, that is, between the first and second phase transitions, the line on which the two sets of prototypes should be is unique, but the distance between the prototypes and the value of γ still maintain a degree of freedom. It is possible to increase one and decrease the other without changing the value of the Lagrangian cost function.

The excessive freedom is very undesirable and increases the likelihood of trapping the algorithm in a poor local minimum. Instead, we propose a regularized objective, namely

$$\begin{aligned} \text{Minimize} \quad & L = \beta P_e + \gamma D - H \\ \text{with respect to} \quad & \gamma, s_j, \lambda_j, \text{and } q_{jk} \\ \text{subject to} \quad & \sum_{k \in \mathcal{K}} q_{jk} = 1 \,, \forall j \in \mathcal{J} \\ & \sum_{j \in \mathcal{J}} \lambda_j = M \\ & v_{nj} = \frac{e^{-\gamma d(x_n, s_j)}}{\sum_{i \in \mathcal{J}} \lambda_j e^{-\gamma d(x_n, s_i)}} \,. \end{aligned} \tag{12}$$

It can be shown that as $\beta \to \infty$, not only $\gamma \to \infty$, but also $\frac{\gamma}{\beta} \to 0$, which is consistent with the original aim of minimizing the misclassification rate P_e. The excessive degrees of freedom are eliminated by the addition of the expected structural cost term. For example, when $\beta = 0$, the set of identical prototypes must be at the global centroid of the training set $\mathcal{T}$ (assuming the squared-distance measure $d(x, y) = ||x - y||^2$.)

The objective (12) is a convex optimization problem in terms of q_{jk}. So the solution for q_{jk} is *analytically* found as

$$q_{jk} = \frac{e^{-\beta D_{jk}}}{\sum_m e^{-\beta D_{jm}}} \,, \tag{13}$$

where

$$D_{jk} = \frac{\sum_n v_{nj} \rho(c_n, k)}{\sum_n v_{nj}}, \tag{14}$$

which is easily interpreted as the expected misclassification rate given that prototype s_j is associated with class label k. Substituting q_{jk} back in the Lagrangian of (12), we obtain

$$L = -\sum_n \log\big(\sum_{j \in \mathcal{J}} \lambda_j e^{-\gamma d(x_n, s_j)}\big) - \sum_n \sum_{j \in \mathcal{J}} \lambda_j v_{nj} \log\big(\sum_{k \in \mathcal{K}} e^{-\beta D_{jk}}\big), \tag{15}$$

which is a natural generalization of the *free energy* (or strictly speaking the *potential*) derived in [3] (and [6] with the mass constraint on codevectors) for the case of clustering or vector quantization.

Necessary Conditions for Optimality

At each value of β, the solution of (15), i.e., the optimal locations of the prototypes $\{s_j\}$, their masses $\{\lambda_j\}$, and γ should satisfy

$$\nabla_{s_j} L = 0 = \gamma \sum_n \lambda_j v_{nj} \nabla_{s_j} d(x_n, s_j)(1 + L_n - L_{nj}), \tag{16}$$

and

$$\frac{\partial L}{\partial \gamma} = 0 = \sum_n \sum_{j \in \mathcal{J}} \lambda_j v_{nj} d(x_n, s_j)(1 + L_n - L_{nj}), \tag{17}$$

and

$$\frac{N}{M} = \sum_n v_{nj}(1 + L_n - L_{nj}), \tag{18}$$

where

$$L_{nj} = \beta \sum_{k \in \mathcal{K}} q_{jk}\{\rho(c_n, k) - D_{jk}\} - \log\big(\sum_{k \in \mathcal{K}} e^{-\beta D_{jk}}\big) - \log\big(\sum_{i \in \mathcal{J}} \lambda_i e^{-\gamma d(x_n, s_i)}\big) \tag{19}$$

and

$$L_n = \sum_{i \in \mathcal{J}} \lambda_i v_{ni} L_{ni}. \tag{20}$$

Note also that

$$L = \sum_n \sum_{j \in \mathcal{J}} \lambda_j v_{nj} L_{nj} = \sum_n L_n. \tag{21}$$

EXPERIMENTAL RESULTS

We performed experiments to compare the performances of the LVQ method [1], the fixed-label DA method [2], and the proposed random-label DA method. As a training set, we used the Finnish phoneme data set that accompanies the standard LVQ software package. The training set consists of 1962 vectors which represent 20-dimensional cepstral coefficients of the corresponding phoneme uttered by the speaker. There are 20 classes (phonemes) in the training set. The experiments comparing LVQ and the fixed-label DA method have already appeared in [2]. In both LVQ and fixed-label DA approaches, the number of prototypes that are associated with a particular class is proportional to the relative frequency of occurrence of that class in the training set. The random-label DA algorithm, circumvents the need for such an *ad hoc* decision.

M (# of prototypes)	20	30	40	50	80	100
P_e (LVQ)	13.25	12.44	11.47	10.96	10.09	8.17
P_e (fixed-label DA)	11.67	9.99	8.36	5.55	4.83	4.23
P_e (random-label DA)	7.65	7.18	6.88	5.86	4.63	4.17

Table 1: A comparison of P_e values (in percent) for the NP-classifiers designed by LVQ, fixed-label DA, and random-label DA methods. The training set is the 20-dimensional, 20 class Finnish phoneme data set.

The experiments are performed using various values of total number of prototypes. The results are presented in Table 1. As seen from the table, in most cases, the random-label DA algorithm outperforms both the fixed-label DA and the LVQ algorithms. The achieved performance gain is more significant when the total number of prototypes is small. For example, the 20-prototype classifier designed by the random-label DA method is even better than the 40-prototype classifier designed by the fixed-label DA, and 100-prototype classifier designed by LVQ. So, one can achieve either substantially improved performance at the same classifier complexity, or similar performance at substantially reduced complexity. Although it is known that deterministic annealing methods avoid many poor local minima, there is no guarantee of finding the global minimum. This is indeed the case of 50 prototypes, where the fixed-label DA performs slightly better than the random-label DA.

CONCLUSIONS

This paper presents an extension of fixed-label DA, where all mappings, including the mapping from prototypes to class labels, are randomized, and the level of randomness is measured by the *joint entropy* which is gradually reduced to zero.

The results of the experiments on the Finnish phoneme set indicate that further significant gains, in terms of better classification performance or reduced classifier complexity, are achieved by randomization and optimization of the mapping $Q(\cdot)$ between the prototypes and the class labels.

REFERENCES

[1] T. Kohonen, G. Barna and R. Chrisley, "Statistical pattern recognition with neural networks: Benchmarking studies", IEEE Proc. ICNN, vol. 1, pp 61-68, 1988.

[2] D. Miller, A. V. Rao, K. Rose and A. Gersho, "A Global Optimization Technique for Statistical Classifier Design", IEEE Trans. on Signal Processing, vol. 44, pp 3108-3122, 1996.

[3] K. Rose, E. Gurewitz and G. C. Fox, "Vector Quantization by Deterministic Annealing", IEEE Trans. on Information Theory, vol. 38, pp 1249-1257, 1992.

[4] K. Rose, "Deterministic annealing for clustering, compression, classification, regression, and related optimization problems", Proceedings of IEEE, vol. 86, no. 11, pp 2210-2239, 1998.

[5] R. O. Duda and P. E. Hart, "Pattern Classification and Scene Analysis", John Wiley & Sons, 1973.

[6] K. Rose, E. Gurewitz and G. C. Fox, "Constrained Clustering as an Optimization Method", IEEE Trans. on Pattern Analysis and Machine Intelligence, vol. 15, no. 8, pp 785-794, 1993.

TIME SERIES CLASSIFICATION USING ADAPTIVE DYNAMIC TARGETS

Ernst Haselsteiner
Department of Medical Informatics
Technical University Graz
Inffeldgasse 16a, 8010 Graz, Austria
http://www-dpmi.tu-graz.ac.at/~hasi/

Abstract. To train a classifier with supervised learning appropriate targets have to be provided. In the case of time series this can be complicated if there is only one target for the whole time series, but the learning algorithm needs a target at each time step. In this paper a new technique is introduced, which is able to provide appropriate targets at each time step. As a result of this technique the impact on classification of each time step is determined, which is very useful in applying the trained classifier to new data. The paper describes this technique in detail and the basic findings of experiments on artificial data and real world data are given.

INTRODUCTION

Most of the literature on supervised learning in neural networks does not pay attention to the targets used for training the network [9, 4, 6]. As it is one of the characteristics of supervised learning that the targets are known, there is nothing to bother about them.

This is not true in the case of time series classification. If a learning algorithm is used, which can use a target at each time step (e.g [10, 1]), it is not obvious how to provide these targets. For classification purpose often there is only one target for the whole time series. Thus, this single target is used for all time steps or it is only applied to some time steps [7]. In the former case for some time steps (e.g. the first ones) this target is much too strong, which results in inappropriate weight updates. In the later case one needs a strategy to find out which time steps should be given a target. In many cases finding such a strategy is not possible as one does not know a priori, which time steps are best suited for classification of the time series.

In this paper a new technique, called *adaptive dynamic targets*, is introduced, which solves the stated problem. The paper is organized as follows. First, the motivation for doing this research is given. This is followed by a brief review of the standard way of using targets. The new technique of

0-7803-5673-X/99/$10.00

adaptive dynamic targets is introduced in the sequel. Two experiments are described and the results on artificial data as well as on real world data are given. The conclusions complete the paper.

MOTIVATION

This research is motivated by classification problems of a Brain Computer Interface (BCI) project at Technical University Graz [8]. The time series to be classified are preprocessed electroencephalograms (EEG) of patients, which imagine movement of left or right hand, without doing any movements. Each time series is assigned the class "left" or "right". Thus, for the supervised training of the classifier two different targets $\mathbf{t_1}$ and $\mathbf{t_2}$ are used. An obvious way to encode the two classes c_1 and c_2 is to set up a network with two units in the output layer and to represent the two classes with targets like

$$\mathbf{t_1} = \begin{pmatrix} 1 \\ 0 \end{pmatrix} \quad \text{and} \quad \mathbf{t_2} = \begin{pmatrix} 0 \\ 1 \end{pmatrix}$$

If the activation functions limit the output values to the range $[0, 1]$, the vector $\mathbf{o}$, consisting of the output values of the two output neurons represents a single point in the unit square, with the first component of the vector on the x axes and the second component on the y axes (see Figure 1).

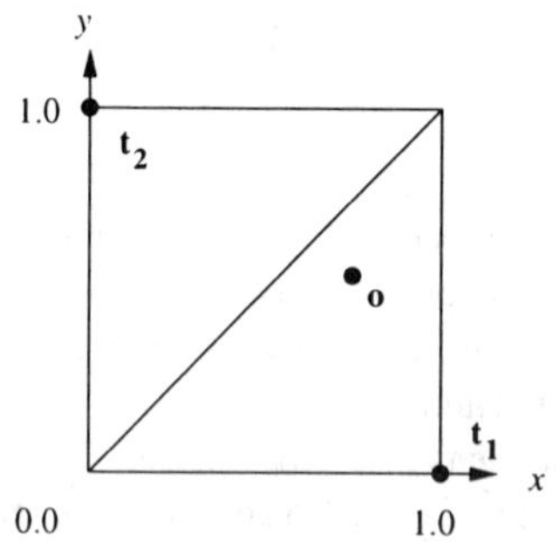

Figure 1: Output space for the two dimensional case. All outputs $\mathbf{o}$ as well as both targets $\mathbf{t}_1$ and $\mathbf{t}_2$ lie within the unit square.

Both targets $\mathbf{t_1}$ and $\mathbf{t_2}$ also lie within this square. To classify a point the square is divided by the line $x - y = 0$. Each point is assigned the class of the target which is next to it. During training one tries to adapt the weights of the network in a way that for each example belonging to c_i the nearest target is $\mathbf{t_i}$. This works fine in general, but the following observations can lead to some improvement.

(a) For training a classifier on time series, although supervised training is used, it is not clear how to use the targets, as often there is only one target for the whole series, but at each time step a target is needed. For example one cannot expect the network to recognize the whole series after processing the first time step, so using the overall target in the first time step seems to be a too strong teacher for the network. To overcome this problem one could provide constraints for the output units instead of actual targets like it is done in [7]. But it is often hard to find such constraints. Thus, an automatic procedure for creating the targets could support the learning.

(b) For high classification rates the output vector $\mathbf{o}$ does not have to reach the value of either $\mathbf{t_1}$ or $\mathbf{t_2}$. It only has to be in the proper half of the unit square. Of course, for good generalization on the one hand having all the outputs far away form the border is good, but on the other hand this implies that the given output space is not used at all, if all the examples are concentrated around $\mathbf{t_1}$ and $\mathbf{t_2}$. If there are different unknown subclasses within one of the two known classes it might be useful to allow them to have different representations within the proper half of the unit square. Thus, we are looking for a way to allow different clusters, which means not pushing all examples exactly towards the targets, but only on the right side of the border.

(c) Regarding networks with recurrent connections as dynamical systems shows, that these networks are able to create desired trajectories in their output spaces. There are a lot of papers which show how to learn and analyze such behaviors of recurrent networks [2, 5, 11, 3]. Unfortunately most of these algorithms lack of application in classification because no desired trajectories are known.

These three points are the motivation to look for a procedure of generating targets during the training to solve the problem described in observation (a) and to satisfy observation (b) by not forcing the outputs to the exact targets. Observation (c) suggests that such a procedure would be able to use the capabilities of recurrent nets more effectively.

ADAPTIVE DYNAMIC TARGETS

To explain the idea of adaptive dynamic targets the conventional way of using targets in supervised learning, which is called fixed targets in the following, is reviewed briefly.

For each sample data a target $\mathbf{t}$ is given and the error $\mathbf{e}$ is defined as

$$\mathbf{e} = \mathbf{t} - \mathbf{o} \tag{1}$$

where $\mathbf{o}$ is the output vector. By using this error vector the outputs are forced into the corners of the unit square, which is illustrated in Figure 2. As stated above this is actually not necessary and can even harm the learning process.

This kind of supervised learning causes some problems, when applied on time series classification, which are described in Section 2.

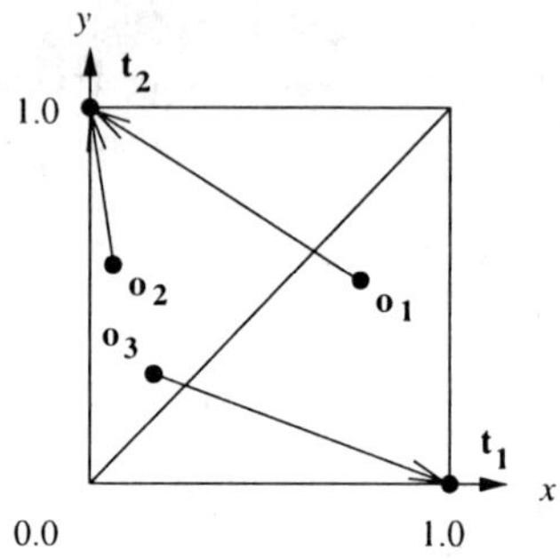

Figure 2: With fixed learning all outputs are forced to the exact position of their targets, which is not necessary.

Therefore a new way of using the given targets is proposed. The new method, called adaptive dynamic targets, tries to find an appropriate target at each time step. The new targets depend on the fixed target $\mathbf{t}$ and on the current output $\mathbf{o}$ of the network. The new targets $\mathbf{t_{new}}$ have to be calculated at every time step during the training and are used to modify the learning procedure in the desired way. At every time step instead of using the fixed target $\mathbf{t}$, the calculated target $\mathbf{t_{new}}$ is used to calculate the error.

$$\mathbf{e} = \mathbf{t_{new}} - \mathbf{o}$$

The determination of the new targets is illustrated in Figure 3.

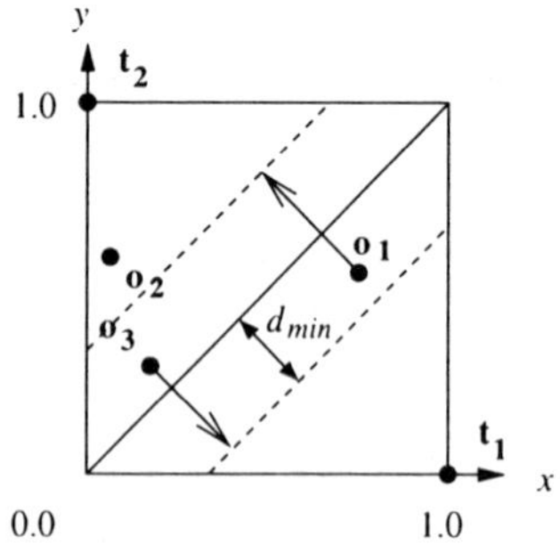

Figure 3: With adaptive dynamic targets the outputs are only forced to be in the right sub space and to have a distance of at least d_{min} to the border. In this case only $\mathbf{o}_1$ and $\mathbf{o}_3$ are moved. The error of $\mathbf{o}_2$ becomes zero. Compare with Figure 2.

If the output lies within the proper subspace and the distance to the border is equal or greater than some parameter d_{min}, no error is desired and therefore the new target becomes the actual output. If the output lies in the wrong sub space or in the right sub space but too near to the border, the new target is the point in the right sub space, having distance d_{min} to the border and being the nearest point to $\mathbf{o}$. This ensures right classification of the calculated new target, a minimal distance to the decision border d_{min} and yields the minimal error vector possible.

Calculating Adaptive Dynamic Targets

The complexity of the calculation of the adaptive dynamic targets depends on the dimensionality and on the positions of the fixed targets within the output space. A very simple case is the one illustrated in Figure 3. The dimension of the output space is two and the targets are made up by the canonical base ($\mathbf{t_1} = (1,0)^T, \mathbf{t_2} = (0,1)^T$). In the following $\mathbf{o}$ is the output of the network, $\mathbf{t}$ is the target for output $\mathbf{o}$, d is the distance of the output to the line $x - y = 0$, d_{min} is the minimal desired distance and $\mathbf{t_{new}}$ is the target to be calculated. The following three cases can be distinguished:

- $\mathbf{t}$ *is the nearest target and* $d \geq d_{min}$

  In this case no error is desired, so the new target is the current output.

$$\mathbf{t_{new}} = \mathbf{o}$$

- $\mathbf{t}$ *is the nearest target and* $d < d_{min}$

  In this case the output is corrected to ensure the desired distance of the output to the decision border.

$$\mathbf{t_{new}} = \begin{cases} \mathbf{o} + \frac{d-d_{min}}{\sqrt{2}} \begin{pmatrix} 1 \\ -1 \end{pmatrix} & \text{if} \quad \mathbf{t} = \begin{pmatrix} 1 \\ 0 \end{pmatrix} \\ \mathbf{o} + \frac{d-d_{min}}{\sqrt{2}} \begin{pmatrix} -1 \\ 1 \end{pmatrix} & \text{if} \quad \mathbf{t} = \begin{pmatrix} 0 \\ 1 \end{pmatrix} \end{cases}$$

- $\mathbf{t}$ *is not the nearest target*

  In this case the output is corrected to make it lie within the proper half of the unit square with the desired distance d_{min} to the decision border. Also in this case the minimal error vector satisfying these conditions is generated by choosing the new target as follows.

$$\mathbf{t_{new}} = \begin{cases} \mathbf{o} + \frac{d+d_{min}}{\sqrt{2}} \begin{pmatrix} 1 \\ -1 \end{pmatrix} & \text{if} \quad \mathbf{t} = \begin{pmatrix} 1 \\ 0 \end{pmatrix} \\ \mathbf{o} + \frac{d+d_{min}}{\sqrt{2}} \begin{pmatrix} -1 \\ 1 \end{pmatrix} & \text{if} \quad \mathbf{t} = \begin{pmatrix} 0 \\ 1 \end{pmatrix} \end{cases}$$

The calculations given are very easy as they apply to a two dimensional problem with the canonical base as target vectors. Using canonical bases for target vectors is very common but not necessary. If other target vectors or higher dimensional output spaces are used, the calculation of the dynamic targets will become more difficult, but the general idea behind dynamic targets will always be applicable.

Adapting Dynamic Targets

In the previous section it is shown that the parameter d_{min} determines the "strength of the teacher". This means if d_{min} gets large, the output is forced very near to its proper target, whereas for $d_{min} = 0$ the output is only forced to lie within the proper subspace of the unit square. Therefore the meaning of the parameter d_{min} is comparable to the strength of the teacher in supervised learning.

Considering time series classification it is unknown, which time steps in a given time series are useful and which may even harm the classification. The learning as well as the classification can be improved if the learning stresses those time steps, which are useful for classification and if the learning does not pay too much attention to time steps, which are not useful for classification. To make the learning emphasize some time steps the parameter d_{min} can be used. This is the basic idea of *adaptive dynamic targets.*

During the training of the classifier there is not one global parameter d_{min}, but a vector $\mathbf{d_{min}}$ of parameters $d_{min,i}$, $i \in \{1, 2, \ldots, n\}$ where n is the number of time steps in one time series. At the beginning all $d_{min,i}$ get initialized to some value $d_{min,init}$. This value should be larger than 0 and smaller than $\sqrt{2}/2$. During the training the error on some independent validation set is measured periodically (e.g. every 10 iterations) for each time step i. If the error at time step i did decrease the value of $d_{min,i}$ will be increased by some small $d_{min,delta}$. If the error did increase, $d_{min,i}$ will be decreased. The possible range of $d_{min,i}$ is $[-\sqrt{2}/2, \sqrt{2}/2]$. It does make sense to allow also negative values of $d_{min,i}$, which are interpreted as follows. If $d_{min,i} < 0$, an error is only generated if the distance of $\mathbf{o}$ to the border is greater than $|d_{min,i}|$ and $\mathbf{o}$ lies on the wrong side of the border. Therefore no error may be generated even if the output lies within the wrong sub space.

The result of this procedure is the vector $\mathbf{d_{min}}$ telling which time steps of the time series are useful for classification and which are not. The value of $d_{min,i}$ directly corresponds to the importance of time step i for classification of the time series. This information is extremely useful, when applying a trained classifier to new data. Doing so assigns a class to each time step of a time series. To find one class for the whole time series a weighted sum of all time steps is used. The weight for each time step can be determined from $\mathbf{d_{min}}$. Therefore a weight vector $\mathbf{w}$ is obtained by normalizing $\mathbf{d_{min}}$.

$$\mathbf{w} = \frac{\mathbf{d_{min}}}{\sqrt{2}} + \frac{1}{2}$$

The class information obtained for every time step is collected in a vector $\mathbf{c}$ and the class information of the whole time series is $\mathbf{w} \cdot \mathbf{c}^T / n$, where n is the number of time steps in the time series.

EXPERIMENTS AND RESULTS

To demonstrate the improvement in supervised learning due to dynamic targets, two experiments were done. In the first experiment artificial data has been used whereas in the second experiment EEG data from the BCI (see Section 2) was used. In both experiments the speed of learning and the achieved generalization capabilities were measured. The trained classifier is an FIR network [10, 6, 1] with two hidden units, two output units and FIR synapses of order two. In the case of the artificial data one input neuron is used and in the case of the EEG data 6 input neurons are used.

Comparing Fixed and Dynamic Targets

To compare the speed of learning the number of epochs needed to fulfill a given criteria is counted. The criteria is the convergence of the error rate on the training data during the training. Unfortunately this value strongly depends on the learning rate used. As the lengths of the error vectors for the dynamic targets are generally much shorter than the lengths of the error vectors calculated using fixed targets it would not be objective to compare these algorithms by using the same learning rate and counting the number of epochs. But there seems to be no way to correct or adjust the learning rate to make the comparison fair.

Therefore in this work the standard algorithm with fixed targets is modified slightly. The error vector $\mathbf{e}$ is calculated according to (1) but its length is adjusted. It is made equal to the length of the error vector, if adaptive dynamic targets would have been used (see Section 3.1). This scheme is called short fixed targets in the sequel. Figure 4 illustrates the three different error vectors.

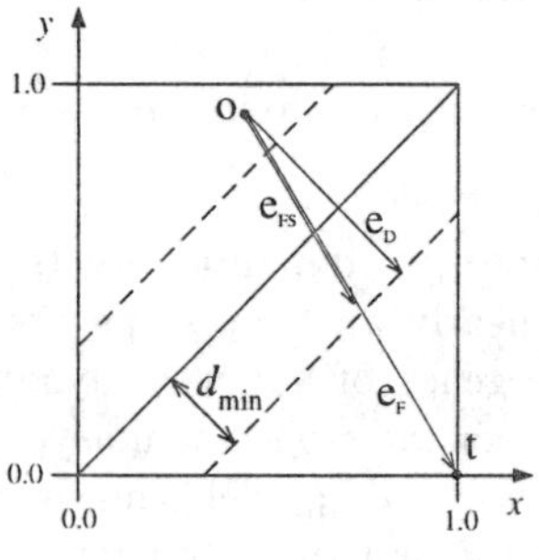

Figure 4: The three different error vectors obtained from using fixed targets ($\mathbf{e_F}$) and adaptive dynamic targets ($\mathbf{e_D}$) and short fixed targets ($\mathbf{e_{FS}}$). The output of the network is $\mathbf{o}$ and its target is $\mathbf{t}$.

To measure the generalization capabilities of the trained classifier the error rates on unseen data is given.

Classification of Artificial Data

In this experiment an FIR MLP network was trained on a temporal version of the well known exclusive or (XOR) problem. The input was a randomly generated binary sequence (e.g. s=(0,1,1,1,0,1,0,1)) added to white noise with $\mu = 0$ and some standard deviation σ. In this experiment targets for all time steps, but the first one, were known. If n is the number of the current time step and $\mathbf{s}(n)$ is the n-th element of $\mathbf{s}$, the target $\mathbf{t}(n)$ at time step n will be

$$\mathbf{t} = \begin{pmatrix} XOR(s(n-1), s(n)) \\ 1 - XOR(s(n-1), s(n)) \end{pmatrix}$$

Due to the described implementation of dynamic targets it is necessary to encode the target in that way. In the case of $n = 1$ $\mathbf{t}$ is set to $(0,1)^T$ or $(1,0)^T$ randomly. The output of the network is considered correct, if the output vector $\mathbf{o}$ is in the same sub space as the target vector $\mathbf{t}$. The error rate on a set of time series is defined as the average over all time steps.

In the experiment 10 classifiers are trained on a set of 50 time series each consisting of 30 time steps. White noise with $\mu = 0$ and $\sigma = 0.1$ is added to the training data. Table 1 gives the number of epochs needed for convergence of the error rate on the training data and the error rate achieved on new unseen data with some given σ and $\mu = 0$. The averages over the 10 classifiers as well as the standard deviations are given.

Learning Scheme	*Number of Epochs*	*Error Rate* $\sigma = 0.1$	*Error Rate* $\sigma = 0.2$
Fixed Targets	141 ± 157	9.5 ± 11.8	15.4 ± 10.0
Short Fixed Targets	60 ± 34	4.6 ± 7.5	10.4 ± 6.4
Adaptive Dynamic Targets	72 ± 20	1.8 ± 0.1	7.1 ± 0.9

Table 1: Results of fixed targets, short fixed targets, and adaptive dynamic targets on artificial data.

These results show that using dynamic targets improves the learning process even if targets for nearly all time steps are known. The number of epochs needed for convergence of adaptive dynamic targets is not significantly smaller, but the error rates on new noisy data are much better. The analysis of the generated vector $\mathbf{d_{min}}$ yields no significant differences between different time steps, only the first time step gets a very low value as it is not possible to classify correctly if only the first time step is given.

Classification of EEG Data

In this experiment real world data is used. Preprocessed EEG signals from patients are classified according to the BCI experiment. The time series consist of 33 time steps, which represent the time from second 3.0 to second 7.0 in a single BCI experiment. 20 classifiers are trained on a set of 160 time series. The number of epochs needed for convergence of the error rate on the training data and the error rate achieved on new unseen test data is given in Table 2. The averages over the 20 classifiers as well as the standard deviations are given.

Learning Scheme	*Number of Epochs*	*Error Rate*
Fixed Targets	1353 ± 1588	16.3 ± 3.1
Short Fixed Targets	833 ± 901	15.8 ± 3.1
Adaptive Dynamic Targets	471 ± 166	14.3 ± 2.1

Table 2: Results of fixed targets, short fixed targets, and adaptive dynamic targets on real world (EEG) data.

In this case the adaptation of the parameters $d_{min,i}$ turned out to be very important, as not all time steps of the given time series were useful for classification. The final values of the vector $\mathbf{d_{min}}$ for one of the trained classifiers are plotted in Figure 5.

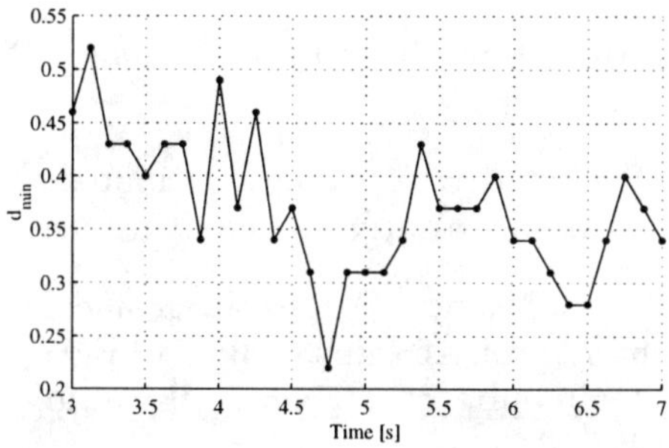

Figure 5: Resulting vector $\mathbf{d_{min}}$ after the training of EEG time series. Only the positive range of $d_{min,i}$ is used. Most of the time seem to be very useful for classification.

CONCLUSION

In this paper a new and simple technique, called *adaptive dynamic targets*, has been introduced. It applies to time series classification with supervised learning. As only the targets get modified, the procedure is not restricted to any specific learning algorithm and can easily be integrated into every existing learning procedure. The proposed technique does better exploit the capabilities of neural networks, which results in faster convergence and better generalization of the trained classifier.

REFERENCES

[1] A. D. Back and A. C. Tsoi, "FIR and IIR synapses, a new neural network architecture for time series modeling," **Neural Computation**, vol. 3, pp. 375–385, 1991.

[2] R. Bakker, J. C. Schouten, C. M. van den Bleek and C. L. Giles, "Neural learning of chaotic dynamics: The error propagation algorithm," Tech. Rep. UMIACS-TR-97-77 and CS-TR-3843, Institute for Advanced Computer Studies, University of Maryland, 1997.

[3] M. Bauer, **Raum-zeitliches Chaos in neuronalen Netzen**, Harri Deutsch, 1994.

[4] C. M. Bishop, **Neural Networks for Pattern Recognition**, Oxford: Clarendon Press, 1996.

[5] E. K. Blum and X. Wang, "Stability of fixed points and periodic orbits and bifurcations in analog neural networks," **Neural Networks**, vol. 5, pp. 577–587, 1992.

[6] S. Haykin, **Neural Networks**, Macmillan College Publishing Company, 1994.

[7] M. I. Jordan, "Serial order: A parallel distributed processing approach," Tech. rep., Institute for Cognitive Science, University of California, 1986.

[8] G. Pfurtscheller, C. Neuper, D. Flotzinger and M. Pregenzer, "EEG-based discrimination between imagination of right and left hand movement," **Electroenceph. clin. Neurophysiology**, vol. 103, no. 6, pp. 642–651, 1997.

[9] D. E. Rumelhart and J. L. McClelland, **Parallel Distributed Processing**, vol. 1, MIT Press, 1986.

[10] E. A. Wan, "Time series prediction by using a connectionist network with internal delay lines," in A. S. Weigend and N. A. Gershenfeld, eds., **Time series prediction: Forecasting the future and understanding the past**, Addison-Wesley Publishing, 1993, pp. 195–217.

[11] X. Wang and E. K. Blum, "Dynamics and bifurcation of neural networks," in M. A. Arbib, ed., **The Handbook of Brain Theory and Neural Networks**, MIT Press, pp. 339–353, 1995.

Some Analytical Results On Critic-driven Ensemble Classification *

David J. Miller and Lian Yan
Department of Electrical Engineering
The Pennsylvania State University
University Park, PA 16802-2701
Phone: (814) 865-6510 E-mail: *miller@bilbo.ee.psu.edu*

Abstract

In recent work [14], we proposed a framework for ensemble classification wherein auxiliary networks, dubbed *critics*, are used to provide reliability information on the ensemble's individual classifiers/experts. We showed experimentally that critic-driven combining schemes extend the applicability of ensemble methods by overcoming the usual requirement that the individual classifier error rate p must be less than 0.5. Here, we support our previous work by proving, under an independence assumption, that performance for a particular critic-driven voting scheme improves with increasing ensemble size N, *so long as* $p+q<1$, with q the critic's error rate in predicting accuracy of expert decisions. While this independence analysis gives significant insight into the conditions for success of critic-based schemes, it does not accurately predict the ensemble performance curve. We thus also develop an analytical approach for predicting the curve, by modeling dependence between experts.

1 Critic-driven Ensemble Classification

In recent years, ensemble classification has emerged as one of the most promising and one of the most actively investigated classification paradigms, see e.g. [6] for a survey. In this approach, probabilistic class estimates or decisions from each classifier in a collection are pooled, using some rule of combination, in order to make the ultimate decision for a new datum. The estimates may be combined either through some form of (soft) averaging or (hard) voting. Ensemble techniques overcome the suboptimality of individual classifiers (due to local minimum traps that limit training effectiveness, poorly chosen features, or deficient training data) via collective decisionmaking – while some classifiers may err, the ensemble decision will still be correct if a sufficient number of classifiers make correct decisions for a

*This work was supported in part by National Science Foundation Career Award IIS-9624870.

given datum. Some understanding of the conditions for success of these methods is gained by considering majority-based voting. Assuming voters make statistically independent errors, the *ensemble* error rate is based on the binomial distribution and decreases for increasing (odd/even) number of voters (N) so long as the individual *voter's* error rate (p) is less than 0.5. If p is greater than 0.5, then the error rate *increases* with (odd/even) N[1]. In practice, it is difficult to form an ensemble of independent classifiers, even for small N. Still this simple analysis does identify what can be hoped for – that performance should improve with increasing N. It also identifies important properties – classifier diversity (independence) and sufficient reliability ($p < 0.5$) – required for voting and other ensemble techniques as well[6]. While $p < 0.5$ appears to be crucial for the success of voting and other simple combining schemes, some recently proposed methods, e.g. stacking [4], boosting[15], and mixture of experts models [8],[11] may overcome this requirement by jointly optimizing the individual classifiers that comprise the ensemble and/or by optimizing the network that combines their decisions. While these methods achieve good performance and may overcome the condition $p < 0.5$, their training requirements may preclude their use for a number of important scenarios. Consider, for example: 1. Large-scale problems: when there are numerous classes, high feature dimensionality, and a large training set, e.g. in character recognition, joint optimization techniques may be too complex to practically implement, even for off-line training. 2. Adaptive/non-stationary settings: here, a sequential, on-line version of the training procedure must be used. Even if one assumes classifiers are separately trained (and adapted), on-line adaptation of the combiner network may be both too computationally complex, as well as ineffective. 3. Multi-sensor classification: e.g., target classification based on multiple sensing modalities, multimedia classification [12], and classification with distributed databases. In this case, labelled training sets may be separately extracted (possibly at different times, and at different sampling rates) for each sensor/source. Thus, the different sensor training sets may be *incommensurate*, i.e. there may be no known correspondence between exemplars from each of the different sets – the different sensors may not have even been applied to the same target examples. This situation prohibits pooling data to form a single, common training set, based on exemplar "total" feature vectors. Since there is no common training data, individual classifiers must be trained and combiner optimization is not possible. 4. Existing or proprietary designs.

Each of these circumstances rules out use of complex ensemble techniques. Instead, simpler schemes which make less assumptions about available data and which reduce the training burden are required. The main techniques in this category are (hard) voting [7], [17] and soft averaging [17]. These methods do not require a common training set; further, they use simple combining rules that do not require training. Such schemes are applicable to all of the aforementioned scenarios. However, this general applicability is achieved by sacrificing performance – methods such as voting are fundamentally limited in that they are only successful when the individual voter's error rate p is less than 0.5. While this requirement is likely easily met when there are few (e.g. less than 10) classes, meeting this requirement appears substantially difficult for real-world problems with tens, hundreds, or even thousands of classes. Within-scene object recognition, where the set of possible objects may be vast, is one such problem. In recent work[14] we developed a

[1]This result is attributed to the Marquis de Condorcet (18th century) [1],[3], who addressed the problem in a jury context.

new combining paradigm that is generally applicable like voting and averaging, but which overcomes performance limitations of these approaches via the use of *critics* – auxiliary networks trained to provide reliability information on expert decisions . For each expert in the ensemble, there is an associated critic. Given the same input feature vector used by the expert, the critic tries to predict whether the expert's decisions are *valid* or *bogus*. The critic is trained after its expert, on the expert's training set, with binary supervising labels indicating whether or not the expert's decisions agree with the true class labels – essentially, the critic is an expert *on* the expert[2]. Since the expert is assumed to tackle a multiclass (> 2 class) problem and the critic only needs to solve a two-class problem, the critic's estimates should be more accurate than the expert's on average [3]. The suggestion is then to incorporate the critic's opinion within the rule of combination so as to achieve more reliable decisions. In [14] we proposed critic-based extensions of both hard voting and soft averaging and presented experimental results to validate the paradigm's effectiveness. Here we further validate this approach with some analytical results. Toward this end, we first introduce a particular form of critic-driven voting, one quite amenable to analysis.

2 Analysis of Voting

2.1 Critic-driven Voting with an Odd-sized Ensemble

In the most straightforward critic-driven extension of majority-based voting, each expert votes for the class which it predicts if the critic assesses the expert prediction is valid. Otherwise, the expert *abstains* from voting. The total number of votes (K) equals N minus the number of abstentions. All votes are given equal weight. If any class receives $l > \frac{K}{2}$ votes, that class is the one predicted by the ensemble. Otherwise the datum is rejected. An alternative rule is motivated by the following fact: the probability of correct decision for majority-based voting with N voters, N even, is *less than* the probability of correct decision with $N-1$ voters. This suggests a modification of the critic-based scheme just described wherein, when K is even, one voter is *dropped* prior to vote tallying. In practice, the voter with least confidence from its critic could be dropped. For analysis purposes, we will assume the dropped voter is randomly selected. This modified scheme has two advantages over the former one. First, it achieves a greater correct decision rate, with less rejections. Second, this modified rule greatly simplifies the performance analysis that we next develop.

2.2 Critic-driven Analysis Under Independence

Assume experts make independent errors, with common rate p. Then, for standard majority-based voting, the probability of l correct votes from a total of N experts is

[2]While this approach does require extra training (for the critics), it is applicable to the aforementioned scenarios where joint optimization is precluded. For example, for the "proprietary" scenario, we are simply assuming each company designs a critic, along with its expert.

[3]The critic's two-class problem may be difficult, as it requires identifying the expert's errors. Still, we have found experimentally that accurate critics, with lower error rate than their experts, can be effectively trained.

given by a binomial distribution. The correct decision rate for the majority-based scheme is

$$P_{c_m}(N) = \sum_{l > \frac{N}{2}}^{N} \binom{N}{l} (1-p)^l p^{N-l}. \tag{1}$$

By Condorcet's theorem, if $p < 0.5$, P_{c_m} increases with odd(even) N[4]. Likewise, if $p > 0.5$, P_{c_m} *decreases* with increasing odd(even) N[1]. Now, consider critic-driven majority voting, wherein a voter is dropped, if required, to achieve an odd-sized voting ensemble. Assume critics make errors at common rate q, independent of their experts and independent of other experts and critics. For convenience, assume q is the rate both for false "valid" and false "bogus" critic decisions. The probability of M voters ($N - M$ abstentions) is binomial, i.e.

$$P_N[M] = \binom{N}{M} (pq + (1-p)(1-q))^M ((1-p)q + p(1-q))^{N-M}. \tag{2}$$

Further, the distribution over the number of correct voters is also binomial. Thus, the probability of correct decision given a critic-driven voting ensemble of size M is

$$P[\text{correct}|M] = \sum_{l > \frac{M}{2}}^{M} \binom{M}{l} (1-\tilde{p})^l \tilde{p}^{M-l}, \tag{3}$$

with $\tilde{p} = \frac{pq}{pq+(1-p)(1-q)}$. Averaging over all ensemble sizes, we get the critic-based correct decision rate:

$$P_{c_c}(N) = \sum_{M=1}^{N} P_N(M) P[\text{correct}|\text{odd}(M)]. \tag{4}$$

Here, odd(M) returns M if M is odd; else, it returns $M - 1$. Similar to P_{c_m}, P_{c_c} increases with (this time increasing *integer*) N for $p < 0.5$, assuming $q < p$. However, unlike P_{c_m}, if q is sufficiently small, then even if $p > 0.5$, P_{c_c} *still* increases with increasing integer N. Essentially, the critics force abstentions so as to reduce the individual *voter's* error rate below 0.5, *even when the expert rate p is above 0.5.* In particular, we have the following result which shows that, under independence, critics extend achievable voting performance:

Theorem: The critic-driven correct decision rate increases with increasing integer N if $p + q < 1$.

Proof: We first write $P_{c_c}(N+1)$ starting from (4), by conditioning on the $(N+1)$st expert's inclusion in the voting. This yields

$$P_{c_c}(N+1) = \sum_{M=0}^{N} \Big(P_N(M) \cdot P[\text{correct}|odd(M)] \cdot \text{Prob}[(N+1)\text{st expert abstains}] + P_N(M) \cdot P[\text{correct}|odd(M+1)] \cdot (1 - \text{Prob}[(N+1)\text{st expert abstains}]) \Big). \tag{5}$$

[4] For N even, $P_{c_m}(N-2) < P_{c_m}(N) < P_{c_m}(N-1)$. Thus, while $P_{c_m} \to 1$ as $N \to \infty$, the performance curve is jagged, dipping on odd-to-even transitions.

Note that the lower limit of the sum in (5) is zero rather than one, as in (4). This reflects the fact that when N voters abstain, a correct decision can still be reached by the $(N+1)$st voter. When the $M = 0$ term is excluded from $P_{c_c}(N+1)$, we can define the lower bound $P_L(N+1)$:

$$P_{c_c}(N+1) > P_L(N+1) \equiv$$

$$\sum_{M=1}^{N} \big(P_N(M) \cdot P[\text{correct}|odd(M)] \cdot \text{Prob}[(N+1)\text{st expert abstains}] +$$

$$P_N(M) \cdot P[\text{correct}|odd(M+1)] \cdot (1 - \text{Prob}[(N+1)\text{st expert abstains}])\big). \quad (6)$$

Now, if M is odd, then $P[\text{correct}|odd(M)] = P[\text{correct}|odd(M+1)] = P[\text{correct}|M]$. So

$$P_L(N+1) = \sum_{M odd} P_N(M) \cdot P[\text{correct}|M] +$$

$$\sum_{M even} \big(P_N(M) \cdot P[\text{correct}|M-1] \cdot \text{Prob}[(N+1)\text{st expert abstains}] +$$

$$P_N(M) \cdot P[\text{correct}|M+1] \cdot (1 - \text{Prob}[(N+1)\text{st expert abstains}])\big). \quad (7)$$

Now, subtracting $P_{c_c}(N)$ from this lower bound $P_L(N+1)$, we obtain

$$P_L(N+1) - P_{c_c}(N) = \sum_{M even} \big(P_N(M) \cdot (P[\text{correct}|M+1] -$$

$$P[\text{correct}|M-1]) \cdot (1 - \text{Prob}[(N+1)\text{st expert abstains}])\big). \quad (8)$$

But, from [1], it is known that the probability of correct decision for an odd number of voters satisfies the recursion:

$$P[\text{correct}|M+1] = P[\text{correct}|M-1] + (1-2\tilde{p}) \cdot B_{M-1}(1-2\tilde{p}; \frac{M}{2}). \quad (9)$$

Here, $B_K(\alpha; L)$ is the binomial probability of L successes in K trials with α the correct decision rate for each voter. Thus, if $\tilde{p} < 0.5$, or equivalently, if $p + q < 1$, we have $P[\text{correct}|M+1] > P[\text{correct}|M-1]$, and then $P_L(N+1) > P_{c_c}(N)$. Finally, we then get $P_{c_c}(N+1) > P_{c_c}(N)$.□
In summary, the critic-driven approach gives a new, less stringent condition for successful voting. Whenever we are able to train critics that are more accurate than their experts (with $q < \frac{1}{2}$), this critic-based scheme will allow relaxation of the condition $p < \frac{1}{2}$.

At this point, one may reasonably ask under which conditions we might expect to find $p < 0.5$. We can envision this occurring 1) when there are numerous classes to discriminate; 2) when there is inadequate training data for building the ensemble; 3) when training and/or use features are unreliable or noisy; 4) when weak individual classifiers are used to limit ensemble complexity [10]; 5) for distributed detection, where the individual classifiers/detectors each base decisions on only one or several sensor outputs [2]; 6) in *context-adaptive* situations, e.g. speaker-adaptive speech recognition or multi-font text recognition [16]. Here, a classifier designed for a particular context will work effectively on within-context data, but may perform poorly outside this scope; 7) when the ensemble is applied as the final "arbiter" to examples rejected by (simpler) previous-stage classifiers.

2.3 Analysis of Voting with Dependent Experts

The theorem of the last section does provide significant insight and is qualitatively consistent with our experimental results [14]. However, the independence assumption does not allow accurate modeling of the actual ensemble performance curve. Predicted curves obtained under independence are *grossly* inaccurate, both for critic-driven and standard majority-based voting. Thus, in addition to performing analysis under independence, we also aim here to provide an analysis tool for more accurately modeling ensemble performance. This could be helpful to the designer in choosing the ensemble size N.

In the social sciences, some results have been achieved, extending the Condorcet theorem to dependent voters[1],[13],[3]. This work provides the important qualitative characterization that voting performance can improve with increasing N even when there is (significant) voter dependence. Here we consider an alternative approach and specifically aim to predict the ensemble performance curve. We will consider the case of majority-based voting in detail, with a similar methodology applying for critic-driven voting. As for the independent case, we assume all experts are statistically identical. However, now, rather than characterizing the system with the single parameter p, we will use three – suppose we know probabilities associated with the decisions of *any pair* of classifiers: $p_1 \equiv$ Prob[both voters correct], $p_2 \equiv$ Prob[both voters in error], $p_3 \equiv$ Prob[1 correct, 1 in error]. We would like to use this (symmetrical dependence) information to obtain improved performance modeling. In the independent case, the probability of correct decision was calculated by considering the joint distribution over the ensemble of voters and summing up the probability over events that correspond to ensemble correct decisions. In the dependent case, we follow a similar procedure; however, we can no longer use the simple joint distribution arising from independence. In fact, the joint distribution consistent with the above pairwise probabilities is not unique. We therefore appeal to methods of statistical inference – in particular the principle of maximum entropy – to help obtain the desired probabilities. Extensive description and justification for this principle can be found in [9]. Maximum entropy was previously used to characterize voting in [7], but only under an independence assumption.

Let $E_i \in \{0,1\}$, with $E_i = 1$ denoting the event that voter i is in error. Then, according to the principle of maximum entropy, we seek the joint pmf $P[\underline{E} = \underline{e}] \equiv P[E_1 = e_1, E_2 = e_2, \ldots, E_N = e_N]$ that maximizes Shannon's entropy while satisfying marginalization constraints. In particular, for the probability of any pair of experts i and j both being correct, we have the constraint:

$$\sum_{l_1}\sum_{l_2}\cdots\sum_{l_N} P[E_1 = l_1, E_2 = l_2, \ldots, E_N = l_N] = p_1, \tag{10}$$

with l_k denoting the set of events summed over for the kth expert. These sets are specified as:

$$l_k = \begin{cases} \{0,1\} & \text{if } k \not\subset \{i,j\}. \\ \{0\} & \text{if } k \subset \{i,j\}. \end{cases} \tag{11}$$

Such constraints are defined for *all* $\frac{N(N-1)}{2}$ voting pairs, (i,j). The constraints associated with p_2 and p_3 can also be defined in a similar way. In the most general case, the maximum entropy joint distribution satisfying constraints for all $L =$

$\frac{N(N-1)}{2}$ pairs takes the form [5]:

$$P[\underline{E} = \underline{e}] = \frac{\exp\left(\sum_{l=1}^{L} \gamma_{\underline{e},l}\right)}{Z}. \tag{12}$$

Here, $\gamma_{\underline{e},l}$ is the Lagrange multiplier associated with the lth voting pair, which depends upon the particular realization $\underline{e}$. Also, Z is the partition function, which achieves the proper normalization.

While the general expression (12) involves $\frac{N(N-1)}{2}$ *sets* of Lagrange multipliers (one set for each pairwise pmf), this expression and subsequent calculations greatly simplify in our case because voters are statistically identical. Thus, while our joint distribution does have the form (12), there are only 3 *distinct* Lagrange multipliers, common to each set – i.e., $\gamma_{\underline{e},l} \in \{\gamma_1, \gamma_2, \gamma_3\}$, with γ_i corresponding to p_i, $i = 1$, 2, and 3. By capitalizing on this fact, it can be shown that, for any realization $\underline{e}_k$ with k correct voters, the form of the joint probability specializes to:

$$P[\underline{E} = \underline{e}_k] = \frac{\exp\left(f(N,k)\right)}{Z}, \tag{13}$$

with $f(N,k) = \gamma_1 \binom{k}{2} + \gamma_2 \binom{N-k}{2} + \gamma_3 \cdot k \cdot (N-k)$, and with $Z = \sum_{l=0}^{N} \binom{N}{l}$ $\exp\left(f(N,l)\right)$. Here, only the parameters γ_1,γ_2, and γ_3 are undetermined. These should be chosen to satisfy the pairwise constraints. For given γ_1, γ_2, and γ_3, we can sum over the joint distribution to obtain the distribution's *estimates* of the pairwise probabilities p_1,p_2,p_3, for example, $\hat{p}_1 = \frac{1}{Z} \sum_{k=2}^{N} \binom{N-2}{k-2} e^{f(N,k)}$. $\hat{p}_2$ and $\hat{p}_3$ can be obtained in a similar way. The Lagrange multipliers should be chosen so that $\hat{p}_i = p_i$. This will be achieved if we minimize (by zeroing) the Kullback distance between them:

$$\min_{\gamma_1,\gamma_2,\gamma_3} \sum_{i=1}^{3} p_i \log \frac{p_i}{\hat{p}_i}. \tag{14}$$

In practice, we perform this minimization using a gradient descent procedure. Note that this minimization is for a given number of voters, N. Finally, the probability of correct decision based on this joint pmf is:

$$\text{Prob[correct]} = \frac{1}{Z} \sum_{k > \frac{N}{2}} \binom{N}{k} \exp\left(f(N,k)\right). \tag{15}$$

Here, we have only described the method for majority-based voting. For critic-based voting the methodology is the same. However, now, we choose the maximum entropy distribution consistent with six distinct pairwise constraints. The larger constraint set accounts for pairwise events where experts in the pair are forced by their critics to abstain from voting. In the next section, we will compare the predictions of our dependence analysis with empirical voting results and with predictions based on independence.

3 Experimental Results

In Figure 1, we compare the hard majority voting approaches on the UC Irvine repository's *glass* data set. We used 10 even training/test splits for designing RBF (expert, critic) pairs with (10, 20) components. Here, $p = 0.37$ and $q = 0.33$. The performance trend of the standard majority voting is almost constant, while the critic-driven majority voting approaches have a decreasing error rate[5] trend and achieve large improvement over standard voting. Note that modified critic-driven majority voting achieves a lower error rate than simple critic-driven majority voting.

We empirically evaluated the performance modeling approach developed in section 2.3 and the independence analysis by comparing predictions obtained for conventional majority voting on the Peterson-Barnsley vowel data set (Figure 2) and for critic-based majority voting on the *glass* data set (Figure 3). The pairwise probabilities were first estimated based on the (designed) classifier ensemble. We then generated performance predictions for each size N by first minimizing the cross entropy cost as we have described, then plugging the resulting Lagrange multipliers into the probability of correct decision expression (15). We can see that the maximum entropy approach accurately predicts standard and critic-based majority voting performances. The main use of this analysis tool is to give the designer some notion of the performance/complexity (N) tradeoff associated with an ensemble scheme, prior to designing the entire ensemble of classifiers[6].

References

[1] S. Berg. Condorcet's jury theorem, dependency among jurors. *Social Choice and Welfare*, 10:87–95, 1993.

[2] R. S. Blum, S. A. Kassam, and H. V. Poor. Distributed detection with multiple sensors. *Proc. of the IEEE*, 85(1):64–79, 1997.

[3] P. J. Boland. Majority systems and the Condorcet jury theorem. *The Statistician*, 38:181–189, 1989.

[4] L. Breiman. Stacked regression. *Mach. Learning*, 24:41–64, 1996.

[5] P. Cheeseman. A method of computing generalized Bayesian probability values for expert systems. In *Proc. of the Eighth Intl. Joint Conf. on Art. Intell.*, volume 1, pages 198–202, 1983.

[6] T. G. Dietterich. Machine-learning research: four current directions. *AI Magazine*, pages 97–136, Winter 1997.

[7] L. K. Hansen and P. Salamon. Neural network ensembles. *IEEE Trans. on Patt. Anal. and Mach. Intell.*, 12(10):993–1001, 1990.

[8] R. A. Jacobs, M. I. Jordan, S. J. Nowlan, and G. E. Hinton. Adaptive mixtures of local experts. *Neural Comp.*, 3:79–87, 1991.

[5] Errors include rejections in our experiments. Each point represents error rates averaged over the 10 runs.

[6] While for this experiment the pairwise probabilities were obtained after designing the entire ensemble, we have found that fairly accurate probabilities can be estimated from very small ensembles ($N = 5$). These probabilities can then be used to predict the entire performance curve (e.g., up to $N = 60$, as in the figure).

[9] E. T. Jaynes. *Papers on Probability, Statistics and Statistical Physics.* Reidel, Dordrecht, 1982.

[10] C. Ji and S. Ma. Combined weak classifiers. In *NIPS 9*, pages 494–500, 1997.

[11] M. I. Jordan and R. A. Jacobs. Hierarchical mixtures of experts and the EM algorithm. *Neural Comp.*, 6:181–214, 1994.

[12] S. Kung and J. Hwang. Neural networks for intelligent multimedia processing. *Proc. of the IEEE*, 86(6):1244–1272, 1998.

[13] K. K. Ladha. Condorcet's jury theorem in light of de Finetti's theorem. *Social Choice and Welfare*, 10:69–85, 1993.

[14] D. J. Miller and L. Yan. Ensemble classification by critic-driven combining. In *Proc. of the IEEE ICASSP*, volume 2, pages 1029–1032, 1999.

[15] R. E. Schapire. Using output codes to boost multiclass learning problems. In *Machine Learning: Proc. of the Fourteenth Intl. Conf.*, pages 313–321, 1997.

[16] J. B. Tenenbaum and W. T. Freeman. Separating style and content. In *NIPS 8*, pages 662–668, 1996.

[17] L. Xu, A. Krzyzak, and C. Y. Suen. Methods of combining multiple classifiers and their applications to handwriting recognition. *IEEE Trans. on Sys., Man, and Cyb.*, 22(3):418–435, 1992.

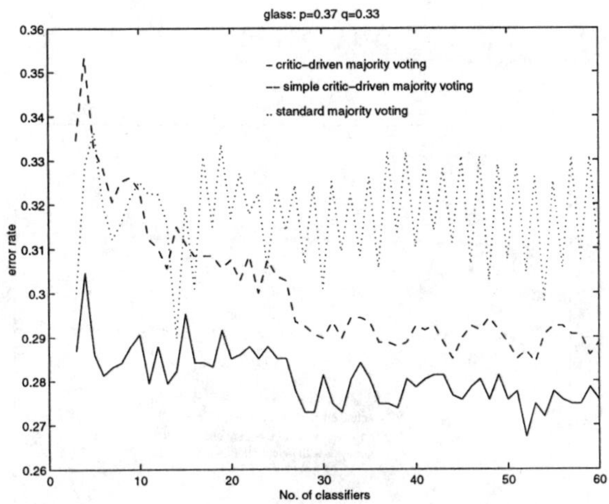

Figure 1: Standard and critic-driven majority voting performance.

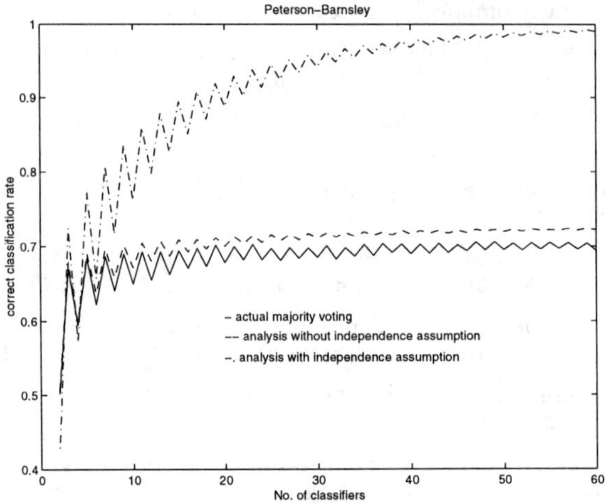

Figure 2: Prediction of standard majority voting performance.

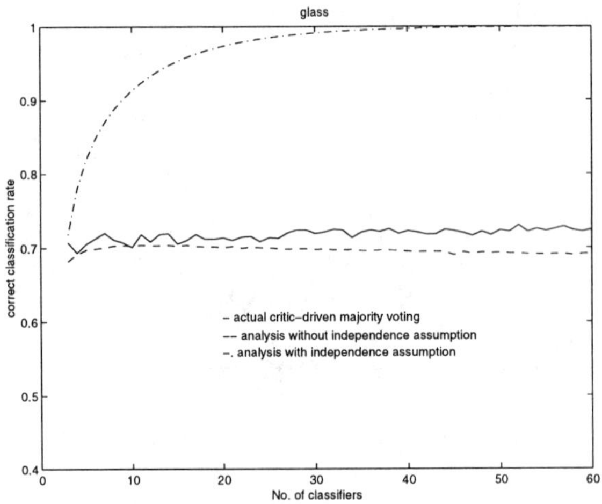

Figure 3: Prediction of critic-driven majority voting performance.

NEURAL ARCHITECTURES FOR PARAMETRIC ESTIMATION OF A POSTERIORI PROBABILITIES BY CONSTRAINED CONDITIONAL DENSITY FUNCTIONS

Juan Ignacio Arribas[†], Jesus Cid-Sueiro[†], Tülay Adali[††] and Anibal R. Figueiras-Vidal[†††]

[†]Dept. of Teoría de la Señal, Comunicaciones e Ing. Telemática, Universidad de Valladolid, Valladolid, Spain.

[††] Dept. of Computer Science and Electrical Engineering, University of Maryland Baltimore County, Maryland, USA.

[†††] Dept. of Tecnologías de las Comunicaciones, Universidad Carlos III de Madrid, Madrid, Spain.

Voice: +34 983 423000, Fax: +34 983 423667, email: jarribas@tel.uva.es

Abstract. A new approach to the estimation of 'a posteriori' class probabilities using neural networks, the Joint Network and Data Density Estimation (JNDDE), is presented in this paper. It is based on the estimation of the conditional data density functions, with some restrictions imposed by the classifier structure; the Bayes' rule is used to obtain the 'a posteriori' probabilities from these densities. The proposed method is applied to three different network structures: the logistic perceptron (for the binary case), the softmax perceptron (for multi-class problems) and a generalized softmax perceptron (that can be used to map arbitrarily complex probability functions). Gaussian mixture models are used for the conditional densities. The method has the advantage of establishing a distinction between the network architecture constraints and the model of the data, separating network parameters and the model parameters. Complexity on any of them can be fixed as desired. Maximum Likelihood gradient-based rules for the estimation of the parameters can be obtained. It is shown that JNDDE exhibits a more robust convergence characteristics than other methods of a posteriori probability estimation, such as those based on the minimization of a Strict Sense Bayesian (SSB) cost function.

INTRODUCTION

It is well known that when a neural network is trained in order to minimize a mean square error [1] or the cross entropy [2] between the targets and the network

0-7803-5673-X/99/$10.00 © 1999 IEEE

outputs, the network provides, after convergence, estimates of the 'a posteriori' probabilities of the classes; many applications in pattern recognition can take advantage of this property: medical diagnosis, financial data analysis, digital communications, among others. In [3], [5], [6], any cost function providing *a posteriori* class probabilities is called *Strict Sense Bayesian* (SSB). General conditions for SSB cost functions can be found in [3], [4] and [6] .

When the *a posteriori* probabilities are estimated using SSB cost functions, no previous assumptions about the data distribution are required: probabilities are estimated without computing the conditional density functions of the classes. It is known, however, that this poses some problems:

- The estimation of *a posteriori* probabilities in well-separated data requires very large training sets.
- Prior knowledge about the data distribution can not be used.
- It is difficult to use unlabeled data to improve learning.

These difficulties are of major importance in many real-world applications. In fact, although the probability estimation property is usually considered as a potential advantage of neural networks, it is rarely used in practice. An alternative to *a posteriori* probability estimation with SSB cost functions is Density Estimation (DE): it consists on estimating the conditional density functions of the data samples corresponding to different classes. Using Bayes' rule:

$$p(j|\mathbf{x}) = \frac{p(j)f(\mathbf{x}\,|\,j)}{\sum_{i=1}^{N} p(i)f(\mathbf{x}\,|\,i)} \tag{1}$$

where N is the number of classes, $p(j)$ is the *a priori* probability of class j, $\mathbf{x}$ is an arbitrary point of the sample space and f denotes a density function; the computation of *a posteriori* probabilities from conditional densities is straightforward. Unlike SSB cost-based methods, the DE approach requires to specify some parametric model for the data densities; in spite of this, we will see that we can take some advantages of it.

In this paper we try to establish a link between DE and SSB-based methods: a density model is assumed, but it can be made as general as desired, without increasing the complexity of the classifier after learning. In the following, we will refer to this method as the *Joint Network and Data Density Estimation* (JNDDE).

The structure of the paper is as follows: first, we describe the basic ideas behind the new method, defining the *implicit set* concept; second, we consider the binary logistic problem, which is going to help us to gain some insight in understanding the nature of the JNDDE. Third, we discuss more complex networks, to make it suitable to more general problems and show two general structures: in the first one, we employ a simple multi-output architecture and the second one is a general classification structure that can map arbitrarily complex probability maps. Fourth, we describe the results obtained through computer simulations. The last section states some conclusions.

FUNDAMENTALS

Consider the N-output neural network whose output is given by $\mathbf{y}=\mathrm{g}(\mathbf{x},\mathbf{w})$, where $\mathbf{x}$ is the input and $\mathbf{w}$ is the parameter set. In order to interpret outputs as probabilities, we assume that

$$0 \le y_i \le 1, \quad i = 1,\ldots,N \quad \sum_{i=1}^{N} y_i = 1 \tag{2}$$

This restriction will be adopted in the following, but it is not strictly necessary. In a binary case, a single output network can be used to estimate the *a posteriori* probability of one of the classes, the other one is readily available by a simple subtraction. Also, a network with N independent outputs can be interpreted as a collection of N independent binary classifiers. According to Eq.(1), if the network is able to estimate the *a posteriori* class probabilities in some particular problem, it should be verified that

$$y_j = g_j(\mathbf{x},\mathbf{w}) = \frac{p(j) f(\mathbf{x}|j)}{\sum_{i=1}^{N} p(i) f(\mathbf{x}\,|\,i)} = p(j\,|\,\mathbf{x}) \quad j = 1,\ldots,N \tag{3}$$

DEFINITION: *Conditional density functions $\{f(\mathbf{x}|j), j=1...N\}$ form an implicit set for a network given by outputs $g_j(\mathbf{x},\mathbf{w})$ if equation (3) are satisfied for some parameter vector $\mathbf{w}$.*

From this definition, it is evident that a neural network can compute exactly the *a posteriori* probabilities of the classes for some parameter vector $\mathbf{w}$ if and only if the conditional density functions form an implicit set. For binary problems, word *set* may be substituted by *pair*, resulting an *implicit pair*.

The main idea of this paper is to establish a parametric model for the family of the possible implicit sets and afterwards to find the implicit set in this family that best fits the data in a Maximum Likelihood sense. The conditional densities will depend on the network parameters, but they may depend on an additional parameter set $\mathbf{v}$, hence we write $f(\mathbf{x}|\mathrm{i})=f_{\mathbf{w},\mathbf{v}}(\mathbf{x}|\mathrm{i})$. The use of over-parametric densities (i.e. with more parameters than the network itself) is essential to allow us to estimate complex data density models without increasing the complexity of the classifier structure. This will be made clear later.

Note that this approach is network-dependent: for each neural network a parametric model of their implicit density sets must be found. For networks with several hidden layers, this may be a difficult task; in this paper, we restrict our analysis to a wide family of networks based on the logistic and the softmax functions.

THE BINARY LOGISTIC PROBLEM (LOP)

Sigmoidal perceptrons

We will describe our method starting with a simple case: the JNDDE binary *Logistic Problem* (Fig. 1). Let us consider a binary problem, with class labels 0 and 1, and a network that computes the logistic or sigmoid function given by:

$$y = \frac{1}{1+e^{-\mathbf{w}'\mathbf{x}-w_0}} \tag{4}$$

where **x** is the sample vector, and $\mathbf{w}^t$ indicates the transposed vector of **w**. In Fig. 1, we show the binary perceptron, consisting of a soft decision step (given by the sigmoid) and a hard decision step (given by the WTA), which assigns one to the winning class and zero to the other (arg max y_i), forming the output vector **d**.

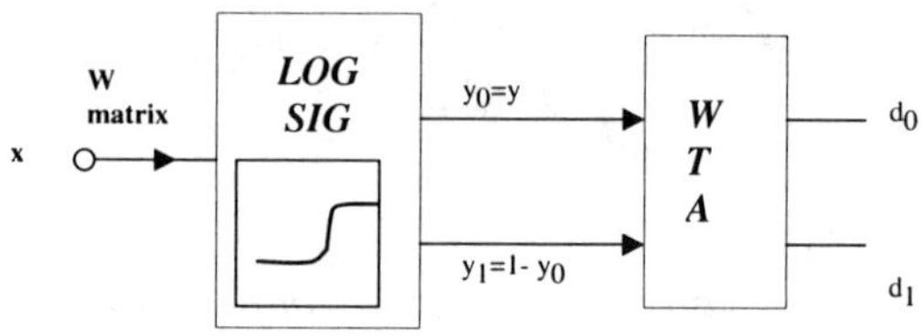

Fig. 1.- The binary Logistic network

We are interested here in training the logistic network in such a way that $y=p(1|\mathbf{x})$; The logistic non-linearity has been found by some authors ([8] is a good example) to be a specially suitable function for the case of Exponential Family distributions, (which includes the Gaussian distribution): it can be shown that the densities of the exponential family form implicit sets of the logistic network [2]. Here, we are interested in the class of all the implicit density sets for the logistic perceptron. It is provided by the following Theorem.

***THEOREM 1**: Density functions $f_0(\mathbf{x})$ and $f_1(\mathbf{x})$ form an implicit set for the logistic perceptron network, if and only if they can be expressed in the form*

$$f_0(\mathbf{x}) = \frac{q_0}{p_0} e^{-\frac{\mathbf{w}^T\mathbf{x}+w_0}{2}} f_c(\mathbf{x})\,; \quad f_1(\mathbf{x}) = \frac{q_1}{p_1} e^{\frac{\mathbf{w}^T\mathbf{x}+w_0}{2}} f_c(\mathbf{x}) \tag{5}$$

where f_c is an arbitrary density function that will be called the Central Density.

Proof: The sufficient condition is straightforward: replacing (3) and (5) in the Bayes' rule, we get a logistic function of the input. The necessity condition can also be proved easily. If (5) holds, we can write

$$p_0 f_0(\mathbf{x}) e^{+\frac{\mathbf{w}^T\mathbf{x}+w_0}{2}} = p_1 f_1(\mathbf{x}) e^{-\frac{\mathbf{w}^T\mathbf{x}+w_0}{2}} \tag{6}$$

where p_1 and p_0 denote the '*a priori*' class probabilities, $f_0(\mathbf{x})=f(\mathbf{x}|0)$ and $f_1(\mathbf{x})=f(\mathbf{x}|1)$ are the conditional class probability density functions. Equation (6) shows the conditions that any implicit pair must verify. Defining

$$f_c(\mathbf{x}) \equiv \frac{p_0 f_0(\mathbf{x})}{q_0} e^{\frac{\mathbf{w}^T\mathbf{x}+w_0}{2}} = \frac{p_1 f_1(\mathbf{x})}{q_1} e^{-\frac{\mathbf{w}^T\mathbf{x}+w_0}{2}} \tag{7}$$

where q_0 , q_1 are some normalizing factors ensuring that $f_c(\mathbf{x})$ is a p.d.f. and. expressing f_0 and f_1 as functions of f_c, the proof is completed. ■

Selecting different central densities, we can generate arbitrary pairs of implicit density functions. For instance, if $f_c(\mathbf{x})$ is a zero-mean Gaussian function, we obtain a Gaussian pair.

Estimating implicit pairs

The JNDDE method is based on making hypotheses about the implicit pair and estimating its parameters using the data. It can be summarized as follows:

1. Assume a model the central p.d.f. $f_c(\mathbf{x},\mathbf{v})$ where $\mathbf{v}$ is the parameter vector to be estimated.
2. Express f_0 and f_1 as a function of $\mathbf{w}$ and $\mathbf{v}$
3. Obtain MLE of $\mathbf{w}$, $\mathbf{v}$. (Gradient-based learning rules can be obtained, but the details are not shown in this paper).
4. Use estimated $\mathbf{w}$ as the perceptron weights, in such a way that y are estimates of *a posteriori* probabilities.

The first step is a key point of the method. The central density should satisfy two important requirements:

1. Since the computation of the conditional densities from the central density requires to integrate the product of the central density by a exponential function, we must select a model f_c providing closed form solutions for these integrand operations.
2. The central density model should be general enough to map arbitrarily complex densities.

The first requirement is satisfied if $f_c(\mathbf{x},\mathbf{v})$ is a Gaussian function. It is easy to show that, in such a case, the class densities are also Gaussian functions, whose parameters depend on **w** and **v**. To satisfy the second requirement, we will assume that $f_c(\mathbf{x})$ is a *Gaussian Mixture*,

$$f_c(\mathbf{x}) = q_1 N(\mathbf{x}, \bar{v}_1) + ... + q_m N(\mathbf{x}, \bar{v}_m) \tag{8}$$

In such a case, both classes are Gaussian mixtures.

THE SOFTMAX PERCEPTRON (SP)

A similar analysis can be done for the multi-class problem. The softmax nonlinearity is the natural generalization of the logistic function for multiple-output problems, see [9] and [10]; consider the *Softmax Perceptron* (SP) in Fig. 2, whose outputs are given in (9).

$$y_j = softmax(\mathbf{w}_j^T \mathbf{x}) = \frac{e^{\mathbf{w}_j^T \mathbf{x}}}{\sum_{i=1}^{N} e^{\mathbf{w}_i^T \mathbf{x}}} = \frac{1}{1 + \sum_{i \neq j}^{N} e^{(\mathbf{w}_i - \mathbf{w}_j)^T \mathbf{x}}} \qquad j = 1,..., N \tag{9}$$

where N is the number of outputs, i.e., number of classes, y_j is the output of our network for class j, **x** its input vector and $\mathbf{w}_j$ is the weight vector.

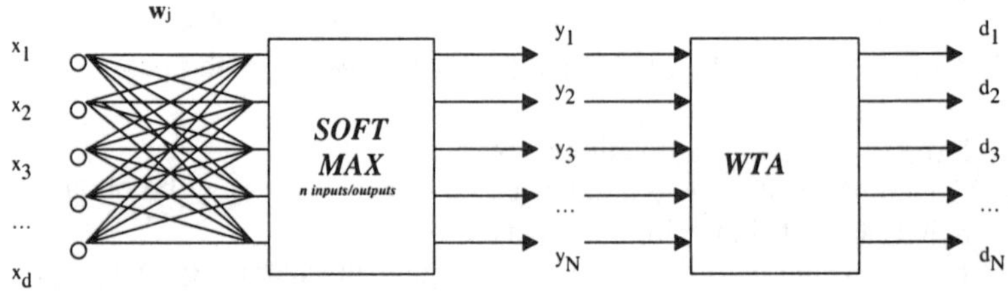

Fig. 2.- The multi-class SP network (**d** is the dimensionality of input data)

We wish y_j to be the *a posteriori* class probability of input sample **x** belonging to class j, so $f_j(\mathbf{x})$ would constitute an *implicit set*.

THEOREM 2: *Density functions $f_j(\mathbf{x})$ form an implicit set for the SP network, if and only if they can be expressed in the form*

$$f_j(\mathbf{x}) = \frac{f_c(\mathbf{x}) e^{\mathbf{w}_j^T \mathbf{x}}}{\int f_c(\mathbf{x}) e^{\mathbf{w}_j^T \mathbf{x}} d\mathbf{x}} \quad \forall j \tag{10}$$

*where j is class index, **x** input sample, **w** network weights and $f_c(\mathbf{x})$ an arbitrary probability density function.*

Proof: The proof is easy and follows the same steps than those in Theorem 1. ■

Thus, in (10) we have the conditional probability density functions for each class in terms of the *Central Density*, constrained by the chosen network structure, SP in this case. Later on, one would compute the partial derivatives for the gradient based learning rules for all the parameters, details not shown in this paper, to obtain Maximum Likelihood Estimates (MLE).

THE GENERALIZED SOFTMAX PERCEPTRON (GSP)

The SP network is not a universal classifier. A more general structure can be obtained by computing the class outputs as the sum of several softmax functions (see Fig. 3).

Again, applying Bayes' formula and defining outputs Z_i as estimates of *a posteriori* class probabilities, constituting densities f(**x**|i) an *implicit set*, we obtain:

$$Z_j \equiv \frac{\sum_{k=1}^{M_j} e^{\mathbf{w}_{j,k}^T \mathbf{x}}}{\sum_{i=1}^{N} \sum_{k=1}^{M_i} e^{\mathbf{w}_{i,k}^T \mathbf{x}}} = \frac{p_j f(\mathbf{x} \mid i)}{\sum_{i=1}^{N} p_i f(\mathbf{x} \mid j)} \quad j = 1, \ldots, N \tag{11}$$

where M_j are the number of elements within *j*-th class, also called the subclass number, and *N* is the number of classes.

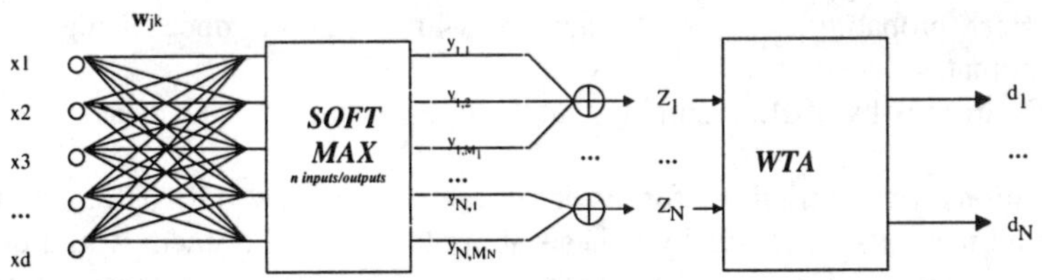

Fig. 3.- The GSP Network

THEOREM 3: Density functions $f_j(\mathbf{x})$ form an implicit set for the GSP network, if and only if

$$f_j(\mathbf{x}) = \frac{f_c(\mathbf{x})\sum_{k=1}^{M_j} e^{\mathbf{w}_{j,k}{}^T\mathbf{x}}}{\int f_c(\mathbf{x})\sum_{k=1}^{M_j} e^{\mathbf{w}_{j,k}{}^T\mathbf{x}} d\mathbf{x}} \qquad j = 1,\ldots,N \tag{12}$$

where j is class index, M_j is the number of elements within j-th class, $\mathbf{x}$ the input sample, $\mathbf{w}$ the network weights and $f_c(\mathbf{x})$ an arbitrary probability density function.

Proof: The proof is easy and also omitted. Again, gradient based learning rules can be obtained which are use to recursively adapt all the parameters, both network and data for the GSP net. ■

RESULTS

We tested JNDDE through computer simulation. Here we consider just a simple example to illustrate the main difference between this method and those based on minimizing a SSB cost. Other simulation examples analyzing convergence speed of stochastic gradient algorithms can be found in [5], [6] and [11].

Consider a zero-mean Gaussian two-dimensional central density f_c with variance matrix $\sigma^2\mathbf{I}$. We took σ=1. Consider the implicit pair of conditional densities resulting from this central density and a logistic perceptron with parameter vector **w**=(-1,3). Data were generated according to this density pair. The simulation work followed three steps

1) Estimate **w** minimizing a cross entropy cost function. Stochastic gradient learning was used. The algorithm was stopped after 16384 iterations. At this time the algorithm has almost reached the final convergence values.
2) Estimate **w** by MLE of the conditional densities. Since classes are Gaussian, the JNDDE approach reduces, in this case, to estimate mean, variances and a priori probabilities of each classes, which can be done using standard estimates.
3) Estimate **w** by MLE, assuming that the true variance of each class is known.

Simulations were carried out for different training set sizes: 4,8,16,32,... up to 2^{13}. Since all methods make true hypothesis about the data, all of them would obtain a perfect estimate of **w** if unlimited data were available. For these finite size examples, we compared the estimation performance using the *Weight Relative Square Error* (WRSE) measure given by

$$WRSE = \frac{\left| \mathbf{w}_{opt} - \mathbf{w}_{est} \right|^2}{\left| \mathbf{w}_{opt} \right|^2} \tag{13}$$

where $\mathbf{w}_{opt}$ is the optimal value of network vector parameter, and $\mathbf{w}_{est}$ the estimated value.

The results obtained, are shown in Fig. 4, which is the average of 50 simulations. As we could expect, JNDDE gets a lower WRSE for the same amount of training data. This is due to the fact that it makes true hypothesis about the data distribution, thus reducing the space search. Note that the difference in WRSE, in a log-log scale, remains constant no matter what is the size of the training set.

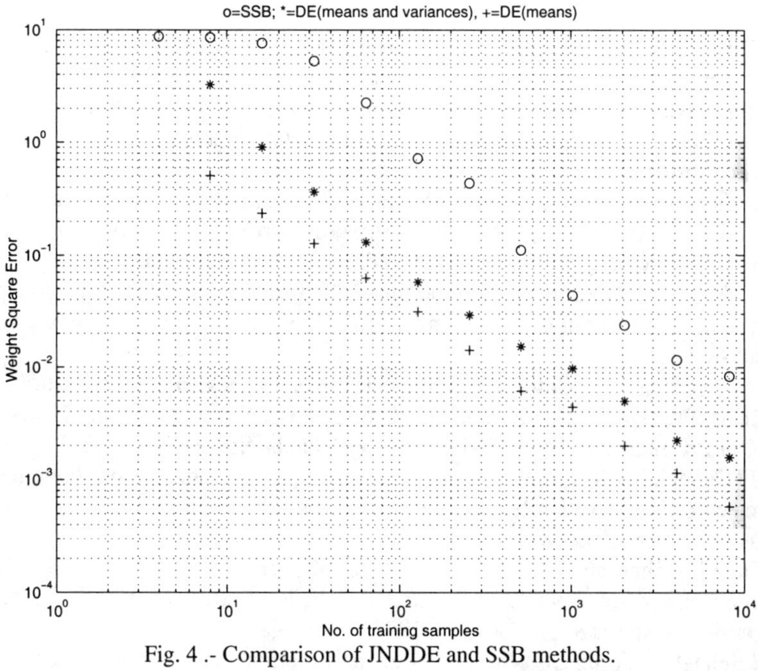

Fig. 4 .- Comparison of JNDDE and SSB methods.

CONCLUSIONS

This paper presents a novel approach to the estimation of 'a posteriori' probabilities with neural networks which is based on estimating the conditional data densities. The JNDDE can reduce the data requirements of the estimation process, at the cost of making assumptions about the sample distribution. The method is based on using two different models, one for the posteriori probabilities (which is given by the network structure) and the other for the conditional data densities; both of them can be made arbitrarily complex. This states an important question for future work: for a given sample, how to identify the complexity of both models. On the other hand, since JNDDE uses a data model, it has the potential capability to extract information from unlabeled data. Use of JNDDE in hybrid learning, both labeled an unlabeled data, is a matter for future research.

ACKNOWLEDGEMENTS: This work has been partially supported by the Spanish Government under C.I.C.Y.T. grant numbers TIC96-0500-C10-03, TIC96-0500-C10-09 and TIC97-0772 . T. Adali's work was partially supported by the National Science Foundation Career Award, NSFNCR-9703161.

REFERENCES

[1] D.W.Ruck, S.K.Rogers, M.Kabrisky, M.E.Oxley and B.W.Suter, "The multilayer perceptron as an approximation to a Bayes optimal discriminant function", **IEEE Trans. on Neural Networks**, vol. 1, no. 4, pp. 296-298, 1990.
[2] S. Amari, "Backpropagation and stochastic gradient descent method", **Neurocomputing**, No.5, pp. 185-196, 1993.
[3] J.Cid-Sueiro, A.R. Figueiras-Vidal, "Cost Functions to Estimate Class Probabilities", **Proceedings of the European Conference on Signal Analysis and Prediction (ECSAP'97)**, pp. 113-116, Praha, 24-27 June, 1997.
[4] P. Smyth, J.W. Miller, R. Goodman, "Objective functions for probability estimation", **Proceedings of the Int. Joint Confefernce on Neural Networks**, pp. 881-886, 1991.
[5] J. I. Arribas , J. Cid-Sueiro, "Bayesian Approaches to the Estimation of A Posteriori Probabilities", **Procs of the IASTED Int. Conf. on Signal Processing and Communications**, pp. 107-110, Canary Islands, Spain, Feb. 1998.
[6] J. Cid-Sueiro, J. Ignacio Arribas, A. R. Figueiras-Vidal, "Cost Functions to Estimate A Posteriori Probability in Multi-Class Problems", to appear in **IEEE Trans. on Neural Networks**.
[7] David Miller and Hasan S. Uyar, "A Mixture of Experts Classifier with Learning Based on Both Labelled and Unlabelled Data", **Neural Information Processing Systems 9**, pp. 571-577, 1996.
[8] Michael I. Jordan, "Why the logistic function? A turotial discussion on probabilities and neural networks", M.I.T., *Computational Cognitive Science Technical Report 9503*, August 13, 1995.
[9] C.M. Bishop, Neural Networks for pattern recognition, Oxford Press,1995.
[10] J.S. Bridle, "Probabilistic interpretation of feedforward classification network outputs with relationship to statistical pattern recognition". In F. Fogelman Soulie and J. Herault (Eds.), **Neurocomputing: Algorithms, Architectures and Applications**, pp. 227-236. New York: Springer Verlag, 1990.
[11] Juan Ignacio Arribas, Jesus Cid-Sueiro, Tulay Adali and Anibal R. Figueiras-Vidal, "Neural networks to estimate ML multi-class constrained conditional probability density functions", **International Joint Conference on Neural Networks**, Washington D.C., USA, July 1999.

ALLPASS VS. UNIT-NORM CONSTRAINTS IN CONTRAST-BASED BLIND DECONVOLUTION

Scott C. Douglas[1] and S.-Y. Kung[2]

[1]Department of Electrical Engineering, Southern Methodist University
Dallas, Texas 75275 USA

[2]Department of Electrical Engineering, Princeton University
Princeton, New Jersey 08544 USA

Abstract — **Many contrast-based independent component analysis (ICA) algorithms can be directly applied to single-channel blind deconvolution tasks without modification. For prewhitened signals, however, an allpass constraint on the filter impulse response is more appropriate than a unit-norm constraint. In this paper, we show how a finite-impulse-response filter can be adapted to maximize or minimize an arbitrary contrast function while approximately maintaining an allpass characteristic in a self-stabilizing fashion. Both on-line time domain and block-based frequency domain algorithms are described. Simulations verify the superior performance behaviors of the new algorithms over those of standard ICA algorithms in prewhitened blind deconvolution tasks.**

I. INTRODUCTION

The blind deconvolution task figures prominently in many practical signal processing problems, in which the goal is to estimate an independent, identically-distributed (i.i.d.) source signal $s(k)$ from a measured signal

$$x(k) = \sum_{j=-\infty}^{\infty} a_j s(k-j), \qquad -\infty < k < \infty, \tag{1}$$

where a_j, $-\infty < j < \infty$ is the unknown channel impulse response. Many techniques have been proposed for this task, among them homomorphic deconvolution, channel estimation followed by sequence estimation, and inverse filtering or blind equalization [1, 2]. In the latter methods, an estimated source sequence $y(k)$ is created from a time-varying FIR filter as

$$y(k) = \sum_{j=0}^{M} m_j(k) x(k-j), \tag{2}$$

[0]This work was supported in part by the Office of Research and Development under Contract No. 98F135700-000.

and the goal is to adjust each $m_j(k)$ for $0 \leq j \leq M$ over time so that

$$y(k) \quad \approx \quad cs(k - \Delta), \tag{3}$$

where c and Δ are arbitrary scaling and integer delay factors, respectively.

Related to blind deconvolution is instantaneous blind source separation, in which a sequence of n-dimensional unknown source signal vectors $\mathbf{s}(k) = [s_0(k) \ \cdots \ s_{n-1}(k)]^T$ with independent components is mixed by a physical process to produce a measured vector sequence

$$\mathbf{x}(k) \quad = \quad [x_0(k) \cdots x_{n-1}(k)]^T \quad = \quad \mathbf{As}(k), \tag{4}$$

where $\mathbf{A}$ is an unknown $(n \times n)$ mixing matrix. Recent efforts have uncovered several independent component analysis (ICA) techniques for extracting each $s_i(k)$ from $\mathbf{x}(k)$ [3]–[10], in which a prewhitened signal vector is computed as

$$\mathbf{v}(k) \quad = \quad \mathbf{Px}(k), \tag{5}$$

where $\mathbf{P}$ is a prewhitening matrix chosen such that $E\{\mathbf{v}(k)\mathbf{v}^{\dagger}(k)\} = \mathbf{I}$, $E\{\cdot\}$ denotes statistical expectation, and $\dagger$ denotes conjugate-transpose. Then,

$$y(k) \quad = \quad \mathbf{w}^T(k)\mathbf{v}(k) \tag{6}$$

is computed, and $\mathbf{w}(k) = [w_0(k) \ \cdots \ w_{n-1}(k)]^T$ is adjusted either to

$$\text{maximize or minimize} \qquad E\{\phi(y(k))\} \tag{7}$$

$$\text{such that} \qquad ||\mathbf{w}(k)|| = 1, \tag{8}$$

where $\phi(y)$ is a real-valued symmetric function and $||\cdot||$ denotes the L_2-norm. Choosing $\phi(y) = |y|^4$ for (7) guarantees that one independent source in $y(k)$ can be extracted, and repeated application of the procedure with deflation allows all non-zero-kurtosis source signals to be extracted [7, 8]. Moreover, although other $\phi(y)$ are not guaranteed to work [9], different $\phi(y)$ can provide more efficient convergence behavior in some cases [10].

Several authors have noted that ICA and blind deconvolution are related problems, to the extent that ICA methods can be translated to the blind deconvolution task [11]–[13]. In this case, $x(k)$ is prewhitened by a prewhitening filter with impulse response p_j, $0 \leq j \leq N$, to produce the signal $v(k)$ such that $E\{v(k)v^*(k+j)\} \approx \delta_j$, $-N \leq j \leq N$. Then, $y(k)$ is computed as

$$y(k) \quad = \quad \sum_{l=0}^{L} w_l(k)v(k-l), \tag{9}$$

and each $w_l(k)$ is adjusted over time to satisfy (7)–(8) with $\mathbf{w}(k) = [w_0(k) \ \cdots \ w_L(k)]^T$. This methodology, however, ignores an inherent structure within prewhitened deconvolution: the equalizer forms an allpass filter at convergence, such that

$$\sum_{l=-\infty}^{\infty} w_l(k)w_{l+j}^*(k) \approx \delta_j, \qquad -L \leq j \leq L, \tag{10}$$

where $w_l(k) = 0$ for $l < 0$ or $l > L$. Note that this constraint includes $||\mathbf{w}(k)|| = 1$ and thus is more stringent than (8). Little is understood about the performance improvements, if any, that would be obtained if (10) is imposed instead of (8) in the prewhitened blind deconvolution task. Moreover, although the theory of allpass filter design appears to be well developed [14], few simple approaches to gradient adaptive allpass filters have been proposed [15], and these IIR-based approaches do not necessarily enjoy the inherent stability properties of existing gradient adaptive FIR filters.

In this paper, we develop a formalism for gradient adaptive allpass filters and apply the resulting techniques to prewhitened blind deconvolution tasks. Our methods extend algorithms for gradient adaptation in the Stiefel manifold of orthonormal matrices [16]–[21] to the space of single-channel dispersive systems. Our methods attempt to adapt the filter coefficients $w_l(k)$ in the space of allpass systems, such that (10) is approximately imposed. Both on-line adaptive and efficient off-line batch versions are described, and simulations verify the useful behaviors of the proposed methods in deconvolving linearly-filtered source signals.

II. BLIND SOURCE SEPARATION WITH ORTHOGONALITY CONSTRAINTS

The allpass-constrained algorithms derived in this paper are temporal extensions of recently-developed gradient adaptive source separation algorithms for the Stiefel manifold of orthonormal matrices [21], the latter of which are briefly reviewed. Let $\mathbf{x}(k)$, $\mathbf{v}(k)$, $\mathbf{A}$, and $\mathbf{P}$ be defined as previously, and let

$$\mathbf{y}(k) = \mathbf{W}(k)\mathbf{v}(k) \tag{11}$$

where $\mathbf{W}(k)$ is an $(n \times n)$ demixing matrix. It is straightforward to show that

$$\mathbf{v}(k) = \mathbf{PAs}(k) = \mathbf{\Gamma s}(k), \tag{12}$$

where $\mathbf{\Gamma}$ is orthonormal; *i.e.* $\mathbf{\Gamma\Gamma}^\dagger = \mathbf{\Gamma}^\dagger\mathbf{\Gamma} = \mathbf{I}$. Since $\mathbf{\Gamma}$ is orthonormal, $\mathbf{W}(k)$ in (11) can be constrained to be orthonormal as well. When each $s_i(k)$ in $\mathbf{s}(k)$ has the same distribution, the optimization problem given by

$$\text{maximize/minimize} \qquad \mathcal{J}(\mathbf{W}) = \sum_{i=0}^{n-1} E\{\phi(y_i(k))\} \tag{13}$$

$$\text{such that} \qquad \mathbf{W}(k)\mathbf{W}^\dagger(k) = \mathbf{I}. \tag{14}$$

forms the basis of generalized parallel extraction procedures [21].

Recently, interest has grown in gradient algorithms that attempt to impose (14) via their adaptive characteristics. [16]–[21]. Matrices $\mathbf{W}$ that satisfy (14) for cost functions $\mathcal{J}(\mathbf{W})$ in which $\mathcal{J}(\mathbf{W}) \neq \mathcal{J}(\mathbf{QW})$ for any Hermitian matrix $\mathbf{Q}$ lie in the *Stiefel manifold* of orthonormal matrices. The differential update

$$\frac{d\mathbf{W}}{dt} = \mathbf{W}\mathbf{W}^{\dagger}\mathbf{G} - \mathbf{W}\mathbf{G}^{\dagger}\mathbf{W}, \qquad \mathbf{G} = \beta\frac{\partial\mathcal{J}(\mathbf{W})}{\partial\mathbf{W}} \tag{15}$$

attempts to maximize (for $\beta > 0$) or minimize (for $\beta < 0$) $\mathcal{J}(\mathbf{W})$ while implicitly maintaining $\mathbf{W}\mathbf{W}^{\dagger} = \mathbf{I}$ for all $t \geq t_0$ if $\mathbf{W}(t_0)\mathbf{W}^{\dagger}(t_0) = \mathbf{I}$ [18]. Moreover, straightforward calculus shows that $\mathbf{G} = \beta\mathbf{f}(\mathbf{y})\mathbf{v}^{\dagger}$ for $\mathcal{J}(\mathbf{W})$ in (13), where $\mathbf{f}(\mathbf{y}) = [f(y_0) \; \cdots \; f(y_{n-1})]^T$ and $f(y) = \partial\phi(y)/\partial y$.

When developing discrete-time approximations of (15), one must address the accumulation of numerical errors that cause $\mathbf{W}(k)$ to deviate from the Stiefel manifold over time, as these errors can eventually cause catastrophic failure of the algorithm. Similar difficulties have been addressed in principal and minor component analysis using *self-stabilized* algorithm forms [19]. A self-stabilized algorithm is locally-stable to the matrix manifold in its operation, such that any deviations of $\mathbf{W}(k)$ away from orthogonality at time $k = k_0$ decay to zero over time. For source separation, the update

$$\mathbf{W}(k+1) = \mathbf{W}(k) + \beta(k)\left[\mathbf{f}(\mathbf{y}(k))\mathbf{v}^{\dagger}(k) - \mathbf{y}(k)\mathbf{f}^{\dagger}(\mathbf{y}(k))\mathbf{W}(k)\right] \tag{16}$$

with $\beta(k) > 0$ provides self-stabilizing behavior to the Stiefel manifold when maximizing (13). To minimize (13), a simple self-stabilized procedure is

$$\mathbf{W}(k+1) = \mathbf{W}(k) + \beta(k)\left[\mathbf{W}(k)\mathbf{W}^{\dagger}(k)\mathbf{f}(\mathbf{y}(k))\mathbf{y}^{\dagger}(k) - \mathbf{y}(k)\mathbf{f}^{\dagger}(\mathbf{y}(k))\right]\mathbf{W}(k) \tag{17}$$

with $\beta(k) < 0$. Details concerning these algorithms' performances and stability properties can be found in [21].

III. ALLPASS-CONSTRAINED ALGORITHMS FOR BLIND DECONVOLUTION

To extend the algorithms in (16) and (17) to blind deconvolution tasks, we use a procedure employed in other similar extensions [12, 13]:

1. Assume a circulant form for $\mathbf{A}$, $\mathbf{P}$, and $\mathbf{W}(k)$.
2. Express the algorithm forms that result from the circulant structure.
3. Let the size of the matrices and vectors tend to infinity.

This procedure produces algorithms that adapt the impulse response $w_l(k)$, $-\infty < l < \infty$ within the space of allpass filters to maximize or minimize an inhomogeneous cost function $\mathcal{J}(\{w_p\})$ [22]. Applying this procedure to (16) yields the allpass-constrained contrast maximization procedure as

$$w_l(k+1) = w_l(k) + \beta(k)\left[f(y_o(k))v_o^*(k-l) - y_o(k)u_o^*(k-l)\right] \tag{18}$$

$$y_o(k) = \sum_{p=-\infty}^{\infty} w_p(k)v(k-p), \qquad u_o(k) = \sum_{q=-\infty}^{\infty} w_{-q}^*(k)f(y(k-q)), \tag{19}$$

where $-\infty < l < \infty$. The corresponding allpass-constrained contrast minimization procedure can be obtained from (17) and is omitted for brevity.

As (18)–(19) involve infinitely-long non-causal filters, we approximate this algorithm by (i) truncating the filter impulse response to $(L+1)$ samples and (ii) delaying the updates in (18) by L samples so that (19) can be causally-implemented. Similar methods have been used in other filtered-gradient algorithms [2]. The resulting algorithm for $0 \leq l \leq L$ and $\beta(k) > 0$ is

$$\begin{aligned} w_l(k+1) &= w_l(k) + \beta(k)\left[f(y(k-L))v^*(k-L-l)\right. \\ &\quad \left. -y(k-L)u^*(k-l)\right] \end{aligned} \tag{20}$$

$$u(k) = \sum_{q=0}^{L} w_{L-q}^*(k)f(y(k-q)), \tag{21}$$

where $y(k)$ is as computed in (9). This algorithm is particularly simple, requiring about $4L$ complex multiply/adds (MACs) at each time instant.

The above procedure can be applied to (17) to obtain an allpass-constrained algorithm for contrast minimization. The details of the derivation are omitted, and only the final algorithm is given: for $0 \leq l \leq L$ and $\beta(k) < 0$,

$$w_l(k+1) = w_l(k) + \beta(k)\left[\psi(k)\xi^*(k-l) - y(k-L)u^*(k-l)\right] \tag{22}$$

$$\xi(k) = \sum_{q=0}^{L} w_{L-q}^*(k)y(k-q), \qquad \psi(k) = \sum_{p=0}^{L} w_p(k)u(k-p), \tag{23}$$

where $y(k)$ and $u(k)$ are given by (9) and (21), respectively. This algorithm requires about $6L$ complex multiply/adds (MACs) at each time instant.

The choice of L for the above algorithms is governed by the characteristics of the prewhitened channel impulse response; that is, L should be large enough to provide adequate modeling of the prewhitened channel inverse. As for the choice of $f(y)$, the local stability properties of (20) and (22) are identical to the spatial algorithms from which they are derived. For real-valued signals and coefficients, the source signal must satisfy

$$\beta(k)\left[E\{s(k)f(s(k))\} - E\{s^2(k)\}E\{f'(s(k))\}\right] \quad > \quad 0. \tag{24}$$

Similar results hold for complex-valued signals if the distribution of $s(k)$ is circularly-symmetric [5]. Eqn. (24) typically includes all source distributions of interest for some non-zero $\beta(k)$ and phase-preserving nonlinearity $f(y)$. Moreover, certain choices of $f(y)$ provide better local convergence behavior than others for a given source distribution [10]. Additionally, the algorithms are observed in simulation to possess similar self-stabilizing properties as their spatial counterparts; that is, they attempt to adjust the impulse response $\{w_i(k)\}$ in the space of allpass filters satisfying (10). Simulations in Section V indicate that these algorithms possess superior convergence properties as compared to those of standard contrast-based ICA algorithms due to their implicit use of the allpass filter constraint.

IV. FREQUENCY DOMAIN IMPLEMENTATIONS

In some situations, the filter length L required by the previously-derived algorithms to provide adequate deconvolution performance can be large, making these algorithms challenging to implement in hardware or software. To simplify their implementation in such cases, we can develop block versions in which the update terms are accumulated over K time instants $0 \leq k \leq K-1$ and are applied at every Kth iteration. The resulting algorithms can then be efficiently implemented using FFT-based fast convolution methods. Similar approaches have been used in traditional adaptive filtering tasks [23]. The frequency domain versions also provide a distinct advantage over their time domain counterparts: the allpass constraint can be easily implemented by normalizing the amplitude of each complex weight of the deconvolution filter frequency response after updating by using the complex sign operation [13]:

$$\mathrm{sgn}(y) \quad = \quad y/|y|. \tag{25}$$

This approach is similar in spirit to the coefficient normalization procedure used in some principal and minor component analysis algorithms [24], and it can be used for both $\beta(k) > 0$ and $\beta(k) < 0$, respectively.

Table 1 lists the equations for the resulting algorithm family, where we have defined $M = 2L$, $\mathcal{E} = [1\ -1\ 1\ \cdots\ -1]^T$, and $\odot$ denotes point-by-point multiplication. Note that several algorithms are possible depending on which time domain data windowing constraints are imposed within the various convolution/correlation operations. We have listed four such constraints (output, filtered output, gradient, and normalization). Although the relative performances of the various versions are currently under study, the best-performing version typically employs all four constraints and requires about $14\log_2 L+26$ complex multiply/adds per data sample per update iteration when $K \gg L$. For long filter lengths, the frequency domain versions save significant amounts of computation over both time domain versions.

Table 1: Family of block-update frequency domain allpass-constrained blind deconvolution algorithms.

Equation	*# of Complex Mult.*
Gradient Calculation	
$\mathcal{T}_i(0) = \mathbf{0}$	0
for $k = 0$ **to** $K-1$ **step** L **do**	
$\mathbf{v}'(k) = [v(k-2L+1) \cdots v(k)]^T$	0
$\mathcal{V}(k) = F\{\mathbf{v}'(k)\}$	$2K \log_2 M$
$\widehat{\mathcal{Y}}(k) = \mathcal{W}(i) \odot \mathcal{V}(k)$	$2K$
$[\bullet \cdots \bullet\ y(k-L+1) \cdots y(k)]^T = F^{-1}\{\widehat{\mathcal{Y}}(k)\}$	$2K \log_2 M$
if {output constraint} **then**	
$\mathbf{y}'(k) = [0 \cdots 0\ y(k-L+1) \cdots y(k)]^T$	0
$\mathcal{Y}(k) = F\{\mathbf{y}'(k)\}$	$2K \log_2 M$
else	
$\mathcal{Y}(k) = \widehat{\mathcal{Y}}(k)$	0
end	
$\mathbf{f}'(k) = [0 \cdots 0\ f(y(k-L+1)) \cdots f(y(k))]^T$	0
$\mathcal{F}(k) = F\{\mathbf{f}'(k)\}$	$2K \log_2 M$
$\widehat{\mathcal{G}}(k) = \mathcal{F}(k) \odot \mathcal{V}^*(k)$	$2K$
$\widehat{\mathcal{U}}(k) = \mathcal{W}^*(i) \odot \{\mathcal{F}(k) + \mathcal{E} \odot \mathcal{F}(k-L)\}$	$4K$
if {filtered output constraint} **then**	
$[u(k-L+1) \cdots u(k)\ \bullet \cdots \bullet]^T = F^{-1}\{\widehat{\mathcal{U}}(k)\}$	$2K \log_2 M$
$\mathbf{u}'(k) = [u(k-2L+1) \cdots u(k)]^T$	0
$\mathcal{U}(k) = F\{\mathbf{u}'(k)\}$	$2K \log_2 M$
$\widehat{\mathcal{T}}(k) = \widehat{\mathcal{G}}(k-L) - \mathcal{Y}(k-L) \odot \mathcal{U}^*(k)$	$2K$
else	
$\widehat{\mathcal{T}}(k) = \widehat{\mathcal{G}}(k) - \mathcal{Y}(k) \odot \widehat{\mathcal{U}}^*(k)$	$2K$
end	
if {gradient constraint} **then**	
$[t_{L-1}(k) \cdots t_0(k)\ \bullet \cdots \bullet]^T = F^{-1}\{\widehat{\mathcal{T}}(k)\}$	$2K \log_2 M$
$\mathbf{t}'(k) = [t_{L-1}(k) \cdots t_0(k)\ 0 \cdots 0]^T$	0
$\mathcal{T}(k) = F\{\mathbf{t}'(k)\}$	$2K \log_2 M$
else	
$\mathcal{T}(k) = \widehat{\mathcal{T}}(k)$	0
end	
$\mathcal{T}_i(k+L) = \mathcal{T}_i(k) + \mathcal{T}(k)$	$2K$
end	
Coefficient Updates	
$\widetilde{\mathcal{W}}(i+1) = \mathcal{W}(i) + \frac{\beta(i)}{K}\mathcal{T}_i(K-1)$	M
$\widehat{\mathcal{W}}(i+1) = \mathrm{sgn}(\widetilde{\mathcal{W}}(i+1))$	M ($M \div$, $M \surd$)
if {normalization constraint} **then**	
$[w_{L-1}(i+1) \cdots w_0(i+1)\ \bullet \cdots \bullet]^T = F^{-1}\{\widehat{\mathcal{W}}(i+1)\}$	$M \log_2 M$
$\mathbf{w}'(i+1) = [w_{L-1}(i+1) \cdots w_0(i+1)\ 0 \cdots 0]^T$	0
$\mathcal{W}(i+1) = F\{\mathbf{w}'(i+1)\}$	$M \log_2 M$
else	
$\mathcal{W}(i+1) = \widehat{\mathcal{W}}(i+1)$	0
end	
increment i	

V. SIMULATIONS

We now verify the performance of the proposed algorithms via simulations. We generate the received signal $x(k)$ from $s(k)$ using the allpass filter

$$A(z) \quad = \quad e^{-j\pi/4}\frac{0.6 - z^{-1}}{1 - 0.6z^{-1}} \tag{26}$$

Since $A(z)$ is allpass, we may assign $v(k) = x(k)$. In all cases, $w_l(0) = \delta_{l-L/2}$ was chosen initially. To evaluate each scheme's performance, the impulse response of the combined channel $C(z,k) = W(z,k)A(z)$ was calculated after each update, and the intersymbol interference (ISI) given by

$$\mathrm{ISI}(k) \quad = \quad \frac{\displaystyle\sum_{j=0}^{3L}|c_j(k)|^2}{\displaystyle\max_{0\leq i\leq 3L}|c_i(k)|^2} - 1 \tag{27}$$

was computed for each algorithm, where $\{c_j(k)\}$ is the combined channel's impulse response. Averages of fifty different simulations runs have been computed in each case.

For the first set of simulations, $s(k)$ is a negative-kurtosis source in which both the real and imaginary parts of $s(k)$ are i.i.d. uniform-$[-\sqrt{6}, \sqrt{6}]$-distributed, such that $E\{|s(k)|^2\} = 1$. Figure 1 shows the evolution of the averaged values of ISI(k) for $\{(9), (20), (21)\}$ with $f(y) = y/|y|$, $\{(9), (21)–(23)\}$ with $f(y) = |y|^2y$, as well as the two corresponding unit-norm-constrained ICA algorithms with $f(y) = y/|y|$ and $f(y) = |y|^2y$, respectively [10, 13], for FIR equalizers with $L = 10$. In this case, the chosen $\beta(k)$ values of 0.0075, −0.006, 0.001, and −0.001, respectively, provide a steady-state ISI of −33dB. As can be seen, the allpass-constrained algorithms provide superior convergence behavior as compared to their unit-norm-constrained counterparts. Also shown for comparison is the constant modulus algorithm (CMA) with $\beta(k) = 0.008$, which provides similar behavior to $\{(9), (21)–(23)\}$ for these signal statistics.

Figure 2 shows the behaviors of the frequency domain versions of the various blind deconvolution algorithms when the real and complex components of $s(k)$ are Laplacian random variables with $E\{|s(k)|^2\} = 1$. We have chosen $K = 10000$, $L = 64$, and $\beta(i)$ values of −1.1, 0.2, −1.5, and 0.55, respectively, to provide similar system convergence rates for this data. As can be seen, the allpass-constrained algorithms provide superior behaviors, and the choice $f(y) = y/|y|$ yields the best-performing algorithm in this case.

VI. CONCLUSIONS

In this paper, we have derived single-channel blind deconvolution algorithms that enforce allpass constraints on the deconvolution system for

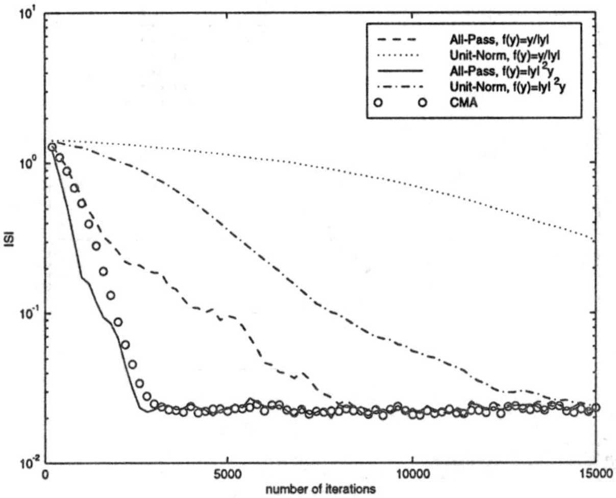

Figure 1: Evolution of $E\{\text{ISI}(k)\}$ for the various on-line algorithms.

prewhitened signal mixtures. Both on-line time domain and block-based frequency domain versions of the algorithms are developed. The former methods are self-stabilizing to the space of deconvolution systems without divides or square roots, whereas the latter methods employ projections back to the space of deconvolution systems in the frequency domain. Simulations verify the excellent performances of the proposed methods as compared to unit-norm-constrained ICA methods for blind deconvolution tasks.

REFERENCES

[1] S. Haykin, ed., *Blind Deconvolution* (Englewood Cliffs, NJ: Prentice-Hall, 1994).

[2] S.C. Douglas, A. Cichocki, and S. Amari, "Self-whitening algorithms for adaptive equalization and deconvolution," *IEEE Trans. Signal Processing,* vol. 47, pp. 1161-1165, Apr. 1999.

[3] P. Comon, "Independent component analysis: A new concept?" *Signal Processing,* vol. 36, no. 3, pp. 287-314, Apr. 1994.

[4] N. Delfosse and P. Loubaton, "Adaptive blind separation of independent sources: a deflation approach," *Signal Processing,* vol. 45, no. 1, pp. 59-83, July 1995.

[5] J.-F. Cardoso and B. Laheld, "Equivariant adaptive source separation," *IEEE Trans. Signal Processing,* vol. 44, pp. 3017-3030, Dec. 1996.

[6] A. Cichocki, R. Thawonmas, and S. Amari, "Sequential blind signal extraction in order specified by stochastic properties," *Electron. Lett.,* vol. 33, no. 1, pp. 64-65, 2 Jan. 1997.

[7] S.-Y. Kung, "Independent component analysis in hybrid mixture: KuicNet learning algorithm and numerical analysis," *Proc. Int. Symp. Multimedia Inform. Processing,* Taipei, Taiwan, pp. 368-381, Dec. 1997.

[8] A. Hyvärinen and E. Oja, "Independent component analysis by general nonlinear Hebbian-like learning rules," *Signal Processing,* vol. 64, no. 3, pp. 301-313, Feb. 1998.

[9] S.-Y. Kung, "Independent component analysis: Extrema-correspondence properties for higher-order moment functions," *Proc. IEEE Workshop Neural Networks Signal Processing,* Cambridge, UK, pp. 53-62, Aug. 1998.

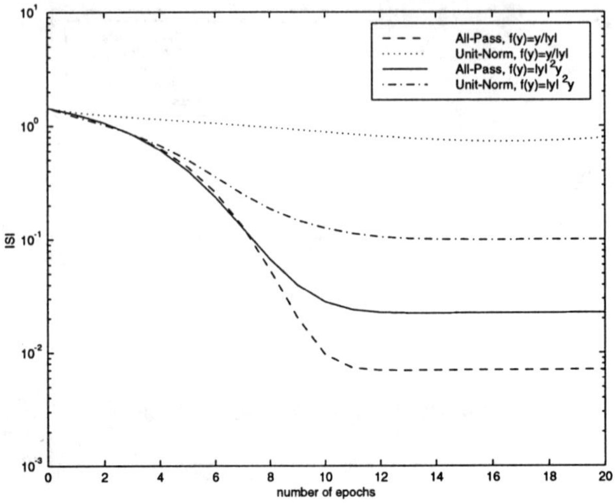

Figure 2: Evolution of $E\{\mathrm{ISI}(i)\}$ for the various block-update algorithms.

[10] S.C. Douglas and S.-Y. Kung, "Design of estimation/deflation approaches to independent component analysis," *Proc. 32nd Asilomar Conf. Signals, Syst., Comput.,* Pacific Grove, CA, vol. 1, pp. 707-711, Nov. 1998.

[11] R.H. Lambert, "Multichannel blind deconvolution: FIR matrix algebra and separation of multipath mixtures," Ph.D. dissertation, Univ. Southern California, Los Angeles, CA, May 1996.

[12] S.C. Douglas and S. Haykin, "On the relationship between blind deconvolution and blind source separation," *Proc. 31st Asilomar Conf. Signals, Syst., Comput.,* Pacific Grove, CA, vol. 2, pp. 1591-1595, Nov. 1997.

[13] S.C. Douglas and S.-Y. Kung, "KuicNet algorithms for blind deconvolution," *Proc. IEEE Workshop Neural Networks Signal Processing,* Cambridge, UK, pp. 3-12, Aug. 1998.

[14] M. Lang, "Allpass filter design and applications," *IEEE Trans. Signal Processing,* vol. 46, pp. 2505-2514, Sept. 1998.

[15] B. Farhang-Boroujeny and S. Nooshfar, "Adaptive phase equalization using allpass filters," *IEEE Int. Conf. Commun.,* vol. 3, pp. 1403-1407, June 1991.

[16] A. Edelman, T. Arias, and S.T. Smith, "The geometry of algorithms with orthogonality constraints," preprint.

[17] U. Helmke and J.B. Moore, *Optimization and Dynamical Systems* (New York: Springer-Verlag, 1994).

[18] T. Chen, S. Amari, and Q. Lin, "A unified algorithm for principal and minor components extraction," *Neural Networks,* vol. 11, pp 385-390, Apr. 1998.

[19] S.C. Douglas, S.-Y. Kung, and S. Amari, "A self-stabilized minor subspace rule," *IEEE Signal Processing Lett.,* vol. 5, pp. 328-330, Dec. 1998.

[20] S.C. Douglas, S.-Y. Kung, and S. Amari "On the numerical stabilities of principal, minor, and independent component analysis algorithms," *Proc. IEEE Workshop Indep. Compon. Anal. Signal Separation,* Aussois, France, pp. 419-424, Jan. 1999.

[21] S.C. Douglas, "Self-stabilized gradient algorithms for blind source separation with orthogonality constraints," submitted to *IEEE Trans. Neural Networks,* Mar. 1999.

[22] S.C. Douglas, S. Amari, and S.-Y. Kung, "Adaptive paraunitary filter banks for spatio-temporal principal and minor subspace analysis," *Proc. IEEE Int. Conf. Acoust., Speech, Signal Processing,* Phoenix, AZ, vol. 2, pp. 1089-1092, Mar. 1999.

[23] J.J. Shynk, "Frequency domain and multirate adaptive filters," *IEEE Signal Processing Mag.,* vol. 9, no. 1, pp. 14-37, Jan. 1992.

[24] K.I. Diamantaras and S.-Y. Kung, *Principal Component Neural Networks: Theory and Applications* (New York: Wiley, 1996).

A STABLE AND ROBUST ICA ALGORITHM BASED ON T-DISTRIBUTION AND GENERALIZED GAUSSIAN DISTRIBUTION MODELS

Jianting CAO [1,2] and Noboru MURATA[2]
[1]Dept. of Electrical and Electronics Engineering, Sophia University,
7-1 Kioicho, Chiyoda-ku, Tokyo 102-8554, JAPAN.
[2]Brain-style Information Systems Group, BSI, RIKEN,
2-1 Hirosawa, Wako-shi, Saitama 351-0198, JAPAN.
E-mails: {cao,mura}@brain.riken.go.jp
Phone: +81-48-467-9665, Fax: +81-48-467-9693.

Abstract. In this paper, we propose a novel Independent Component Analysis (ICA) algorithm which enables to separate mixtures of sub-Gaussian, super-Gaussian and Gaussian primary source signals. Alternative activation functions in the algorithm are derived by using parameterized t-distribution and generalized Gaussian distribution density models. The functions are self-adaptive based on estimating the high-order moments of extracted signals. Moreover, a stability condition of the proposed algorithm for separating the true solution is given. Simulation experiment results are presented to illustrate the effectiveness and performance of the proposed algorithm.

INTRODUCTION

Blind source separation is of considerable interest due to many applications in science and technology, e.g. in recovering of biomedical signals (EEG, MEG)[1-10]. In the problem of blind separation of instantaneously mixed sources, let us assume that $\mathbf{s}(k) = [s_1(k), \cdots, s_n(k)]^T$ denotes a column vector of n zero-mean, mutually independent original source signals generated at discrete time k. The observations $\mathbf{x}(k) = [x_1(k), \cdots, x_n(k)]^T$ are propagated from the original sources through a linear matrix. Here we discuss the instantaneous mixture model which is written as

$$\mathbf{x}(k) = A\mathbf{s}(k), \qquad k = 1, 2, \cdots \tag{1}$$

0-7803-5673-X/99/$10.00 © 1999 IEEE

where $A \in \Re^{n \times n}$ is an unknown non-singular mixing matrix. In order to recover the unknown sources, we use the linear transform from observations,

$$\mathbf{y}(k) = W\mathbf{x}(k), \tag{2}$$

where $\mathbf{y}(k) = [y_1(k), \cdots, y_n(k)]^T$ and W is an $n \times n$ separating (de-mixing) matrix.

The task of ICA is to find a de-mixing matrix W in the way that reestablishes the independent condition

$$p_y(\mathbf{y}) = \prod_{i=1}^{n} p_i(y_i), \tag{3}$$

where $p_y(\mathbf{y})$ is the joint probability density function (pdf) and $p_i(y_i)$ are marginal pdfs of $\mathbf{y}$. The solutions for output $\mathbf{y}$ recovering the sources $\mathbf{s}$ are under the condition of $WA = P\Lambda$ or $WA = I$, where P, Λ and I are permutation, scaling diagonal and identity matrices, respectively.

Minimizing the mutual information (MI) is an ICA approach in which the mutual stochastic independence of output signals $y_i(k)$ is measured by the Kullback-Leibler divergence between the joint pdf $p_y(\mathbf{y})$ and the marginal pdfs $p_i(y_i)$. It is defined as :

$$\begin{aligned} D(\mathbf{y}|W) &= \int p_y(\mathbf{y}) \log \frac{p_y(\mathbf{y})}{\prod_{i=1}^{n} p_i(y_i)} d\mathbf{y} \\ &= -H(\mathbf{y}|W) + \sum_{i=1}^{n} H(y_i|W), \end{aligned} \tag{4}$$

where $H(\mathbf{y}|W) = -\int p_y(\mathbf{y}) \log p_y(\mathbf{y}) d\mathbf{y}$ is the entropy of $\mathbf{y}$ and $H(y_i|W) = -\int p_i(y_i) \log p_i(y_i) dy_i$ are the entropies of each y_i.

Applying the stochastic descent method of the natural gradient version, the rule of updating W is derived as (see for example Amari et al. 1996)

$$\begin{aligned} \Delta W(k) &= -\eta(k) \frac{\partial D(\mathbf{y}|W)}{\partial W} W^T W \\ &= -\eta(k) \left[-\frac{\partial H(\mathbf{y}|W)}{\partial W} + \sum_{i=1}^{n} \frac{\partial H(y_i|W)}{\partial W} \right] W^T W \\ &= \eta(k) \left[I - \varphi(\mathbf{y}(k)) \mathbf{y}^T(k) \right] W(k), \end{aligned} \tag{5}$$

where $\eta(k) > 0$ is the learning rate, $\varphi(\cdot)$ is a vector of activation functions, whose components for each output y_i are

$$\varphi_i(y_i) = -\frac{d}{dy_i} \log p_i(y_i) = -\frac{\dot{p}_i(y_i)}{p_i(y_i)}, \tag{6}$$

where $\dot{p}_i(y_i) = dp_i(y_i)/dy_i$.

The above function is optimal if we know the original source probability distributions $p_i(s_i)$. However, the knowledge of $p_i(s_i)$ is not available in various cases. One practical solution is to approximate the pdfs using a stochastic expansion, such as the Edgeworth expansion (see Comon 1994) or the Gram-Charlier expansion (see Amari et al. 1996). Another approach is to use hypothesized probability density model, e.g. the hyperbolic Cauchy probability distribution density model is used by Mackay [10], the mixtures of Gaussian density model (Pearson model) is used by Girolami [8,9], the generalized Gaussian density model is used by Choi et al [5]. Many other researchers used switching activation function method (e.g. choice either cubic or hyperbolic-tangent function) for separating the mixtures of sub- and super-Gaussian sources according to the estimated kurtosis of the extracted outputs [3,6]. However, most of those reports does not show the stability of developed algorithms.

This paper presents a locally stable algorithm which is derived by using both generalized Gaussian probability density model for sub- and Gaussian signals and **t**-distribution density model for super- and Gaussian signals. The main purpose for introducing the **t**-distribution density model is to guarantee the algorithm remains stable. Also, the activation function derived by **t**-distribution density model has a robust property.

INDEPENDENT COMPONENT ANALYSIS ALGORITHM

Let us firstly introduce a quantity which can measure the sharpness of distribution of a random variable so called the kurtosis. The normalized kurtosis is defined as

$$\kappa(y) = \frac{m_4(y)}{m_2^2(y)} - 3, \tag{7}$$

where $m_2(y) = E[y^2]$ and $m_4(y) = E[y^4]$ denote the 2nd-, 4th-order moments. The kurtosis with positive, zero and negative values corresponds to super-Gaussian, Gaussian and sub-Gaussian, respectively. The relationship between the kurtosis and freedom degree parameters which in the generalized Gaussian distribution and **t**-distribution will be determined in next two sections.

Generalized Gaussian distribution

The probability density function for the generalized Gaussian distribution with α degrees of freedom (having α number of distributions) and scaling factor λ_α is represented by

$$p_\alpha(y) = \frac{\alpha\lambda_\alpha}{2\Gamma(\frac{1}{\alpha})}\exp(-|\lambda_\alpha y|^\alpha), \tag{8}$$

where $\Gamma(\cdot)$ is the Gamma function defined by

$$\Gamma(x) = \int_0^\infty t^{x-1}\exp(-t)dt. \tag{9}$$

By changing the value of α, a family of distributions with different sharpness will be given. When $\alpha = 2$, the distribution is reduced to the normal distribution. If $\alpha < 2$, the distribution curves are much peaky than the normal one, otherwise $\alpha > 2$, the distributions are much fat than the normal distribution. This property is corresponding to the kurtosis of a signal with sub-, super- or Gaussian distribution.

The relationship between the 2nd-, 4th-order moments, kurtosis and scaling constant λ_α is derived as follows. From the definition formula, the 2nd-order moment is obtained by

$$m_2 = \int_{-\infty}^{\infty} y^2 p_\alpha(y)dy = \frac{\Gamma(\frac{3}{\alpha})}{\lambda_\alpha^2 \Gamma(\frac{1}{\alpha})}, \tag{10}$$

and the 4th-order moment is given by

$$m_4 = \int_{-\infty}^{\infty} y^4 p_\alpha(y)dy = \frac{\Gamma(\frac{5}{\alpha})}{\lambda_\alpha^4 \Gamma(\frac{1}{\alpha})}. \tag{11}$$

The constant λ_α can be derived by using one of above equations, e.g. from the 2nd-order moment,

$$\lambda_\alpha = \left[\frac{\Gamma(\frac{3}{\alpha})}{m_2 \Gamma(\frac{1}{\alpha})}\right]^{\frac{1}{2}}. \tag{12}$$

Using the solutions of moments, we found the relative formula between the kurtosis and parameter α as

$$\kappa_\alpha = \frac{m_4}{m_2^2} - 3 = \frac{\Gamma(\frac{5}{\alpha})\Gamma(\frac{1}{\alpha})}{\Gamma^2(\frac{3}{\alpha})} - 3. \tag{13}$$

In Eq. (13), given a value of κ_α, the parameter α can be determined.

t-distribution

The probability density function of y for **t**-distribution with β degrees of freedom and scaling factor λ_β is defined by

$$p_\beta(y) = c_\beta \left(1 + \frac{(\lambda_\beta y)^2}{\beta}\right)^{-\frac{1}{2}(\beta+1)}, \tag{14}$$

$$c_\beta = \frac{\lambda_\beta \Gamma(\frac{\beta+1}{2})}{\sqrt{\pi\beta}\,\Gamma(\frac{\beta}{2})}, \tag{15}$$

where freedom degrees within $0 < \beta < \infty$ represents a family of distributions with changing sharpness referred to the Gaussian distribution, specially when $\beta = 1$ the distribution is called the Cauchy distribution, and when β approaches to infinity, it reduces to the normal distribution.

The 2nd-order moment for the **t**-distribution is

$$\begin{aligned} m_2 &= \int_{-\infty}^{\infty} y^2 p_\beta(y) dy \\ &= 2c_\beta c^{\frac{1}{2}(\beta+1)} \left[\frac{1}{c^{q-1}} \sum_{r=1}^{q-1} \frac{(-1)^{r+1}}{2r+1} {}_{q-2}C_{r-1} \left(\frac{y}{\sqrt{y^2+c}} \right)^{2r+1} \right]_0^\infty, \\ &= \frac{\beta\Gamma(\frac{\beta-2}{2})}{2{\lambda_\beta}^2\Gamma(\frac{\beta}{2})}, \end{aligned} \tag{16}$$

where $c = \beta/{\lambda_\beta}^2$ and $q = \beta/2$, we used the formula of the finite series including the binomial coefficients

$$\sum_{r=0}^{n} \frac{(-1)^r}{r+a} {}_nC_r = \frac{\Gamma(n+1)\Gamma(a)}{\Gamma(n+a+1)}, \tag{17}$$

where ${}_nC_r$ denotes the combination of n things taken r at a time. Note that when $\beta > 2$ the variance of **t**-distribution is exist. The scaling constant λ_β can be derived as

$$\lambda_\beta = \left[\frac{\beta\Gamma(\frac{\beta-2}{2})}{2m_2\Gamma(\frac{\beta}{2})} \right]^{\frac{1}{2}}. \tag{18}$$

Similarly the 4th-order moment can be derived as :

$$\begin{aligned} m_4 &= \int_{-\infty}^{\infty} y^4 p_\beta(y) dy \\ &= 2c_\beta c^{\frac{1}{2}(\beta+1)} \left[\frac{1}{c^{q-2}} \sum_{r=2}^{q-1} \frac{(-1)^{r+1}}{2r+1} {}_{q-3}C_{r-2} \left(\frac{y}{\sqrt{y^2+c}} \right)^{2r+1} \right]_0^\infty \\ &= \frac{3\beta^2\Gamma(\frac{\beta-4}{2})}{4{\lambda_\beta}^4\Gamma(\frac{\beta}{2})}, \qquad \beta > 4. \end{aligned} \tag{19}$$

Using the solutions of the 2nd-, 4th-order moments, the kurtosis can be derived as

$$\kappa_\beta = \frac{m_4}{m_2^2} - 3 = \frac{3\Gamma(\frac{\beta-4}{2})\Gamma(\frac{\beta}{2})}{\Gamma^2(\frac{\beta-2}{2})} - 3, \tag{20}$$

where the condition of $\beta > 4$ is necessary for the kurtosis in existence. Unlike κ_α has arbitrary number such as positive, zero and negative, κ_β is always positive for arbitrary value of β. κ_β approach to zero if β approach to infinity. For the reason of stability, we restrict the freedom degrees of generalized Gaussian distribution in the case of $\kappa_\alpha > 0$ and instead of it, we employ **t**-distribution.

Activation functions

In order to separate mixtures of sub-Gaussian and super-Gaussian sources, we use an ICA algorithm with an automatic switching activation functions based on the generalized Gaussian distribution and **t**-distribution density models. From these two models, the functions are derived as

$$\varphi_i(y_i) = \alpha\lambda_\alpha \operatorname{sgn}(y_i)|\lambda_\alpha y_i|^{\alpha-1}, \qquad \kappa_\alpha = \hat{\kappa}_i \leq 0 \tag{21}$$

and

$$\varphi_i(y_i) = \frac{(1+\beta)y_i}{y_i^2 + \frac{\beta}{\lambda_\beta^2}}, \qquad \kappa_\beta = \hat{\kappa}_i > 0, \tag{22}$$

respectively, where $\hat{\kappa}_i = \hat{m}_4/\hat{m}_2^2 - 3$ is an estimate of the kurtosis obtained by using the following 2nd-, 4th-order moments estimation formula

$$\hat{m}_j(k) = [1-\eta(k)]\hat{m}_j(k-1) + \eta(k)y_i^j(k), \qquad j = 2, 4, \tag{23}$$

In Eq. (23), $0 < \eta(k) < 1$ is learning rate. The selection of function in Eq. (21) or Eq. (22) can be determined if given a $\hat{\kappa}_i$. Note that Eq. (22) has a robust property in which it can against the larger outliers (see figure 1).

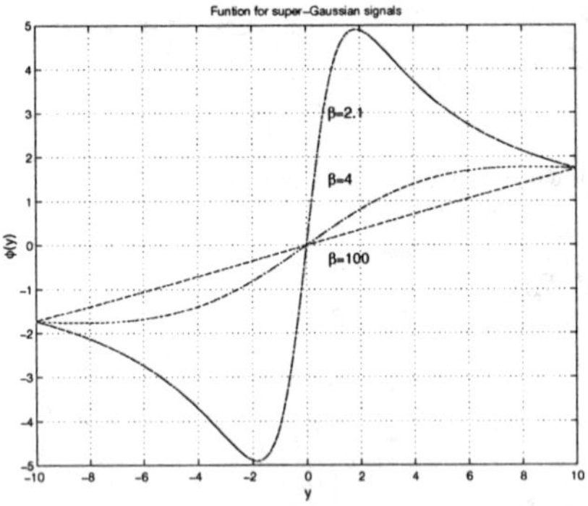

Fig. 1 A family of activation functions based on **t**-distribution model used for super-Gaussian signals.

Implementation

The implementation of proposed ICA algorithm is simply summarized as :

- Calculating the output $\mathbf{y}$ by given observations and an initial W .
- Estimating the high-order moments of the extracted signals $\mathbf{y}$.
- Normalizing the variance of output y_i to 1 by using the estimate of the 2nd-order moment, e.g. $y_i' = y_i/\sqrt{\hat{m}_2}$ and calculating the scaling constants λ_α and λ_β.
- Selecting α or β corresponding to the estimate of $\hat{\kappa}_i$.
- Updating W by using a suitable activation function.

STABILITY ANALYSIS OF ICA ALGORITHM

In the stability analysis, we seek conditions that guarantee the behavior of the learning system approaching to the equilibrium. When the Jacobian of the time averaged weight update $\partial\left(E\left[I-\varphi(\mathbf{y})\mathbf{y}^T\right]W\right)/\partial W$ has negative real parts, the equilibrium is asymptotically stable (see Amari et. al. 1997). Hence, the necessary and sufficient conditions are

$$\begin{aligned} E[y_i^2\dot{\varphi}_i(y_i)]+1 &> 0, & (24)\\ E[\dot{\varphi}_i(y_i)] &> 0, & (25)\\ E[y_i^2]E[y_j^2]E[\dot{\varphi}_i(y_i)]E[\dot{\varphi}_j(y_j)]-1 &> 0, & (26)\end{aligned}$$

for all $i, j(i \neq j)$, where $E[\cdot]$ denotes expectation and $\dot{\varphi} = d\varphi/dy$.

Applying these conditions, we illustrate the stability for all of the cases such as only sub-Gaussian or super-Gaussian signals, mixtures of sub-, super-Gaussian signals or/and Gaussian signal by using generalized Gaussian and t-distribution density models.

For the case of sub-Gaussian signals, the differential of the function in Eq. (21) can be derived by

$$\dot{\varphi}_i(y_i) = \lambda_\alpha^2\alpha(\alpha-1)|\lambda_\alpha y_i|^{\alpha-2}. \qquad (27)$$

All of the stability conditions are satisfied for sub-Gaussian signals ($\alpha > 2$) since

$$E[y_i^2\dot{\varphi}_i(y_i)]+1 = \int_{-\infty}^{\infty} y_i^2\dot{\varphi}_i(y_i)p_\alpha(y_i)dy_i + 1 = \alpha > 0, \qquad (28)$$

$$E[\dot{\varphi}_i(y_i)] = \int_{-\infty}^{\infty} \dot{\varphi}_i(y_i)p_\alpha(y_i)dy_i = \frac{\lambda_\alpha^2\alpha(\alpha-1)\Gamma(\frac{\alpha-1}{\alpha})}{\Gamma(\frac{1}{\alpha})} > 0, \qquad (29)$$

and

$$E[y_i^2]E[\dot{\varphi}_i(y_i)] = \frac{\alpha(\alpha-1)\Gamma(\frac{3}{\alpha})\Gamma(\frac{\alpha-1}{\alpha})}{\Gamma^2(\frac{1}{\alpha})} > 1, \qquad (30)$$

where $E[y_i^2] = m_2$ is the same one as in Eq. (10).

For the case of a Gaussian distribution signal, when $\alpha = 2$, then in Eq. (30), $E[y_i^2]E[\dot{\varphi}_i(y_i)] = 1$. If there are two Gaussian signals then the condition in Eq. (26) is not satisfied. From stability point of view, this is a reason which explains why we cannot separate more than one Gaussian source signal. If two Gaussian sources are non-white, by introducing the AR model it is possible to separate them (Amari personal memo 1998). However, If one signal is Gaussian and another signal is sub-Gaussian, it is easy to show the condition in Eq. (26) is satisfied since $E[y_j^2]E[\dot{\varphi}_j(y_j)] > 1$.

Note that the case of super-Gaussian signals, $\alpha < 2$, the stability conditions are not always satisfied. For example, when $\alpha = 0.5, 0.2$, etc., the conditions in Eqs. (25) and (26) are not satisfied since Eqs. (29) and (30) have no solution.

Now we discuss the case of super-Gaussian signals by using **t**-distribution density model. The differential of the function in Eq. (21) can be derived as

$$\dot{\varphi}_i(y_i) = \frac{(1+\beta)(c-y_i^2)}{(y_i^2+c)^2}, \tag{31}$$

where $c = \beta/\lambda_\beta^2$.

The conditions in Eqs. (24) and (25) are always satisfied since

$$E[y_i^2\dot{\varphi}_i(y_i)] + 1 = \int_{-\infty}^{\infty} y_i^2\dot{\varphi}_i(y_i)p_\beta(y_i)dy_i + 1 = \frac{2\beta}{\beta+3} > 0 \tag{32}$$

and

$$E[\dot{\varphi}_i(y_i)] = \int_{-\infty}^{\infty} \dot{\varphi}_i(y_i)p_\beta(y_i)dy_i = \frac{\lambda_\beta^2(\beta+1)}{\beta+3} > 0 \tag{33}$$

are positive.

For the condition in Eq. (26), let us recall the Eq. (16), the 2nd-order moment

$$E[y_i^2] = \frac{\beta\Gamma(\frac{\beta-2}{2})}{2\lambda_\beta{}^2\Gamma(\frac{\beta}{2})} > 0 \tag{34}$$

is positive when $\beta > 2$. Using Eqs. (33) and (34), we have

$$E[y_i^2]E[\dot{\varphi}_i(y_i)] = \frac{\beta(\beta+1)\Gamma(\frac{\beta-2}{2})}{2(\beta+3)\Gamma(\frac{\beta}{2})} > 1. \tag{35}$$

Hence, all of the stability conditions for super-Gaussian signals are satisfied by using the **t**-distribution density model.

$E[y_i^2]E[\dot{\varphi}_i(y_i)]$ approaches to 1 if and only if β approaches to infinity. This is the case for a Gaussian signal. As we have already discussed, when the sources are more than one Gaussian source, the algorithm does not guarantee the stability.

For the case of mixtures of super-Gaussian and sub-Gaussian (or Gaussian) signals, it is easy to show all of the conditions are satisfied by using the above results.

SIMULATION RESULTS

In order to confirm the validity and the performance of the developed algorithm we illustrate one ill-condition example in which not only mixtures of super- and sub-Gaussian sources, but also Gaussian source. Three orignal source signals with different kurtosis such as $\kappa_1 = -0.1060$, $\kappa_2 = 0.0073$ and $\kappa_3 = 3.3748$ (see Fig.2a) were mixed by a random matrix

$$A = \begin{bmatrix} -0.1106 & 0.8436 & -0.1886 \\ 0.2309 & 0.4764 & 0.0709 \\ 0.5839 & -0.6475 & 0.8338 \end{bmatrix}$$

by $\mathbf{x}(k) = A\mathbf{s}(k)$. It is assumed that three sources and mixing matrix A are unknown but the mixture signals $\mathbf{x}(k)$ are accessible (see Fig.2b). The initial condition is $W(0) = I$, the learning rate $\eta = 0.001$. The recovered signals is shown in Fig.2c. It can be seen from the result that the sources are well separated by using the proposed algorithm.

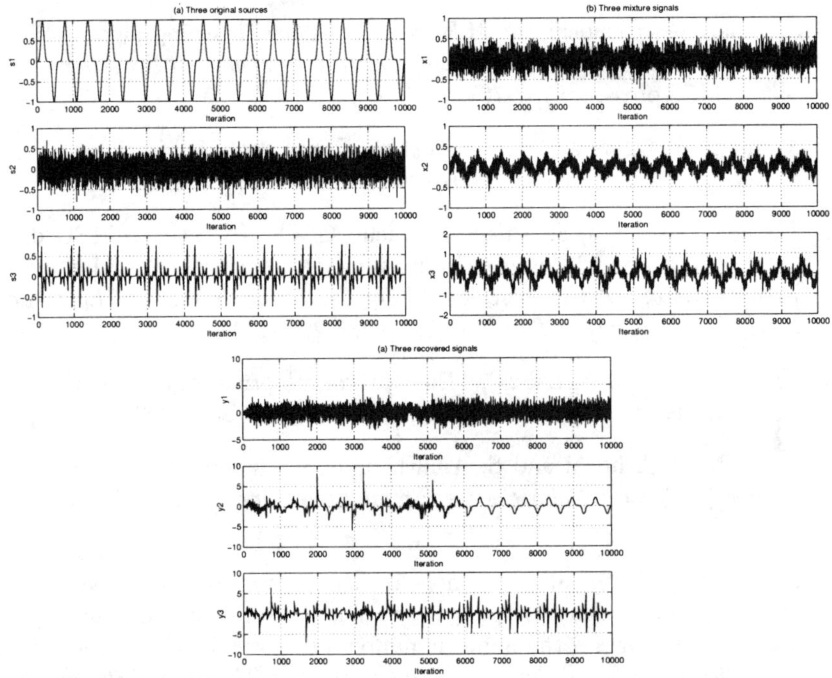

Fig. 2 Experimental results showing the separation of mixed super-, sub- and Gaussian sources.

CONCLUSIONS

In this paper we proposed a novel locally stable and robust ICA algorithm which enables to separate mixtures of sub-Gaussian, super-Gaussian and Gaussian original source signals. The flexible activation functions in the algorithm were derived by using parameterized **t**-distribution and generalized Gaussian distribution density models. The functions are self-adaptive based on estimating the high-order moments of extracted signals. The main advantages are that the proposed algorithm satisfies the stability conditions and also it can separate mixtures of sub-Gaussian, super-Gaussian and Gaussian original source signals which is very useful in the EEG and MEG data analysis.

Acknowledgments: The authors gratefully acknowledge Prof. Shun-ichi Amari and Prof. Andrzej Cichocki at RIKEN Brain Science Institute for constructive comments and useful discussions. The first author acknowledge partial support from Tateisi Science and Technology Foundation (Japan).

REFERENCES

[1] S. Amari, A. Cichocki and H.H. Yang, "A new learning algorithm for blind signal separation", *Advances in Neural Information Processing System 8*, MIT press, pp.757-763, 1996.

[2] S. Amari, T. Chen, and A. Cichocki, "Stability analysis of adaptive blind source separation", *Neural Networks*, Vol. 10, **8**, pp.1345-1351, 1997.

[3] J. Cao, N. Murata, S. Amari, A. Cichocki and T. Taketa, "MEG data analysis based on ICA approach with pre- & post-processing techniques", *Proceedings of 1998 International Symposium on Nonlinear Theory and its Applications (NOLTA)*, Vol. 1, pp.287-290, 1998.

[4] J.F. Cardoso and B. Laheld, "Equivariant adaptive source separation", *IEEE Trans. on Signal Processing*, vol. 44, **12** pp 3017-3030, 1996.

[5] S. Choi, A. Cichocki and S. Amari "Flexible independent component analysis", *Neural Networks for Signal Processing*, Vol. VIII, pp.83-93, 1998.

[6] A. Cichocki, I, Sabala, S. Choi, B. Orsier and R. Szupiluk "Self-adaptive independent component analysis for sub-Gaussian and super-Gaussian mixtures with unknown number of source signals", *Proceedings of 1997 International Symposium on Nonlinear Theory and its Applications (NOLTA)*, Vol. 1, pp.731-734, 1997.

[7] P. Comon, "Independent component analysis, a new concept?", *Signal Processing*, Vol. 36, **3**, pp.287-314, 1994.

[8] M. Girolami and C. Fyfe "Negentropy and kurtosis as projection pursuit indices provide generalized ICA algorithms", *NIPS'96 Workshop: Blind Signal Processing*, Snowmaas, Colorado, Dec, 1996.

[9] T-W. Lee, M. Girolami and T. Sejnowski "Independent component analysis using an extended infomax algorithm for mixed sub-Gaussian and super-Gaussian sources", *Neural Computation*, Vol.11(2): 417-441, 1999.

[10] D. Mackay "Maximum likelihood and covariant algorithms for independent component analysis", *Phd. thesis, University of Cambridge*, 1996.

SEPARATION OF AN OVERLAPPED SIGNAL USING SPEECH PRODUCTION MODELS

Hideyuki WATANABE †, *Satoru FUJITA* †,
and Shigeru KATAGIRI † ‡
† ATR Human Information Processing Research Laboratories
‡ NTT Communication Science Laboratories
2-2 Hikaridai, Seika-cho, Soraku-gun, Kyoto 619-0288, Japan
Tel.: +81-774-95-1095
Fax.: +81-774-95-1008
E-mail: hwatana@hip.atr.co.jp

Abstract

In this paper, we propose a novel approach to separating an overlapped signal using speech production models. The speech production models are used as top-down constraints for solving a one-channel signal separation, which is a one-to-many mapping problem. The goal of our approach is to not only realize high-accuracy separation systems but to also construct a computational model of human auditory mechanisms based on perception-production interaction. The utility of the proposed method is demonstrated in a task of separating a mixture of two-speaker vowel utterances using primitive linear speech production models.

1 INTRODUCTION

The technology of separating an overlapped acoustic signal into its component source signals has been extensively studied in recent years. This technology is very important for various kinds of real-world signal processing applications, including cross-talk speech recognition, noisy speech recognition, sound monitoring, and automatic music transcription.

Currently, much of the research is directed at multi-channel (microphone-array) techniques, such as separation based on source localization [1] and blind signal separation (independent component analysis) which extracts statistically-independent signal components by multi-channel signal transformations [2] [3]. Typically, in these methods, no prior knowledge of component signals (except their statistical independence) is assumed (or some probability distributions are assumed at most). This lack of prior information actually

0-7803-5673-X/99/$10.00 © 1999 IEEE

increases the applicability of the methods, but it limits the separation capability, so that a large number of microphone units are required to achieve an acceptable performance. On the other hand, many researchers are vigorously studying an approach that mimics human auditory processes as a more realistic technique that only requires a few channels [4] [5]. Most implementations based on this approach, however, do not use prior information on the production mechanisms of component signals, but instead utilize the "rules of thumb" of human audition in time-frequency analyses. Obviously, such empirical rules are not necessarily sufficient for achieving a satisfactory level of performance.

One of the convincing theories concerning human auditory processes is the articulatory reference theory of speech perception, which is also referred to as the motor theory of speech perception [6]. This theory states that when a human being listens to a speech sound, he/she perceives it by referring to its corresponding production structure (articulatory structure). Actually, the theory has not been fully proven yet in the psychological and/or neurophysiological sense. Nevertheless, the validity of the theory is supported by many scientific findings (e.g., see [7]). Naturally, we are brought to the challenge of developing sound separation technologies using *a priori* knowledge of speech production mechanisms.

Considering this theoretical viewpoint, we propose in this paper a new formalization method of speech sound separation, which is based on top-down prior knowledge of speech production mechanism. There are many possible formalization methods, and we specially employ linear all-pole (autoregressive) modeling for the speech production process. By using linguistic category-dependent production models, one can also simultaneously perform speech separation and speech recognition. The method is evaluated in a task of separating overlapped vowel sounds, each spoken by a male speaker and a female speaker.

2 FORMALIZATION OF SOUND SEPARATION USING SPEECH PRODUCTION MODELS

Let $s_i(t)$ $(i = 1, 2, ..., I)$ be the waveform of an original speech uttered by speaker $\#i$, where I denotes the number of speakers. Assume that the overlapped speech waveform $s(t)$ is observed:

$$s(t) \quad = \quad s_1(t) + s_2(t) + \cdots + s_I(t). \qquad (1)$$

The task here is to reconstruct the component signal sources, $\{s_i(t)\}_{i=1}^{I}$, from the observed signal $s(t)$.

In order to solve this one-to-many mapping problem, we must inevitably impose some appropriate constraints on the component signals. Considering that we are interested in a mathematical model of a human speech segregation based on the motor theory (and the application of the model to signal separation techniques), we use speech production models as the constraints. For a concrete example, a speech production model is prepared for each speaker and the model parameters are estimated so that the summation of the signals synthesized by these models can approximate the observed overlapped speech. We then regard each synthesized signal as a separated, reconstructed speech of its corresponding speaker. Let $\overline{s}_{i,\theta_i}(t)$ $(i = 1, 2, ..., I)$ be the signal synthesized by the production models of the speaker $\#i$, where θ_i denotes the adjustable model parameter set. The speech separation problem then leads to the minimization problem:

$$\{\hat{\theta}_1, .., \hat{\theta}_I\} = \arg \min_{\theta_1,\ldots,\theta_I} J[s|\overline{s}_{1,\theta_1} + \cdots + \overline{s}_{I,\theta_I}], \tag{2}$$

where $J[x|y]$ denotes a criterion that measures the discrepancy between the two signals: $x(t)$ and $y(t)$. The separated, reconstructed signals are given by the synthesized signals from the optimized models: $\overline{s}_{i,\hat{\theta}_i}(t)$ $(i = 1, 2, ..., I)$. Fig. 1 illustrates the separation procedure for two speakers.

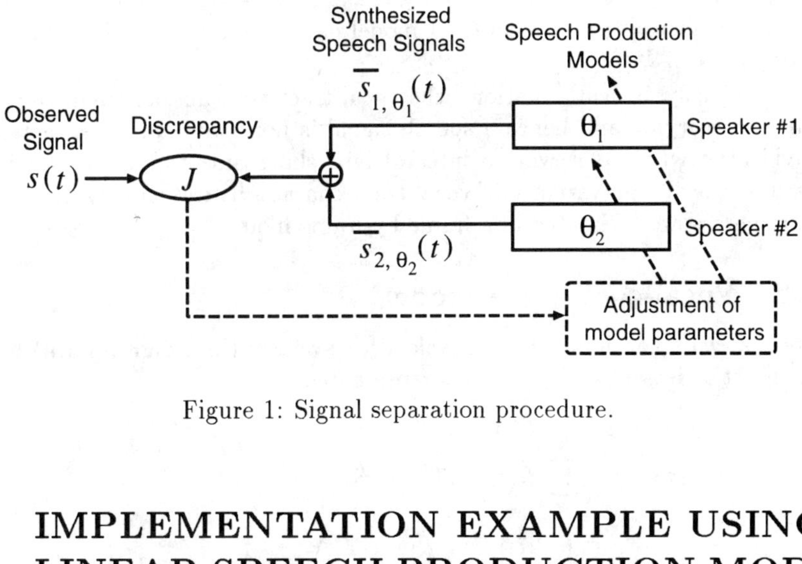

Figure 1: Signal separation procedure.

3 IMPLEMENTATION EXAMPLE USING LINEAR SPEECH PRODUCTION MODELS

We focus our attention here on the separation of overlapped voiced speech, which is accompanied by vocal-cord vibrations. As a primitive speech production model, we consider the linear terminal-analog model (Fig. 2) consisting of the all-pole (autoregressive) vocal tract model [8] and Klatt's (Rosenberg's) vocal-cord-wave model [9].

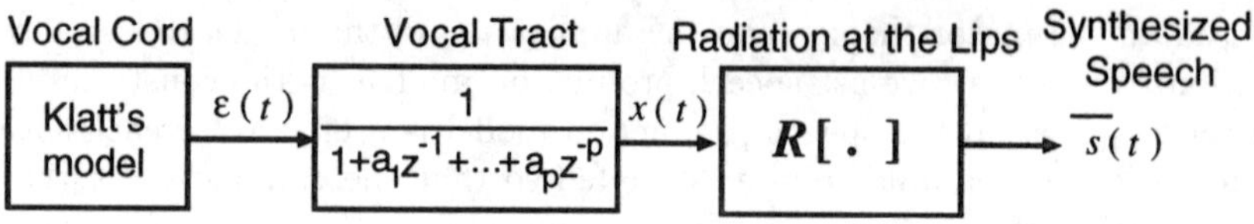

Figure 2: A linear terminal-analog speech production model.

3.1 Vocal Tract Model

The synthesized speech signal of speaker $\#i$ ($i = 1, 2, ..., I$) based on the all-pole vocal tract model is expressed as:

$$\bar{s}_{i,\theta_i}(t) = \mathcal{R}\left[x_{i,\theta_i}(t)\right], \tag{3}$$

$$x_{i,\theta_i}(t) + a_1^{(i)} x_{i,\theta_i}(t-1) + \cdots + a_M^{(i)} x_{i,\theta_i}(t-M) = \varepsilon_i(t). \tag{4}$$

$\mathcal{R}[\cdot]$ denotes a high-pass filter corresponding to the characteristics of the sound radiation at the lips, $\{a_m^{(i)}\}_{m=1}^{M}$ denote the coefficients of the M-order all-pole vocal tract filter of speaker $\#i$, and $\varepsilon_i(t)$ is the vocal-cord waveform of the speaker $\#i$. The vocal-tract model parameters to be adjusted are the coefficients $\{a_m^{(i)}\}_{m=1}^{M}$.

Obviously, in real situations, the vocal-tract transmission characteristics are time-varying and hence a speech signal is non-stational. We, therefore, divide the whole observation interval into short-time frames in which we assume the time-invariance of vocal-tract characteristics, namely, the model parameters are estimated in a frame-by-frame mode.

3.2 Vocal-cord-wave Model

Speaker $\#i$'s vocal-cord volume-velocity waveform (in a digital form) based on Klatt's (Rosenberg's) model is represented as

$$\varepsilon_i(t) = \tilde{\varepsilon}_i(t/f_s), \tag{5}$$

$$\tilde{\varepsilon}_i(\tau) = \sum_n f_i(\tau - nT^{(i)} - \tau_0^{(i)}), \tag{6}$$

$$f_i(\tau) = \begin{cases} \dfrac{27A^{(i)}}{4{T_o^{(i)}}^2}\tau^2\left(1 - \dfrac{\tau}{T_o^{(i)}}\right) & (0 \le \tau \le T_o^{(i)}) \\ 0 & \text{(otherwise)} \end{cases}. \tag{7}$$

$\tilde{\varepsilon}_i(\tau)$ represents the vocal-cord waveform in an analog form where τ (sec) denotes the continuous time, f_s (Hz) is a sampling frequency, $T^{(i)}(> 0)$ (sec) denotes a pitch period of speaker $\#i$'s utterance, $\tau_0^{(i)}$ (sec) corresponds to a phase, $A^{(i)}(> 0)$ denotes an amplitude (a gain factor), and $T_o^{(i)}$ $(0 < T_o^{(i)} < T^{(i)})$ (sec) represents the duration of a glottis-open period. The vocal-cord waveform is illustrated in Fig. 3. The vocal-cord-wave parameters to be adjusted are $A^{(i)}$, $T^{(i)}$, $T_o^{(i)}$ and $\tau_0^{(i)}$.

According to auditory psychoacoustics findings, a human being refers to the difference of pitch patterns among source signals as a key to segregation. Based on this theory, we estimate in advance, for each speaker, the average speaker-specific pitch period and utilize it as an inital value for the optimization of $T^{(i)}$.

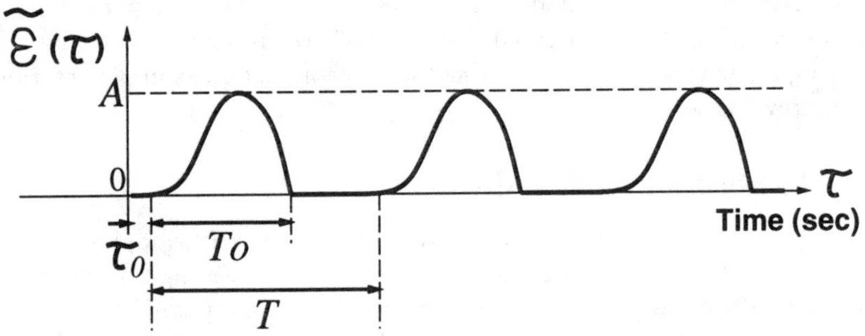

Figure 3: Klatt's (Rosenberg's) vocal-cord-wave model

3.3 Discrepancy Measure to be Minimized

Since the restoration of linguistic units such as phonemes is the most important task here, we apply, as a discrepancy measure (minimization criterion), the mean square error in a low-frequency region in which linguistic features are prominent.

$$J_1[s|\bar{s}_\theta] = \sum_t \{\mathcal{L}[s(t)] - \mathcal{L}[\bar{s}_\theta(t)]\}^2, \quad (8)$$

where $\bar{s}_\theta(t) = \bar{s}_{1,\theta_1}(t) + \bar{s}_{2,\theta_2}(t) + \cdots + \bar{s}_{I,\theta_I}(t)$, $\theta = (\theta_1, \theta_2, ..., \theta_I)$, and $\mathcal{L}[\cdot]$ denotes a low-pass filter. The above measure is a rather simple mean square error calculated in the time domain and hence does not reflect the human auditory characteristics. We use this simple measure for its computational simplicity.

3.4 Minimization Procedure

To efficiently minimize J_1 with respect to the I parameter sets: $\theta_1, \theta_2, ..., \theta_I$, we optimize each parameter set by turns. Specifically, we first optimize θ_1 while fixing the other parameter sets, then optimize θ_2, and so forth. We repeat this alternating minimization several times.

For the optimization of vocal-tract filters, we prepare a codebook of filter-coefficient vectors for each speaker and for each linguistic category (such as a phoneme), and, for each speaker, search the codebook to find the optimal filter. This way, by referring to the linguistic category that the optimal filter belongs to, we can perform speech recognition as well as speech separation.

On the other hand, we optimize the vocal-cord parameters using the gradient descent method.

4 EXPERIMENTAL EVALUATION

To evaluate the proposed model-based separation method, we performed an experiment in which we separated a mixture of two-speaker (a male and a female) Japanese vowel utterances using the implementation example described in the previous section.

4.1 Experimental Conditions

Each utterance was individually recorded at 22.050 kHz sampling, down-sampled at 12 kHz and quantized in 16 bits. The two digital waveforms were manually superposed; we used the resulting superposed waveform as an observed signal. The frame-by-frame estimation of the speech production models was executed by moving a 20-msec Hanning time window with a 10-msec shift length. The order of each all-pole vocal tract filter was 16. The reconstructed speech of each speaker was computed by concatenating the frame-based synthesized signals with the overlap-add method [10]. The above-mentioned procedure is summarized in Fig. 5.

We simulated the sound radiation at the lips ($\mathcal{R}[\cdot]$) by using the first-order pre-emphasis filter: $1 - 0.98z^{-1}$. For the low-pass filter ($\mathcal{L}[\cdot]$) used in the discrepancy measurement (8), we applied the pole-zero Butterworth-type filter with a (4,4)-th order and 2000 Hz cut-off frequency. The codebook size was set to 8 in the codebook searching to find the optimal vocal-tract parameters. For the vocal-cord-wave parameters, the amplitude $A^{(i)}$ was calculated in a closed form so that the power of the synthesized signal became equal to that of the observed signal. The other parameters, $T^{(i)}$, $T_o^{(i)}$ and $\tau_0^{(i)}$, were adjusted with the gradient method. The male pitch period $T^{(1)}$ was first set to 8.33 msec (120 Hz) and the female one $T^{(2)}$ to 3.33 msec (300 Hz). In addition, there were 30 iterations of gradient descent and 5 alternating minimizations.

4.2 Experimental Results

We processed three kinds of mixtures of utterances: (1) /a/ for the male and /u/ for the female, (2) /a/ for the male and /i/ for the female, and (3) /aiueo/ for the male and /aeiou/ for the female.

Fig. 4 shows the frame-based phoneme recognition results obtained from the codebook searching with the frame-by-frame speech separation process. For the single-vowel cases (/a/+/u/ and /a/+/i/), we could obtain a fairly good recognition. For the more difficult case of connected vowels (i.e., /aiueo/ +/aeiou/), on the other hand, the performance was less satisfactory; we are left with much room for further improvement in the codebook designs and the searching procedures.

An informal listing indicated that the speaker characteristics and the linguistic contents were largely restored for single-vowel utterances (/a/+/u/ and /a/+/i/) and comparatively restored for connected-vowel utterances (/aiueo/+/aeiou/). Fig. 6 shows the original waveforms in (a) (male) and (b) (female) and the reconstructed waveforms in (c) (male) and (d) (female) in the third case (/aiueo/+/aeiou/). The reconstructed waveforms look different from their corresponding original ones. However, this difference is mainly due to phase difference in the waveform synthesis stage, which is basically irrelevant to the speech intelligibility. Experimental results suggest that the proposed method has a fundamental potential for developing a new model-based speech sound separation technique. This method also can be quite useful for researchers who wish to elucidate the interactive mechanism of human speech perception and production.

(1) Male: /a/ + Female: /u/

Male: aaaaaaaaaaaaaaaaaaaoaaaaaaaaaoaaoaaaaaaaaaaaaaaaa

Female: uuuuuuuuuuuuuuuuuuuuuuuuuiuoauuuuuuuuuuuuauuuuuuuuu

(2) Male: /a/ + Female: /i/

Male: aaaaaaaauaaaaaoaoaoaaoooaooaaaaeoaooaaooaoooaooaa

Female: iuiiiiiiuiiiiiuiuiiiiieuieoiiiiuuioaiiuaiiuiioaii

(3) Male: /aiueo/ + Female: /aeiou/

Male:

aaaaaaaaaaaaaaaaaaaoaaaaooeuioiiooiooiiuuiuiueuiiiuou
uuouiiuouuuuiiuoeoooooooouooooaaoooooooooooooooooouoouo
ooeoooooooo

Female:

aaaoaaaaaaaaaauaaaaaieeieeiiiiiieiiiiieiuuiuiiiiiiiii
iiiiiiiiiiiiiioieuueuoooooooaiaaiaauuiieiiieiaaiieiii
ioiiieeiii

Figure 4: Frame-by-frame phoneme recognition results.

5 CONCLUSION

We have proposed a new approach to the separation of an overlapped signal using speech production models as top-down knowledge of sources. We demonstrated the fundamental utility of this approach in a preliminary experiment of vowel utterance separation using the linear all-pole speech production models. The next step in our research would be to apply a more sophisticated and realistic production model to replace the simple linear model. We also plan to investigate the separation of speech utterances and other types of sounds. Research efforts along the model-based approach should be useful for not only producing new sound separation techniques but also for analyzing the interactive mechanism between speech perception and production in the human auditory system.

References

[1] T. Yamada, S. Nakamura, and K. Shikano, "Robust speech recognition with speaker localization by a microphone array," *Proc. ICSLP,* pp. 1317-1320, Oct. 1996.

[2] A. J. Bell and T. J. Sejnowski, "An information-maximization approach to blind separation and blind deconvolution," *Neural Computation,* Vol. 7, pp. 1129-1159, 1995.

[3] J.-F. Cardoso, "Blind signal separation: statistical principles," *Proc. IEEE,* Vol. 86, No. 10, pp. 2009-2025, Oct. 1998.

[4] I. Masuda-Katsuse, H. Kawahara, and K. Aikawa, "Speech segregation based on continuity of spectral shapes," *Working Notes of the Computational Auditory Scene Analysis Workshop, held with 15th Intl. Joint Conf. on Artificial Intelligence (IJCAI'97),* pp. 39-45, Aug. 1997.

[5] M. Unoki and M Akagi, "Signal extraction from noisy signal based on auditory scene analysis," *Proc. ICSLP,* Vol. 4, pp. 1515-1518, 1998.

[6] A. M. Liberman, "Some results of research on speech perception," *J. Acoust. Soc. Am.,* Vol. 29, No. 1, pp. 117-123, Jan. 1957.

[7] R. Akahane-Yamada, E. McDermott, T. Adachi, H. Kawahara, and J. S. Pruitt, "Computer-based second language production training by using spectrographic representation and HMM-based speech recognition scores," *Proc. ICSLP,* 1998.

[8] J. D. Markel and A. H. Gray, *Linear Prediction of Speech,* Springer-Verlag, New York, 1976.

[9] D. H. Klatt and L. C. Klatt, "Analysis, synthesis, and perception of voice quality variations among female and male talkers," *J. Acoust. Soc. Am.,* Vol. 87, No. 2, pp. 820-857, Feb. 1990.

[10] E. B. George and M. J. T. Smith, "Analysis-by-synthesis/overlap-add sinusoidal modeling applied to the analysis and synthesis of musical tones," *J. Audio Eng. Soc.,* Vol. 40, No. 6, pp. 497-516, Jun. 1992.

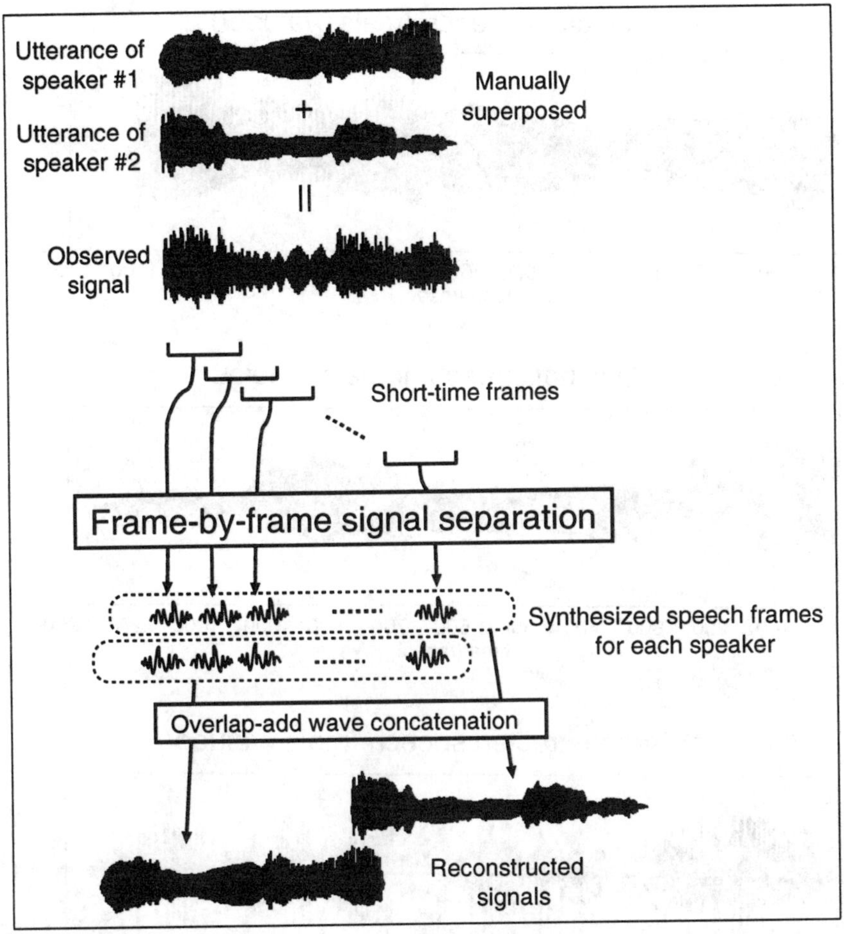

Figure 5: Procedure of signal separation experiments

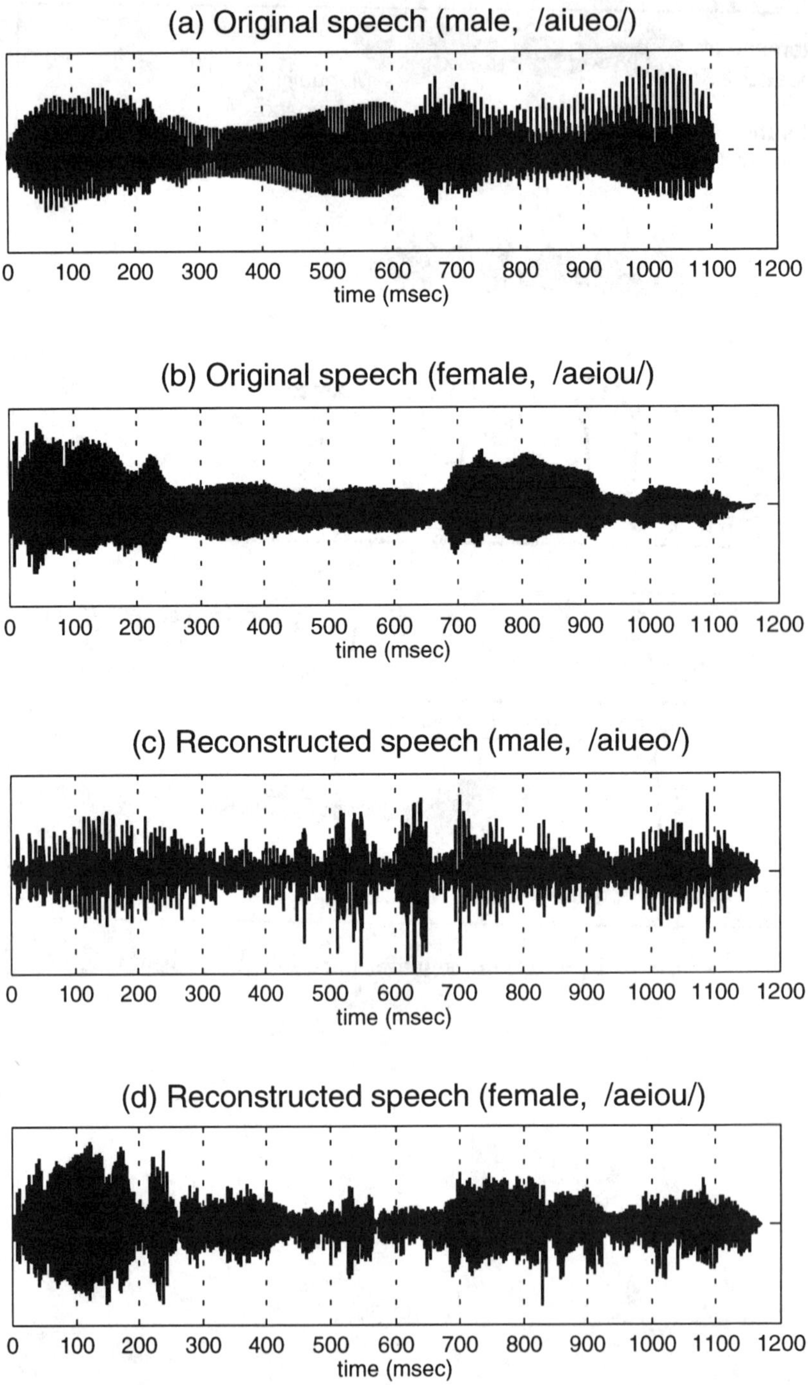

Figure 6: Original and reconstructed speech waveforms.

GEOMETRICAL STRUCTURES OF FIR MANIFOLD AND THEIR APPLICATION TO MULTICHANNEL BLIND DECONVOLUTION

L.-Q. Zhang, A. Cichocki and S. Amari
Brain-style Information Systems Research Group, BSI
The Institute of Physical and Chemical Research
Saitama 351-0198, Wako shi, JAPAN
Phone: +81 48 4679665
Fax: +81 48 4679694
E-mail: {zha, cia}@open.brain.riken.go.jp
Web: www.open.brain.riken.go.jp

Abstract. In this paper we study geometrical structures on the manifold of FIR filters and their application to multichannel blind deconvolution. First we introduce the Lie group and Riemannian metric to the manifold of FIR filters. Then we derive the natural gradient on the manifold using the isometry of the Riemannian metric. Using the natural gradient, we present a novel learning algorithm for blind deconvolution based on the minimization of mutual information. We also study properties of the learning algorithm, such as equivariance and stability. Simulations are given to illustrate the effectiveness and validity of the proposed algorithm.

INTRODUCTION

Recently blind separation of independent sources has become an increasing important research area due to its theoretical interest, as well as its rapidly growing applications in various fields, such as telecommunication systems, image enhancement and biomedical signal processing. Refer to review papers [4] and [9] for details.

It has been shown that the natural gradient improves dramatically the learning efficiency in instantaneous blind separation and blind deconvolution. For doubly infinite filter mixtures, the natural gradient algorithm has been developed by Amari, Douglas and Cichocki [6]. The main objective of this paper is to develop an efficient learning algorithm for training causal finite filter demixing models. In contrast to doubly infinite filters, the causal finite filters do not have the self closed multiplication and inverse operations in the manifold of filters with a fixed length. In general the multiplication of two

0-7803-5673-X/99/$10.00

filters with a given length makes a new filter with extended length, so does the inverse operation.

In this paper, we derive the explicit form of the natural gradient for FIR filters by using the isometric property of the Lie group on the manifold. The natural gradient descent scheme is employed to train the FIR filter and a novel learning algorithm is developed to update the parameters in the demixing model. Computer simulations are given to illustrate the effectiveness and validity of the proposed algorithm. For the space limitation, the detailed proofs will be presented in a full paper.

The natural gradient approach to FIR filters is of theoretical and practical interest since under a simple condition, any doubly finite filters can be decomposed into a cascade form of two FIR filters, one is a delay filter and the other is a forward filter [26].

PROBLEM FORMULATION

In this paper, as a convolutive mixing model, we consider a multichannel linear time-invariant(LTI) systems, with no poles on the unit circle, of the form

$$\mathbf{x}(k) = \mathbf{H}(z)\mathbf{s}(k), \tag{1}$$

where $\mathbf{H}(z) = \sum_{p=0}^{\infty} \mathbf{H}_p z^{-p}$, is an unknown mixing filter, $\mathbf{H}_p$ is an $n \times n$-dimensional matrix of mixing coefficients at time-lag p, which is called the impulse response at time p, $\mathbf{s}(k)$ is an n-dimensional vector of source signals with mutually independent components at time k, and $\mathbf{x}(k)$ is an $n-$dimensional sensor vector at time k, $k = 1, 2, \cdots$. The goal of multichannel blind deconvolution is to retrieve source signals $\mathbf{s}(k)$ only using sensor signals $\mathbf{x}(k), k = 1, 2, \cdots$, and certain knowledge of the source signal distributions and statistics. We carry out blind deconvolution by using a causal finite multichannel equalizer of the form

$$\mathbf{y}(k) = \mathbf{W}(z)\mathbf{x}(k), \tag{2}$$

where $\mathbf{W}(z) = \sum_{p=0}^{N} \mathbf{W}_p z^{-p}$, N is the length of FIR filter $\mathbf{W}(z)$, $\mathbf{y} = [y_1(k), \cdots, y_n]^T$ is an n-dimensional vector of the outputs and $\mathbf{W}_p$ is an $n \times n$-dimensional coefficient matrix at time lag p.

The set of all FIR filters $\mathbf{W}(z)$ of length N, having the constraint that $\mathbf{W}_0$ is nonsingular, is denoted by $\mathcal{M}(N)$,

$$\mathcal{M}(N) = \left\{ \mathbf{W}(z) \mid \mathbf{W}(z) = \sum_{p=0}^{N} \mathbf{W}_p z^{-p},\ det(\mathbf{W}_0) \neq 0 \right\}. \tag{3}$$

$\mathcal{M}(N)$ is a manifold of dimension $n^2(N+1)$. In general, multiplication of two filters in $\mathcal{M}(N)$ will enlarge the filter length and the result does belong to $\mathcal{M}(N)$ anymore. This makes it difficult to introduce the Riemannian structure to the manifold of finite multichannel filters. In order to explore possible geometrical structures of $\mathcal{M}(N)$ which will lead to effective learning algorithms for $\mathbf{W}(z)$, we define algebraic operations of filters in the Lie group framework.

When we apply $\mathbf{W}(z)$ to the sensor signal $\mathbf{x}(k)$, the global transfer function from $\mathbf{s}$ to $\mathbf{y}$ is defined by $\mathbf{G}(z) = [\mathbf{W}(z)\mathbf{H}(z)]_N$. The goal of the blind deconvolution task is to find $\mathbf{W}(z)$ such that

$$\mathbf{G}(z) = \mathbf{P}\mathbf{\Lambda}\mathbf{D}(z), \tag{4}$$

where $\mathbf{P} \in \mathbf{R}^{n\times n}$ is a permutation matrix, $\mathbf{D}(z) = \mathrm{diag}\{z^{-d_1}, \cdots, z^{-d_n}\}$, and $\mathbf{\Lambda} \in \mathbf{R}^{n\times n}$ is a nonsingular diagonal scaling matrix. In other words, the objective of multichannel blind deconvolution is to recover the source vector $\mathbf{s}(k)$ from the observation vector $\mathbf{x}(k)$, up to scaling, reordering and delays.

Assume that $p(\mathbf{y}, \mathbf{W})$, $p_i(y_i, \mathbf{W})$ are the joint probability density function of $\mathbf{y}$ and marginal pdf of y_i, $(i = 1, \cdots, n)$ respectively. In order to separate independent sources by a demixing model, we formulate the blind deconvolution problem into an optimization problem. Our target is to make the components of $\mathbf{y}$ as both mutually independent and identically independently distributed as possible. To this end, we employ the Kullback-Leibler divergence as a cost function [5]

$$l(\mathbf{y}, \mathbf{W}) = -H(\mathbf{y}, \mathbf{W}) + \sum_{i=1}^{n} H(y_i, \mathbf{W}), \tag{5}$$

where $H(\mathbf{y}, \mathbf{W}) = -\int p(\mathbf{y}) \log p(\mathbf{y}) d\mathbf{y}$, $H(y_i, \mathbf{W}) = -\int p_i(y_i) \log p_i(y_i) dy_i$. The divergence $l(\mathbf{y}, \mathbf{W})$ is a nonnegative functional, which measures the mutual independence of the output signals $y_i(k)$. The output signals $\mathbf{y}$ are mutually independent if and only if $l(\mathbf{y}, \mathbf{W}) = 0$.

GEOMETRICAL STRUCTURE FOR MANIFOLD $\mathcal{M}(N)$

In this section we introduce geometrical structures, such as the Lie group and Riemannian metric, to the manifold of FIR filters. Such structures are useful for the derivation of the natural gradient [1].

Lie Group

In the manifold $\mathcal{M}(N)$, Lie operations *multiplication* $*$ and *inverse* $\dagger$ are defined as follows: For $\mathbf{B}(z)$, $\mathbf{C}(z) \in \mathcal{M}(N)$,

$$\mathbf{B}(z) * \mathbf{C}(z) = \sum_{p=0}^{N} \sum_{q=0}^{p} \mathbf{B}_q \mathbf{C}_{(p-q)} z^{-p}, \tag{6}$$

$$\mathbf{B}^{\dagger}(z) = \sum_{p=0}^{N} \mathbf{B}_p^{\dagger} z^{-p}, \tag{7}$$

where $\mathbf{B}_p^\dagger$ are recurrently defined by $\mathbf{B}_0^\dagger = \mathbf{B}_0^{-1}$, $\mathbf{B}_p^\dagger = -\sum_{q=1}^{p} \mathbf{B}_{p-q}^\dagger \mathbf{B}_q$, $p = 1, \cdots, N$. With these operations, both $\mathbf{B}(z) * \mathbf{C}(z)$ and $\mathbf{B}^\dagger$ still remain in the manifold $\mathcal{M}(N)$. It is easy to verify that the manifold $\mathcal{M}(N)$ with the above operations forms a Lie Group. The identity element is $\mathbf{E}(z) = \mathbf{I}$. Moreover, the Lie group possesses the following properties

$$\begin{array}{ll} \text{1) Associative Law :} & \mathbf{A}(z) * (\mathbf{B}(z) * \mathbf{C}(z)) = (\mathbf{A}(z) * \mathbf{B}(z)) * \mathbf{C}(z), \\ \text{2) Inverse Property :} & \mathbf{B}(z) * \mathbf{B}^\dagger(z) = \mathbf{B}^\dagger(z) * \mathbf{B}(z) = \mathbf{I}. \end{array} \tag{8}$$

In fact the Lie multiplication of two $\mathbf{B}(z)$, $\mathbf{C}(z) \in \mathcal{M}(N)$ is the truncated form of the ordinary multiplication up to order N, that is

$$\mathbf{B}(z) * \mathbf{C}(z) = [\mathbf{B}(z)\mathbf{C}(z)]_N \tag{9}$$

where $[\mathbf{B}(z)]_N$ is such a truncated operator that any terms with orders higher than N in the polynomial $\mathbf{B}(z)$ are omitted.

Riemannian Metrics

The Lie Group has an important property that it admits an invariant Riemannian metric. Let $T_{\mathbf{W}(z)}$ be the tangent space of $\mathcal{M}(N)$, and $\mathbf{X}(z)$, $\mathbf{Y}(z) \in T_{\mathbf{W}(z)}$ be the tangent vectors. We introduce the inner product with respect to $\mathbf{W}(z)$ as $< \mathbf{X}(z), \mathbf{Y}(z) >_{\mathbf{W}(z)}$. Since $\mathcal{M}(N)$ is a Lie group, any $\mathbf{B}(z) \in \mathcal{M}(N)$ defines an onto-mapping: $\mathbf{W}(z) \to \mathbf{W}(z) * \mathbf{B}(z)$. The multiplication transformation maps a tangent vector $\mathbf{X}(z)$ at $\mathbf{W}(z)$ to a tangent vector $\mathbf{X}(z) * \mathbf{B}(z)$ at $\mathbf{W}(z) * \mathbf{B}(z)$. Therefore we can define a Riemannian metric on $\mathcal{M}(N)$, such that the right multiplication transformation is isometric, that is, it preserves the Riemannian metric on $\mathcal{M}(N)$. Explicitly we write it as follows

$$< \mathbf{X}(z), \mathbf{Y}(z) >_{\mathbf{W}(z)} = < \mathbf{X}(z) * \mathbf{B}(z), \mathbf{Y}(z) * \mathbf{B}(z) >_{\mathbf{W}(z) * \mathbf{B}(z)} . \tag{10}$$

If we define the inner product at the identity $\mathbf{E}(z)$ by $< \mathbf{X}(z), \mathbf{Y}(z) >_{\mathbf{E}(z)} = \sum_{p=0}^{N} tr(\mathbf{X}_p \mathbf{Y}_p^T)$, then $< \mathbf{X}(z), \mathbf{Y}(z) >_{\mathbf{W}(z)}$ is automatically induced by

$$< \mathbf{X}(z), \mathbf{Y}(z) >_{\mathbf{W}(z)} = < \mathbf{X}(z) * \mathbf{W}(z)^\dagger, \mathbf{Y}(z) * \mathbf{W}(z)^\dagger >_{\mathbf{E}(z)} . \tag{11}$$

NATURAL GRADIENT

Stochastic gradient optimization methods for parameterized systems suffer from slow convergence due to the statistical correlations of the processes signals. While quasi-Newton and related methods can be used to improve convergence, they also suffer from the mass computation and numerical instability, as well as local convergence.

The natural gradient search scheme is an efficient technique for solving iterative estimation problems [1]. For a cost function $l(\mathbf{W}(z))$ defined on the

Riemannian manifold $\mathcal{M}(N)$, the natural gradient $\tilde{\nabla}l(\mathbf{W}(z))$ is the steepest ascent direction of the cost function $l(\mathbf{W}(z))$ as measured by the Riemannian metric on $\mathcal{M}(N)$, which is the contravariant form of partial derivatives (ordinary gradient), denoted by $\frac{\partial l(\mathbf{W}(z))}{\partial \mathbf{W}(z)} = \sum_{p=0}^{N} \frac{\partial l(\mathbf{W}(z))}{\partial \mathbf{W}_p} z^{-p}$, where $\frac{\partial l(\mathbf{W}(z))}{\partial \mathbf{W}_p} = \left(\frac{\partial l(\mathbf{W})}{\partial \mathbf{W}_{pij}}\right)_{n\times n}$. The natural gradient $\tilde{\nabla}l(\mathbf{W}(z))$ of the cost function $l(\mathbf{W}(z))$ is calculated as follows [1]

$$\langle \mathbf{X}(z), \tilde{\nabla}l(\mathbf{W}(z))\rangle_{\mathbf{W}(z)} = \langle \mathbf{X}(z), \frac{\partial l(\mathbf{W}(z))}{\partial \mathbf{W}(z)}\rangle_{\mathbf{E}(z)}, \tag{12}$$

for any $\mathbf{X}(z) \in T_{\mathbf{W}(z)}$. Comparing the both side of (12), we have

$$\tilde{\nabla}l(\mathbf{W}(z)) = \frac{\partial l(\mathbf{W}(z))}{\partial \mathbf{X}(z)} * \mathbf{W}(z). \tag{13}$$

where $\mathbf{X}(z)$ is a nonholonomic base, which is defined by

$$d\mathbf{X}(z) = d\mathbf{W}(z) * \mathbf{W}^{\dagger}(z) = \left[d\mathbf{W}(z)\mathbf{W}^{-1}(z)\right]_N. \tag{14}$$

It should be noted that $d\mathbf{X}(z) = \left[d\mathbf{W}(z)\mathbf{W}^{-1}(z)\right]_N$ is a nonholonomic basis, which has a definite geometrical meaning and proves to be useful in blind separation algorithms [3].

LEARNING ALGORITHM

We apply the stochastic gradient descent method to obtain a learning algorithm. First, we simplify the cost function $l(\mathbf{y}, \mathbf{W})$ for the FIR demixing model as follows,

$$l(\mathbf{y}, \mathbf{W}(z)) = -\frac{1}{2}\log(det(\mathbf{W}_0\mathbf{W}_0^T)) - \sum_{i=1}^{n} \log p_i(y_i(k), \mathbf{W}(z)), \tag{15}$$

where $det(\mathbf{W}_0\mathbf{W}_0^T)$ is the determinant of matrix $\mathbf{W}_0\mathbf{W}_0^T$.

For the gradient of $l(\mathbf{y}, \mathbf{W}(z))$ with respect to $\mathbf{W}(z)$, we calculate the total differential dl of $l(\mathbf{y}, \mathbf{W}(z))$ when we takes a differential $d\mathbf{W}(z)$

$$dl(\mathbf{y}, \mathbf{W}(z)) = l(\mathbf{y}, \mathbf{W}(z) + d\mathbf{W}(z)) - l(\mathbf{y}, \mathbf{W}(z)). \tag{16}$$

Following the procedure for deriving the natural gradient algorithm for the instantaneous mixing case [5], we have

$$dl(\mathbf{y}, \mathbf{W}(z)) = -tr(d\mathbf{W}_0\mathbf{W}_0^{-1}) + \varphi(\mathbf{y})^T d\mathbf{y}, \tag{17}$$

where tr is the trace of a matrix and $\varphi(\mathbf{y})$ is a vector of nonlinear activation functions, $\varphi_i(y_i) = -\frac{d\log p_i(y_i)}{dy_i} = -\frac{p_i'(y_i)}{p_i(y_i)}$. Using the notation (14), we have

$$dl(\mathbf{y}, \mathbf{W}(z)) = -tr(d\mathbf{X}_0) + \varphi(\mathbf{y})^T d\mathbf{W}(z)\mathbf{W}^{-1}(z)\mathbf{y}. \tag{18}$$

From this equation we obtain the partial derivatives of $l(\mathbf{y}, \mathbf{W}(z))$ with respect to $\mathbf{X}(z)$,

$$\frac{\partial l(\mathbf{W}(z))}{\partial \mathbf{X}_p} = -\delta_{0,p}\mathbf{I} + \varphi(\mathbf{y})\mathbf{y}^T(k-p), \; p = 0, \cdots, N \tag{19}$$

Using the natural gradient descent learning rule we present a novel learning algorithm as follows

$$\begin{aligned} \Delta \mathbf{W}_p &= -\eta \sum_{q=0}^{p} \frac{\partial l(\mathbf{W}(z))}{\partial \mathbf{X}_q} \mathbf{W}_{p-q} \\ &= \eta \sum_{q=0}^{p} \left(\delta_{0,q}\mathbf{I} - \varphi(\mathbf{y})\mathbf{y}^T(k-q)\right) \mathbf{W}_{p-q}, \end{aligned} \tag{20}$$

for $p = 0, 1, \cdots, N$, where η is a learning rate. In particular, the learning algorithm for $\mathbf{W}_0$ is described by

$$\Delta \mathbf{W}_0 = \eta \left(\mathbf{I} - \varphi(\mathbf{y})\mathbf{y}^T(k)\right) \mathbf{W}_0. \tag{21}$$

It should be noted that the algorithm looks similar but in fact it is not identical to the Amari et al algorithm [6]. The essential difference is that the update rule for $\Delta \mathbf{W}_p$ in this paper only depends on $\mathbf{W}_q$, $q = 0, \cdots, p$, while in [6] it depends on all parameters $\mathbf{W}_q$, $q = 0, \cdots, N$.

The algorithm (20) has two important properties, uniform performance (the equivariant property) and nonsingularity of $\mathbf{W}_0$. In the multichannel deconvolution problem, an algorithm is equivariant if its dynamical behavior depends on the global transfer function $\mathbf{G}(z)$ but not on the specific mixing filter $\mathbf{H}(z)$. In fact the learning algorithm (20) has equivariant property in the Lie group sense. Writing the learning algorithm in the Lie group form and multiplying both sides by the mixing filter $\mathbf{H}(z)$ in the Lie group sense, we obtain

$$\Delta \mathbf{G}(z) = -\eta \frac{\partial l(\mathbf{W}(z))}{\partial \mathbf{X}(z)} * \mathbf{G}(z). \tag{22}$$

where $\mathbf{G}(z) = \mathbf{W}(z) * \mathbf{H}(z)$. From equation (19) we know $\frac{\partial l(\mathbf{W}(z))}{\partial \mathbf{X}(z)}$ is formally independent of the mixing channel $\mathbf{H}(z)$. This means that the algorithm (20) is equivariant.

Another important property of the learning algorithm (21) is that it keeps the nonsingularity of $\mathbf{W}_0$ provided the initial $\mathbf{W}_0$ is nonsingular [22]. This means that the learning algorithm (20) keeps the filter $\mathbf{W}(z)$ on the manifold $\mathcal{M}(N)$ if the initial filter is on the manifold. The condition implies that the equilibrium points of the learning algorithm satisfy the following equations

$$E\left\{\varphi(\mathbf{y}(k))\mathbf{y}^T(k-p)\right\} = 0, \; \textit{for } p = 1, \cdots, N, \tag{23}$$

$$E\left\{\mathbf{I} - \varphi(\mathbf{y}(k))\mathbf{y}^T(k)\right\} = 0. \tag{24}$$

This means that we can make the separated signals $\mathbf{y}$ as mutually independent as possible if the nonlinear activation function $\varphi(\mathbf{y})$ are suitably chosen.

STABILITY OF LEARNING ALGORITHM

Since the learning algorithm for updating $\mathbf{W}_k$, $k = 0, 1, \cdots, N$, is a linear combination of $\mathbf{X}_k$, $k = 0, 1, \cdots, N$, the stability of $\mathbf{X}_k$, $k = 0, 1, \cdots, N$ implies the stability of the learning algorithm. In order to analyze the stability of the learning algorithm, we suppose that the separating signals $\mathbf{y} = (y_1, \cdots, y_n)^T$ are not only spatially mutually independent but also temporally i.i.d. for each component . Now consider the learning algorithm for updating $\mathbf{X}_p$

$$\frac{d\mathbf{X}_p}{dt} = \eta(\delta_{0,p}\mathbf{I} - \varphi(\mathbf{y}(k)\mathbf{y}(k-p)^T), \quad p = 0, 1, \cdots, N. \tag{25}$$

To analyze the asymptotic properties of the learning algorithm, we take expectation on equation (25)

$$\frac{d\mathbf{X}_p}{dt} = \eta\left[\delta_{0,p}\mathbf{I} - E\{\varphi(\mathbf{y})\mathbf{y}^T(k-p)\}\right], \quad p = 0, 1, \cdots, N. \tag{26}$$

Then the stability conditions for (26) are

$$m_i + 1 > 0, \ for\ i = 1, \cdots, n, \tag{27}$$

$$\kappa_i > 0, \ for\ i = 1, \cdots, n, \tag{28}$$

$$\kappa_i\kappa_j\sigma_i^2\sigma_j^2 > 1, \ for\ i, j = 1, \cdots, n, \tag{29}$$

where $m_i = E(\varphi'(y_i(k))y_i^2(k)]$, $\kappa_i = E[\varphi_i'(y_i)]$, $\sigma_i^2 = E[|y_i|^2]$, $i = 1, \cdots, n$. The conditions are identical to the ones derived by Amari et al [2] for instantaneous mixtures. Detailed derivation is left in a full paper.

SIMULATIONS

When we implement the natural gradient algorithm, first we have to estimate the length of the demixing filter. The choice of the length N of the demixing filter usually depends on the mixing filter and error tolerance for recovered signals. Although we do not know the mixing filter in the blind deconvolution, we can estimate roughly the length N of the demixing filter. The overestimate of N will not affect the outcome of the natural gradient approach, but will increase the computing cost. Computer simulations show that the parameters in the overestimate range will automatically converge to zero provided the mixing model is minimum phase and stable.

To evaluate the performance of the proposed learning algorithms, we employ the multichannel intersymbol interference, denoted by M_{ISI}, as a criteria,

$$\begin{aligned} M_{ISI} &= \sum_{i=1}^{n} \frac{\sum_{j=1}^{n}\sum_{p=0}^{N}|\mathbf{G}_{pij}| - \max_{p,j}|\mathbf{G}_{pij}|}{\max_{p,j}|\mathbf{G}_{pij}|} \\ &+ \sum_{j=1}^{n} \frac{\sum_{i=1}^{n}\sum_{p=0}^{N}|\mathbf{G}_{pij}| - \max_{p,i}|\mathbf{G}_{pij}|}{\max_{p,i}|\mathbf{G}_{pij}|} \end{aligned} \tag{30}$$

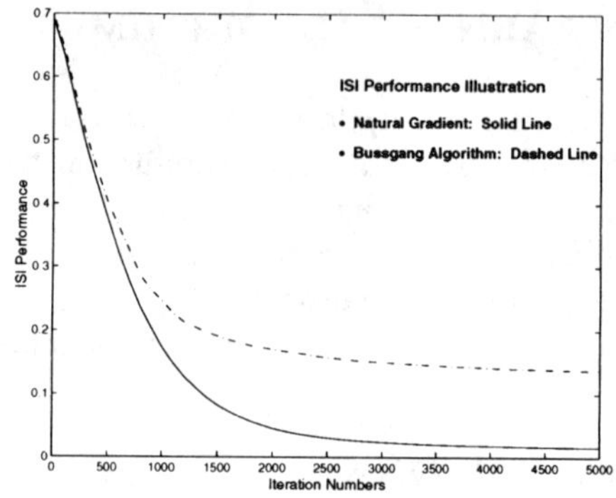

Figure 1: M_{ISI} performance of the natural gradient algorithm

It is easy to show that $M_{ISI} = 0$ if and only if $\mathbf{G}(z)$ is of the form (4).

We give two examples to demonstrate the behavior and performance of the algorithm (20). In both examples the mixing model is multichannel ARMA model as follows

$$\mathbf{x}(k) + \sum_{i=1}^{10} \mathbf{A}_i \mathbf{x}(k-i) = \mathbf{B}_0 \mathbf{s}(k) + \sum_{i=1}^{10} \mathbf{B}_i \mathbf{s}(k-i) + \mathbf{v}(k), \tag{31}$$

where $\mathbf{x}, \mathbf{s}, \mathbf{v} \in \mathbf{R}^3$. The matrices $\mathbf{A}_i$ and $\mathbf{B}_i$ are randomly chosen such that the mixing system is stable and minimum phase. The sources $\mathbf{s}$ are chosen to be i.i.d signals uniformly distributed in the range (-1,1), and $\mathbf{v}$ are the Gaussian noises with zero mean and a covariance matrix $0.1\mathbf{I}$. The nonlinear activation function was chosen $\varphi(y) = y^3$.

Example 1. We use AR model as a mixing system, which can be exactly inverted by a FIR filter. A large number of simulations show that the natural gradient learning algorithm can easily recover source signals in the sense of $\mathbf{W}(z)\mathbf{H}(z) = \mathbf{P}\mathbf{\Lambda}$.

Figure 1 illustrates 100 trial ensemble average M_{ISI} performance of the natural gradient learning algorithm and the Bussgang algorithm. It is observed that the natural gradient algorithm usually needs less than 2000 iterations to obtain satisfactory results, while the Bussgang algorithm needs more than 20000 iterations to obtain satisfactory results since there is a long plateau in the Bussgang learning.

Example 2. We choose ARMA model as a mixing system such that the system is stable and minimum phase. A large number of simulations have been performed. Figure 2 and 3 illustrate a typical example of the coefficients of global transfer function $\mathbf{G}(z) = \mathbf{W}(z)\mathbf{H}(z)$ at initial state and after 3000 iterations respectively, where the $(i,j)-$th sub-figure plots the coefficients of the transfer function $G_{ij}(z) = \sum_{k=0}^{\infty} g_{ijk} z^{-k}$ up to order of 80.

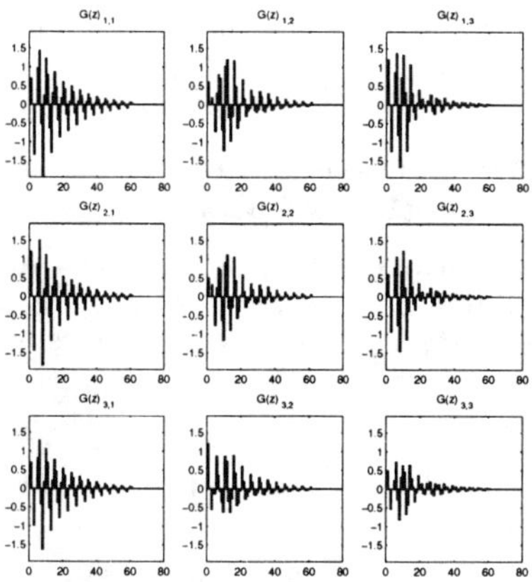

Figure 2: $\mathbf{G}(z)$ at the initial state

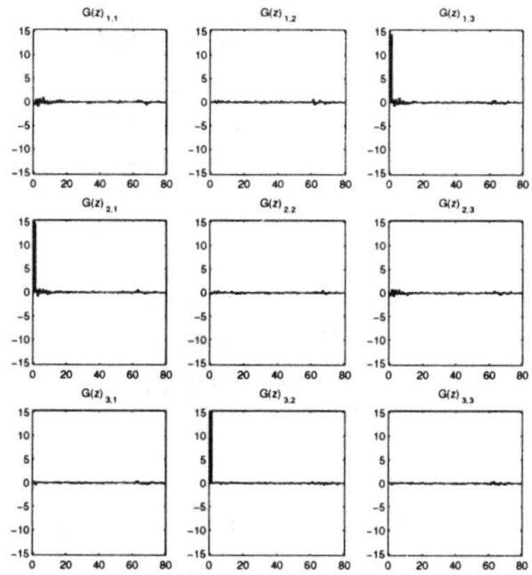

Fig. 3. $\mathbf{G}(z)$ after 3000 iterations

REFERENCES

[1] S. Amari. Natural gradient works efficiently in learning. *Neural Computation*, 10:251–276, 1998.

[2] S. Amari, T. Chen, and A. Cichocki. Stability analysis of adaptive blind source separation. *Neural Comput.*, 10:1345–1351, 1997.

[3] S. Amari, T. Chen, and A. Cichocki. Nonholonomic orthogonal contraints in blind source separation. *Neural Comput.*, to be published.

[4] S. Amari and A. Cichocki. Adaptive blind signal processing– neural network approaches. *Proceedings of the IEEE*, 86(10):2026–2048, 1998.

[5] S. Amari, A. Cichocki, and H.H. Yang. A new learning algorithm for blind signal separation. In G. Tesauro, D.S. Touretzky, and T.K. Leen, editors, *Advances in Neural Information Processing Systems 8 (NIPS*95)*, pages 757–763, Cambridge, MA, 1996. The MIT Press.

[6] S. Amari, S. Douglas, and A. Cichocki. Multichannel blind deconvolution and source separation using the natural gradient. *IEEE Trans. on Signal Processing*, to appear.

[7] S. Amari, S. Douglas, A. Cichocki, and H. Yang. Novel on-line algorithms for blind deconvolution using natural gradient approach. In *Proc. 11th IFAC Symposium on System Identification, SYSID'97*, pages 1057–1062, Kitakyushu, Japan, July, 8-11 1997.

[8] A.J. Bell and T.J. Sejnowski. An information maximization approach to blind separation and blind deconvolution. *Neural Computation*, 7:1129–1159, 1995.

[9] J.-F Cardoso. Blind signal separation: Statistical principles. *Proceedings of the IEEE*, 86(10):2009–2025, 1998.

[10] J.-F. Cardoso and B. Laheld. Equivariant adaptive source separation. *IEEE Trans. Signal Processing*, SP-43:3017–3029, Dec 1996.

[11] A. Cichocki, S. Amari, and J. Cao. Neural network models for blind separation of time delayed and convolved signals. *Japanese IEICE Transaction on Fundamentals*, E-82-A(9):1057–1062, Sept 1997.

[12] A. Cichocki and R. Unbehauen. Robust neural networks with on-line learning for blind identification and blind separation of sources. *IEEE Trans Circuits and Systems I : Fundamentals Theory and Applications*, 43(11):894–906, 1996.

[13] A. Cichocki, R. Unbehauen, and E. Rummert. Robust learning algorithm for blind separation of signals. *Electronics Letters*, 30(17):1386–1387, 1994.

[14] S. Douglas, A. Cichocki, and S. Amari. Multichannel blind separation and deconvolution of sources with arbitrary distributions. In *Proceeding of 1998 Int'l IEEE Workshop on Neural Networks for Signal Processing(NNSP'97)*, pages 436–445, Florida, US, September 1997.

[15] D. N. Godard. Self-recovering equalization and carrier tracking in two-dimensional data communication systems. *IEEE trans. Comm.*, COM-28:1867–1875, 1980.

[16] M. I. Gurelli and C. L. Nikias. Evam: An eigenvector-based algorithm for multichannel blind deconvolution of input colored signals. *IEEE Trans. Signal Processing*, 43:134–149, 1995.

[17] Y. Hua. Fast maximum likelihood for blind identification of multiple fir channels. *IEEE Trans. Signal Processing*, 44:661–672, 1996.

[18] Y. Sato. Two extensional applications of the zero-forcing equalization method. *IEEE Trans. Commun.*, COM-23:684–687, 1975.

[19] L. Tong, R.W. Liu, V.C. Soon, and Y.F. Huang. Indeterminacy and identifiability of blind identification. *IEEE Trans. Circuits, Syst.*, 38(5):499–509, May 1991.

[20] L. Tong, G. Xu, and T. Kailath. Blind identification and equalization base on second-order statistics: A time domain approach. *IEEE Trans. Information Theory*, 40:340–349, 1994.

[21] K. Torkkola. Blind separation of convolved sources based on information maximization. In S. Usui, Y. Tohkura, S. Katagiri, and E. Wilson, editors, *Proc. of the 1996 IEEE Workshop Neural Networks for Signal Processing 6 (NNSP96)*, pages 423–432, New York, NY, 1996. IEEE Press.

[22] H. Yang and S. Amari. Adaptive on-line learning algorithms for blind separation: Maximum entropy and minimal mutual information. *Neural Comput.*, 9:1457–1482, 1997.

[23] D. Yellin and E. Weinstein. Criteria for multichannel signal separation. *IEEE Trans. Signal Processing*, 42:2158–2168, 1994.

[24] L. Zhang and A. Cichocki. Blind deconvolution/equalization using state-space models. In *Proceeding of 1998 Int'l IEEE Workshop on Neural Networks for Signal Processing(NNSP'98)*, pages 123–131, Cambridge, UK, August 31-September 2 1998.

[25] L. Zhang and A. Cichocki. Blind separation of filtered source using state-space approach. In *Neural Information Processing Systems, (NIPS'98) Vol 11*, page to appear, Denver, December 1-3 1998. MIT press.

[26] L. Zhang, A. Cichocki, and S. Amari. Multichannel blind deconvoltion of nonminimum phase systems using information backpropagation. In *Proceedings of the Fifth International Conference on Neural Information Processing(ICONIP'99)*, page to appear, Perth, Australia, Nov. 16-20 1999.

APPLICATIONS

CLUSTERING APPROACH TO SQUARE AND NON-SQUARE BLIND SOURCE SEPARATION

Marc M. Van Hulle
K.U.Leuven, Laboratorium voor Neuro- en Psychofysiologie
Campus Gasthuisberg, Herestraat, B-3000 Leuven, BELGIUM
Tel.: + 32 16 34 59 61, Fax: + 32 16 34 59 93
E-mail: marc@neuro.kuleuven.ac.be
Web: simone.neuro.kuleuven.ac.be

Abstract. Recently, a number of heuristic techniques, mostly based on topographic maps, have been introduced in order to overcome (some) of the limitations of the Blind Source Separation (BSS) algorithms that are rooted in the theory of Independent Component Analysis. Here, we introduce a new heuristic that relies on the tendency of the mixture samples to *cluster* around the source directions in mixture space. We use linear mixtures of speech signals and consider BSS problems with not only square but also non-square mixing matrices (less mixtures than sources).

INTRODUCTION

The majority of the Blind Source Separation (BSS) algorithms are based on the theory of Independent Component Analysis (ICA) [1]. The idea is to find the statistically independent source signals $\mathbf{s}(t) = [s_i(t)]$, $i = 1, ..., m$, with t a time index, from the mixtures $\mathbf{v}(t) = A_{mm}\mathbf{s}(t)$, with A_{mm} the $m \times m$ mixing matrix, by optimizing a criterion such as maximum likelihood (ML). However, for reasons of the assumed linearity of the mixing, the ensuing increased algorithmic complexity to encompass non-linear mixtures, or the inability to address the case where there are fewer mixtures than sources, a number of heuristic techniques have been devised, mostly based on topographic maps. They meet one or several of the previous concerns and fall in two broad categories by their ability to separate either: 1) linear- or mildly non-linear mixtures of signals with either sub-Gaussian (*i.e.* approximatively uniform) [2, 3, 4, 5, 6], or 2) super-Gaussian (*i.e.* sharply peaked) source densities [7, 8]. The idea behind the second category, and to which we will restrict ourselves in this contribution, is as follows. The column vectors of the mixing matrix A_{mm} define, for the linear case, an independent component coordinate system the basis vectors of which are usually non-orthogonal. For sharply peaked source signals, such as speech signals, when only one signal s_i is

0-7803-5673-X/99/$10.00

active at a time, the mixed signal will be directed along a line parallel to A_{mmi}. The direction of this line can be retrieved by least-squares fitting [7] or by some sort of edge detection procedure *e.g.* the Hough transform [8]. Here, we propose a completely different approach. Basically, it relies on the tendency of the mixture samples to *cluster* around the source directions. We perform density-based clustering with a topographic map trained with the kernel-based Maximum Entropy learning Rule (kMER) [6].

CLUSTERING APPLIED TO BSS

Consider two sharply peaked source signals that are linearly mixed. As an example, we consider speech signals from the TIMIT database [9], namely $dr1/fcjf0/sa2$ (TR_1) and $dr1/ftbr0/sa1$ (TR_2), respectively 34509 and 55092 samples long and sampled at 16 kHz (Fig. 1A,B). We scale these signals so that they have a unit peak amplitude, and add zeros at the end of the first signal so that it also contains 55092 samples. The scaled signals are then linearly mixed using the following mixing matrix:

$$A_{22} = \begin{pmatrix} 0.7 & 0.3 \\ 0.3 & 0.7 \end{pmatrix}. \tag{1}$$

The resulting scatter plot is shown in Fig. 1C. The high-density directions correspond to the coordinate axes of the source signals and thus to the directions defined by the column vectors of the mixing matrix.

In order to separate these speech signals from their mixtures, we adopt a three-stage approach. We first train a topographic map with kMER and use it for generating an estimate of the mixture distribution. We then use this estimate to perform a density-based clustering analysis and, finally, we perform BSS in both a parametric and a non-parametric manner.

Stage 1: Density estimation with kMER

Topographic map formation

Let A be a two-dimensional lattice of N formal neurons. To each neuron i corresponds a circular Receptive Field (RF) region with radius σ_i and center $\mathbf{w}_i = (w_{i1}, w_{i2})$. Both the centers and the radii are trained with our learning rule kMER of which the mathematical details, including a proof of convergence, are discussed elsewhere [6, 10]. In summary, the centers $\mathbf{w}_i$ are updated so that they become the medians of the neurons' RF regions, and the radii σ_i are updated so that the probability for neuron i to be active will be $\frac{1}{N}$, $\forall i$.

We now use kMER for developing a *lin-polar* map of the mixture coordinate frame with amplitude $R = \sqrt{v_1^2 + v_2^2}$ and phase angle:

$$\theta = \min\left(\arctan\frac{v_2}{v_1}, \pi + \arctan\frac{v_2}{v_1}\right). \tag{2}$$

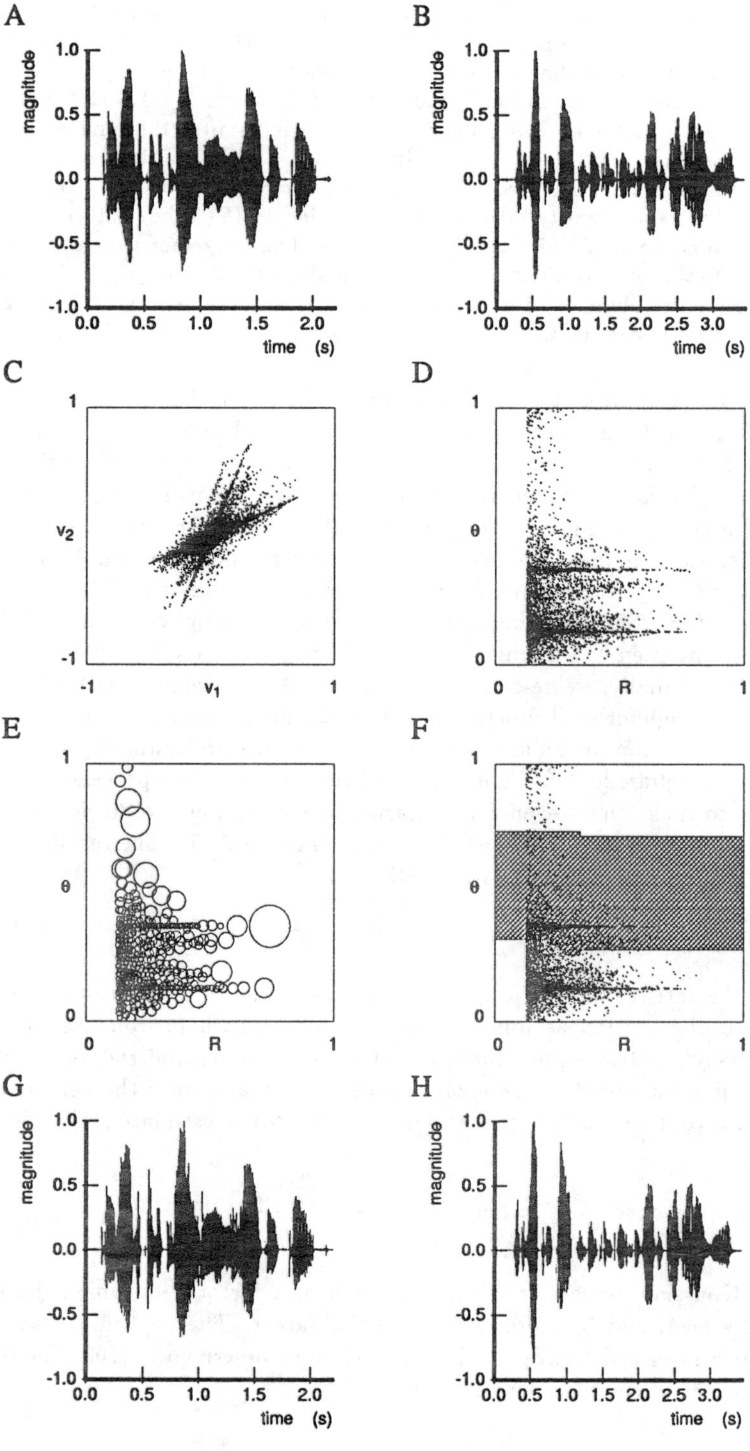

Figure 1: Continues on next page.

Figure 1: Clustering approach to Blind Source Separation. (A,B) Two example speech signals, taken from the TIMIT database, that are subject to linear mixing. (C,D) Scatter plot of the signal mixtures (v_1, v_2) in the original Cartesian coordinate frame (C) and in the lin-polar frame (R, θ) for $R \geq 0.125$ (D). No signal sharpening or whitening has been performed prior or after the mixing process. (E) RF regions of a 24×24 lattice trained with kMER on the data shown in (D). (F) Result of density-based clustering with the SKIZ technique, starting from the variable kernel density estimate corresponding to (E). The "influence zones" of the two clusters, of which one is shaded, are shown together with the training data. Note the effect of the toroidal extension along the θ-axis. (G,H) The speech signal estimates obtained after demixing with the first strategy (CL). The signals are scaled for unit peak amplitude.

Such a map is more suited for density-based clustering than a Cartesian map: the mixture samples now display the tendency to cluster along lines parallel to the R-axis and hence, the clusters will be better defined when the mixture samples have smaller amplitudes. In order to avoid the region with the highest density in mixture space, *i.e.* where the source signal amplitudes are close to their mean values, we only consider a mixture sample when its amplitude $R \geq R_{min}$, with $R_{min} > 0$. Here, we will take $R_{min} = 0.125$ so that $M = 7859$ lin-polar transformed mixture samples remain. For the sake of convenience, we scale the range of the polar axis from $[0, \pi]$ to $[0, 1]$ (Fig. 1D). Finally, we re-consider the $N = 24 \times 24$ lattice, with the same Gaussian neighborhood function, cooling scheme, number of epochs, and radius initialization procedure as in [6]. The weights are randomly initialized in the unit square $[0, 1]^2$. Contrary to others, we do not pre-process the audio signals to make them even more sharply tuned [8], and we do not perform any whitening of the lattice inputs prior to training [5, 7]. The resulting map of the RF regions is shown in Fig. 1E.

Density estimation

Once the lattice has converged, we determine the density estimate of the mixture distribution as follows. We allocate at each neuron weight $\mathbf{w}_i$ a radially-symmetric, equi-volume kernel with center $\mathbf{w}_i$ and radius σ_i. Since the radii are adapted to the local sample density, and since the kernels have an equal volume, we obtain a *variable* kernel density estimate [11]:

$$\widehat{p_{\rho_s}}(\mathbf{v}) = \sum_{i=1}^{N} \frac{\exp(-\frac{\|\mathbf{v}-\mathbf{w}_i\|^2}{2(\rho_s \sigma_i)^2})}{Z_i}, \tag{3}$$

when Gaussian kernels are used, and with ρ_s a factor with which the radii σ_i are scaled, and Z_i a proper normalizing factor. The "optimal" degree of smoothness ρ_s is determined with the technique described in [10]. The result is $\rho_s = 1.3$, when optimizing in steps of 0.1.

Stage 2: Density-based clustering

In order to determine how many clusters there are, and to identify the areas spanned by them (*i.e.* the "influence zones"), we apply the SKeleton by Influence Zones (SKIZ) technique [12]. Basically, this technique considers L equidistant levels at which cross-sections with $\widehat{p_{\rho_s}}$ are taken. From the connected regions obtained in this way, the "influence zones" are identified. For the simulations reported on, we have used $L = 16$ levels. The result for Fig. 1E is shown in Fig. 1F.

Stage 3: Blind source separation

We compare three strategies. First, we consider the mixture samples, which belong to a given influence zone, as samples originating from the same source channel. In other words, we perform a classification of the mixture samples (case labeled CL). The source signals are then reconstructed by supplementing the R-value of the mixing signal with the corresponding sign (by definition of our phase angle θ). Note that, as a result of classification, there is only one "winning" channel at any given time. For source signals with sharply peaked densities, such as speech signals, this is a reasonable assumption since these signals hover around their mean values most of the time.

Second, we classify the weights of the converged lattice, and determine the median along the θ-axis for each weight class: the two medians then serve as regression lines for the corresponding weight classes (case labeled CL_REG_W). (Note that medians are less sensitive to outliers than means and, in this way, a better definition of the regression lines is obtained in practice.) From the resulting θ's, the column vectors of the mixing matrix A_{22} can be determined, up to a scale factor, the inverse of the mixing matrix calculated, A_{22}^{-1}, and the source signals reconstructed.

Finally, we repeat the previous case but now starting from a classification of the mixture samples, instead of the lattice weights (case labeled CL_REG_S).

RESULTS

The source signal estimates obtained with the first strategy (CL case) are shown in Fig. 1G,H; the estimates for the other two strategies cannot be distinguished from the originals by visual inspection. In order to quantify all these results, we first scale the source signal estimates so that their variances equal those of the corresponding source signals s_1 and s_2. We then determine the mean squared error (MSE) between the originals and the estimates, as well as the signal-to-noise-ratio (SNR) (in dB). Furthermore, we also consider two other speech signals for testing purposes, $dr5/fbmh0/sx146$ (TE_1) and $dr1/ftbr0/sa2$ (TE_2), which are, respectively, 41268 and 44852 samples long. The results are summarized in Table 1. Note the high precision achieved with the CL_REG_S and CL_REG_W strategies (sometimes bounded by the single precision representation used in the algorithm). The estimate obtained for the mixing matrix eq. (1) is, *e.g.* for the CL_REG_S case:

$$\widehat{A_{22}} = \begin{pmatrix} 0.6962 & 0.3000 \\ 0.3038 & 0.7000 \end{pmatrix}. \quad (4)$$

The quality of our source estimates (CL case) seems to be superior to that obtained by Lin and co-workers, albeit that the estimates they showed were for a mildly non-linear but still whitened mixing (compare their Fig. 4 in [7] to our Fig. 1 – unfortunately they did not show the results for their linear mixing case or quantify the performance). When a regression analysis on the weights is performed, Lin and co-workers report for their linear case a mismatch of less than 1 % in the orthogonality of the source signal estimates, *i.e.* when the estimates are expressed as linear combinations of the original source signals, whereas we find 2.5 % on average (CL_REG_W strategy). This discrepancy in performance is, in our opinion, due to the fact that Lin and co-workers have whitened their mixture signals: whitening orthogonalizes the independent coordinate axes.

Table 1: Performance for a square BSS problem expressed in terms of the mean squared error (MSE) and signal-to-noise-ratio (SNR) in dB (between brackets). The first two signals (TR_1, TR_2) were used for training; the last two for testing (TE_1, TE_2).

MSE	CL	CL_REG_W	CL_REG_S
TR_1	2.16×10^{-3} (7.52)	5.64×10^{-8} (53.3)	7.41×10^{-13} (102.2)
TR_2	1.07×10^{-3} (9.41)	2.75×10^{-5} (25.3)	1.10×10^{-6} (39.3)
TE_1	4.60×10^{-3} (5.25)	1.12×10^{-7} (51.4)	4.10×10^{-12} (95.7)
TE_2	2.68×10^{-3} (8.39)	3.49×10^{-5} (27.3)	1.39×10^{-6} (41.2)

BSS FROM FEWER MIXTURES

By virtue of our clustering approach, we can also tackle the "non-square" BSS problem where the number of mixtures is less than the number of sources. Additional sources introduce additional density directions in mixture space. As an example, we consider the case of 3 sources and 2 mixtures. The CL strategy stays the same but the other two have to be adapted since the mixing matrix is no longer square. We suggest the following heuristic: we first project the mixture sample onto each of the three source directions that we have identified. We then select the source direction which produces the largest projection result (*i.e.* the smallest angular difference in the lin-polar map) and assign the projection result to the corresponding source channel. For the residue of that projection, we invert the remaining 2×2 mixing matrix and determine the other two source signal estimates. We use the previous

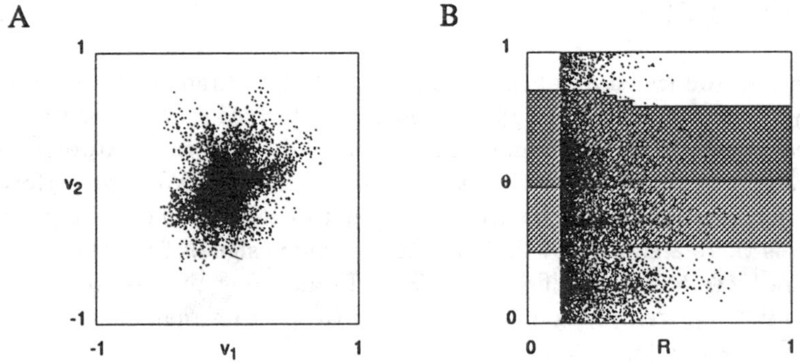

Figure 2: Blind Source Separation from fewer mixtures. Scatter plot produced by a mixture of three signals in the Cartesian coordinate frame (A) and the lin-polar frame (R, θ) for $R \geq 0.125$ (B). In (B), the influence zones of the three clusters are indicated with different gray levels.

two source signals TR_1 and TR_2 and select a third one form the TIMIT database TR_3 ($dr5/fbmh0/si1136$; 71476 samples) for training our lattice. We use the following mixing matrix:

$$A_{23} = \begin{pmatrix} 0.7 & 0.3 & -0.3 \\ 0.3 & 0.7 & 0.7 \end{pmatrix}, \tag{5}$$

and again consider 55092 samples for mixing (Fig. 2A). This results in 13519 lin-polar transformed mixture samples (Fig. 2B). For testing, we use the TIMIT signals $dr1/fcjf0/sa1$ (46797 samples), $dr1/ftbr0/sa2$ (44852 samples), and $dr5/fbmh0/sx146$ (41268 samples) (labeled further as TE_1, TE_2, TE_3). Again, no signal sharpening or whitening has been performed. The clustering result is shown in Fig. 2B; the BSS results are summarized in Table 2. The estimate of the mixing matrix eq. (5), obtained with the CL_REG_S strategy, is:

$$\widehat{A_{23}} = \begin{pmatrix} 0.7133 & 0.2930 & -0.3000 \\ 0.2867 & 0.7070 & 0.7000 \end{pmatrix}. \tag{6}$$

Table 2: Performance for a non-square 2×3 BSS problem.

MSE	CL	CL_REG_W	CL_REG_S
TR_1	6.00×10^{-3} (3.08)	2.49×10^{-3} (6.90)	2.39×10^{-3} (7.07)
TR_2	4.24×10^{-3} (3.42)	4.58×10^{-3} (3.08)	4.11×10^{-3} (3.55)
TR_3	2.93×10^{-3} (4.91)	2.27×10^{-3} (6.02)	2.26×10^{-3} (6.04)
TE_1	6.55×10^{-3} (1.99)	3.68×10^{-3} (4.49)	3.58×10^{-3} (4.61)
TE_2	6.66×10^{-3} (4.44)	7.17×10^{-3} (4.12)	6.31×10^{-3} (4.68)
TE_3	4.44×10^{-3} (5.40)	3.44×10^{-3} (6.52)	3.39×10^{-3} (6.57)

CONCLUSION

We have introduced a new heuristic approach to the Blind Source Separation (BSS) of linear mixtures of signals with sharply peaked source densities, such as speech signals. Our technique relies on the tendency of the mixture samples to *cluster* around the source directions in mixture space. We have performed a density-based clustering analysis in order to determine how many source channels there are, and how to separate the source signals from the observed mixtures. We have quantified the BSS performance in the case of square as well as non-square mixing matrices applied to speech signals taken from the TIMIT database [9].

Acknowledgments

The author wishes to thank T. Gautama for his help in developing the "non-square" BSS example. M.M.V.H. is a research associate of the Fund for Scientific Research – Flanders (Belgium) and is supported by research grants received from the Fund for Scientific Research (G.0185.96), the National Lottery (Belgium) (9.0185.96), the Flemish Ministry of Education (GOA 95/99-06), and the Flemish Ministry for Science and Technology (VIS/98/012).

REFERENCES

[1] P. Comon, "Independent component analysis – a new concept?," Signal Processing, vol. 36(3), pp. 287-314, 1994.

[2] T. Kohonen, K. Raivo, O. Simula, O. Ventä, and J. Henriksson, "Combining linear equalization and self-organizing adaptation in dynamic discrete-signal detection," Proc. IJCNN'96 (San Diego, CA, June 17-21), pp. I223-I228, 1996.

[3] E. Moreau, and O. Macchi, "High-order contrasts for self-adaptive source separation, Int'l J. of Adaptive Control and Signal Processing, vol. 10, pp. 19-46, 1996.

[4] P. Pajunen, A. Hyvärinen, and J. Karhunen, "Nonlinear Blind Source Separation by Self-Organizing Maps," Progress in Neural Information Processing (Proceedings of the International Conference on Neural Information Processing, ICONIP'96), S.-I. Amari, L. Xu, L.-W. Chan, I. King, and K.-S. Leung (Eds.), Vol. 2, pp. 1207–1210, 1996.

[5] M. Herrmann, and H.H. Yang, "Perspectives and limitations of self-organizing maps in blind separation of source signals," Progress in Neural Information Processing (Proceedings of the International Conference on Neural Information Processing, ICONIP'96), S.-I. Amari, L. Xu, L.-W. Chan, I. King, and K.-S. Leung (Eds.), Vol. 2, pp. 1211-1216, 1996.

[6] M.M. Van Hulle, "Kernel-based equiprobabilistic topographic map formation," Neural Computation, vol. 10, pp. 1847-1871, 1998.

[7] J.K. Lin, D.G. Grier, J.D. Cowan, "Faithful representations of separable distributions," Neural Computation, vol. 9, pp. 1305-1320, 1997.

[8] J.K. Lin, D.G. Grier, J.D. Cowan, "Feature extraction approach to blind source separation," Proc. IEEE NNSP97 (Amelia Island Plantation, Florida), pp. 398-405, 1997.

[9] TIMIT, "DARPA TIMIT acoustic-phonetic speech corpus," NIST Speech Disc 1-1.1 (CD-ROM), 1990.

[10] M.M. Van Hulle, "Clustering with kernel-based equiprobabilistic topographic maps," Proc. IEEE NNSP98 (Cambridge, UK), pp. 204-213, 1998.

[11] B.W. Silverman, Density Estimation for Statistics and Data Analysis, Chapman and Hall: London, 1986.

[12] J. Serra, Image analysis and mathematical morphology, New York: Academic Press, 1982.

SEPARATION OF ACOUSTIC SIGNALS USING SELF-ORGANIZING NEURAL NETWORKS

Temujin Gautama & Marc M. Van Hulle
K.U.Leuven, Laboratorium voor Neuro- en Psychofysiologie
Campus Gasthuisberg, Herestraat 49, B-3000 Leuven, BELGIUM
Tel.: + 32 16 34 59 61, Fax: + 32 16 34 59 93
E-mail: {temu,marc}@neuro.kuleuven.ac.be

Abstract— Spectral modeling is an essential component in many signal processing applications, such as speech enhancement and sound monitoring. This paper will demonstrate its use in the separation of acoustic sources from a compound signal that is registered by one sensor. Our technique distinguishes itself from the popular blind source separation procedure by its much higher noise insensitivity and its ability to cope with varying as well as non-square mixing conditions.

INTRODUCTION

In the neural network literature, the separation of acoustic sources is mostly described in the context of *blind source separation* (BSS) (for references, see [4]). The source signals are assumed to be statistically independent and they are separated from the recorded mixture signals without knowing the mixture process. This process is generally assumed to be linear, with as many sensors as there are sources, and with a mixing matrix that is both invertible and constant over time. Under these conditions, the mixing process can be modeled and the source signals can be recovered. However, since the BSS algorithms are essentially based on amplitude information, they tend to be quite sensitive to noise [1].

If there is only a single sensor present, BSS of two sources becomes a non-square problem, and the current BSS algorithms cannot be used. One way to tackle this problem is to model the spectral characteristics of the acoustic sources. *Single microphone speech enhancement* [7] and *sound monitoring* [6] are well-known applications in this field, where acoustic sources are first modeled using spectral information, after which these models are applied either to cancel out noise, to detect changes in the acoustic sources, or to identify the presence of an unknown source.

Modeling the sources using their spectral characteristics is expected to yield an increased resistance to noise and to variabilities in the mixing pro-

0-7803-5673-X/99/$10.00 © 1999 IEEE

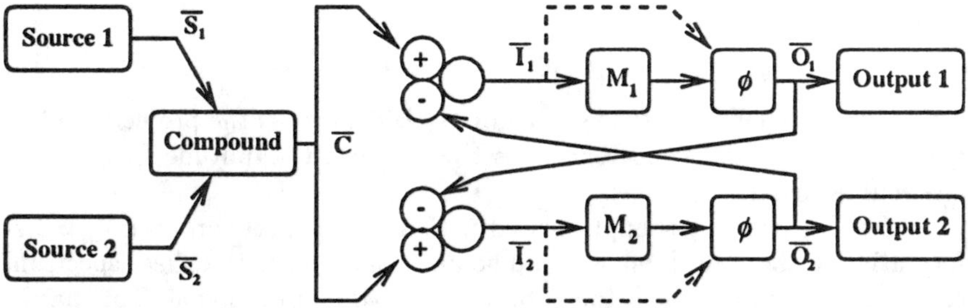

Figure 1: Diagram of the system. All signals are represented by their complex Fourier spectrum.

cess. These reasons form the main motivations to adopt a spectral modeling approach to source separation. The models considered in this paper will be developed using self-organizing learning rules trained on spectral information of music signals.

PROCEDURE

Basic Strategy

Figure 1 shows the overall system setup. The source signals $\bar{\mathbf{S}}_1$ and $\bar{\mathbf{S}}_2$ are linearly mixed so as to produce the compound signal $\bar{\mathbf{C}}$ (note that all signals are represented by their complex short-time Fourier transforms (STFT) $\bar{\mathbf{S}}$, instead of by their amplitude spectra $\mathbf{S}$). The compound signal $\bar{\mathbf{C}}$ is applied to the two branches in the system, one for each source signal. In every branch, the estimated spectrum of the other source is subtracted from $\bar{\mathbf{C}}$ and the result, $\bar{\mathbf{I}}_n$, is quantized by the model $\mathbf{M}_n$, ($n \in \{1, 2\}$). The quantized signal is then phase aligned (ϕ) to the model's input signal $\bar{\mathbf{I}}_n$ (dashed line), so as to produce the source signal estimate $\bar{\mathbf{O}}_n$ after a number of iterations. The exact nature of both the quantization and the phase alignment will be elaborated upon in sections 2.2 and 2.3. The quantization models will be trained using self-organizing learning rules, as will be explained in section 2.3. As an example, we will consider signals taken from two musical instruments (oboe and piano).

Quantization

The spectral model of each instrument, $\mathbf{M}_n$, consists of a set of codebook vectors $\{\bar{\mathbf{Q}}_{n,i}\}$ each of which represents a *"prototypical"* Fourier spectrum. These vectors should be translation (in time) invariant so as to avoid redundancy in the representation. Hence, $\mathbf{M}_n$ will consist of amplitude spectra $\mathbf{Q}_{n,i}$, rather than complex Fourier spectra $\bar{\mathbf{Q}}_{n,i}$. At time t, the winning codebook $\mathbf{Q}_{n,c}$ for input (amplitude) spectrum $\mathbf{I}_n$ will be selected using the dot product criterion, *i.e.*

$$|\mathbf{Q}_{n,c} \cdot \mathbf{I}_n| \geq |\mathbf{Q}_{n,i} \cdot \mathbf{I}_n|, \; \forall \; \mathbf{Q}_{n,i} \in \mathbf{M}_n. \tag{1}$$

The reconstructed amplitude spectrum $\mathbf{O}_n$ is defined as the projection of $\mathbf{I}_n$ onto the winner $\mathbf{Q}_{n,c}$ and represents the estimated amplitude spectrum of one source signal.

Next, $\mathbf{O}_n$ has to be supplemented with a phase spectrum. If not, transformation to the time domain would be impossible, since the phase spectrum contains information about the harmonic structure of the signal. The matching process (eq. 1), however, should not be performed on complex codebook vectors, since these are not translation invariant. Hence, we have opted for a two-step procedure: first the matching is performed on the amplitude spectrum only (*Amplitude Quantization*), after which the reconstruction is combined with the corresponding phase spectrum (*Complex Quantization* and *Phase Alignment*). This two-step procedure and the training of the models will be explained in section 2.3.

Building and Applying the Quantization Models

Amplitude Quantization Model. The codebook for model $\mathbf{M}_n$, $\{\mathbf{Q}_{n,i}\}$, is determined, for every instrument separately, using self-organizing learning rules. We will compare three such rules: kMER [5], Kohonen's SOM [2] and the k-means algorithm [3].

In the kMER (kernel-Based Maximum Entropy Rule, [5]), every neuron i is determined by its traditional weight vector $\mathbf{w}_i$, supplemented by a radius σ_i. A given neuron is active if the Euclidean distance between its weight vector and the input pattern is smaller than the neuron's radius. The rule is aimed at building a topographic map in which all neurons have the *same* activation probability, the exact level of which is determined by the parameter ρ. This is achieved by adjusting both the weight vectors and their radii in an iterative manner (note that the activation regions are allowed to overlap). The update rule incorporates a neighborhood function (in lattice space) whose width decreases over time in order to yield topographic maps.

Training patterns for the learning rules consist of amplitude spectra coming from a single instrument. They are thresholded in energy in order to avoid training on irrelevant spectra by using the following criterion: input pattern μ will be considered for training when:

$$E^{\mu} \geq \frac{< E >_{training\ set}}{4}, \tag{2}$$

where E^{μ} is the signal energy of input pattern μ. Since the actual quantization will be performed by means of projections, all patterns are normalized prior to training.

The result of this stage is a quantized amplitude spectrum for each instrument (computed as the projection of $\mathbf{I}_n$ onto the winner $\mathbf{Q}_{n,c}$). We now have to extend this representation with a phase spectrum in order to restore the harmonic structure of the time signal.

Complex Quantization Model. The prototype amplitude spectra $\mathbf{Q}_{n,i}$ will be extended with representative phase spectra so as to preserve the phase relationships between frequency components. This will be done only once, after training, by combining every prototype spectrum $\mathbf{Q}_{n,i}$ with the phase spectrum of its closest matching input spectrum $\bar{\mathbf{I}}^{\gamma}$:

$$\frac{|\mathbf{Q}_{n,i} \cdot \mathbf{I}^{\gamma}|}{||\mathbf{I}^{\gamma}||} \geq \frac{|\mathbf{Q}_{n,i} \cdot \mathbf{I}^{\mu}|}{||\mathbf{I}^{\mu}||}, \ \forall \mu. \tag{3}$$

The result of this stage is a reconstruction of the complex spectrum. However, the exact positioning in the time slice still has to be determined.

Phase Compensation. The reconstructed, complex spectrum $\bar{\mathbf{O}}_n$ is the Fourier transform of a time slice $o_n(t)$. If we assume the phase relations between the frequency components to be correct, then $o_n(t)$ represents a time slice that corresponds to $i_n(t)$, except for a time shift within the time slice, Δt (the phase spectrum determines both the phase relations between components and the exact positioning in time). This Δt can be computed as the period of time over which $o_n(t)$ has to be circularly shifted over the time slice, so as to optimally correlate with input signal $i_n(t)$ (circularly shifted, since the discrete Fourier transform assumes the signals to be periodic over the given time slice):

$$\Delta t = \arg\max[|i_n(t) \circ o_n(\Delta t)|] = \arg\max[|\mathcal{F}^{-1}\{\bar{\mathbf{I}}_n \cdot \bar{\mathbf{O}}_n^*\}|], \tag{4}$$

where $\circ$ denotes the correlation function, $\mathcal{F}^{-1}$ the inverse Fourier transform, $\bar{\mathbf{I}}_n$ is the Fourier transform of $i_n(t)$ and $\bar{\mathbf{O}}_n^*$ is the complex conjugate of the Fourier transform of $o_n(t)$.

As a result, these three stages yield time signals that approximate the original source signals in frequency content, phase relations and positioning in time.

Source Separation

Once the spectral models have been trained, the separation of the acoustic sources can be performed by means of a process that iteratively refines the spectral estimates of the sources at every time step t. Figure 1 shows that the input to quantization model $\mathbf{M}_1$ is formed by the subtraction of the estimated spectrum of source 2 at iteration $(k-1)$ from the compound spectrum $\bar{\mathbf{C}}(t)$, and *vice versa*.

To improve performance, a competitive element has been included: at iteration k, the model $\mathbf{M}_m$ $(m \neq n)$ that finds the worse match to its input, will quantize $(\bar{\mathbf{C}}(t) - \bar{\mathbf{O}}_n^k(t))$, rather than $(\bar{\mathbf{C}}(t) - \bar{\mathbf{O}}_n^{(k-1)}(t))$. The separation will now be explained at iteration k.

1. The inputs to the quantization models, $\bar{\mathbf{I}}_n(t)$, are set to $(\bar{\mathbf{C}} - \bar{\mathbf{O}}_m^{(k-1)})$, with $m \neq n$.

2. *Amplitude quantization.* The closest amplitude matches $\mathbf{Q}_{n,c}$ are found for every model $\mathbf{M}_n$ using eq. (1). The cosines of the angles between $\mathbf{Q}_{n,c}$ and the input amplitude spectrum $\mathbf{I}_n(t)$ are determined. $\mathbf{M}_{win}$ denotes the model that yields the highest cosine. $\mathbf{I}_{win}(t)$ is projected onto the model's best matching vector $\mathbf{Q}_{win,c}$. The length of this projection corresponds to a gain factor that is applied to $\mathbf{Q}_{win,c}$.
3. *Complex quantization.* $\mathbf{Q}_{win,c}$ is combined with its corresponding phase spectrum.
4. *Phase alignment* on this complex spectrum yields $\bar{\mathbf{O}}^k_{win}$, using eq. (4).
5. The complex residue spectrum $(\bar{\mathbf{C}}(t) - \bar{\mathbf{O}}^k_{win}(t))$ is quantized by the remaining model $\mathbf{M}_m$ and phase aligned resulting in $\bar{\mathbf{O}}^k_m(t)$ $(m \neq win)$.

At iteration 0, the inputs to the quantization models $\mathbf{M}_n$ are all set to $\bar{\mathbf{C}}$ (*i.e.* the initial spectral estimates of the models are set to the null vector). The number of iterations is set to 3 in all simulations described in this paper. When the competitive element was not included, the performance decreased by 0.2 dB and more than 3 iterations did not lead to better results.

SIMULATIONS

Signal Generation and Evaluation

The signals that were used in this section have been generated on a Crystal 4232 audio controller (Yamaha OPL3 FM synthesizer) at a sampling rate of 11025 Hz, using the simulated sounds of an oboe and a piano. The compound signal was created by linearly mixing two separate time signals (9 seconds of music): $c(t) = 0.5\, s_1(t) + 0.5\, s_2(t)$.

The STFT is computed every 256 samples, resulting in a 3/4 overlap; the length of one time window is 1024 samples. At this stage, no windowing is applied so as not to interfere with the correlation strategy of the phase alignment (eq. 4). For the inverse transform, the consecutive complex spectra are transformed to the time domain. Every time window is then Hamming windowed and offset in time (spectrum i will have an offset of $(256 \cdot i)$ samples), after which they are summed so as to yield the reconstructed time signal, $o_n(t)$.

The two training sets consist of the short-time Fourier spectra of time signals of two octaves (15 notes, 10 seconds) played on the respective instruments (the same octaves for both oboe and piano). The resulting spectra are then thresholded using eq. 2 and normalized, resulting in training sets of 364 and 387 complex spectra for oboe and piano respectively. The compound signal consists only of notes included in the training sets.

In all simulations, the signal-to-noise ratio (SNR) has been used to evaluate performances. It is defined for the original time signal $s_n(t)$ and its reconstruction $o_n(t)$ as follows:

$$SNR(s_n, o_n)[dB] = -10 \log_{10} \sum_{t=0}^{T-1} (\frac{s_n(t)}{\sigma_{s_n}} - \frac{o_n(t)}{\sigma_{o_n}})^2, \quad (5)$$

where σ_{s_n} and σ_{o_n} are the standard deviations of s_n and o_n, and T is the length of the shorter signal. The SNR of a round transformation (STFT, followed by its inverse) is 14.88 dB for the oboe and 22.19 dB for the piano training signal.

Dimensionality Reduction

In order to facilitate and speed up the training process of the quantization models, a dimensionality reduction has been performed on the amplitude spectra of both training signals separately, by means of the Karhunen-Loève transform. First, the principal components of the amplitude spectra (512 dimensions) are computed, after which every data point is projected onto the first N (≤ 512) principal components.

To monitor the quality of the reconstruction of the training signals as a function of N, the amplitude spectra coming from the STFT were first reduced in dimensionality, after which they were reconstructed, extended with their original phase spectra and, finally, transformed back to the time domain. Figure 2A shows the quality of the reconstructed time signals as a function of N. The piano model reaches its maximum for $N \geq 300$ and the oboe model for $N \geq 100$. The difference in performance between instruments corresponds to the quality of the above mentioned round transformation (14.88 dB and 22.19 dB for oboe and piano) and hence is not due to the Karhunen-Loève transform. To monitor the influence of the dimensionality reduction on the overall separation performance, all learning rules were trained for five different values of N (20, 40, 100, 200 and 300).

Training the Models

The goal of training is the generation of a codebook of 49 vectors (kMER and SOM are configured as 7x7 lattices). During all training sessions, the mean squared error on the training set is monitored, so as to find good parameter sets for the different learning rules (learning rates, length of training and, for the kMER and the SOM algorithm, the neighborhood cooling schemes). Both the kMER and the SOM algorithm are initialized with the sane random configurations, while k-means is initialized with vectors randomly taken from the training set.

Figure 2B shows the SNRs of the different models' reconstructed piano training signals, evaluated in function of the dimensionality reduction (oboe results not shown). The solid lines show the quality of the reconstruction when the original phase spectrum is used to reconstruct the complex Fourier spectrum (*i.e. Amplitude Quantization* only). Dashed lines represent

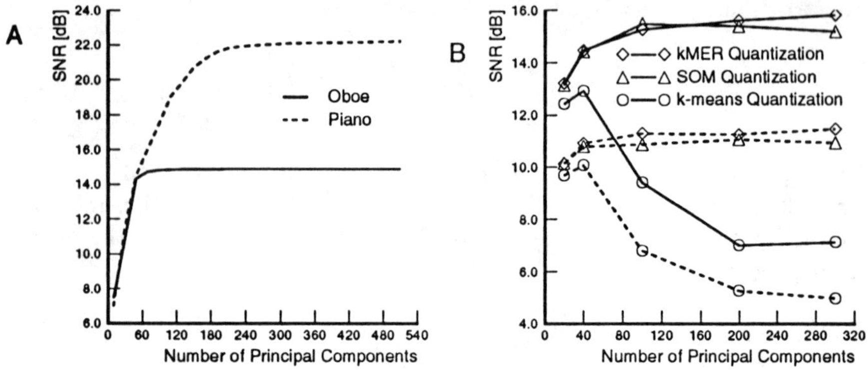

Figure 2: Performance (S/N in dB) in function of the number of principal components of A) the reconstruction of the training signals after dimensionality reduction (the phase is untouched); B) the reconstruction of the piano training signal after amplitude quantization (solid line) and complex quantization (dashed line) for the different learning rules.

the SNRs of the reconstruction where the *Complex Quantization* and *Phase Alignment* of section 2.3 have been used. Figure 2B also demonstrates the robustness of the different learning rules with respect to the dimensionality of the training data. The k-means performance is prone to the sparseness that is due to a higher dimensionality. Thus, it cannot take advantage of an increasing number of eigenvectors which allows for better spectral reconstructions, whereas the kMER and the SOM can (*cfr.* Fig. 2A).

Separation

The complete system is tested and evaluated for the different models (five different sets for all three learning rules: $N = 20, 40, 100, 200, 300$). Figure 3A shows the performances for the different models, evaluated in function of the number of principal components. Again, kMER and SOM's performances do not degrade with increasing dimensionality.

In order to better evaluate the performances, we considered two more models: $\mathbf{M}_{tr}$ and $\mathbf{M}_{sep}$. $\mathbf{M}_{tr}$ corresponds to a *"perfect"* quantization of the training set: the models consist of all normalized spectra of the training sets. $\mathbf{M}_{sep}$ corresponds to optimal codebooks with respect to the separation: they consist of all normalized spectra of the separate instruments in the compound signal. The results are $\mathbf{M}_{tr}$ (7.97 dB and 11.89 dB) and $\mathbf{M}_{sep}$ (9.07 dB and 14.06 dB).

CONCLUDING REMARKS

In this paper, spectral models have been applied in the context of the separation of acoustic sources. The models themselves have been developed

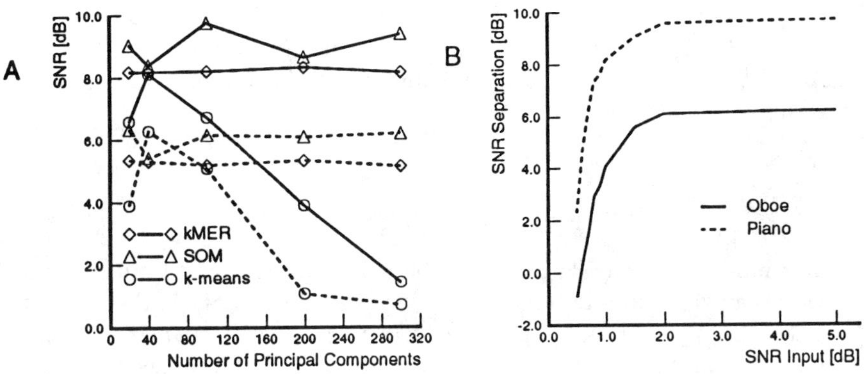

Figure 3: A) Performances of the system for the different models, evaluated in function of the dimensionality reduction. Solid lines represent the performances for the piano, dashed lines the ones for the oboe. B) Performance of the system in a noisy environment

using self-organizing learning rules. The recognition/separation is based on the projection of the compound signal onto the models' prototypical spectra. It is due to this projection that our models are invariant to noise: when uniform noise is added to the compound signal (Fig. 3B), the performance is still reasonable (5.59 dB) for a SNR of 1.5 dB of the compound signal (this corresponds to a noise amplitude of 30%). In applications in the field of sound monitoring, these models would be able to recognize the expected acoustic events, even in a noisy environment. The complete system could also be used to detect and separate unknown events from the modeled ones and even perform on-line modeling.

Contrary to blind source separation, our system does not assume a fixed mixing. Simulations have been performed where the mixing was variable. The mixing coefficients for the two instruments varied as a sine and a cosine, both with an offset of 1.1, with one cycle over the total signal length. Such a variable mixing could be caused by moving acoustic sources. When using the SOM models trained on 100 principal components, the separation performance hardly degraded and stayed at 5.83 dB for the oboe and 9.78 dB for the piano signal, compared to 6.17 dB and 9.75 dB when the mixing is constant (Fig. 3A). If two sensors were present, the sources could easily be located and tracked through time, since the mixing coefficients can be determined.

All these are topics for further investigation.

ACKNOWLEDGMENTS

T.G. is supported by a scholarship from the Flemish Institute for the promotion of Scientific Technological Research in Industry (I.W.T.). M.M.V.H. is a research associate of the Fund for Scientific Research – Flanders (Belgium) and is supported by research grants received from the Fund for Scientific Research (G.0185.96), the National Lottery (Belgium) (9.0185.96), the Flemish Ministry of Education (GOA 95/99-06), and the Flemish Ministry for Science and Technology (VIS/98/012).

REFERENCES

[1] Herrmann, M., and Yang, H.H. (1996). Perspectives and limitations of self-organizing maps in blind separation of source signals. In S.-I. Amari, L. Xu, L.-W. Chan, I. King, and K.-S. Leung (Eds.), *Progress in Neural Information Processing* (Vol. 2, pp. 1211-1216). Proceedings of the International Conference on Neural Information Processing (ICONIP'96), London: Springer-Verlag.

[2] Kohonen, T. (1982). Self-organized formation of topologically correct feature maps. *Biol. Cybern.*, **43**, pp. 59-69.

[3] Krishnaiah, P.R., and Kanal, L.N. (1982). *Classification, Pattern Recognition, and Reduction of Dimensionality*, Handbook of Statistics, vol. 2, Amsterdam: North Holland.

[4] Van Hulle, M.M. (1998). Kernel-based equiprobabilistic topographic map formation. *Neural Computation*, **10**, pp. 1847-1871.

[5] Van Hulle, M.M. (1998). Clustering with kernel-based equiprobabilistic topographic maps. *Proc. IEEE NNSP98* (Cambridge, U.K.), pp. 204-213.

[6] Watanabe, H., Matsumoto, Y., Tanaka, S., & Katagiri, S. (1998). Sound Monitoring based on the generalized probabilistic descent method. *Proc. IEEE NNSP98* (Cambridge, U.K.), pp. 383-392.

[7] Xie, F., and Van Compernolle, D. (1994). A family of MLP based non-linear spectral estimators for noise reduction. *Proc. ICASSP'94*, vol. II, pp. 53-56.

NEURO BAYESIAN BLIND EQUALIZATION WITH BER ESTIMATION IN DIGITAL CHANNELS

L. M. San José-Revuelta and J. Cid-Sueiro
Dpto. Teoría de la Señal y Comunicaciones e Ingeniería Telemática
E.T.S. Ingenieros de Telecomunicación, Universidad de Valladolid
C^o del Cementerio, s/n. 47011 - Valladolid, SPAIN
Phone: +34 (9)83 42.30.00, Fax: +34 (9)83 42.36.67
E-Mail: luismi@atenea.tel.uva.es, jescid@tel.uva.es

Abstract. The implementation of an optimal Bayesian algorithm for digital equalization is infeasible due to its computational complexity. In this paper we present a new approach to Bayesian blind equalization which is based on a hybrid architecture involving Neural Networks and Evolutionary Computation concepts. We develop a theoretical analysis which leads to recursive formulas to estimate the Probability Density Functions (PDFs) of both the channel and the received samples. These parameters are used directly by the algorithms to perform equalization. Beginning with a brief revision of previous neural, genetic and Radial Basis Function (RBF) networks-based approaches, we outline the theoretical algorithm that will serve as a reference for future neuro-evolutionary derivations. We also show the capability of these structures to perform blind Bit Error Rate (BER) estimation in reception. Finally, several computer simulations and comparative results are exposed.

INTRODUCTION

Adaptive blind channel equalization consists in compensating intersymbol interference (ISI) in order to decide which is the right sequence of transmitted symbols, using the only available information consisting of the received noisy samples, with the capability of also tracking slow variations in the channel during transmission [1, 2, 3, 4].

In the equalizer design, the conflicting objectives of performance and complexity need to be taken into account. The Viterbi Algorithm (VA) equalizer suffers from a high computational complexity, and the performance of the conventional linear and hard-decision feedback equalizer sometimes appears to be unsatisfactory [1, 5].

0-7803-5673-X/99/$10.00

Multilayer Perceptrons and RBF networks have been tested, exhibiting better detection capabilities than linear structures [6, 7]. In [8, 9], it is proved that RBF networks with the adequate feed-back loop (*RRBFs, Recurrent-RBFs*) compute infinite memory detection, though they have two important drawbacks: the convergence to suboptimal solutions and the complexity. In [8, 9], we suggest a strategy to limit this complexity by means of merging all the conditional estimates of the channel into one function which represents the estimate of the conditional channel PDF. In 1994, Iltis proposed an alternative blind equalizer derived from Kalman filter structures [10].

Chen and Wu exposed in [11] a new approach to blind equalization using a Genetic Algorithm (GA) scheme based on maximum likelihood (ML) joint channel and data estimation.

In the present paper, we extend the Bayesian approach worked out in [8, 9] taking into account a more complete formula for the calculus of the received samples PDF. This way we get a symbol-by-symbol blind equalizer focusing on the main inconvenient found on previous work: that of convergence. The developed algorithm allows us to adjust the level of complexity of the receiver (this parameter affects directly to the convergence properties of the algorithm). The receiver can also estimate BER rates without any reference signal using the capability of RRBF-based equalizers so as to calculate the *a posteriori* probabilities associated to every possible channel state [3, 8, 9]. We will see that even in the cases where the channel is not accurately estimated, the equalizer obtains accurate estimates of the BER.

In the case of tracking time-variant channels in non-stationary scenarios, the equalizer shows the capability of blind tracking with no previous knowledge of either any reference signal or the parameters of the time variation model ([10] assumes these parameters to be known).

BAYESIAN CHANNEL ESTIMATION ALGORITHM

Digital Channel Model

An ideal model for base-band channel representation that is widely used for the analysis and optimization of conventional detectors is

$$r_k = h(x_k, x_{k-1}, \ldots, x_0) + n_k = y_k + n_k \tag{1}$$

where x_k and r_k represent the transmitted and received symbol at instant k, respectively. We will suppose equiprobable binary bipolar transmission ($x_k = 1$ or $x_k = -1$), with $Pr\{x_k = 1\} = Pr\{x_k = -1\} = 1/2$ (the generalization to more complex channel models does not lead to special difficulties). All transmitted symbols are assumed to be statistically independent as well. Considering this model of causal dispersive channel, every received sample r_k is related to the transmitted sequence in accordance with (1). Function h is not known and must be estimated, and $\{n_k\}$ represents a stationary random noise process whose statistical parameters will be estimated.

Estimation of the Conditional PDF of h

As mentioned before, in this section we extend the analysis carried out in [8] by considering the PDF of **h** based on all received samples.

For reasons of computational efficiency, we consider that function **h** has a Gaussian PDF distribution and is not based on the concrete values of **x** that have been transmitted. After several algebraic operations, whose mathematical details can be seen in [3], the expression of the conditional PDF of **h** takes the form of a weighted sum of Gaussian functions

$$f_h(\mathbf{h}|\mathbf{r}_k) = \sum_{i=0}^{L_k} \lambda_i(k) N\left(\mathbf{h} - \mathbf{h_{e,i}}(k), \mathbf{S_i}(k)\right) \quad (2)$$

where $L_k = 2^{m+k+1}$ and $\mathbf{r}_k = (r_k, r_{k-1}, \ldots, r_0)$. $\lambda_i(k)$ represents the conditional probabilities $Pr\{\mathbf{x_i}^{(k)} | \mathbf{r}_k\}$ (the probability of the i-th possible sequence of symbols, based on the whole set of received samples); $N\left(\mathbf{h} - \mathbf{h_{e,i}}(k), \mathbf{S_i}(k)\right)$ is the PDF of the channel based on the received samples: $f_h(\mathbf{h}|\mathbf{r}_k)$; and, finally, $\mathbf{h_{e,i}}$ and $\mathbf{S_i}$ are the mean and the variances matrix when assuming that the i-th possible sequence (*candidate*) of transmitted symbols (i.e., the sequence given by the binary representation of index i) was the real sequence sent by the transmitter. So, every $\mathbf{h_{e,i}}$ represents a filter candidate, each of them being associated to a hypothesis, and with a relative weight given by the corresponding $\lambda_i(k)$.

The expressions used to generate the new population of candidates from the previously existing ones are shown on Table I (see also [3] for a more detailed mathematical description).

The coefficients $\lambda_i(k)$ have been defined as

$$\lambda_{i_0,i_1,i_2,\ldots,i_k}(k) = P_{i_0}(0) P_{i_0,i_1}(1) P_{i_0,i_1,i_2}(2) \cdots P_{i_0,i_1,i_2,\ldots,i_k}(k) \quad (3)$$

Note that in the latter expressions, the decimal index i is equivalent to the value of the binary string $\mathbf{i}_k = i_0, \ldots, i_k$. The coefficients $P_i(j)$ are given by

$$P_i(j) = \alpha \cdot p_i(j) N\left(r_j - \mathbf{h_{e,i}}(j)^T \mathbf{x_i}^{(j)}, \sigma_i^2(j)\right) \quad j = 0, 1, \ldots, k \quad (4)$$

where α is a normalizing constant ensuring that $\sum_i P_i(j) = 1$, and the following recursive relation holds

$$\lambda_{i_0,i_1,i_2,\ldots,i_k}(k) = \lambda_{i_0,i_1,i_2,\ldots,i_{k-1}}(k-1) P_{i_0,i_1,i_2,\ldots,i_k}(k) \quad (5)$$

Step of Decision

Considering the optimal MAP decision rule, we find that the estimate $s_k = \hat{x}_k$ of the transmitted symbol is

$$s_k = g(\mathbf{r}(k)) = \text{sign}\left(\sum_{i=00\ldots0}^{11\ldots1} \lambda_{i,1}(k) - \sum_{i=00\ldots0}^{11\ldots1} \lambda_{i,0}(k) \right) \quad (6)$$

Pruning Step

The complexity of this initial version, -Algorithm 1-, grows exponentially in accordance to time and channel memory size, i.e., the number of candidates (possible sequences of transmitted symbols) at time k is $n_c = 2^{m+k+1}$. Though theoretically relevant, this scheme is obviously infeasible. This leads directly to the necessity of a *pruning procedure* in order to keep n_c constant when increasing k.

Different strategies can be applied in order to generate and select the surviving candidates (also called *centroids* in RBF literature) at instant k from those at $k-1$. In [8] all possible channel candidates are merged into an only Gaussian function according to their respective weight, and [10] gathers together those candidates that have supposed the same transmitted symbol sequences in the last m received symbols. In [3] we used several ideas from Evolutionary Computation to implement the pruning procedure with the goal of reducing complexity. In this paper we will focus on the use of one of these alternatives: *Direct Sampling Selection.* This strategy selects the individuals with the highest weight λ_i in Eq. 2, -Algorithm 2-.

The procedure is as follows: we choose as a parameter the maximum number of centroids up to which we let the equalizing algorithm run. This parameter is called $n_{c,max}$. During the initial time, with k small, the number of centroids, n_c, is less than $n_{c,max}$ and the behavior will be the one previously described. When n_c equals to $n_{c,max}$, the pruning procedure begins to work selecting the n_s sets $\{\mathbf{S_i}, \mathbf{h_{e,i}}, \lambda_i, i\}$ whose λ_i is greater as surviving centroids (possible symbol paths) for next k. The resulting process is schematized in Table I.

In this table, $\mathbf{x}_j^{(k)}$ represents the binary sequence of length $m+k+1$, given by the binary representation of subindex j, $\sigma_j^2(k)$ is the variance if assuming that $\mathbf{x}_j^{(k)}$ was the real sent sequence, and $\sigma_{n,j}^2(k)$ is the noise variance associated to this hypothesis.

Note that this model allows to give more importance to the contribution of the noise in Eq. 1 in order to compensate for a greater ignorance of $\mathbf{h}$. Thus, when the binary sequence given by index j is very different from the real sent sequence, $\mathbf{h_{e}}_{,j}$ will be a poor estimate of the impulse response of the channel; in addition, σ_j^2 will be high (the corresponding term in Eq. 2 should be a Gaussian function with a very wide variance), and, since $\mathbf{h_{e}}_{,j}$ has little reliable information about $\mathbf{h}$, the parameter $\sigma_{n,j}^2$ is large, indicating that, in the model described by Eq. 1, the noise component is relevant.

SIMULATIONS AND RESULTS

Four channels taken from [12] were chosen to make comparisons: $H_1(z) = 0.35 + 0.87z^{-1} + 0.35z^{-2}$, $H_2(z) = 1 + 0.8z^{-1} - 0.81z^{-2}$, $H_3(z) = 0.319 + 0.620z^{-1} + 0.634z^{-2} + 0.323z^{-3} + 0.087z^{-4}$ and $H_4(z) = 0.548 + 0.789z^{-1} + 0.273z^{-2} - 0.044z^{-3} + 0.012z^{-4} + 0.017z^{-5} - 0.017z^{-6} + 0.007z^{-7} + 0.009z^{-8}$.

1. **Initialization**

   $\mathbf{h_{e,i}}(-1) \approx (0,\ldots,0)^T$. $\mathbf{S_i}(-1) = \sigma_h^2 \mathbf{I} \quad \sigma_h^2 \gg 0 \quad \lambda_i(-1) = 2^{-m}$

   $ncmax = M^p$, p integer, $M = 2$ with binary transmission.

2. **Estimation of the PDF of h**

   $n_c = 2^{m+k+1}$, centroids number, $n_s = 2^m$, previous k surviving centroids.
   FOR $l = 1$ TO n_s

   FOR $i = 0, 1$

   $$\mathbf{d}_{2l-i}(k) = \mathbf{S}_l(k-1)\mathbf{x}_{2l-i}^{(k)} \qquad \sigma_{2l-i}^2(k) = \sigma_{n,l}^2 + {\mathbf{x}_{2l-i}^{(k)}}^T \mathbf{d}_{2l-i}(k)$$
   $$\epsilon_{2l-i}(k) = r_k - \mathbf{h}_{\mathbf{e},l}(k-1)^T \mathbf{x}_{2l-i}^{(k)} \qquad \sigma_{n,2l-i}^2(k-1) = (1-\mu)\cdot\sigma_{n,l}^2(k) + \mu\cdot\epsilon_{2l-i}^2(k)$$

   $\mathbf{h_{e,i}}$ and $\mathbf{S_{h,i}}$ associated to every surviving centroid of the previous k.

   $$\begin{cases} \mathbf{h}_{\mathbf{e},2l-i}(k) = \mathbf{h}_{\mathbf{e},l}(k-1) + \frac{1}{\sigma_{2l-i}^2(k)}\epsilon_{2l-i}(k)\mathbf{d}_{2l-i}(k) \\ \mathbf{S}_{2l-i}(k) = \mathbf{S}_l(k-1) - \frac{1}{\sigma_{2l-i}^2(k)}\mathbf{d}_{2l-i}(k)\mathbf{d}_{2l-i}(k)^T \end{cases}$$

   $$f_r(r_k|\mathbf{r}_{k-1}) = \sum_{\mathbf{i}_k=0,\ldots,0}^{1\ldots1} \lambda_{\mathbf{i}_{k-1}}(k-1)p_{\mathbf{i}_k}(k)\cdot N\left(r_k - \mathbf{h}_{\mathbf{e},\mathbf{i}_k}(k)^T\mathbf{x}_{\mathbf{i}_k}^{(k)}, \sigma_{\mathbf{i}_k}^2(k)\right)$$

   FOR $i_0, i_1, \ldots, i_k = 00\ldots0$ TO $(2ns) - 1$

   $$P_{i_0,i_1,\ldots,i_k}(k) = \frac{p_{i_0,\ldots,i_k}(k)N\left(r_k - \mathbf{h}_{\mathbf{e},i_0,\ldots,i_k}(k)^T\mathbf{x}_{i_0,\ldots,i_k}^{(k)}, \sigma_{i_0,\ldots,i_k}^2(k)\right)}{f_r(r_k|\mathbf{r}_{k-1})}$$
   $$\lambda_{i_0,i_1,\ldots,i_k}(k) = \lambda_{i_0,i_1,\ldots,i_{k-1}}(k-1)P_{i_0,i_1,\ldots,i_k}(k)$$

   $$f_h(\mathbf{h}|\mathbf{r}_k) = \sum_{i=0}^{2^{m+k+1}-1} \lambda_i(k)\cdot N\left(\mathbf{h} - \mathbf{h_{e,i}}(k)^T, \mathbf{S_i}(k)\right)$$

3. **Decision Step.** $s_k = \text{sign}\left(\sum_{i=00\ldots0}^{11\ldots1}\lambda_{i,1}(k) - \sum_{i=00\ldots0}^{11\ldots1}\lambda_{i,0}(k)\right)$

4. **Pruning Procedure.**

   IF $nc \geq ncmax$ THEN (Calculus of the ns sets $\{\mathbf{S_i}, \mathbf{h_{e,i}}, \lambda_i, i\}$ for next iteration)

   - $indexes = \{i\}$ so that $\lambda_i(k)$ is within the ns highest λ_i's.
   - $\lambda_i(k+1) = \lambda_{indexes}(k)$, $\quad \mathbf{h_{e,i}}(k+1) = \mathbf{h}_{\mathbf{e},indexes}(k)$,
     $\mathbf{S_i}(k+1) = \mathbf{S}_{indexes}(k)$
   - $ns = ncmax/2$

   ELSE $\quad ns = nc$

5. $\mathbf{k = k+1}$. **Go to Step 2.**

Table I: Blind Equalization Algorithm No. 2

Figure 1 shows the symbol error probability against SNR for three different equalization methods [1] : (1) PRLS *(Probabilistic Recursive Least Squares)*, derived in [8], an algorithm for detection and unsupervised learning of the channel response based on RRBF networks, (2) PLMS *(Probabilistic Least Mean Squares)*, which is a simplified version of the former where the adaptation step is computed using the noise variance and the expected variance of the channel estimate, (3) [10] method, and (4) Algorithm 2 (Table I) with $n_{c,max} = 64$ (512 for $H_4(z)$, though the graphic displays 64).

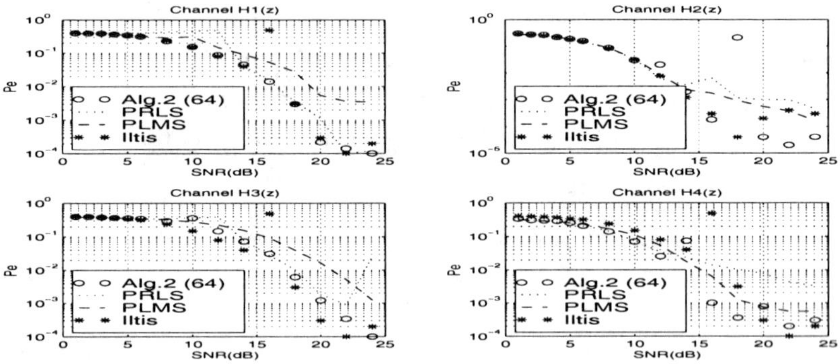

Figure 1: BER versus SNR for different blind equalization methods: PRLS, PLMS, [10] and Algorithm 2 with $n_{c,max}$ equal to 64, 64, 64 and 512 respectively.

It can be observed that complex non-linear structures (such as Algorithm 2, [10] and PRLS) exhibit a very similar performance with respect to BER. PLMS has a worse performance and the differences with the other methods are greater as the SNR increases.

PRLS and PLMS show a worse convergence behavior, especially with low SNR ratios (though this is not evidenced by this graphical representation).

Next, we study the influence of the SNR on the convergence of the algorithm proposed (Algorithm 2, Table I). In order to quantify this, let us define the *Least Square Error* as

$$LSE(k) = \min\{\| \mathbf{h}(k) - \mathbf{h_e}(k) \|, \| \mathbf{h}(k) + \mathbf{h_e}(k) \|\} \quad (7)$$

where $\mathbf{h_e}$ represents the best channel filter candidate at k. The minimum of two different error measures is given by the fact that, for a binary symmetric transmission, there is no way to discriminate between **h** and **-h** without supervision. This is a well known limitation that is independent from the estimation algorithm.

Figure 2 shows the evolution of $LSE(k)$ for three different qualities of the channel: $SNR = 4, 14$ and 24 dB. And for each one, the evolution of $LSE(k)$ for several values of the parameter $n_{c,max}$ is computed.

This figure displays the evolution of the channel impulse response function estimate, $\mathbf{h_e}$, which the algorithm gives for the real channel response, **h**. In

[1] The blind equalizers chosen in this paper to compare with, are those directly related with the use of probabilities (Bayesian approaches): PRLS and PLMS, from [8], and a method derived from Kalman filter structures, [10].

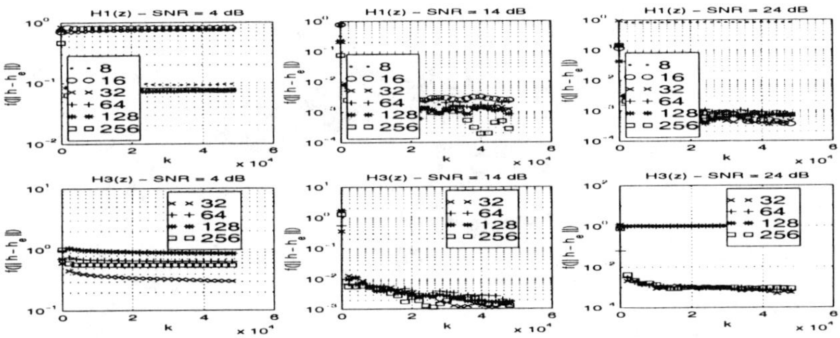

Figure 2: Evolution of $LSE(k)$ for three different qualities of the channel: SNR = 4, 14 and 24 dB. For each graph, several values of the parameter $n_{c,max}$ were considered. Simulations belong to $H_1(z)$ and $H_3(z)$.

the case of SNR=4 dB (very low SNR), the convergence and quality of the estimate is much poorer than with higher values. Note, also, that the estimate of **h** becomes more accurate (smaller LSE) as the SNR is increased (notice the vertical axis scale).

In order to evaluate the performance of Algorithm 2 with respect to the speed of convergence, we run simulations for $H_2(z)$ with our equalizer, taking different values of the maximum number of candidates: $n_{c,max} = 128$ and 256, and we compare them with the LSE evolution corresponding to [10].

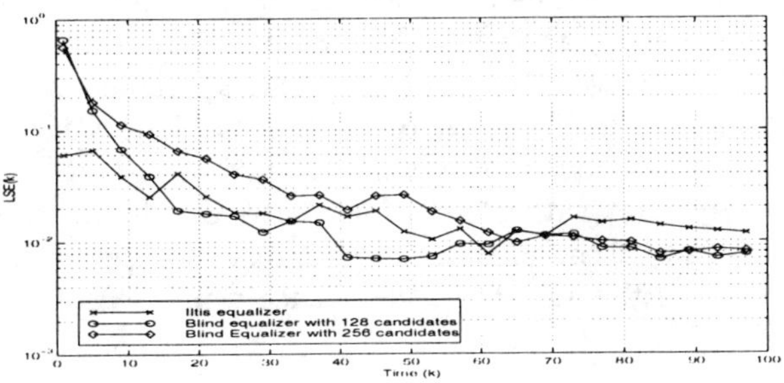

Figure 3: Evolution of $LSE(k)$ for stationary channel $H_2(z)$ with SNR = 18 dB.

Figure 3 shows that these three schemes manifest similar speed of convergence with alike accuracy in the estimate of **h**. Also, notice that when the equalizer converges to the optimal solution, the maximum size of the equalizing network does not affect the speed of convergence, but has great influence on the probability of good convergence.

The 3D representation of Figure 4 shows a combination of both concepts previously described. $LSE(k)$ decreases with both k and the SNR (the z-axis will represent the logarithm of LSE). For a given SNR, $LSE(k)$ as a function of k should be a decreasing function, and, conversely, for a fixed instant k, $LSE(k)$ should also be a decreasing function of the channel quality.

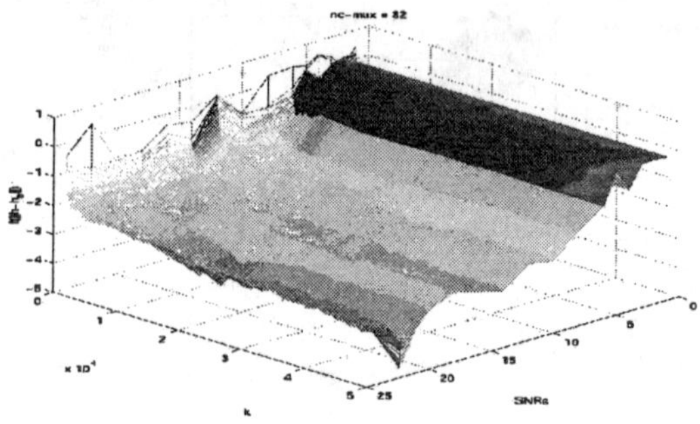

Figure 4: Evolution of $LSE(k)$ vs time k and SNR. $H_2(z)$ with $n_{c,max} = 32$.

Now, let us consider the question of convergence. One of the main problems of the PRLS algorithm exposed in [8] is its quite frequent convergence to wrong or suboptimal solutions (suboptimal $\mathbf{h}$ estimates). As we emphasized previously, one of the most important advantages of the proposed algorithm for blind equalization (Algorithm 2) is that it permits the selection of the complexity level, by using the parameter $n_{c,mac}$. This parameter determines the maximum size of the equalizing network, since it represents the maximum number of centroids (or filter candidates), being each one associated to a possible transmitted sequence of symbols ($\mathbf{x_i}$).

The size of the equalizing network greatly affects the convergence properties of the algorithm. Figure 5 shows the number of faults (convergence to bad solutions) for channel $H_2(z)$ with SNR = 8, 10, 12 dB. These results prove the importance of the selection of the right size for the equalizing network. If we consider $H_2(z)$ with SNR = 8 dB, we can see that with $n_{c,max} = 8$, 16 or 32 centroids, the number of errors is very high and Algorithm 2 seldom converges properly. But with a size of 64 centroids, the behavior improves significantly. In fact, if this number of centroids is duplicated, the performance remains quite the same and this choice would waste computational resources.

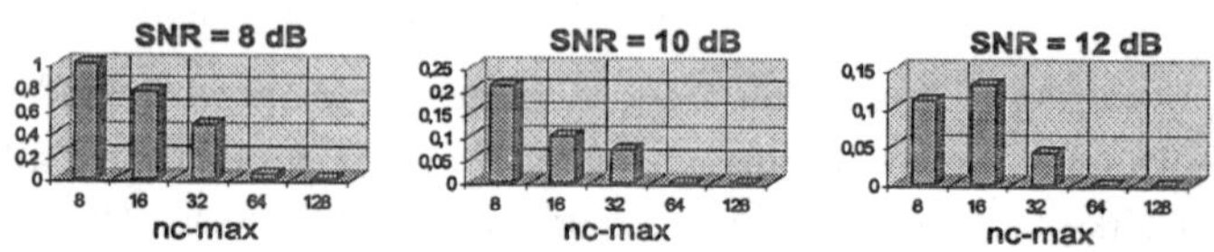

Figure 5: Percentage of bad convergence for channel $H_2(z)$. Number of samples: 25000. Threshold: 0.02. Number of iterations for each SNR: 100.

Next, the problem of blind tracking of digital time-variant channels with no previous knowledge of the time variations process is dealt with. Our algorithm performs tracking tasks with no previous knowledge about the time variation model of $\mathbf{h}$. (For instance, [10] assumes these parameters to be known, as well as the noise variance). Since Bayesian equalizers require a noise variance estimate to compute conditional data densities, by computing them as the average of $\widehat{e}_k = r_k - \widehat{h}(\widehat{\mathbf{s}}_k)$, (see Table I), where $\widehat{\mathbf{s}}_k$ is a decision vector, the channel variations can be taken into account.

Fig. 6 is an example of this, using Algorithm 2. Initially, a perfect knowledge of linear channel $H_3(z)$, with SNR = 18 dB, is assumed. The channel response is modified at each symbol time according to a simple random walk model: $\mathbf{h}_k = \mathbf{h}_{k-1} + \mathbf{dh}_k$, where vector $\mathbf{h}_k$ contains the channel coefficients and $\mathbf{dh}_k$ is a random zero-mean Gaussian vector with variance σ_h^2.

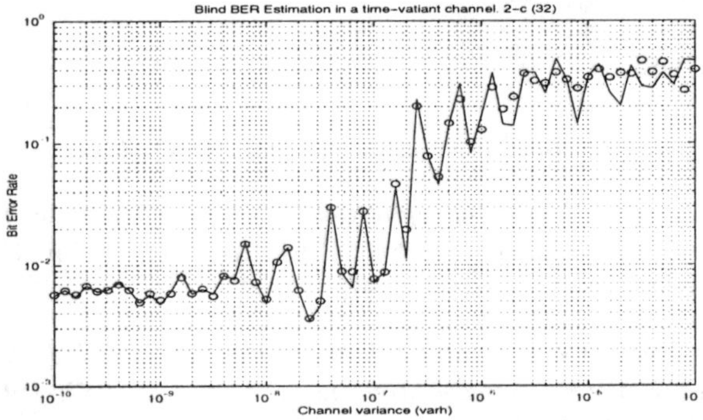

Figure 6: BER estimation in a time variant linear channel $H_3(z)$. Random walk model. The continuous line shows the BER estimates; the dots represent the real BER. Direct Sampling Selection is used as evolutionary pruning strategy.

The influence of the different sampling selection strategies on the convergence properties is analyzed in [3], proving that implementing a self-adapting procedure which selects the appropriate pruning method, we can reduce the non-converging situations.

Finally, we analyze the problem of blind BER estimation in the receiver with no reference signal, considering non-stationary channel $H_1(z)$ ($\sigma_h^2 \neq 0$). Simulations show that when the equalizer does not converge to the optimal channel function, the blind BER estimate is right, but when the equalizer converges properly, the BER estimate sometimes fails. At the end of the process, 85 % of the cases where properly estimated. Future work will focus on this problem of blind estimation for time-variant channels.

CONCLUSIONS

We have presented the theoretical analysis that leads to an algorithm for blind digital channel equalization in a non-supervised mode, which allows blind BER estimation and time-variant channel tracking. Due to the exponentially increasing complexity of the theoretical algorithm, the necessity of a pruning procedure is introduced. This stage selects the set of paths of transmitted symbols which has a higher probability of being the real transmitted sequence. The final implementation of the algorithm shows the capability of on-line BER estimation and permits to define the required level of complexity, which is related to the BER and convergence properties. The equalizer developed also shows the capability of tracking time-variant channels with no previous knowledge of the time-variation parameters.

ACKNOWLEDGMENTS

The authors wish to acknowledge financial support from: CICYT grants TIC96-0500-09 and TIC 97-0772; and Programa FPI de la Junta de Castilla y León (Conv: Orden de 30/10/96).

REFERENCES

[1] J. G. PROAKIS, *Digital Communications.* 2nd ed. NY: McGrawHill, 1989.

[2] S. CHEN, *Genetic Algorithm Optimization for Blind Channel Identification with Higher Order Cumulant Fitting,* IEEE Trans. on Evolut. Comput., vol. 1, No. 4, pp. 259-265, November 1997.

[3] L. M. SAN JOSÉ-REVUELTA AND J. CID-SUEIRO, *Robust Blind Channel Equalization Based on Evolutionary Strategies and On-Line BER Estimation,* International COST 254 Workshop on Intelligent Communication Technologies and Applications, Neuchatel, Switzerland, May 1999.

[4] S. T. ALEXANDER, *Adaptive Signal Processing: Theory and Applications,* Springer-Verlag, NJ, 1986.

[5] J. THIELECKE, *A Soft-Decision State-Space Equalizer for FIR Channels,* IEEE Trans. on Comm., vol. 45, No. 10, pp. 1208-1217, October 1997.

[6] G. J. GIBSON, S. SIU AND C. COWAN, *The Application of Nonlinear Structures to the Reconstruction of Binary Signals,* IEEE Trans. on Signal Processing, vol. 39, No. 8, 1991.

[7] B. MULGREW, *Applying Radial Basis Functions,* IEEE Signal Processing Magazine, pp. 50-65, March 1996.

[8] J. CID-SUEIRO, A. ARTÉS-RODRÍGUEZ AND A. R. FIGUEIRAS-VIDAL, *Recurrent Radial Basis Function Networks for Optimal Symbol-by-Symbol Equalization,* Signal Processing, vol. 40, No. 2, November 1994.

[9] J. CID-SUEIRO, AND A. R. FIGUEIRAS-VIDAL, *Recurrent Radial Basis Function Networks for Optimal Blind Equalization,* Proc. 1993 IEEE-SP Workshop on Neural Networks for Sig. Proc., Linthicum, MA, pp. 562-571, June 1993.

[10] R. A. ILTIS, J. J. SHYNK AND K. GIRIDHAR, *Bayesian Algorithms for Blind Equalization Using Parallel Adaptive Filtering,* IEEE Trans. on Communications, v. 42, No. 2/3/4, pp. 1017-1032, Feb./Mar./Apr. 1994.

[11] S. CHEN AND Y. WU, *Maximum Likelihood Joint Channel and Data Estimation Using Genetic Algorithms,* IEEE Trans. on Signal Processing, vol. 46, No. 5, pp. 1469-1473, May 1998.

[12] A. P. CLARK, *Equalizers for Digital Modems,* Pentech Press, London, 1985.

MAXIMUM LIKELIHOOD BLIND SOURCE SEPARATION IN GAUSSIAN NOISE

Joaquín Míguez, Luis Castedo
Departamento de Electrónica e Sistemas, Universidade da Coruña
Campus de Elviña s/n, 15071 A Coruña, Spain
Tel.: ++34 981 167150 ext. 1212
Fax: ++34 981 167160
E-mail: miguez@des.fi.udc.es

Abstract. This paper presents a new Maximum Likelihood (ML) approach to the separation of convolutive mixtures of unobserved sources in the presence of Additive Temporally White Gaussian Noise (ATWGN). The proposed method proceeds in two steps. First, the ML estimate of the mixing system is computed and, afterwards, this estimate is employed to obtain the ML estimates of the sources. The proposed algorithms rely on the knowledge of the sources probability density function (p.d.f.) and the noise second order statistics. These are fairly realistic assumptions in applications such as digital communications.

INTRODUCTION

Blind Source Separation (BSS) is an important problem in signal processing that consists of the recovery of an unobserved set of signals (sources) from observations of linear mixtures of them without *a priori* knowledge of the mixing coefficients. Separation is often carried out with a linear system that can be interpreted as the synaptic weights of a single layer neural network. Thus, it is recognized that adaptive algorithms for BSS adequately work as unsupervised learning rules for neural networks.

Since the pioneering work by Jutten and Herault [7] many algorithms for BSS have been proposed. Special attention has been devoted to algorithms based on the Maximum Likelihood (ML) estimation of the inverse of the linear mixture model [8, 4]. These algorithms are statistically optimum and can be shown to be equivalent to other information theoretic approaches such as the infomax principle [2]. Furthermore, the ML estimation of the mixing model does not require the statistical independence of the source signals, although most authors make this assumption for the sake of simplicity [5]. Instead, some knowledge of the sources joint probability density function (p.d.f.) is required. This is not the typical statement of the BSS problem but it is reasonable in many contexts [5] (e.g., digital communications).

0-7803-5673-X/99/$10.00 © 1999 IEEE

However, most existing approaches to BSS focus on the separation of instantaneous linear mixtures of the sources with no noise contaminating the observed data and only some recent papers adress the problem of separating convolutive mixtures or dealing with noisy observations (see [10, 6] and references therein). In this work, we present an ML based approach for blind separation of convolutive mixtures of the source signals that is also specifically devised to account for the presence of Additive Temporally White Gaussian Noise (ATWGN). The proposed method proceeds in two steps. First, the ML estimate of the convolutive mixing system is computed. The ML estimation algorithm exploits the knowledge of the sources and noise statistics to achieve optimum performance (in the likelihood sense) even when the Signal-to-Noise Ratio (SNR) is low. In a second step, the ML estimate of the mixing system is employed to obtain the corresponding ML estimates of the sources.

The remaining of this paper is organized as follows. Section 2 introduces the signal model. In section 3 we derive an algorithm that computes the ML estimate of the mixing system. Section 4 briefly describes how to obtain the ML estimates of the sources. Illustrative computer simulations are presented in section 5 and, finally, section 6 is devoted to the conclusions.

SIGNAL MODEL

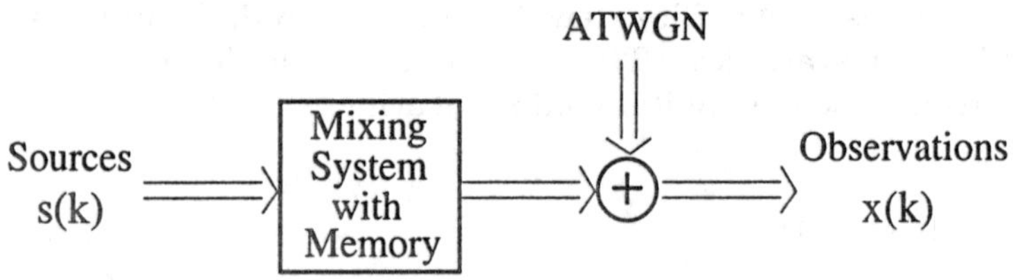

Figure 1: Block diagram of a convolutive mixing system with Additive Temporally White Gaussian Noise (ATWGN).

Figure 1 plots the block diagram of a discrete-time mixing model that consists of a Multiple Input Multiple Output (MIMO) system with N inputs, L outputs, a memory of size m and ATWGN. The observed signals can be written as the convolution sum

$$\mathbf{x}(k) = \sum_{i=0}^{m-1} \mathbf{A}^H(i)\mathbf{s}(k-i) + \mathbf{n}(k) \tag{1}$$

where $\mathbf{s}(k) = [s_1(k), \ldots, s_N(k)]^T$ is the $N \times 1$ source vector at time k, $\mathbf{x}(k) = [x_1(k), \ldots, x_L(k)]^T$ is the observation vector, $\mathbf{n}(k) = [n_1(k), ..., n_L(k)]^T$ is an $L \times 1$ vector of zero-mean temporally white complex Gaussian components with covariance matrix $\mathbf{\Sigma}$, the sequence of $N \times L$ matrices $\mathbf{A}(0), ..., \mathbf{A}(m-1)$ represents the mixing system and ^{H} denotes Hermitian transposition. We address the general case where the signals and the parameters in the model are complex valued.

In the subsequent derivation of the ML algorithm we will make use of the

$Nm \times L$ overall mixing matrix

$$\mathbf{A} \triangleq \left[\mathbf{A}^T(m-1) \quad \cdots \quad \mathbf{A}^T(0)\right]^T \tag{2}$$

and the $Nm \times 1$ *redundant* source vector

$$\tilde{\mathbf{s}}(k) \triangleq \left[\mathbf{s}^T(k-m+1) \quad \cdots \quad \mathbf{s}^T(k)\right]^T \tag{3}$$

Using (2) and (3), the observations can be rewritten as

$$\mathbf{x}(k) = \mathbf{A}^H\tilde{\mathbf{s}}(k) + \mathbf{n}(k). \tag{4}$$

We will assume that $\mathbf{A}$ is full row rank in order to ensure that it is identifiable.

Finally, note that when the mixing system memory size is $m = 1$ eq. (4) reduces to an instantaneous linear mixture model.

ML ESTIMATION OF THE MIXING MATRIX

In this section we introduce an algorithm that computes the ML estimate of the overall channel matrix from the received signals using only the knowledge of the transmitted symbols and noise statistics. For simplicity, we will drop, when possible, the time index k when referring to the symbols and the observations (i.e., we will write $\tilde{\mathbf{s}}$ and $\mathbf{x}$ instead of $\tilde{\mathbf{s}}(k)$ and $\mathbf{x}(k)$).

Estimation Problem

The ML estimate of the mixing matrix is the solution to the following problem [9]

$$\widehat{\mathbf{A}} = \arg \max_{\mathbf{A}} \{\mathcal{J}(\mathbf{A}) = E_{\mathbf{x}} \log(f_{\mathbf{x};\mathbf{A}}(\mathbf{x})\} \tag{5}$$

where $f_{\mathbf{x};\mathbf{A}}(\cdot)$ is the joint p.d.f. of the observations that depends on the mixing matrix $\mathbf{A}$, $E_{\mathbf{x}}$ denotes mathematical expectation with respect to $\mathbf{x}$ and $\log(\cdot)$ is the natural logarithm. Function $f_{\mathbf{x};\mathbf{A}}(\cdot)$ is the likelihood of the mixing matrix and $\mathcal{J}(\mathbf{A})$ will be referred as the *expected log-likelihood* of $\mathbf{A}$. In order to solve (5) we need to express $f_{\mathbf{x};\mathbf{A}}(\cdot)$ in terms of the transmitted symbols and the ATWGN joint p.d.f.'s ($f_{\mathbf{s}}(\cdot)$ and $f_{\mathbf{n}}(\cdot)$, respectively). Both functions are assumed to be known, including the noise covariance matrix $\mathbf{\Sigma}$.

Let us define the $(Nm + L) \times 1$ *extended* vectors

$$\tilde{\mathbf{s}}_e = \left[\tilde{\mathbf{s}}^T \quad \mathbf{n}^T\right]^T \quad \mathbf{x}_e = \left[\tilde{\mathbf{s}}^T \quad \mathbf{x}^T\right]^T \tag{6}$$

It is well known that the joint p.d.f.'s of $\mathbf{s}_e$ and $\mathbf{x}_e$ are related by [9]

$$f_{\mathbf{x}_e;\mathbf{A}}(\mathbf{x}_e) = \frac{f_{\tilde{\mathbf{s}}_e}(t_{\mathbf{A}}^{-1}(\mathbf{x}_e)}{|J_t|} \tag{7}$$

where $f_{\tilde{\mathbf{s}}_e}(\cdot)$ and $f_{\mathbf{x}_e;\mathbf{A}}(\cdot)$ are the joint p.d.f.'s of $\tilde{\mathbf{s}}_e$ and $\mathbf{x}_e$, respectively, $t_{\mathbf{A}}(\tilde{\mathbf{s}}_e) = \left[\tilde{\mathbf{s}}^T \quad (\mathbf{A}^H\tilde{\mathbf{s}} + \mathbf{n})^T\right]^T$ is an invertible transformation and $|J_t|$ is the absolute value of the Jacobian of $t_{\mathbf{A}}(\cdot)$. It is straightforward to show that $|J_t| = 1$. Thus, assuming the sources and the ATWGN components are statistically independent, the joint p.d.f. $f_{\mathbf{x}_e;\mathbf{A}}(\cdot)$ can be factorized as

$$f_{\mathbf{x}_e;\mathbf{A}}(\mathbf{x}_e) = f_{\tilde{\mathbf{s}}}(\tilde{\mathbf{s}}) f_{\mathbf{n}}(\mathbf{x} - \mathbf{A}^H \tilde{\mathbf{s}}). \tag{8}$$

and the joint p.d.f. of the observations is the marginal p.d.f.

$$f_{\mathbf{x};\mathbf{A}}(\mathbf{x}) = E_{\tilde{\mathbf{s}}} f_{\mathbf{n}}(\mathbf{x} - \mathbf{A}^H \tilde{\mathbf{s}}). \tag{9}$$

that is readily obtained from (8). Finally, substituting the complex Gaussian multivariate p.d.f. into (9) and neglecting constants that do not affect the solution, problem (5) can be written as

$$\widehat{\mathbf{A}} = \arg \max_{\mathbf{A}} \left\{ \mathcal{J}(\mathbf{A}) = E_{\mathbf{x}} \log \left(E_{\tilde{\mathbf{s}}} e^{-||\mathbf{x} - \mathbf{A}^H \tilde{\mathbf{s}}||_{\mathbf{\Sigma}^{-1}}} \right) \right\} \tag{10}$$

where $||\mathbf{v}||_{\mathbf{M}} = \mathbf{v}^H \mathbf{M} \mathbf{v}$. In practice, expectation $E_{\mathbf{x}}$ has to be estimated from the available observations, $\mathbf{x}(0), ..., \mathbf{x}(K-1)$, and the optimization problem to be solved becomes

$$\widehat{\mathbf{A}} = \arg \max_{\mathbf{A}} \left\{ \mathcal{J}(\mathbf{A}) = \frac{1}{K} \sum_{k=0}^{K-1} \log \left(E_{\tilde{\mathbf{s}}} e^{-||\mathbf{x}(k) - \mathbf{A}^H \tilde{\mathbf{s}}||_{\mathbf{\Sigma}^{-1}}} \right) \right\}. \tag{11}$$

To ease notation, we keep using $\mathcal{J}(\mathbf{A})$ to denote the estimate of the expected log likelihood obtained from the observations.

Estimation Algorithm

A simple way to solve problem (11) is to use the natural gradient algorithm [1]

$$\widehat{\mathbf{A}}_{i+1} = \widehat{\mathbf{A}}_i + \mu \widehat{\mathbf{A}}_i \widehat{\mathbf{A}}_i^H \nabla_{\widehat{\mathbf{A}}_i} \mathcal{J}(\widehat{\mathbf{A}}_i) \tag{12}$$

where μ is the step-size parameter, $\widehat{\mathbf{A}}_i$ is the mixing matrix estimate at the ith iteration and $\nabla_{\mathbf{A}}$ is the gradient operator with respect to $\mathbf{A}$. The gradient of the expected log-likelihood can be calculated using the chain rule

$$\begin{aligned} \nabla_{\mathbf{A}} \mathcal{J}(\mathbf{A}) &= -\frac{1}{K} \sum_{k=0}^{K-1} \frac{1}{E_{\tilde{\mathbf{s}}} e^{-||\mathbf{x}(k) - \mathbf{A}^H \tilde{\mathbf{s}}||_{\mathbf{\Sigma}^{-1}}}} \\ &\quad \times E_{\tilde{\mathbf{s}}} e^{-||\mathbf{x}(k) - \mathbf{A}^H \tilde{\mathbf{s}}||_{\mathbf{\Sigma}^{-1}}} \nabla_{\mathbf{A}} ||\mathbf{x}(k) - \mathbf{A}^H \tilde{\mathbf{s}}||_{\mathbf{\Sigma}^{-1}} \end{aligned} \tag{13}$$

and, resorting to the properties of the gradient operator [9],

$$\nabla_{\mathbf{A}} ||\mathbf{x}(k) - \mathbf{A}^H \tilde{\mathbf{s}}||_{\mathbf{\Sigma}^{-1}} = \tilde{\mathbf{s}} (\mathbf{A}^H \tilde{\mathbf{s}} - \mathbf{x}(k))^H \mathbf{\Sigma}^{-1}. \tag{14}$$

Substituting (14) into (13) and the resulting expression into (12) we arrive at

$$\widehat{\mathbf{A}}_{i+1} = \widehat{\mathbf{A}}_i + \mu \widehat{\mathbf{A}}_i \widehat{\mathbf{A}}_i^H - \frac{1}{K} \sum_{k=0}^{K-1} \frac{E_{\tilde{\mathbf{s}}} e^{-||\mathbf{x}(k) - \mathbf{A}^H \tilde{\mathbf{s}}||_{\mathbf{\Sigma}^{-1}}} \tilde{\mathbf{s}} (\mathbf{A}^H \tilde{\mathbf{s}} - \mathbf{x}(k))^H \mathbf{\Sigma}^{-1}}{E_{\tilde{\mathbf{s}}} e^{-||\mathbf{x}(k) - \mathbf{A}^H \tilde{\mathbf{s}}||_{\mathbf{\Sigma}^{-1}}}} \tag{15}$$

that serves as an optimization rule to solve (11).

Discussion

The practical implementation of algorithm (15) depends on the ability to compute the mathematical expectations $E_{\tilde{\mathbf{s}}}$. On the one hand, when the sources are discrete this calculation can be made analitically as a simple summation. Indeed, assuming there are P different values of the redundant source vector, $\tilde{\mathbf{s}}_i, \quad i = 1, ..., P$, with corresponding probability masses $f_{\tilde{\mathbf{s}}}(\tilde{\mathbf{s}}_i), \quad i = 1, ..., P$, then

$$E_{\tilde{\mathbf{s}}}\phi(\tilde{\mathbf{s}}) = \frac{1}{P}\sum_{i=1}^{P}\phi(\tilde{\mathbf{s}}_i)f_{\tilde{\mathbf{s}}}(\tilde{\mathbf{s}}_i)$$

where $\phi(\cdot)$ is an arbitrary function of the sources. On the other hand, when the sources are continuous it is not possible, in general, to find an analytic expression of the corresponding expectations and they must be approximated either in a deterministic (numerical integration) or stochastic (e.g., Monte Carlo integration) way [3]. For the simulation results presented in section 5 we have chosen to approximate $E_{\tilde{\mathbf{s}}}$ deterministically but Monte Carlo integration is also a valid approach and it may provide a reduced complexity for more practical implementations of algorithm (15).

The main drawback of likelihood maximization algorithms, however, arises from the existence of local maxima in the expected log-likelihood function, $\mathcal{J}(\mathbf{A})$. As a consequence, a gradient algorithm like (15) is not guaranteed to converge to the ML estimate of the mixing matrix when started from an arbitrary initial estimate $\widehat{\mathbf{A}}_0$. In the simplest case (instantaneous mixture of two sources, i.e., $N = 2$ and $m = 1$) simulations indicate that the choice of the initial estimate is not critical but for more complex situations a proper selection of the algorithm initial conditions is required. A sensible strategy is to resort to a suboptimum blind system identification algorithm (see [5] and references therein) to obtain the initial estimate. For instance, an algorithm devised for an instantaneous mixing model without noise could yield a rough estimate of $\mathbf{A}$ that would be useful to initialize (15).

It must also be noticed that the ML estimate of the mixing matrix, $\widehat{\mathbf{A}}$, may have its rows disordered with respect to the actual matrix $\mathbf{A}$. This ambiguity comes from the equality

$$\mathbf{x} = \mathbf{A}^H\tilde{\mathbf{s}} + \mathbf{n} = (\mathbf{PA})^H\mathbf{P}\tilde{\mathbf{s}} + \mathbf{n} \tag{16}$$

where $\mathbf{P}$ is an arbitrary $Nm \times Nm$ permutation matrix. As a consequence, $\mathbf{A}$ cannot be distinguished from $\mathbf{PA}$ using the observations $\mathbf{x}$ alone. The output $\widehat{\mathbf{A}}$ provided by algorithm (15), therefore, may be an estimate of $\mathbf{PA}$ and the corresponding sources may be also permuted as $\tilde{\mathbf{s}}_p = \mathbf{P}\tilde{\mathbf{s}}$. In order to perform optimum source separation we need to find a permutation matrix $\widehat{\mathbf{P}}$ such that $\widehat{\mathbf{P}}\tilde{\mathbf{s}}_p = \widehat{\mathbf{P}}\mathbf{P}\tilde{\mathbf{s}}$ has the same subvector structure as $\tilde{\mathbf{s}}$ in eq. (3) up to an $N \times N$ permutation of the source subvectors $\mathbf{s}(k-m+1), ..., \mathbf{s}(k)$ (the same permutation for the m subvectors). This is ensured by any $\widehat{\mathbf{P}}$ that satisfies the set of equations

$$\widehat{\mathbf{P}}\ \ E_{\tilde{\mathbf{s}}_p}\tilde{\mathbf{s}}_p(k)\tilde{\mathbf{s}}_p^H(k-l)\ \ \widehat{\mathbf{P}}^T = E_{\tilde{\mathbf{s}}}\tilde{\mathbf{s}}(k)\tilde{\mathbf{s}}^H(k-l), \ \ l = 1, ..., m-1. \tag{17}$$

The autocorrelation matrices $E_{\tilde{\mathbf{s}}}\tilde{\mathbf{s}}(k)\tilde{\mathbf{s}}^H(k-l)$ are known but those corresponding to $\tilde{\mathbf{s}}_p$ must be estimated. To do that, $\tilde{\mathbf{s}}_p(k)$ can be estimated using the observations $\mathbf{x}(k)$ and the disordered estimate $\widehat{\mathbf{A}}$. The simulations presented in section 5 correspond to discrete sources and, in this case, the permuted source vectors can be accurately estimated as

$$\widehat{\mathbf{s}}_p(k) = \arg \min_{\tilde{\mathbf{s}}} \left\{ \Omega(\tilde{\mathbf{s}}, k) = ||\mathbf{x}(k) - \widehat{\mathbf{A}}^H \tilde{\mathbf{s}}||_{\Sigma^{-1}} \right\}, \quad k = 0, ..., K-1. \quad (18)$$

Note that the metrics $\Omega(\tilde{\mathbf{s}}, k)$ must have been calculated for every possible value of $\tilde{\mathbf{s}}$ at the last iteration of algorithm (15). In the sequel, we assume that $\widehat{\mathbf{A}}$ stands for the *reordered* estimate $\widehat{\mathbf{P}}\widehat{\mathbf{A}}$.

Finally, note that, besides the permutation of $\widehat{\mathbf{A}}$, a sign ambiguity may exist for sources s_i with a symmetric p.d.f. since the random variables s_i and $-s_i$ have the same probability distribution. This may be a problem or not depending on the application.

ML ESTIMATION OF THE SOURCES

Since the linear mixture is convolutive a realization of the source vector $\mathbf{s}(k)$ contributes to m consecutive observation vectors $\mathbf{x}(k)$, ..., $\mathbf{x}(k+m-1)$. Therefore, observations containing information from $\mathbf{s}(k)$ overlap with those containing information from $\mathbf{s}(k)$ through $\mathbf{s}(k+m-1)$ and the whole frame of available obervations, $\mathbf{x}(0)$, ..., $\mathbf{x}(K-1)$, must be used in order to perform ML estimation of the sources. Stacking the K available observation vectors yields

$$\mathsf{x} = \mathsf{A}^H \mathsf{s} + \mathsf{n} \quad (19)$$

where

$$\mathsf{x} = \left[\mathbf{x}^T(0) \;\; \mathbf{x}^T(1) \;\; \cdots \;\; \mathbf{x}^T(K-1)\right]^T \quad (20)$$

$$\mathsf{s} = \left[\mathbf{s}^T(-m+1) \;\; \cdots \;\; \mathbf{s}^T(0) \;\; \cdots \;\; \mathbf{s}^T(K-1)\right]^T \quad (21)$$

$$\mathsf{n} = \left[\mathbf{n}^T(0) \;\; \cdots \;\; \mathbf{n}^T(K-1)\right]^T \quad (22)$$

$$\mathsf{A} = \begin{bmatrix} \mathbf{A}(m-1) & \mathbf{0} & \cdots & \mathbf{0} \\ \mathbf{A}(m-2) & \mathbf{A}(m-1) & \cdots & \mathbf{0} \\ \vdots & \mathbf{A}(m-2) & \ddots & \vdots \\ \mathbf{A}(0) & \vdots & \ddots & \mathbf{A}(m-1) \\ \vdots & \mathbf{A}(0) & \ddots & \mathbf{A}(m-2) \\ \vdots & \vdots & \ddots & \vdots \\ \mathbf{0} & \cdots & \cdots & \mathbf{A}(0) \end{bmatrix}. \quad (23)$$

A is the $N(K+m-1) \times LK$ *joint* mixing matrix constructed from $\mathbf{A}$, s is the $N(K+m-1) \times 1$ *stacked* source vector whereas x and n are the $LK \times 1$ stacked observation and noise vectors, respectively.

Let us denote the ML estimate of the joint mixing matrix as $\widehat{\mathsf{A}}$. This is a matrix constructed as in (23) but using $\widehat{\mathbf{A}}$ instead of $\mathbf{A}$. It is straightforward to show that the ML estimate of the joint source vector is the realization of s that minimizes the Euclidean distance between the joint observation vector and $\widehat{\mathsf{A}}^H\mathsf{s}$, i.e.,

$$\widehat{\mathsf{s}} = \arg \min_{\mathsf{s}} \left\{ \Omega_{joint}(\mathsf{s}) = ||\mathsf{x} - \widehat{\mathsf{A}}^H \mathsf{s}||_{\Sigma_{joint}^{-1}} \right\} \tag{24}$$

where

$$\Sigma_{joint} = \begin{bmatrix} \Sigma & \cdots & 0 \\ \vdots & \ddots & \vdots \\ 0 & \cdots & \Sigma \end{bmatrix}_{LK \times LK} .$$

When the sources are continuous the solution to (24) is simply the linear transformation of the joint observation vector

$$\widehat{\mathsf{s}} = \left(\widehat{\mathsf{A}} \Sigma_{joint}^{-1} \widehat{\mathsf{A}}^H \right)^{-1} \widehat{\mathsf{A}} \Sigma_{joint}^{-1} \mathsf{x} \tag{25}$$

and the Mean Square Error (MSE) between the actual joint source vector and its ML estimate can be written in terms of the actual and estimated joint mixing matrices as

$$\mathrm{MSE}(\mathsf{A}, \widehat{\mathsf{A}}) = \mathrm{Trc}\left[\mathsf{R}_{ss} - \mathsf{R}_{ss}\mathsf{Q}^H - \mathsf{Q}\mathsf{R}_{ss} + \mathsf{Q}\mathsf{R}_{ss}\mathsf{Q}^H + (\widehat{\mathsf{A}}\Sigma_{joint}^{-1}\widehat{\mathsf{A}}^H)^{-1} \right] \tag{26}$$

where $\mathrm{Trc}[\cdot]$ is the matrix trace operator, $\mathsf{R}_{ss} = E_{\mathsf{s}}\mathsf{s}\mathsf{s}^H$ is the autocorrelation of the joint source vector and $\mathsf{Q} = (\widehat{\mathsf{A}}\Sigma_{joint}^{-1}\widehat{\mathsf{A}}^H)^{-1}\widehat{\mathsf{A}}\Sigma_{joint}^{-1}\mathsf{A}^H$. When the estimated and actual joint mixing matrices match, $\mathsf{A} = \widehat{\mathsf{A}}$, the error reduces to

$$\mathrm{MSE}(\mathsf{A}) = \mathrm{Trc}\left[\left(\mathsf{A}\Sigma_{joint}^{-1}\mathsf{A}^H \right)^{-1} \right]. \tag{27}$$

For discrete sources the solution to problem (24) is a nonlinear transformation of the observations. The ML estimate $\widehat{\mathsf{s}}$ is obtained by an exhaustive search on the set of all possible different values of s. The computational complexity of this search may be prohibitively high but, in practice, it can be efficiently implemented by means of a constrained complexity dynamic programming algorithm [11] that yields an estimate of s with a *high* likelihood, although perhaps not the one with the *maximum* likelihood.

COMPUTER SIMULATIONS

In this section we present some computer simulation results that illustrate the validity of the proposed approach. The first computer experiment consists of the separation of two independent identically distributed (i.i.d.) real sources with a triangular p.d.f.

$$f_{s_i} = \begin{cases} 2(1 - s_i), & 0 \leq s_i \leq 1 \\ 0 & \text{otherwise} \end{cases} \quad i = 1, 2. \tag{28}$$

The complex ATWGN covariance matrix is $\mathbf{\Sigma} = \sigma_n^2 \mathbf{I}_L$, where $\mathbf{I}_L$ is the $L \times L$ identity matrix; the SNR in dB of the ith source is defined as

$$SNR_i = 10 \log_{10} \left(\frac{E_{s_i} |s_i|^2}{\sigma_n^2} \right)$$

and the linear mixture is instantaneous ($m = 1$)

$$\mathbf{A} = \begin{bmatrix} -0.43 - j0.33 & 0.13 + j0.19 & -1.15 - j0.59 & 1.19 + j0.14 \\ -1.66 - j0.17 & 0.29 + j0.73 & 1.19 + j2.18 & -0.04 - j0.11 \end{bmatrix} \quad (29)$$

where $j = \sqrt{-1}$. Figure 2 plots $\mathbf{MSE}(\mathsf{A}, \widehat{\mathsf{A}})$ and $\mathbf{MSE}(\mathsf{A})$ for several values of the SNR, that is equal for both sources (SNR $= SNR_1 = SNR_2$). It is apparent that the performance achieved using the ML estimate of the mixing matrix to separate the sources is practically optimum.

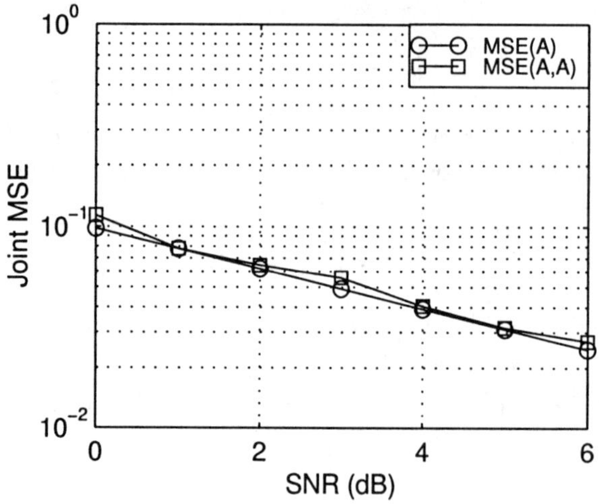

Figure 2: MSE$(\mathsf{A}, \widehat{\mathsf{A}})$ and MSE(A) versus SNR. The sources are two i.i.d. triangular random processes. The number of observation vectors available to estimate $\mathbf{A}$ was $K = 60$. Algorithm (15) was iterated 600 times with a step size $\mu = 0.1$. The expectations $E_{\bar{s}}$ were approximated by numerical integration with an integration step of 0.1. $\widehat{\mathbf{A}}_0$ was chosen randomly.

The second computer experiment consists of the separation of a convolutive mixture of two i.i.d. discrete uniform complex sources $s_i \in \{1, j, -1, -j\}$, $i = 1, 2$. The mixing system memory size is $m = 2$ and the corresponding mixing matrix is

$$\mathbf{A} = \begin{bmatrix} 0.26 + 0.47j & 0.11 + 0.04j & 0.18 + 0.26j & 0.36 - 0.19j \\ -0.88 - 0.45j & -0.16 + 0.34j & 0.51 + 0.53j & -0.29 - 0.36j \\ 0.08 + 0.14j & -0.06 + 0.03j & -0.59 - 0.45j & -0.65 - 0.27j \\ 0.08 + 0.14j & -0.06 + 0.03j & -0.59 - 0.45j & -0.65 - 0.27j \\ 0.66 - 0.14j & 0.06 - 0.04j & -0.05 + 0.48j & -0.07 + 0.13j \\ -0.34 + 0.01j & -0.61 - 0.71j & 0.15 - 0.01j & -0.07 + 0.57j \\ -0.28 + 0.01j & 0.52 + 0.51j & -0.03 - 0.01j & -0.04 + 0.07j \\ 0 & 0.38 - 0.18j & 0 & -0.24 + 0.55j \\ 0 & -0.08 - 0.43j & 0 & -0.87 - 0.39j \end{bmatrix}. \quad (30)$$

The covariance matrix of the ATWGN is $\mathbf{\Sigma} = \sigma_n^2 \mathbf{I}_L$ and the SNR is defined as in the previous experiment. Figure 3 plots the probability of making an error when estimating the sources for several values of the SNR (equal for both sources). We compare the error probability curves obtained when (a) the actual channel is used to estimate the sources and (b) the ML estimate $\widehat{\mathbf{A}}$ is employed. A constrained complexity trellis search algorithm [11] has been used to solve problem (24). This explains the slight loss in performance for low values of the SNR. For intermediate values, however, practically optimum separation is achieved.

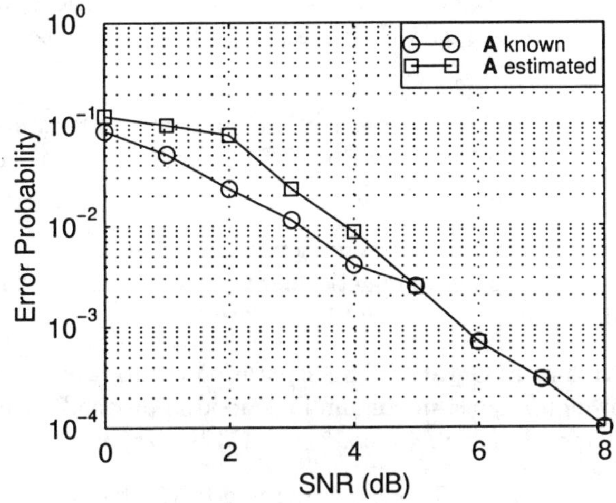

Figure 3: Error probability versus the SNR. The sources are two i.i.d. discrete uniform complex sources. The number of observation vectors available to estimate A was $K = 60$. Algorithm (15) was iterated 600 times with a step size $\mu = 0.08$.

CONCLUSIONS

This paper has introduced a new ML approach to the problem of BSS in Gaussian noise that is valid for both instantaneous and convolutive linear mixture models. The method proceeds in two steps: first, the ML estimate of the mixing system is computed and, afterwards, it is used to calculate the ML estimates of the sources. The proposed estimation algorithm requires knowledge of the sources joint p.d.f.

REFERENCES

[1] S. Amari, "Natural gradient works efficiently in learning," **Neural Comput.**, vol. 10, no. 2, pp. 251–276, 1998.

[2] A. J. Bell and T. J. Sejnowski, "An information-maximization approach to blind separation and blind deconvolution," **Neural Comput.**, vol. 7,

no. 6, pp. 1004–1034, 1995.

[3] O. Cappé, A. Doucet, M. Lavielle and E. Moulines, "Simulation-based methods for blind maximum-likelihood filter identification," **Signal Processing**, vol. 73, no. 1-2, pp. 3–25, January 1999.

[4] J. F. Cardoso, "Infomax and maximum likelihood for blind source separation," **IEEE Signal Processing Letters**, vol. 4, no. 4, pp. 112–114, April 1997.

[5] J. F. Cardoso, "Blind signal separation: Statistical principles," **Proceedings of the IEEE**, vol. 86, no. 10, pp. 2009–2025, October 1998.

[6] A. Hyvarinen, "Noisy independent component analysis by gaussian moments," **Proc. of the 1st Int. Workshop on ICA and Signal Sep.**, pp. 473–478, January 1999.

[7] C. Jutten and J. Herault, "Blind separation of sources i. an adaptive algorithm based on neuromimetic architecture," **Signal Processing**, vol. 24, no. 1, pp. 1–10, July 1991.

[8] D. T. Pham and P. Garat, "Blind separation of mixture of independent sources through a quasi-maximum likelihood approach," **IEEE Trans. Signal Processing**, vol. 45, no. 7, pp. 1712–1725, July 1997.

[9] F. K. Tai and Z. Y. Ting, **Generalized Multivariate Analysis**, Science Press Beijing and Springer-Verlag, 1990.

[10] J. K. Tugnait, "Adaptive blind separation of convolutive mixtures of independent linear signals," **Signal Processing**, vol. 73, no. 1-2, pp. 139–152, January 1999.

[11] S. Verdu, **Multiuser Detection**, Cambridge (UK): Cambdridge University Press, 1998.

A SUBOPTIMAL BAYESIAN EQUALIZER USING AN NONLINEAR MULTILAYER COMBINER

Sooyong Choi and Daesik Hong
Info. and Telecom. Lab., Dept. of Electronic Eng., Yonsei University,
Phone : +82-2-361-2871, Fax : +82-2-312-4887
E-mail : daesikh@sunlight.yonsei.ac.kr

Abstract. In order to reduce the complexity and enhance the performance of the Bayesian equalizer (REQ) using the radial basis function (RBF) network, a new equalizer (RNEQ) using the RBF network with a nonlinear multilayer combiner is proposed. The proposed RNEQ produces the output using nonlinear multilayer combiner. The RNEQ is applied to a digital communication system and a nonlinear magnetic storage system. From computer simulation results, the RNEQ with almost 70% reduced structure over the REQ shows nearly the same performance in a digital communication system. And in a nonlinear digital storage system the RNEQ is superior to the REQ. In addition, the RNEQ converges to lower mean squared error value than the other equalizers.

INTRODUCTION

The radial basis function (RBF) network as a channel equalizer, which is denoted as REQ, is equivalent to the optimal Bayesian equalization solution and shows good results in a stationary channel [1]. However, the complexity of the RBF network is a significant hindrance. Furthermore, the exact assessment of the parameters are required, which are the centers of the basis function in the RBF network and the estimation of the channel through which data symbol pass [1, 2]. Also the RBF network is very sensitive to the state of the initialization of the centers [1, 2]. These problems are the added goods for using RBF.

In this paper, in order to reduce the complexity and enhance the performance of the REQ, a new equalizer using an RBF network with a nonlinear multilayer combiner (RNEQ) is proposed. The conventinal REQ, which is shown in Fig. 1(a), consists of an input layer of source node, a single hidden layer of nonlinear processing units, and an output layer of linear weights. The input layer of the REQ has the same tapped-delay line (TDL) structure as the convetional equalizer. The hidden layer is composed of basis functions and

0-7803-5673-X/99/$10.00 © 1999 IEEE

the nonlinear transform of the input data is undergone in the hidden layer. The output is computed by linear combining the outputs of the RBF units. Therefore the classification of the nonlinearly transformed data is made by means of linear decision boundary.

In contrast with the REQ, the basic structure of the proposed RNEQ has four layers, which are an input layer, two hidden layers and one output layer as shown in Fig. 1(b). The input layer and the output layer of the RNEQ are the same structure as those of the REQ, respectively. Also the first hidden layer of the RNEQ is identical with the hidden layer of the REQ. The second hidden layer of the RNEQ, which is termed nonlinear multilayer combiner (NMC), is composed of multilayer perceptron (MLP). The NMC forms nonlinear decision boundaries. Therefore the output of the RNEQ is classified by nonlinear decision boundary which is established by the NMC. The use of the NMC preserves the performance and simplifies the structure of the REQ.

BAYESIAN EQUALIZER USING AN RBF NETWORK [1, 2]

Consider binary phase shift keying (BPSK) modulated symbols and a typical data transmission system as shown in Fig. 2. It is assumed that the transfer function of the channel is $H(z)$ and the number of taps of a channel is $n+1$. The REQ as the symbol-decision equalizer is shown in Fig. 1(a). When the order and the delay of the REQ are m and τ, respectively. There are $p = 2^{n+m}$ combinations of the channel input sequence

$$\mathbf{s}(t) = \{s(t)s(t-1)\cdots s(t-m+1-n)\}^T. \tag{1}$$

This produces p points of the noise-free channel output vector, $\hat{\mathbf{y}}(t) =$

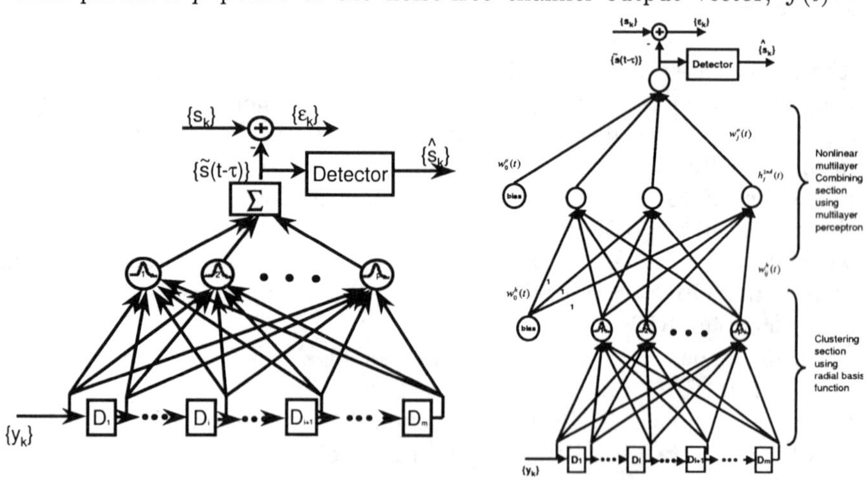

(a) (b)

Figure 1: Schematic diagram of (a) the REQ and (b) the RNEQ.

$[\hat{y}(t) \cdots \hat{y}(t-m+1)]^T$, which is a m-dimensional vector and one of the desired channel states. These p points can be partitioned into two classes according to the output of the REQ, $\tilde{s}(t-\tau)$,

$$\begin{cases} Y^+_{m,\tau} = \{\hat{\mathbf{y}}(t) | \tilde{s}(t-\tau) = 1\}, \\ Y^-_{m,\tau} = \{\hat{\mathbf{y}}(t) | \tilde{s}(t-\tau) = -1\} \end{cases} \tag{2}$$

and n^+ and n^- are the numbers of the states in $Y^+_{m,\tau}$ and $Y^-_{m,\tau}$, respectively.

Additive white Gaussian noise (AWGN) is added to channel output, which is the input vector of the REQ. Then the output of the REQ, $\tilde{s}(t-\tau)$, is

$$\tilde{s}(t-\tau) = \sum_{i=1}^{p} w_i \exp\left(- \frac{\|\mathbf{y} - \mathbf{c_i}\|^2}{\rho_i} \right), \tag{3}$$

where p is the number of the RBF units, $\mathbf{c_i}$ is the ith RBF center, ρ_i is the width of the ith RBF unit and w_i is the ith weight connecting the ith RBF unit to the output node for linear combining. After computing the output of the REQ and sampling at symbol duration T, we can obtain the optimal Bayesian equalization solution as follows:

$$\hat{s}(t-\tau) = sgn(s_k) = \begin{cases} -1, & \text{if } s_k < 0 \\ 1, & \text{if } s_k \geq 0. \end{cases} \tag{4}$$

The adaptation of the weights is achieved using the least mean square (LMS) algorithm [3]. If the weights are trained for a sufficient duration using the LMS algorithm, the weights converge to the optimal weights [2]. The training methods for the centers are generally two ways. One is the supervised k-means clustering procedure and the other is the unsupervised clustering [1, 2].

A SUBOPTIMAL BAYESIAN EQUALIZER USING AN RBF NETWORK WITH A NONLINEAR MULTILAYER COMBINER

Design Method of the RNEQ

In order to overcome several problems of the REQ, the RNEQ as shown in Fig. 1(b) is proposed. The input data of the RNEQ are nonlinearly transformed in the first hidden layer which consists of basis functions. The transformed data

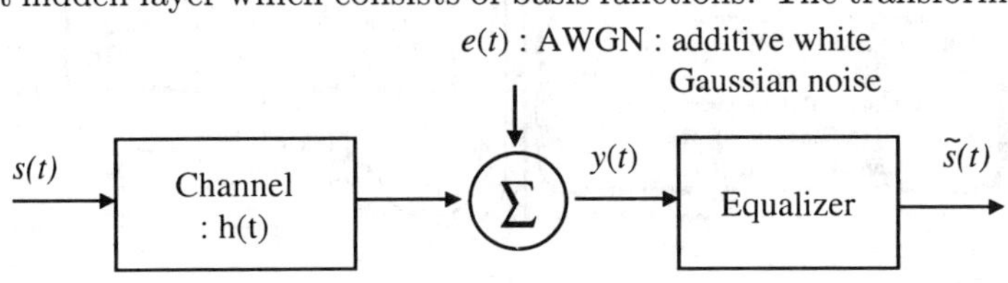

Figure 2: Schematic of data transmission system.

are classified into each corresponding class by nonlinear decision boundaries in the second hidden layer which is composed of perceptrons. The second hidden layer plays a role of a NMC that has the same structure as the MLP.

Through various experiments the design method of the RNEQ can be obtained as shown in Fig. 3. Two things are considered in order to design the RNEQ with the simpler structure than the REQ. The number of operations such as multiplication and addition for the RNEQ has to be smaller than that for the REQ and the number of the NMC nodes is used as small as possible for avoiding the problems of the MLP such as local minima, slow convergence speed, unpredictable training results and etc.

When an equalization problem of BPSK modulated symbols is considered, the channel output must be included one set of the two desired channel output sets as written in (2). At first the REQ is designed and the resultant equalizer is denoted as the (m, p) REQ, which has m inputs, p RBFs and 1 output. One NMC node is only added to the (m, p) REQ and then the $(m, p, 1)$ RNEQ is made, which has m inputs, p RBFs, 1 NMC node and 1 output.

The first simplified equalizer can be designed from the $(m, p, 1)$ RNEQ. Averaging the closest two centers among the centers of the same class, $\{Y^{+}_{m,\tau}\}$ or $\{Y^{-}_{m,\tau}\}$ and doubling the width of the basis function, and then one NMC node is added to the NMC layer. The $(m, p/2, 2)$ RNEQ that is a new simplified equalizer has been constructed. If the same process is repeated, the second simplified equalizer can be obtained. The second simplified equalizer is the $(m, p/4, 3)$ RNEQ. After the ith designing process, the ith simplified equalizer can be obtained with the structure of the $(m, p/2^i, i+1)$ RNEQ. This design process does not stop until the associated class of any original center among the p centers changes from one class to other class. In this paper this event is called a class change.

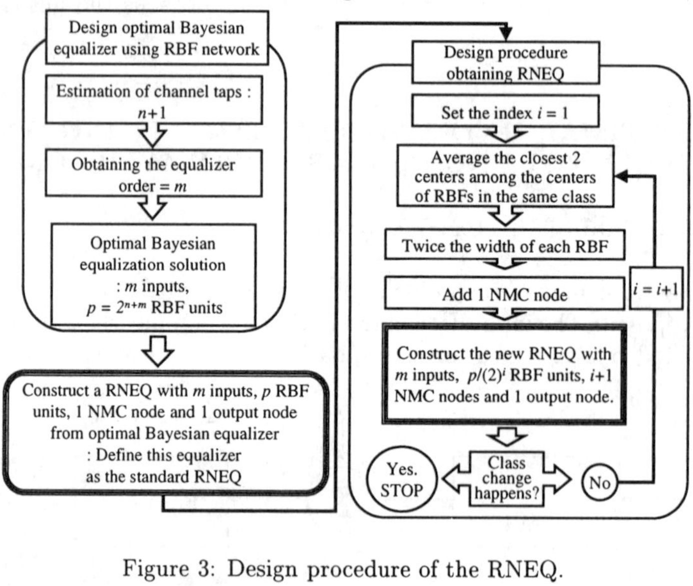

Figure 3: Design procedure of the RNEQ.

Training Method of the RNEQ

Consider the (m, p, l) RNEQ. When the input of the RNEQ is the noisy channel output vector that is the m-dimensional vector, the output of the ith RBF in the first hidden layer is

$$h_i^{1st}(t) = \exp\left(- \frac{\|\mathbf{y} - \mathbf{c_i}\|^2}{\rho_i} \right), \quad 1 \leq i \leq p, \tag{5}$$

where $\mathbf{c_i}$ is the m-dimensional center of the ith RBF and ρ_i is the width of the ith RBF unit. The output of the basis function that is multiplied by weights is the input of the NMC. The input of the jth NMC node is

$$net_j(t) = \sum_{i=1}^{p} w_{ij}^h(t) h_i^{1st}(t) + \theta_j^{1st}(t), \quad 1 \leq j \leq l, \tag{6}$$

where w_{ij}^h is the weight that connects the output of the ith basis function to the input of the jth NMC node and θ_j^{1st} is the input bias of the jth NMC node. Therefore the jth NMC node output in the second hidden layer is

$$h_j^{2nd}(t) = \frac{1 - \exp(net_j)}{1 + \exp(net_j)}, \quad 1 \leq j \leq l, \tag{7}$$

and the output of the RNEQ, $\tilde{s}(t - \tau)$, is

$$\tilde{s}(t - \tau) = \sum_{j=1}^{l} w_j^o(t) h_j^{2nd}(t) + \theta^o, \tag{8}$$

where w_j^o is the weight connecting the jth NMC node to the output node and θ^o is the bias of the output node. After computing the output of the RNEQ and sampling at symbol duration T, we can obtain the $\hat{s}(t - \tau)$ which is the estimation value of the desired value $s(t - \tau)$ by equation (4).

Both the supervised k-means clustering procedure and the unsupervised clustering algorithm can be used to train the centers. And the backpropagation [4] algorithm can be used for the updating method of the weights in the NMC. For the training the RNEQ, after settling down the states of the centers, the weights are trained using the backpropagation algorithm

Comparison of the RNEQ with the REQ

The complexities of each equalizer can be calculated in terms of the numbers of nodes and weights and the number of each operation. Consider the (m, p) REQ and the (m, p', l) RNEQ. The number of operations such as multiplication, division, addition and exp and the number of nodes and weights to be updated are shown in Table I. In Table I, it is assumed that the algorithm for updating centers is the supervised k-means clustering algorithm and the

Table 1: Comparisons of Complexities of the REQ with the RNEQ.

Parameters	(m,p) REQ	(m,p',l) RNEQ
No. of weights	p	$p' \times l + l$
No. of multiplication [DC]	$(m+3) \times p$	$(l+m+2) \times p' + 7l$
No. of division [DC]	p	$p' + l$
No. of addition [DC]	$(2m+1) \times p$	$(m+l) \times p' + 6l$
No. of exp [DC]	p	$p' + l$

training method for the weights uses the LMS algorithm for the REQ and the backpropagation algorithm for the RNEQ, respectively.

The condition that the computational complexity of the RNEQ is less than that of the REQ can be obtained from Table I. When $l = 2$ and $p' = p/2$, the conditions that the numbers of multiplication and addition for the RNEQ is smaller than those for the REQ are $(m+3)p > (m+4)p/2 + 14$ for multiplication and $(2m+1) > (m+2)p/2 + 18$ for addition, respectively. From these equations, we can obtain the following conditions:

$$\begin{cases} (m+2)p > 28, & \text{for multiplication} \\ mp > 8, & \text{for addition.} \end{cases} \tag{9}$$

When $l = 3$ and $p' = p/4$, through the same procedure the conditions are given by

$$\begin{cases} (3m+7)p > 84, & \text{for multiplication} \\ (7m+1) > 72, & \text{for addition.} \end{cases} \tag{10}$$

From (13), (14) and $p = 2^{n+m}$, if the input order is larger than 2, the $(m, p/2, 2)$ RNEQ and the $(m, p/4, 3)$ RNEQ can be designed with the simpler structure than the optimal Bayesian equalizer using the RBF network.

Fig. 4 shows decision boundaries for the optimal $(2,16)$ REQ, the $(2,8)$ REQ and the $(2,8,2)$ RNEQ. The decision boundary is shown in Fig. 4(a) when the optimal Bayesian Equalizer is adopted as a channel equalizer. Fig. 4(b) shows the decision boundary of the $(2,8)$ REQ. The decision boundary of the optimal $(2,16)$ REQ is nonlinear with a curved line as shown in Fig.

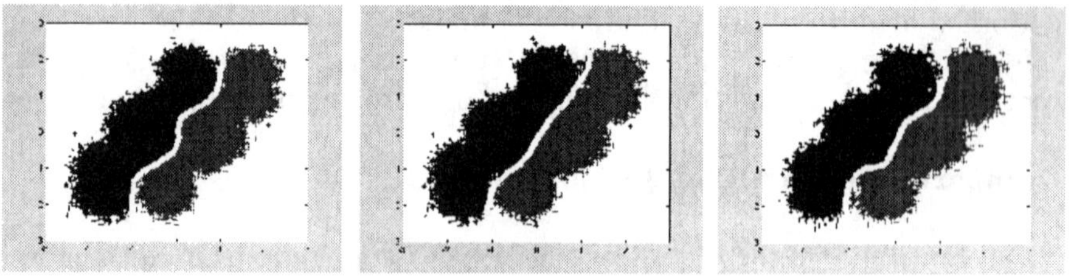

(a) (b) (c)

Figure 4: Decision regions formed by (a) optimal Bayesian equalizer, the (2,16) REQ, (b) the (2,8) REQ and (c) the (2,8,2) RNEQ when the channel $H(z) = 1.0 + 0.8z^{-1} + 0.5z^{-2}$ and SNR = 10dB.

4(a), however, the decision boundary of the $(2, 8)$ REQ is changed into almost a straight line as shown in Fig. 4(b), which is the reason for the performance degradation. Fig. 4(c) shows the decision boundary for the $(2, 8, 2)$ RNEQ. The performance degradation due to the reduction of the number of the centers does not happen, which results from using the NMC. In other words, the NMC compensates for the reduction of the centers. As shown in Fig. 4(c), the decision boundary is nonlinear with a curved line, which is almost the same as the decision boundary of the optimal $(2, 16)$ REQ.

SIMULATION RESULTS

The performance of the RNEQ is compared with the LE (linear equalizer) [5] and the REQ with respects to BER (bit error rate) and MSE (mean squared error). In simulations, the training of the weights for each equalizer does not stop until the iteration of the training become 100,000, after which there is little variation of MSE value. And there is no gain in BER by using momentum parameter, but the momentum parameter having the value of 0.5 is used in order to obtain the fast convergence speed of the RNEQ.

Simulation Results in a Digital Communication Channel

Consider the data transmission system as shown in Fig. 2 when a channel transfer function is given by $H(z) = 0.3482 + 0.8704z^{-1} + 0.3482z^{-2}$ [1] and data symbols are BPSK modulated. The optimal Bayesian equalizer is the $(4, 64)$ REQ with delay 1.

BER graphs for the REQ and RNEQs are shown in Fig. 5. The proposed RNEQs have two kinds of structures such as the $(4, 32, 2)$ RNEQ and the $(4, 16, 3)$ RNEQ. The RNEQs show the nearly same performance as the $(4, 64)$ REQ as shown in Fig. 5(a). Fig. 5(b) show that although the number of NMC nodes increases, there is no performance gain. In other words, there is no use adding more NMC nodes to the RNEQ in terms of BER. On the contrary, the RNEQ with 4 or 5 NMC nodes needs almost twice times as many training period as the RNEQ with 2 or 3 NMC nodes in our simulations.

The complexity for each equalizer is shown in Table II. The relative degree of the complexity for the RNEQ over the optimal REQ is defined as

$$\begin{aligned} &\textit{Degree of complexity } [DC]\ (\%) \\ &= 100 \times \frac{\text{the number of a operation for the RNEQ}}{\text{the number of a operation for the optimal REQ}}. \end{aligned} \tag{11}$$

The (4,32,2) RNEQ and the (4,16,3) RNEQ obtain the reduction of the complexity about 50% and about 70% over the (4,64) REQ, respectively. Therefore it is known from the simulation results in the communication system that the RNEQ having the simpler structure than the REQ can obtain the optimal Bayesian equalization solution. When the signal to noise ratio (SNR) is 18dB and the learning rate of each equalizer is 0.01, Fig. 6 illustrates the initial

Table 2: Comparisons of Complexities in a Digital Communication Channel.

Parameters	(4,64)REQ	(4,32,2)RNEQ	(4,16,3)RNEQ
No. of weights	64	66	51
No. of mul. [DC]	448 [100%]	260 [58.0%]	165 [36.8%]
No. of div. [DC]	64 [100%]	34 [53.1%]	19 [29.7%]
No. of add. [DC]	576 [100%]	204 [35.4%]	130 [22.6%]
No. of exp [DC]	64 [100%]	34 [53.1%]	19 [29.7%]

convergence characteristics of the weights for each equalizer. The LE shows the fastest equalizers among the four equalizers but has the worst MSE value. The equalizers using RBF network show the slower convergence rate than the LE. After settled down the states of the centers of the RBFs, however, the equalizers using the RBF network rapidly converge. The convergence speed of the (4,32,2) RNEQ is faster than that of the (4,16,3) between two RNEQs and the MSE value of the (4,32,2) RNEQ is lowest among four equalizers.

Simulation Results in a Nonlinear Digital Storage Channel

In high-density magnetic recording channel, if there are adjacent transitions, the effective readback signal is partially erased. Using the amount of partial erasure effect value α, the output signal of the magnetic recording channel can be expressed [6]. Reference [6] shows the amount of partial erasure effect value according to normalized recording density. Fig. 7(a) shows BER graphs for the LE, the REQ and the RNEQ in a digital storage system when recording density is 2.0 and α is 0.7. When BER $= 10^{-4}$, the REQ obtains about 4.8dB

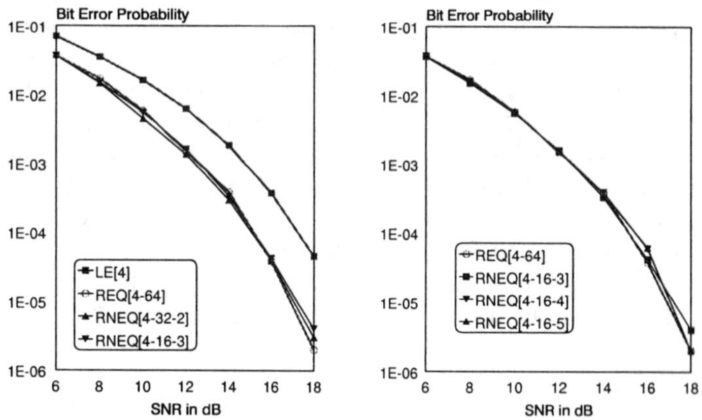

(a) (b)

Figure 5: BER graphs (a) for the RNEQ, the REQ and the LE, (b) for the RNEQs at varying the number of the NMC node and (c) for the RNEQ with 3 input order at varying the number of RBFs in a digital communication system. The optimal REQ is the (4-64) REQ.

performance gain over LE and the RNEQ provides about 5.0dB gain over the LE and about 0.5dB gain over the REQ. When there is nonlinear distortion, the RNEQ with simpler structure than the REQ produces better BER than the other equalizers. The reason is that the RNEQ contains nodes which have the strong capability of nonlinear mapping both in the first and the second hidden layer. Fig. 7(b) show BER graphs for the REQ and the RNEQ in the severe nonlinear data storage channel. It is assumed that recording density is 2.0 and the partial erasure effect, α, is 0.5. As the nonlinear distortion is severer, the performance gain of the RNEQ over the REQ increases.

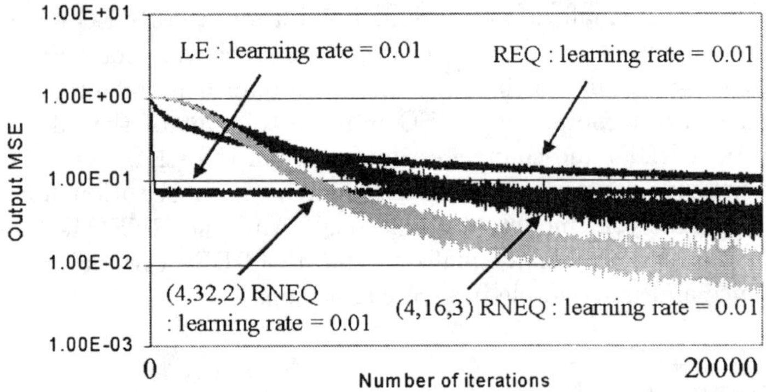

Figure 6: MSE graphs for the (4,16,3) RNEQ, the (4,32,2) RNEQ, the (4,64) REQ and the (4) LE in a digital communication system and SNR = 18dB.

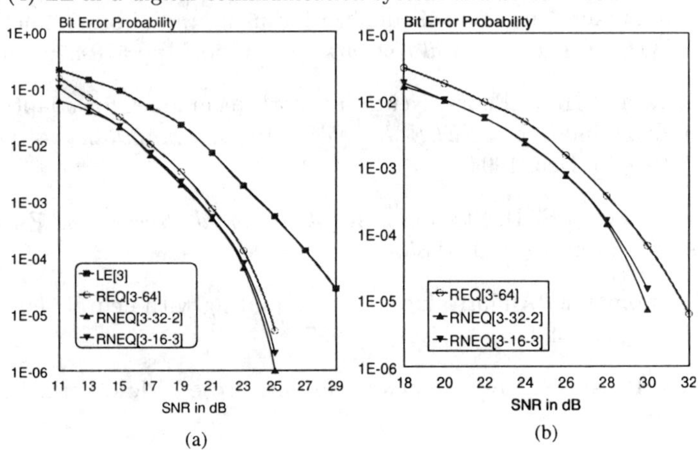

Figure 7: BER graphs for the RNEQ, the REQ and the LE in a nonlinear digital storage system, (a) when recording density is 2.0 and $\alpha = 0.7$, (b) when recording density is 2.0 and $\alpha = 0.5$.

CONCLUSIONS

In this paper, a new neural equalizer that is called as RNEQ is proposed in order to simplify the structure and enhance the performance of the optimal Bayesian equalizer using the RBF network. The RNEQ classify the data symbols by nonlinear decision boundaries which are formed by the nonlinear multilayer combiner. The RNEQ is applied to a digital communication system in which there is linear ISI, the RNEQ with almost 70% reduced structure over the optimal REQ shows nearly the same performance as the optimal Bayesian equalization solution. In addition, the RNEQ has not only the simpler structure but also better in performance than the REQ in a digital storage system in which the dominant impairment factor is nonlinear distortion. Furthermore, the RNEQ is more tolerant of the effects of the parameters of RBFs and channel estimation than the REQ.

In conclusion, the proposed RNEQ can be a substitution for optimal Bayesian equalizer in moderate conditions. And the RNEQ with simpler structure than the Bayesian equalizer using the RBF network are better in performance and more stable in severe situations.

REFERENCES

[1] S. Chen, B. Mulgrew, and P. M. Grant, "A clustering technique for digital communications channel equalization using radial basis functon networks," *IEEE Trans. Neural Networks*, vol.4, no.4, pp.570-579, Jul. 1993.

[2] U. Mitra and H. V. Poor, "Neural network techniques for adaptive multiuser demodulation," *IEEE J. Select. Areas. Commun.*, vol.12, no.8, pp.1460-1470, Dec. 1994.

[3] B. Widrow and S. D. Streams, *Adaptive Signal Processing*, Englewood Cliffs, NJ: Prentice Hall, 1985.

[4] R. P. Lippmann, "An introduction to computing with neural nets," *IEEE ASSP Magz.*, vol.4, pp.4-22, Apr. 1987.

[5] John G. Proakis, *Digital Communications*, 3rd ed. New York: McGraw-Hill, 1995.

[6] S. Choi, S. Ong, C. You and D. Hong, "A quadratic sigmoid neural equalizer for nonlinear digital magnetic recording channels," *IEEE Commun. Letters*, vol.2, no.9, pp.263-265, Sep. 1998.

APPLICATIONS OF SOAR TO MONOCHROMATIC IMAGE RESTORATION

Yusuf Öztürk and Hüseyin Abut
Department of Electrical and Computer Engineering,
San Diego State University,
San Diego, CA 92182-1309

ABSTRACT

This study proposes a novel autoassociator for associative storage of gray scale images. An autoassociator is a system capable of completing an incomplete pattern, when it is presented part of a learned pattern. Patterns in the proposed associative architecture are gray scale images with integer values. The proposed system, based on a recently developed technique called "SOAR", is superior to existing techniques using binary neural networks for associative storage of gray scale images [1] in both the computational complexity and the processing unit requirements. The proposed system is parallel in nature and can be implemented using a synchronous or asynchronous parallel system with no added complexity.

1. INTRODUCTION

Pattern completion is a sub-discipline in image processing dealing with the restoration of missing or incomplete information. In a pattern completion problem, a partial input is presented to a neural network which has been initially trained with a set of complete state vectors. The network is required to complete the missing part with some degree of uncertainty. Associative memories have been demonstrated as effective means of pattern completion in the literature, which allow the storage and recall of information by association with other semantically related information [1,3,5].

Several recent studies on image restoration using associative memories have utilized the ubiquitous Hopfield network [2], which is binary in nature and is not suitable for storing image pixels as noted in [1,3]. The existing methods in literature maps the image in to the binary domain by means of a slicing technique [1] and representing each pixel with a layer of neurons. In [1], the authors have decomposed images into s-gray levels and employed a simple gray code transformation for keeping the bit planes correlated. Every bit plane has been stored separately using a 2-D binary Hopfield network. Recall and restoration are performed on independent bit planes using the Hamming distance as a quality measure. Once restoration is achieved on every individual bit plane, the results are

0-7803-5673-X/99/$10.00 © 1999 IEEE

transformed back to the image space for display purposes. It is not difficult to see that the existing methods using binary Hopfield network for storage and associative recall require n-neural units for representing each pixel, where n is the number of gray levels, which inherently suffer from large number of neural units and memory.

Here we propose a novel solution for image storage and restoration of missing information using System of Associative Relationships (SOAR)[4] which is an extension of a recently developed technique called Multivalued Recursive Network (MAREN)[5]. MAREN is a novel integer-valued, recurrent, and nonlinear associative memory structure inspired from Hopfield and Tank's binary associative memory [2].

2. SYSTEM of ASSOCIATIVE RELATIONS (SOAR)

SOAR makes use of the pair-wise pixel associations. The pixels defined by a token over an image are used for determining associations among pixels. SOAR defines an analysis token as an irregular mask which determines the underlying structure of associations. The size and shape of the token depends on the target application. The token can be simply a 3x3 window centered at a pixel whose associations are explored or it can be of any. In this study the token will be the entire region of interest in the image.

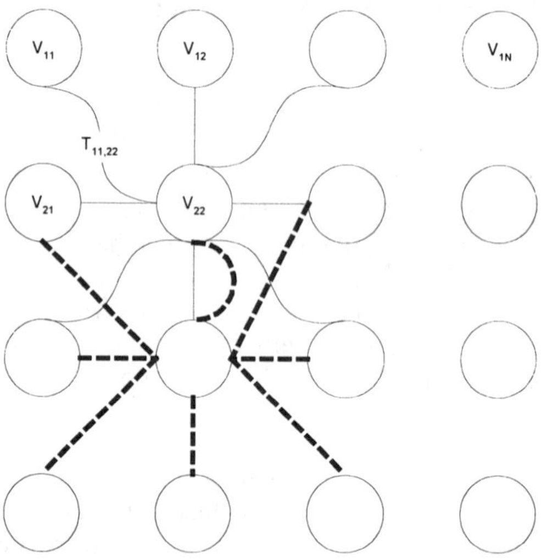

Figure 1: A 3x3 Neigborhood of Association

Once we decide on the token shape, we can then proceed to the definition of the SOAR architecture. To explore the associations among pixel elements defined by the token, each pixel is mapped to a processing node and the internal status of this processing node is set to the intensity of that particular pixel. The association map can change from one application to another depending on the neighborhood of association, which can be as small as a 3x3 window centered on every pixel or set to include an entire image. Here the neighborhood of 3x3 implies inter-pixel correlations outside a neighborhood of 3x3 are negligible or ignored. On the other hand a neighborhood of association covering entire image assumes every pixel is associated with every other. Sample association map with a 3x3 neighborhood of association is shown in figure 1. In Figure 1, the unidirectional links between processing elements –inherently those of pixels-- will store the associations between pixel pairs.

Assuming pixel intensities in a neighborhood of association are: { V_{ij}; i,j=1,2,….q }, SOAR encodes the larger/smaller information called the "interpixel connection strength" between pixels within the neighborhood as

$$\begin{aligned} &\mathrm{T(i,j;k,l)} = \mathrm{sgn}\{\mathrm{V}_{\mathrm{ij}}, \mathrm{V}_{\mathrm{kl}}) \quad \text{where } 1 \le i,j,k,l \le q \\ &\text{where} \quad \mathrm{V}_{\mathrm{ij}} \text{ and } \mathrm{V}_{\mathrm{kl}} \in \mathrm{R}_{\mathrm{n}} \quad \mathit{except}:(i,j) \ne (k,l) \end{aligned} \tag{1}$$

The associations between the pixels outside neighborhood of association are simply zero. The signum function here is the ternary sign function of discrete mathematics:

The prescription formulated in (1) can be extended to encode interpixel relationships within an image can be redefined for an ensemble of images as shown in (2).

$$T_P(i,j;k,l) = \sum_{p=1}^{P} \mathrm{sgn}\,(V_{ij}^{p} - V_{kl}^{p}) \tag{2}$$

As it can be seen from (2) the connection strength $\mathrm{T(i,j;k,l)}$ between the pixel at location *(i,j)* and the one at *(k,l)* increases if $V_{ij} > V_{kl}$; but it decreases if $V_{ij} < V_{kl}$, and remains unchanged if $V_{ij} = V_{kl}$. Furthermore, we would like to point out that the connection strengths are anti-symmetric, i.e., $T(k,l;i,j)=-T(i,j;k,l)$.

Assume an image is stored using the prescription given in (1) and a new image is presented to the system. The basic question will be the degree of similarity of this new image to the previously stored image. In [4] a new similarity measure has been proposed as part of the SOAR and it is given by (3)

$$\mathrm{E} = \sum_{\mathrm{i}}\sum_{\mathrm{j}}\sum_{\mathrm{k}}\sum_{\mathrm{l}} \mathrm{T(i,j;k,l)} * \mathrm{sgn}(\mathrm{V}_{\mathrm{ij}} - \mathrm{V}_{\mathrm{kl}}) \tag{3}$$

If we use (2) for the term $T(i,j;k,l)$ in (3) we can write:

$$\mathrm{E} = \sum_{\mathrm{i}}\sum_{\mathrm{j}}\sum_{\mathrm{k}}\sum_{\mathrm{l}} \mathrm{sgn}(\mathrm{V}_{\mathrm{ij}}^{1} - \mathrm{V}_{\mathrm{kl}}^{1}) * \mathrm{sgn}(\mathrm{V}_{\mathrm{ij}} - \mathrm{V}_{\mathrm{kl}}) \tag{4}$$

It is not difficult to deduce from (4) that E will attain its maximum value if the input image equal to the previously stored image, two patterns are correlated with a correlation coefficient of "1." E will have a value near zero if the input image is not correlated with any one of the stored images. The similarity measure proposed in SOAR is a *correlation indicator function* between two specific patterns.

Assume a set of P images are stored in the system using (2). Next, a new image is presented to the system. Here we will be interested in the similarity of this new image to any one of the images stored in the previous step. In [4] the similarity measure is extended to an ensemble of images stored in connections $T_P(i,j;k,l)$ via:

$$E = \sum_{p}\sum_{i}\sum_{j}\sum_{k}\sum_{l} \operatorname{sgn}(V_{ij}^{p} - V_{kl}^{p}) * \operatorname{sgn}(V_{ij} - V_{kl}) \tag{5}$$

If the test image is equal to one of the stored images, say p_l; the similarity function will be close to its theoretical maximum, which will be a function of neighborhood of association. If the neighborhood is the entire image than the maximum value of E will be

$$E_{max} = q^2 - q \quad \text{where } q \text{ is the number pixels in the image region of interest.} \tag{6}$$

In [4] it has been shown that this similarity measure is analogous to correlation in the case of single pattern comparison. It has also been shown that the similarity measure in (5) is highly correlated with the mean squared error. The similarity measure given in (5) is independent of the mean. That is, it does not matter how much bias exists in the data; the ordering between pixels for similar image regions will be the same.

3. LEARNING AND ASSOCIATIVE RECALL

Learning starts with the definition of the region of associations (ROA). In our experiments we have used a region of association covering the entire image. We have assumed a system capable of restoring the image in cases with, at most, ¼ of the original image data is missing. The learning process is very simple as defined in (2). The images to be stored are presented to the network and the associations are stored in the connections defined by region of association of Equation (2).

Now let us select one of the stored images. One such image is shown in Figure 3-a and suppose that some portions of this image has been removed as shown in Figure 3-b. The problem is to restore the missing region. The recall or restoration process will start with a presentation of the image under study -- image with missing segment -- to the system. We set the internal state of every processing unit V_{ij} equal to the corresponding pixel value. The internal state of processing units corresponding to the missing block will be set to zero.

The next step is to update the processing units as explained below until the system converges to a steady state, i.e., the energy of the system stops changing. The processing units (neurons) must be updated in such a way that the energy of the system never decreases. Let d be the difference between two consecutive discrete values of neuron (i,j) and let us define the energy changes when we update neuron (i,j) as

$$\Delta E_{ij}^{+} = E(V_{ij} + d) - E(V_{ij}) \tag{7}$$
$$\Delta E_{ij}^{-} = E(V_{ij}-d) - E(V_{ij})$$

Here ΔE_{ij}^{+} and ΔE_{ij}^{-} denote the difference in total E when we increment or decrement V_{ij}, respectively. Every neuron update must satisfy $\Delta E_{ij} \geq 0$. This can be achieved by simply incrementing and decrementing V_{ij} and observing the change in total energy.

If the energy increases by increasing or decreasing V_{ij}, we increment or decrement V_{ij}, accordingly. If both decrease the energy no change must be done. These situations are depicted in the Table-1.

ΔE_{ij}^{+}	ΔE_{ij}^{-}	V_{ij}
=0	=0	$V_{ij}\pm d$ or V_{ij}
>0	≤0	$V_{ij}+d$
≤0	>0	$V_{ij}-d$
<0	<0	V_{ij}
=0	<0	$V_{ij}+d$ or V_{ij}
<0	=0	$V_{ij}-d$ or V_{ij}
>0	>0	$V_{ij}\pm d$

Table 1: Neuron Update Rule

The update rule presented in Table-1 can be reformulated by

$$_{ij} = \begin{cases} V_{ij} + d & \text{if } k_{ij} > 0 \\ V_{ij} - d & \text{if } k_{ij} < 0 \\ V_{ij} & \text{if } k_{ij} = 0 \end{cases} \quad \text{Where } k_{ij} = \operatorname{sgn}(\Delta E_{ij}^{+}) - \operatorname{sgn}(\Delta E_{ij}^{-}) \tag{8}$$

Here $\Delta E_{ij} \geq 0$ is guaranteed. Updating V_{ij} changes only the terms including V_{ij} in (5). Using this we can rewrite (7) as follows:

$$\Delta E_{ij}^{+} = \sum T(i,j;k,l)\ [\operatorname{sgn}(V_{ij}+d-V_{kl}) - \operatorname{sgn}(V_{ij}-V_{kl})]$$
$$\Delta E_{ij}^{-} = \sum -T(i,j;k,l)\ [\operatorname{sgn}(V_{ij}-d-V_{kl}) - \operatorname{sgn}(V_{ij}-V_{kl})] \tag{9}$$

The recall process stops when the energy of the system does not change more than a threshold. A decision on mapping the qualitative measure to a quantitative measure is yet to be made. Figure 2-a shows the energy of the system during the recall process as a function of individual neuron update count (x-axis * 10^3). Whereas, Figure 2-b depicts the change in the mean square error (MSE) between the original image and the restored image during the restoration process as a function of individual neuron update count as above. As it can be seen from these plots, the MSE does not change after 2000 individual updates and can be used as a stopping criterion.

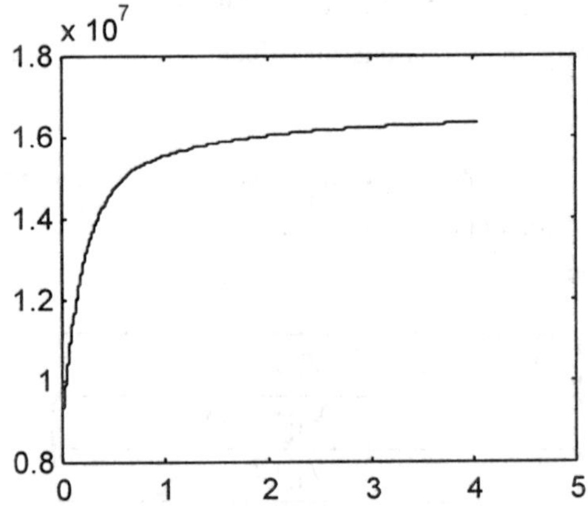

(a) Energy of the system

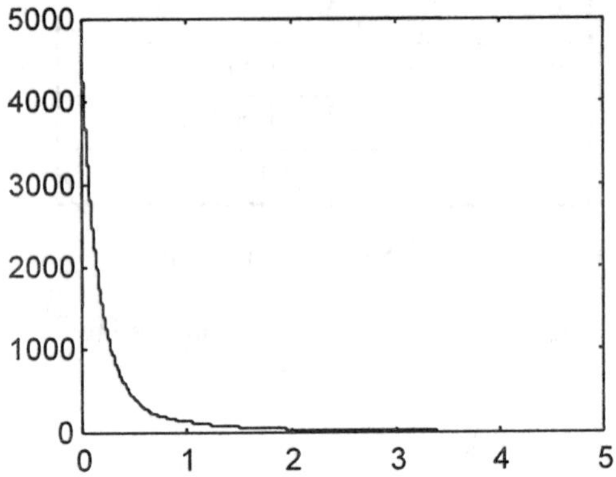

(b) Mean square error

Figure 2: The dynamics of energy surface of the system and mean square error during restoration

In Figure 3, we present our experimental findings from a sample image restoration application. The original image in Figure 3-a is stored in the proposed system using the association prescription given in (2). A large portion of the original image is removed to yield the test pattern of Figure 3-b is. This image with missing data is fed to our system as a starting point. At the end of the process, the system has converged to the result in Figure 3-c. As it can be clearly seen from the results presented in this set of pictures, the system has restored the missing information remarkably well.

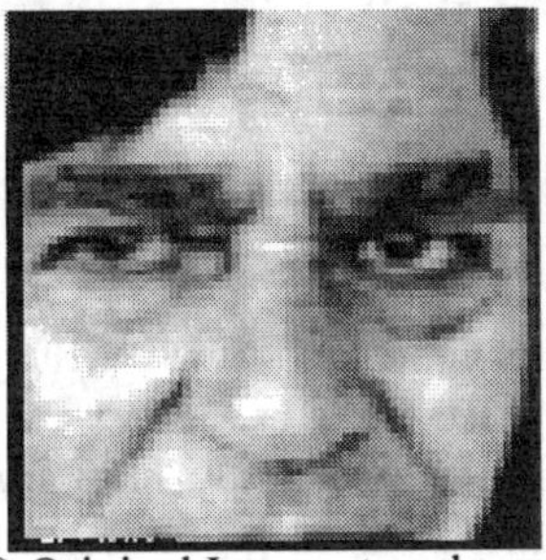

(a) Original Image stored

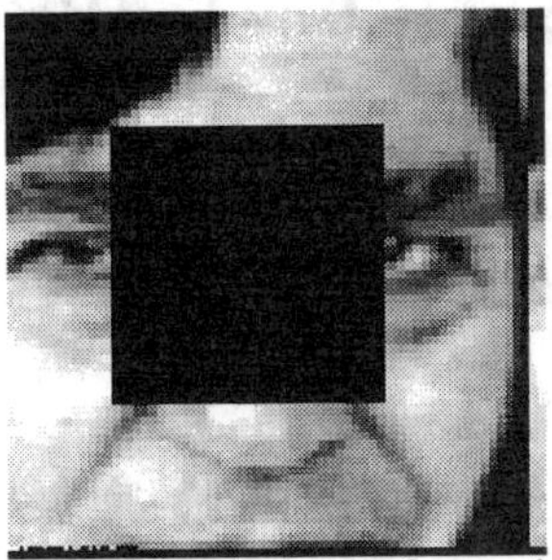

(b) Image with missing information

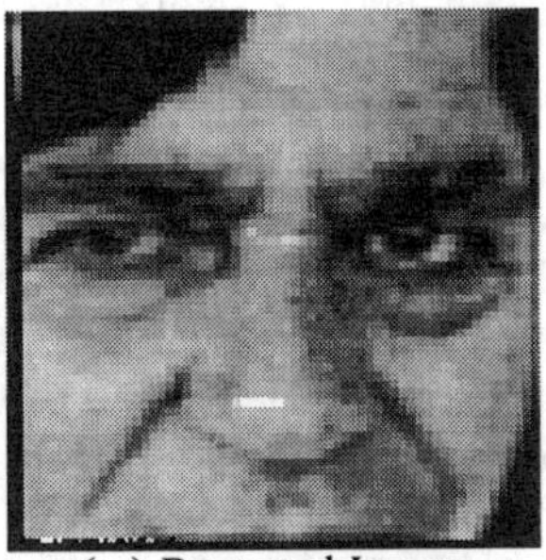

(c) Restored Image

Figure 3: A sample restoration by using the proposed technique

4. CONCLUSION

The novel autoassociative memory discussed here has successfully restored missing parts of a monochromatic image. The system has been tested on several images and resulted identical results.

5. REFERENCES

[1] S. Sinha, S. Biswas, "Associative Memory for Gray-Scale Images", IEEE International conference on Image Processing, October 26-29, 1997, Santa Barbara, CA.

[2] J.J. Hopfield, "Neural Networks and Physical Systems with Emergent Collective Computational Abilities", Proceedings of Nat. Ac. of Sci. USA, Vol. 79, April 1982.

[3] R. P. Lippmann, "An Introduction to computing with neural nets", **IEEE ASSP Mag.**, vol.4, pp. 4-22, 1987.

[4] Y. Öztürk and H. Abut, "A Novel Similarity Measure for Compression and Classification," *ICASSP-98 Proceedings,* May 12-15, 1998, Seattle, WA.

[5] M. Erdem and Y. Öztürk, "A New Family of Multivalued Networks ", **Journal of Neural Networks,** Vol. 9, No.6, pp. 979-989, August 1996.

EFFICIENT APPROXIMATION OF A NEURAL FILTER FOR QUANTUM NOISE REMOVAL IN X-RAY IMAGES

Kenji SUZUKI†, Isao HORIBA†, and Noboru SUGIE‡
†Faculty of Information Science and Technology
Aichi Prefectural University, Nagakute, Aichi, 480-1198 Japan
‡Faculty of Science and Technology
Meijo University, Nagoya 468-8502, Japan
E-mail: k-suzuki@ist.aichi-pu.ac.jp

Abstract. An efficient filter approximating the neural filter (NF) trained to remove quantum noise from X-ray images has been realized. A novel analysis method is proposed for making the characteristics of the trained NF clear. It analyzes a nonlinear system with plural inputs by using its outputs when the specific input signals are fed to it. The experimental results have demonstrated that the approximate filter, which is realized by using the results of the analysis, is sufficient for approximation of the trained NF, and efficient at computational cost.

INTRODUCTION

Nonlinear filters using a neural network (NN) model, called the neural filters (NFs), have been proposed [1, 2, 9, 10, 11, 12, 13]. This class of NFs unifies linear and nonlinear filters, such as finite impulse response filters, microstatistic filters, generalized weighted order statistic filters, an epsilon filter, and generalized stack filters [1, 2, 11, 12, 13]. Moreover, it has been proved that any continuous mapping can be approximately realized by multilayer NNs [4, 6]. These studies have made the capability of the NFs clear. However, the characteristics of the NFs after training has not yet been understood explicitly, because to analyze them is difficult due to the complicated structure composed of a lot of units with non-linear characteristics. For practical use, particularly in the critical applications such as these to medical systems, it is quite important to make their characteristics clear. Moreover, since the computational cost of the NF is exceedingly high, the reduction is a serious issue for making the NF practicable. In applications of filters to medical X-ray images, it is required to process 30 images, consisting of 1024×1024 pixels×10 bit, per a second. Therefore, how to analyze the NF after training and to reduce the computational cost have remained serious issues.

0-7803-5673-X/99/$10.00 © 1999 IEEE

In this paper, we propose a novel method for analyzing the NFs after training to make the characteristics clear and realizing its efficient approximate filter using the results of the analysis. One of basic functions of image-processing filters is a noise reduction. Noises in images are classified into two broad types: one is signal-dependent noise; the other is signal-independent noise. The quantum noise is a representative of the former one, and is generally observed in the photon-limited images such as the images obtained in darkness, those obtained by an infrared camera, X-ray images, and so on. We narrow down the issues to the NF trained to remove quantum noise from X-ray images, because the analysis of the NFs trained to solve various problems is difficult. By experiments on the NF trained with medical X-ray image sequences, the efficiency of both the proposed analysis method and the approximate filter are shown.

NEURAL FILTER FOR QUANTUM NOISE REMOVAL IN X-RAY IMAGE SEQUENCES

Architecture of the Neural Filter [9]

The NF consists of a multilayer NN in which the activation functions of the units in the input, hidden, and output layers are an identity function, a sigmoid function, and an identity function, respectively. The inputs to the NF are an object pixel value $g(x, y, t)$ and spatially/spatiotemporally neighboring pixel values. We explain a three-layered spatiotemporal NF as an example. An output from the NF is represented by

$$f(x, y, t) = NN(\boldsymbol{I}_{x,y,t}) \tag{1}$$

where

$$\boldsymbol{I}_{x,y,t} = \{g(x-i, y-j, t-k)|i, j, k \in R\} \tag{2}$$

is the input vector to the NF, $NN(\mathbf{I})$ denotes an output of the multilayer NN, and R denotes the spatiotemporal support of the filter. The NF is trained by the back-propagation algorithm [8]. Through training, it becomes the optimal filter with desired characteristics by presenting an input image sequence together with a desired teaching image sequence. The error to be minimized by the training is defined as

$$E = \sum_{p} (T_C^p - O^p)^2 \tag{3}$$

where p is a pattern number, T_C^p is the pth pattern in the teaching image, and O^p is the pth pattern in an output image.

Making Input Images and Teaching Images [9]

Medical X-ray images are treated in this paper, because the analysis of new technologies such as the NF is quite important particularly in applications to medical systems. The goals of improving the image quality of medical X-ray image sequences are; (1) removing the quantum noise from low-dose X-ray images to reduce X-ray exposure to patients and (2) enhancing the edges in diagnostic regions in the images to discern them clearly. The quantum noise in X-ray images decreases with increasing X-ray exposure. Using this property, the input and teaching images for training can be made from high-dose X-ray images. Since the X-ray quantum noise is originated from a signal-dependent Poisson-distributed noise source [7], the low-dose X-ray images $g_L(x, y, t)$, the input images to the NF, can be synthesized from the high-dose X-ray image with negligible quantum noise $g_H(x, y, t)$ as the following equation:

$$g_L(x, y, t) = g_H(x, y, t) + X_N(\sigma) \tag{4}$$

where

$$\sigma = k_N \sqrt{g_H(x, y, t)} \tag{5}$$

is the standard deviation, k_N is a parameter related to radiation dosage, and $X_N(\sigma)$ is white Gaussian noise.

The teaching image for the NF, $T_C(x, y, t)$, is synthesized from the high-dose X-ray image by performing a high-pass filtering, the window function of which is expressed as

$$P_H(u, v) = \begin{cases} \frac{\frac{3}{2} + \frac{1}{2}\cos(\pi \frac{f_C - \sqrt{u^2+v^2}}{f_C})}{2} & (0 \leq \left|\sqrt{u^2 + v^2}\right| \leq f_C) \\ & (f_C \leq \left|\sqrt{u^2 + v^2}\right| \leq \frac{1}{2}) \end{cases} \tag{6}$$

where u is the normalized horizontal spatial frequency, v is the normalized vertical spatial frequency, and f_C is a parameter for determining the pass band. We got the parameter k_N of 0.0024 by the actual measurement of noise in the low-dose X-ray images, and used f_C of 1/16.

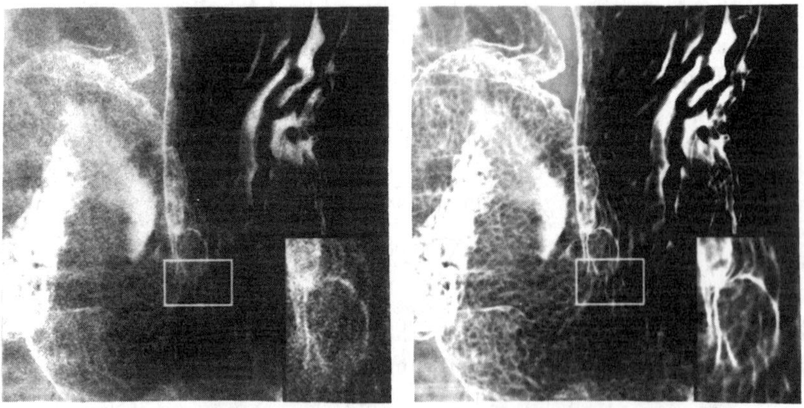

(a) Input image $g_L(x, y, t_0)$. (b) Teaching image $T_C(x, y, t_0)$.

Figure 1: Images used for training.

Training the Neural Filter

The spatiotemporal support of the NF consists of 5 pixels $(x, y-1), (x-1, y), (x, y), (x+1, y)$, and $(x, y+1)$ in each of 5 consecutive frames. The number of units in the input, hidden, and output layers are 25, 20, and 1, respectively (hereafter referred to as 25–20–1). The noisy input image $g_L(x, y, t_0)$ used for training, a radiograph of the stomach consisting of 512×512 pixels, and the teaching image $T_C(x, y, t_0)$ are shown in Figures 1(a), (b), respectively. Here, t_0 denotes the time of the current frame. This kind of images is used for medical examination of stomach cancer. By training with these images, the NF acquires the function of noise reduction with preserved edges. The NF was trained 80,000 epochs with the training sets in the regions indicated by the white rectangular frames in Figure 1. The training converged with the error E of 0.024.

ANALYSIS METHOD FOR THE NEURAL FILTER AFTER TRAINING

If a system is a linear system, it can be analyzed easily and clearly by using its impulse response. However, it is difficult to analyze the NFs after training mathematically because of their non-linearity. We narrow down the issues to the NF trained to remove quantum noise form X-ray images. Since the trained NF can be treated as a nonlinear deterministic system, the proposed method analyzes it by using its outputs when the input signals for analysis are fed to it.

Since the quantum noise is signal-dependent noise, the NF trained to remove it should be realized with the characteristics depending on gray levels in the images. The specific input for analysis is shown in Figure 2. The ramp signal with respect to a parameter s is fed to a target-input unit, and each base signal D, a constant value with respect to the parameter s, is fed to the other input units. The input vector to the NF is defined as the following equation:

$$\forall q, D \quad \boldsymbol{I}_{q,D} = \{I_{1,D}, I_{2,D}, \cdots, I_{q,D}, \cdots, I_{N_I,D}\} \tag{7}$$

where

$$\forall m \quad I_{m,D}(s) = \begin{cases} s & if\ m = q \\ D & if\ m \neq q \end{cases} \quad (0 \leq s \leq 1), \tag{8}$$

N_I is the number of input units of the NF, m is the input unit number, s is a parameter, and q is the target-input unit number. An output from the NF to which the input signals for analysis are fed, can be expressed as

$$O_{q,D}(s) = NN(\boldsymbol{I}_{q,D}(s)). \tag{9}$$

Equation (9) represents an output response of a system when the base signal D, i.e. an average input gray level to the system, is fed. By approximating the output responses to polynomial expressions, an unknown system can be analyzed.

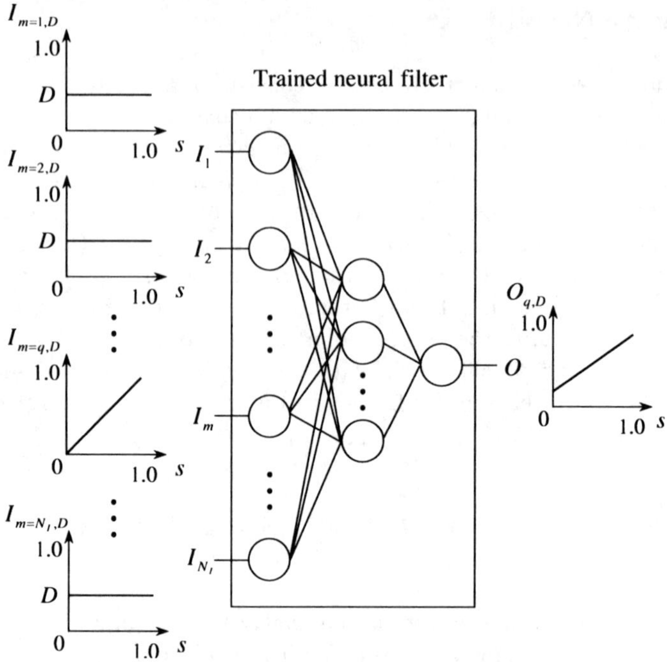

Figure 2: Input signals for analysis.

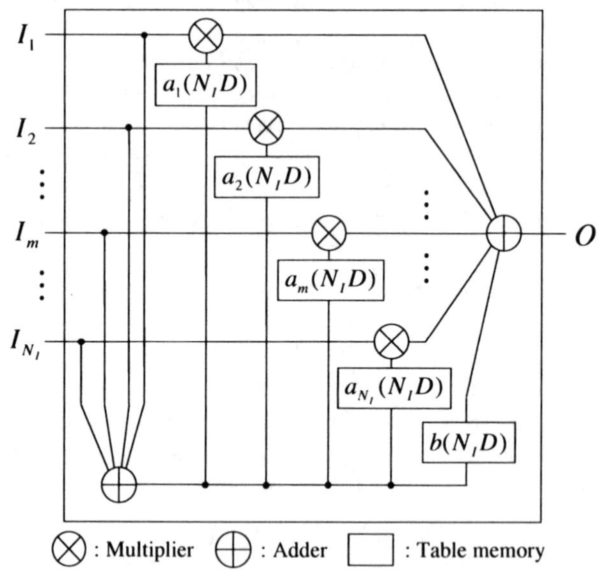

Figure 3: Hardware architecture of the approximate filter.

ANALYZING THE TRAINED NEURAL FILTER

Experimental Results

The proposed analysis method was applied to the trained NF. Since the output responses of the trained NF, to which the input signals for analysis are fed, are nearly straight, they can be approximated by linear functions. By this approximation as well as approximating the base signal D to the summation of the inputs, the trained NF can be represented by the following filter model:

$$O = \sum_{m=1}^{N_I} a_m(D) \cdot I_m + b(D) \tag{10}$$

where $a_m(.), b(.)$ are filter coefficients. The approximate filter model is shown in Figure 3. $a_m(.), b(.)$ are implemented as look-up tables, with the summation of the inputs, $N_I \cdot D$, as an index. The hardware architecture is very simple: it consists of multipliers, adders, and table memories. Using the approximate filter, the cost of hardware of the NF is reduced to about 1/20.

The filter coefficients of the approximate filter, which are obtained from the output responses of the trained NF, are shown in Figure 4. Here, T denotes a time interval between frames. The filter coefficients at the center pixels (c), i.e. those at (x,t), are relatively greater than those at the other pixels are. The filter coefficients decrease in proportion to the distance from the current frame t_0. This characteristic is due to the training to process time-varying images. They are minus in the frames $t_0 - 3T$ and $t_0 - 4T$. This shows that the edge-enhancement is mainly performed in these frames. The filter coefficients at the middle of D are relatively greater than the others are. This is caused by the gray level histogram in the training region: the frequency around the middle of gray levels is relatively high, as Figure 5 shows.

Quantitative Evaluation of Approximate Precision

An output image of the NF and that of its approximate filter are shown in Figures 6(a) and (b), respectively. At first sight, the difference between them can not be distinguished. So that, the image quality was evaluated quantitatively using the improvement in signal-to-noise ratio (ISNR) [3, 5]. The ISNR is defined as:

$$ISNR(t) = 10\log_{10} \frac{\sum_{x,y \in R_E} \{T_C(x,y,t) - g(x,y,t)\}^2}{\sum_{x,y \in R_E} \{T_C(x,y,t) - f(x,y,t)\}^2} \tag{11}$$

where R_E is a region for evaluation. The results are shown in Figure 7. The mean for the signal-to-nose ratio of the input image sequence was 11.6 dB. The frame number 10 is used in the training. The ISNR of the approximate filter is nearly equal to that of the NF in not only the trained frame but also

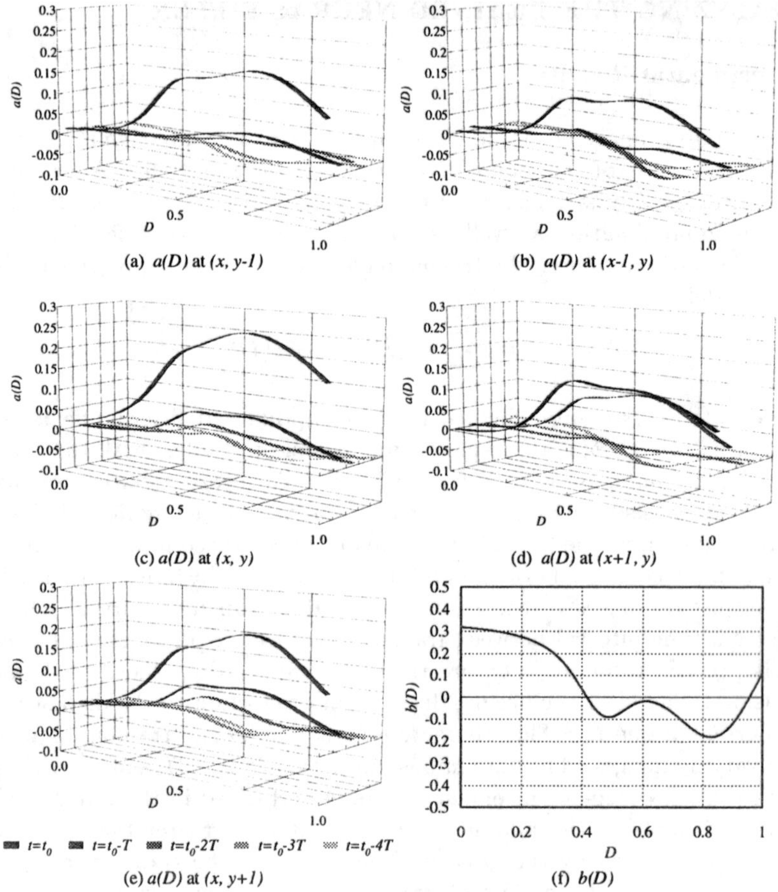

Figure 4: Filter coefficients of the approximate filter.

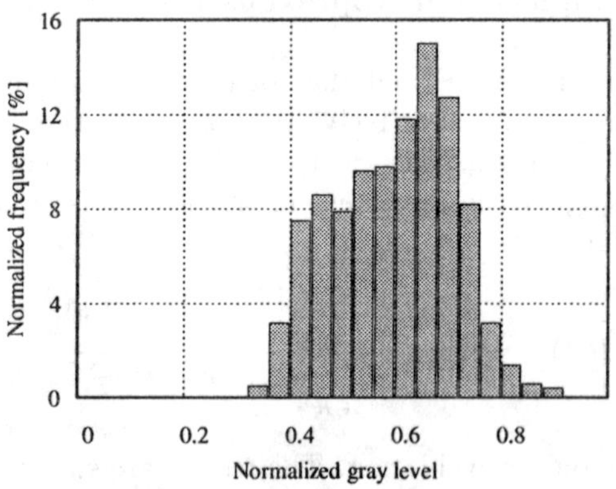

Figure 5: Gray level histogram in the training region.

non-trained frames. The means for the ISNRs of the NF and the approximate filter are 1.866 dB and 1.866 dB, respectively. This shows that the approximate filter is a good approximation of the NF.

In order to evaluate the versatility, two experiments were performed. First, we experimented on the approximate filter by expanding the support of the filter. Table 1 shows quantitative evaluation of approximate precision in terms of the error between output images of the NF and those of the approximate filter. Both the mean absolute error (MAE) and the root mean square error (RMSE), are also small in this case. This shows that the proposed analysis method work well in case of expanding the support. Next, we applied the approximate filter to the other kind of images. Two kinds of images exist in the medical X-ray images: one is a double contrast radiograph of a stomach, which was used in the training; the other is an angiograph of a coronary artery of a heart. The result with the angiographs of a coronary artery is shown in Table 1. The errors in all cases are quite small. It shows that the approximate filter has enough versatility to apply it to medical systems.

Table 1: Quantitative evaluation of the approximate precision between the neural filter and the approximate filter

Structure of the NF	Image	MAE [%]	RMSE [%]
25–20–1	DCR	0.278	0.453
100–20–1	DCR	0.187	0.255
	ACA	0.200	0.260

DCR: Double contrast radiograph,
ACA: Angiograph of a coronary artery.

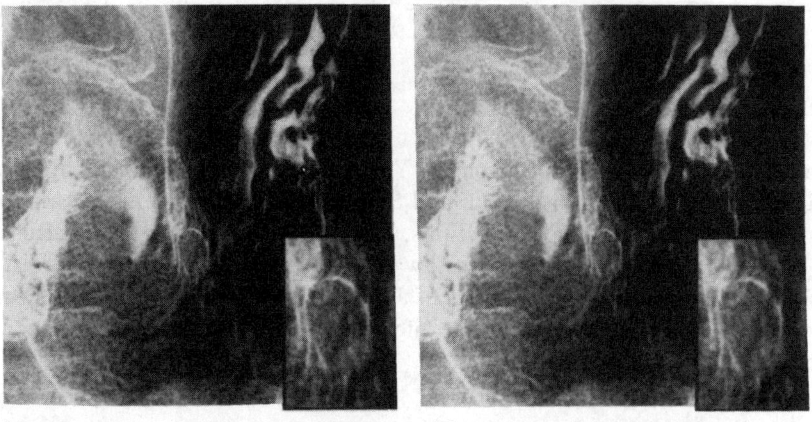

(a) Output image of the neural filter. (b) Output image of the approximate filter.

Figure 6: Comparison of the image quality of an output image of the neural filter with that of the approximate filter.

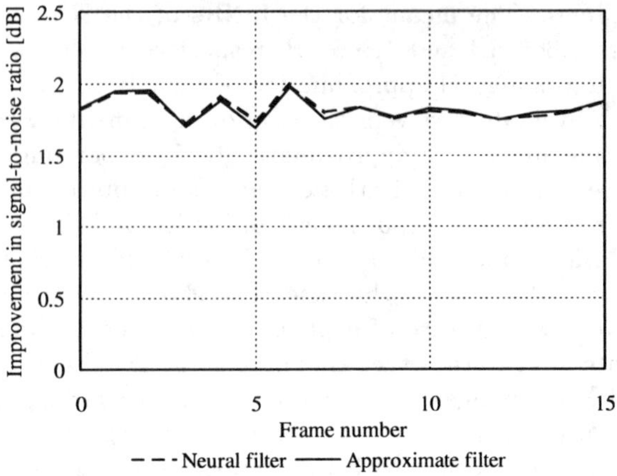

Figure 7: Comparison of the performance of the approximate filter with that of the neural filter.

CONCLUSION

In this paper, we have presented a novel method for analyzing the NF trained to remove quantum noise from X-ray images. The proposed method proved to be useful to analyze an unknown nonlinear system with plural inputs such as the trained NF. The results of the analysis of the NF trained with medical X-ray image sequences have made the characteristics clear, leading to its efficient approximate filter. The experimental results have demonstrated that the approximate filter, constructed of simple hardware, is sufficient for approximation of the trained NFs, and efficient at computational cost. It is also shown that the approximate filter has enough versatility to apply it to medical systems.

Acknowledgement

The authors wish to thank Shigenobu Yanaka, Kouichi Koike, Ken Ishikawa, Shigeyuki Ikeda, Katsumi Suzuki, and Shuji Sugeno of Hitachi Medical Corporation for their valuable suggestions, Michio Nanki of Chubu Rosai Hospital for evaluating images. This work was supported by the Ministry of Education, Science, Sports and Culture of Japan under grant-in-aid for quantum information theoretical approach to life science and grant-in-aid for encouragement of young scientists 10750302.

REFERENCES

[1] K. Arakawa and H. Harashima, "A nonlinear digital filter using multilayerd neural networks," in **Proc. IEEE Int. Conf. Communications '90**, 1990, vol. 2, pp. 424–428.

[2] K. Arakawa, K. Ymamakawa and M. Koyama, "Neural net-type filter with optimization of its nonlinear function," **Trans. IEICE-A**, vol. J78-A, no. 2, pp. 151–160, Feb. 1995.

[3] M. R. Banham and A. K. Katsaggelos, "Digital image restoration," **IEEE Signal Processing Magazine**, vol. 14, no. 2, pp. 24–41, Mar. 1997.

[4] A. R. Barron, "Universal approximation bounds for superpositions of a sigmoidal function," **IEEE Trans. Information Theory**, vol. 39, no. 3, pp. 930–945, Mar. 1993.

[5] J. C. Brailean, R. P. Kleihorst, S. Efstratiadis, A. K. Katsaggelos and R. L. Lagendijk, "Noise reduction filters for dynamic image sequences: A review," **Proc. IEEE**, vol. 83, no. 9, pp. 1270–1291, Sep. 1995.

[6] K. Funahashi, "On the approximate realization of continuous mappings by neural networks," **Neural Networks**, vol. 2, pp. 183–192, 1989.

[7] A. Macovski, **Medical Imaging Systems**, NJ: Englewood Cliffs, PrenticeHall, 1983.

[8] D. E. Rumelhart, G. E. Hinton and R. J. Williams, **Learning internal representations by error propagation**, MA: M.I.T. Press, vol. 1 of **Parallel Distributed Processing**, chap. 8, pp. 318–362, 1986.

[9] K. Suzuki, I. Horiba, N. Sugie and S. Ikeda, "Improvement of image quality of x-ray fluoroscopy using spatiotemporal neural filter which learns noise reduction, edge enhancement and motion compensation," in **Proc. Int. Conf. on Signal Processing Applications and Technology**, Boston, MA, Oct. 1996, vol. 2, pp. 1382–1386.

[10] K. Suzuki, I. Horiba, N. Sugie and M. Nanki, "A recurrent neural filter for reducing noise in medical x-ray image sequences," in **Proc. Int. Conf. on Neural Information Processing**, Kitakyushu, Japan, Oct. 1998, vol. 1, pp. 157–160.

[11] L. Yin, J. Astola and Y. Neuvo, "A new class of nonlinear filters - neural filters," **IEEE Trans. Signal Processing**, vol. 41, no. 3, pp. 1201–1222, Mar. 1993.

[12] L. Yin, J. Astola and Y. Neuvo, "Adaptive multistage weighted order statistic filters based on the back propagation algorithm," **IEEE Trans. Signal Processing**, vol. 42, pp. 419–422, Feb. 1994.

[13] Z. Z. Zhang and N. Ansari, "Structure and properties of generalized adaptive neural filters for signal enhancement," **IEEE Trans. Neural Networks**, vol. 7, no. 4, pp. 857–868, Jul. 1996.

LEARNING GESTALT OF SURFACES IN NATURAL SCENES

Amir Assadi,* Stephen Palmer,† Hamid Eghbalnia‡

INTRODUCTION.

Most surfaces in the world around us are generally not ideal smooth mathematical surfaces. Many rough and non-smooth surfaces, such as rolling hills covered by grass, star fish, sea urchins, hedgehogs, and many other living beings or natural scenes are perceived to have a piecewise smooth "global" geometric form that is quite distinct from the small-scale "local" geometric properties. How do we associate a large-scale piecewise smooth surface geometry to such objects or scenes (what we may call an imaginary model surface), and how do we use the small-scale information to add "texture" to the imaginary surfaces in order to complete our visual understanding? Questions of similar nature have attracted vision scientists for some time. There are a number of different formulations and even proposed solutions to such local-global problems. The most relevant theories come under names Multi-scale Representations, Scale Space Theory and their variations (based on the idea that different characteristics of an image reveal themselves at different levels of resolution.) Our theory complements such existing theories in several ways.

We develop a computational model for scenes with surfaces that have rough and non-smooth small-scale structure but with a perceived global (larger-scale) geometric form. Examples include grass and meadow, surfaces textured with sand-paper, natural scenes having rough texture such as the skin of crocodile, pine cones, a field of sea urchins, forests, ripples and waves on water surfaces, etc. Another domain of examples arise in scientific exploration of microscopic images, such as the atomic force microscopy (AFM) images from alloys in materials science, molecular beam epitaxy (MBE), rough surfaces due to ballistic deposition (BD surfaces) and random deposition surfaces (RD). As a last example, one may translate some outstanding image processing problems of infra-red astronomy to understanding the random tex-

*Center for the Mathematical Sciences and Department of Medical Physics, UW-Madison, WI 53706

†Center of Cognitive Studies and Department of Psychology, UC-Berkeley, CA 94720

‡Center for the Mathematical Sciences and Department of Mathematics, UW-Madison, WI 53706

0-7803-5673-X/99/$10.00 © 1999 IEEE

ture of clouds combined with noise, e.g. to describe algorithms that detect stars within noisy data provided by infra-red imaging devices .

APPROACH.

We propose several complimentary paradigms (each leading to a computational model for a 3D-textured surface), and their experimental implementation in order to compare their outcomes form several points of view: model precision and effectiveness in predicting human visual perception under various circumstances, computational performance, robustness and algorithmic efficiency. One reason for multiple paradigms is the diversity of circumstances under which visual perception occurs. Another reason is the applications to different subjects, as briefly mentioned above. It is often the case that each scenario calls for a different paradigm. Finally, comparison of different paradigms leads to better evaluation of the algorithms.

Our main contributions are: (a) the development of the general geometric theories that associate to a given 3D-textured surface S, a family of piecewise smooth surfaces S(t) that are "perceived-forms" of S from the large-scale view (in the world.) (b) Development of the learning algorithms to recover shape from texture for large families of such surfaces.

We will endow members of S(t) with 2D-texture via an appropriately defined "transfer" (pull-back of the texture mapping, in the jargon of differential topology) of the actual texture of S from the scene. Next, we will adapt an appropriate scheme to define objective functions on these families, whose minima distinguish the 2D-surfaces that best describe the large-scale form of the original surface S in early vision, under the given circumstances (e.g. a particular observer in the scene.) This step is accomplished by means of a neural network with back propagation. For simplicity at this stage, we assume that the figure-ground separation is achieved, e.g. via a suitable segmentation algorithm. We show later how to modify such methods to obtain segmentation algorithms based on Gestalt. The "shape" of S(t) and its other mathematical invariants are studied via standard tools from differential geometry and topology. Our work in progress includes validation of the perceived form of natural surfaces by comparison of the outcome of our computational theory with that of S through psychophysical measurements. An important step (in the algorithms and their experimental implementations) is the mathematical problem of constructing the family S(t), briefly mentioned below.

The Rolling Spheres Paradigm.

Consider a sphere of radius R=t rolling on the given 3D-textured surface S, and let the locus of centers of these spheres be called S(t). As the radius R approaches zero, S(t) approaches S. Through suitable reflection of the spheres, a family S(t) is defined. Here, t is the absolute value of a parameter that lies in an appropriate interval (-a, a), and S(0)= S. Thus, the family approximates S, and for larger values of the parameter, interprets the surface with coarser resolution. The mathematical consideration of this model leads to very interesting and natural problems in studying geometric invariants of surfaces with Lipschitz structure on the one-hand, and controlled topology on the other. From the view-point of pure mathematics, these problems are basic and quite interesting in their own right.

The Foliation Paradigm.

The family S(t) can be also considered as leaves of a foliation (in the sense of differential topology.) To obtain such a foliation, we must describe an integrable distribution (as in Frobenius Theorem). Since we are in 3D, every distribution is determined through its associated unit normal field. An averaging procedure based on a family of meshes that approximate the surface (e.g. as in a one-dimensional analogue of the rolling ball paradigm applied to "noisy coordinate curves" of S cut by the planes passing through the mesh and parallel to the z-axis, if we locally describe the surface as a graph of a real-valued 2D function.)

Statistical Averaging.

This approach lends itself to discrete models, where averages of normals to the discretized surface are performed in multi-step neighborhood. These define the statistical normal vector fields, hence the 2D-distribution that will be verified to be integrable, and integrate to form the family of surfaces S(t).

The Jacobi-Hamilton Equations and Level Surface Theory.

When S is the graph of a function with bounded variation (here, our functions are assumed continuous with realistic variations), we can apply the Jacobi-Hamilton equations, and the family provided by Hopf's solution. The level-surfaces of this family yield S(t). The crux of the matter here is to reduce the problem for a reasonably large collection of surfaces to the case of a graph. The graph case can be also generalized, but here we leave out the technical discussion. This approach may be compared with the Heat Equation approach in which heat diffuses according to a Dirichlet boundary problem (e.g. Hummel 1987[1]: The scale space formulation of pyramid data structures.) Discrete versions corresponding to the heat equations are simulated and solved, e.g. as in Hummel 1987[1]. Similarly, the discrete versions of our models can be simulated and solved. We may remark that high performance computing techniques are available for faster and more extensive numerical experiments (e.g. using the WARP parallel systolic array).

Convolution and Filtering.

Finally, we consider a family of kernels K(t), and perform the convolutions that average the local variation of the surface, and gives rise to the family S(t). This approach is the closest in spirit to multi-resolution or multi-scale theories of vision. On the other hand, unlike the existing theories that lead to blurring the image in order to extract coarser information, we propose to create surfaces embedded in the 3D-space with coarser details but not necessarily blurred images. Thus, in our approach the image processing step will be done at a later stage.

We have also made progress on applying the above-mentioned theory of vision to initiate the mathematical theory of textured surfaces (including natural scenes) that has led us to non-trivial mathematical constructs and more subtle and advanced geometric theories.

DISCUSSION.

It is worthwhile to mention the novel features of our theory, and compare it to the research of others.

(a) We propose to use the theory of foliation from differential topology to create a geometric model for the world with sufficient flexibility, so that at least one locally smooth mathematical surface (i.e. one leaf of the foliation) can be singled out to characterize the perceived geometry of the scene, so that most of these computations happen" in the world." Only at a later stage, the image of the artificial 2D-textured mathematical surface in the 3D-space will be compared with the image of the real scene. In the work of others, all the manipulations (e.g. blurring and other approximations) are applied directly to the 2D-image that represents the real world. Our method has several advantages: for instance, as progress is made on finding more efficient algorithms to describe shape from texture, shape from shading, shape from motion etc., we can apply them to the 2D-images obtained from the leaves of the foliation, and update the output of the theory. Thus, this theory complements the present and future image-based research on perception of shape.

(b) The mathematical theory in the research has potential for experimental verification via virtual reality environments, although at this point we plan to use standard computer graphics to generate the textured surfaces in the family that will be used in the psychophysical measurements. The VR will open a new research direction: merging of the tactile and visual senses to describe the perceived shape (and eventually, merging of auditory, tactile and visual senses.) The image-based algorithms do not seem to have this flexibility.

(c) The theory of foliation will provide a powerful interface between modern mathematical research and biology and a new set of tools and geometric concepts (to our knowledge, foliation theory has not been done in neuroscience in the past.) On the mathematical side, one expects that a powerful and rigorous variational method eventually emerges, that will lead to a PDE whose solution describe the perceived shape. As usual, such PDEs have the advantage of bringing well-understood and standard tools (e.g. numerical solutions) to solve the related problems.

COMPUTATION.

In this paper, we describe the results of computations with the Foliation Method, and comment on the Rolling Ball Method that has a similar flavor. To avoid excessive technicalities, we consider an approximation to both methods that works well in most cases that the variations in the curvature of the Gestalt surface is not too high, corresponding to pieces of surfaces with Gestalt curvatures bounded between two numbers that are bounded in absolute value by a constant multiple of the measure of solid viewing angle. The cases where the principal curvatures of the Gestalt surface varies frequently and rapidly (within the viewing angle) pose technical problems related to emerging cusp and fold singularities for the leaves of foliation formed by the loci of centers of rolling balls. Such singularities need to be smoothed out, in order to restore the smoothness percept of the Gestalt of the surface, and are generally artifacts of the computational method, rather than a natural part of the foliated structure that our theory predicts abstractly. We hope to address in the near future a solution for such technical difficulties of our methods.

An intuitively appealing and computationally accessible method for construction of the leaves of the foliation are as follows.

First, we solve the corresponding problem in the Euclidean plane. Consider a curve C that is the graph of a piecewise smooth function f on an interval [a,b]. Assume that the variation of differences of values of consecutive maxima (correspondingly consecutive minima) are bounded between a small integer constant multiple of b-a. This condition is similar to the condition regarding the roughness of the surface relative to the viewing angle. We can select an average smoothing of C in a variety of ways. For example, we select a uniformly distributed sample of points within [a,b], take the DFT of f using the sample, and filter the high ferequencies. Analogously, we could take a convolution of f with a Gaussian smoothing kernel. In the latter approach, different variances give rise to different smoothing of C, analogous to the Convolution and Filtering Method that we described above. The smaller the variance, the closer the smoothing to C. As the variance grows, the smoothing loses more local features of C and approaches to a flat curve. The Rolling Ball Method becomes the rolling circle method, and intuitively corresponds to the smoothing kernel above. To use these observation for the case of vision, we assume that the observer's eye is located in a fixed location above the plane, say on the z-axis looking down at the xy-plane containing C. In any of the above cases, the family of curves in the plane correspond to the family of smoothing curves after scale and shifts that brings the viewing angle for the graph to be the same as the viewing angle for C.

Next, we the case of the curve above to simplify the problem for the construction of family of smoothing for a surface. Consider the point P on a (natural) surface S. First, construct a piecewise linear neighborhood of the point P that are local maxima, approximating the given surface patch. Next, we select a randomly and uniformly distributed set of points on S, and construct a piecewise linear surface L consisting of simplixes (flat triangular surfaces) that contain P in their center, and so that L is topologically a disk. Next, we take an average of the unit normal vectors for the piecewise linear surface L, and normalize it for the sake of convenience. Call this vector N. For a generic choice of random points and a generic choice of L, the vector N has the following property: Among the perpendicular pairs of planes passing through N, intersecting the surface S in a pair of curves C and C', almost all pairs C and C' have (infinitely many) smoothings at different scales (as described above) that are graphs of functions that are, in particular, at least twice differentiable at P, and in general, piecewise differentiable. (The proofs of these assertions follow from general position and transversality arguments in differential topology.) Any such pair C and C' is then smoothed with a choice of scale, and give rise to candidates for the coordinate curves for a smoothing of the surface at P. That is, these curves interpolate to define a surface patch for a leaf of the foliation at that scale around the point P. In practice, one selects a set of points along C and C' and constructs the system of curves that we may call a system of "rough coordinate curves" near P. Then one proceeds with the choice of "rough normals", and cuts a system of rough coordinate curves on the natural surface. Next, we continue to smooth the system of curves at each scale, and interpolate them to form the surface elements. The surface elements come together to form the leaves of the foliation, as proposed above. For the sake of brevity, we leave out the discussion of smoothing textures. Suffice it to say that texture is constructed as a section of a vector bundle on S, or locally, a set of vector-valued functions. Kernel smoothings of a section representing an original texture are constructed in the same spirit.

The figures (1,2,3 and 4) below illustrate the procedure. More details and discussion of the code is available on the web site

http:://www.cms.edu/~cvg.

RELATED RESEARCH.

Among other related literature, we mention a sample from Scale Space Theory and related approaches. Rosenfeld and Thurston [2] take advantage of simultaneous use of different size operator masks (corresponding to the spatial-frequency channels, as it was shown later) to increase sensitivity to edges of variable resolution, and also to withstand the noise. Burt and Adelson [3] considered the pyramidal structure as the Laplacian of successively blurred

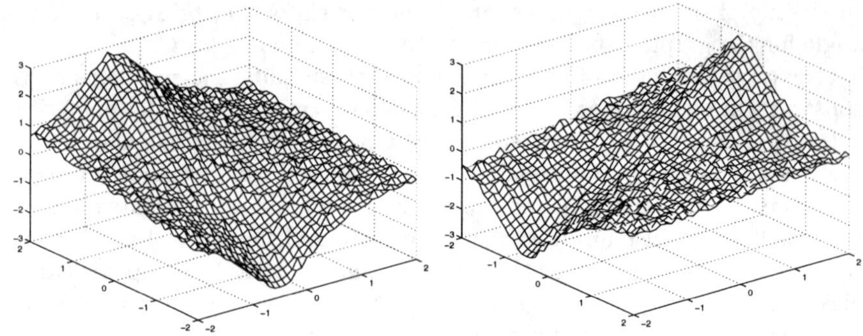

Figure 1: "Rough" surfaces before foliation

images (also known as the Gaussian pyramid.) Other related literature include: Crowley and Parker [4] and Meer, Baugher and Rosenfeld [5] and Mallats[6] There is also a body of literature based on applications of fractals, e.g. Pentland [7] (Fractal-based descriptions) proposes fractal -like models for all texture, and Peleg et. al. [8] refute Pentland's hypothesis, and suggests how to use changes in measurements in different scales to characterize texture. A later suggestion along these lines is due to Keller et. al [9] who introduce the concept of lacunarity (that captures the second order statistics of fractal surfaces.)

On the perception side, Gagalowicz [10] , Julesz and Bergen [11] and Caelli [12] among others, have offered evidence for visual perception as a the local process. Wilson and Bergen [13] and DeValois et. al [14] establish the relevance of multi-channel frequency analysis systems in human vision, and explain/measure quantitatively the relationship to Hubel and Wiesel's classical discovery of receptive fields (RF) of simple cells. Hubel and Wiesel's theory of receptive fields is, on the other hand, interpreted by Koenderink and van Doorn [15] in terms of filtering theory of images (i.e. convolutions with the Laplacian of Gaussians); thus we may regard RFs as multi-channel filters in this context. This is potentially a starting point to search for the physiological basis for our theory, and further relationships to other related

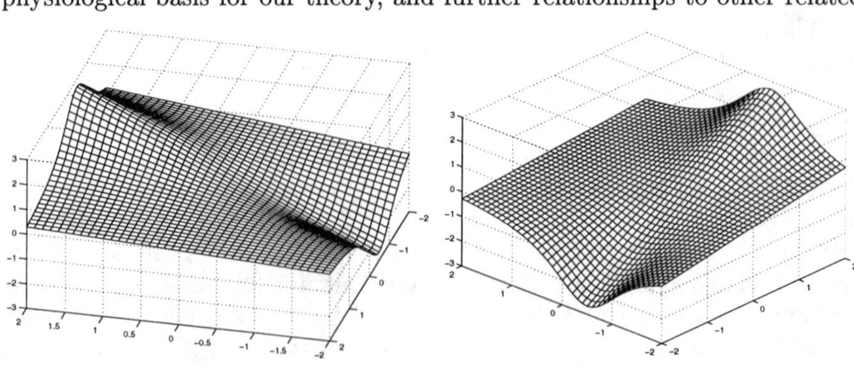

Figure 2: Surfaces after foliation

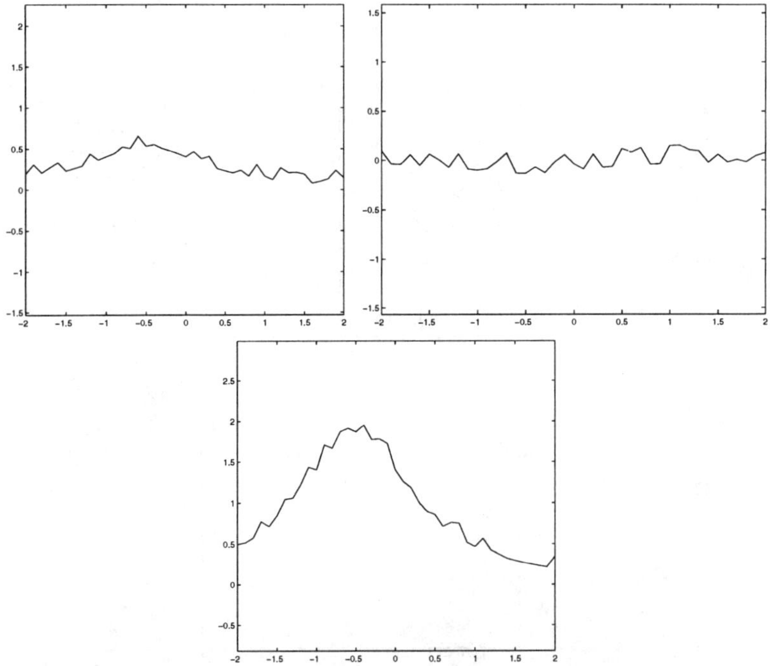

Figure 3: Some coordinate Curves of "rough" surfaces

biologically plausible vision research. Another related direction is Porat and Zeevi [16] and Clark and Bovik [17] on the Gabor representation for early vision. A distributed architecture, made up of multiple spatially and spectrally localized RF and defined as Gabor filters, yields an early low-level representation of the visual input. Watson's [18] theory of cortex transform is conceptually modeled after Gabor-type RF, and provides a distributed representation in terms of both spatial and spectral localization, and it can be potentially integrated into a computational model of our proposed theory.

Applications to Symmetry and Structural Regularity. Symmetry is a much-encountered theme in science. Here, We use the term symmetry in reference to all kinds of transformations that leave invariant some form of geometry, together with the related concepts such as harmony. Therefore, similarity in Euclidean plane geometry is a form of symmetry (in the so-called conformal geometry, where angles are preserved) although it is not necessarily a rigid motion. The term quasi-symmetry can be used for perceived regularity of structure that is compelling in its organization, but fails to be a strict symmetry in the mathematical sense above. This aspect of our research has a long-term history in psychology as the investigation of cognitive processes that underlie human perception of geometric forms, much the same way that Henri Poincare' posed in his 19-th Century treatise Science and Hypothesis and led to his discovery of non-Euclidean worlds long before the Einstein's theory of gravity. We believe that our theory to address such issues in the

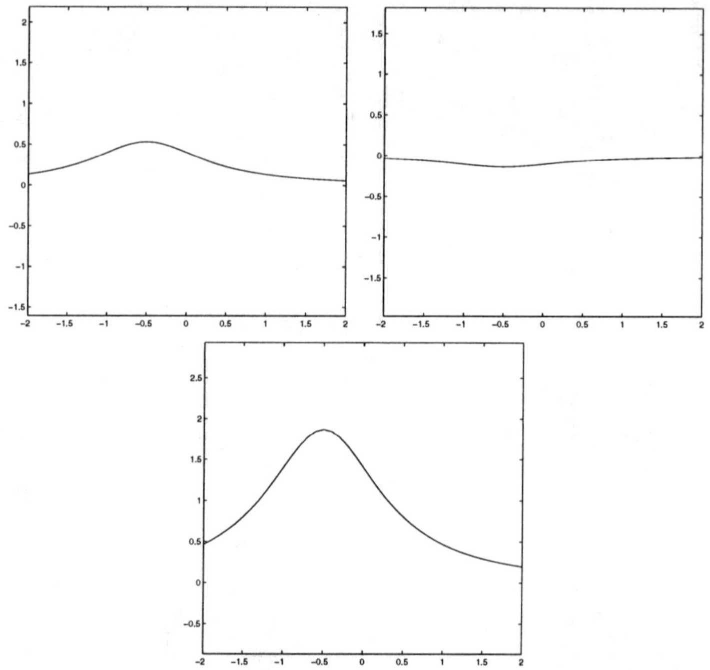

Figure 4: Coordinate Curves after foliation

realm of cognitive neuroscience, with computational models to support the cognitive theories, and investigation of physiological evidence to establish the low-level computations of visual, tactile motor and auditory processes that contribute to our perception of symmetry/regularity in structure. As an example, our theory predicts that the human perception of symmetry could be modeled via the mathematical symmetry of the Gestalt of the surface s (bounding the object) as discussed above. Thus, the rotational symmetry of a pine cone or a sea urchin does not have a counterpart in its original physical shape, but it is measurable when we apply the Gestalt theory of surfaces. Such observations agree with our daily experience, and can be measured in psychophysics.

REFERENCES

1..Hummel, R., The Scale-Space Formulation of Pyramid Data Structures, PCV88(107-123).

2..Thurston, M., and Rosenfeld, A., Edge and Curve Detection for Visual Scene Analysis, TC(20), No. 5, May 1971, pp. 562-569.

3..Burt, P.J., and Adelson, E.H., The Laplacian Pyramid as a Compact Image Code, Commun(31), No. 4, April 1983, pp. 532-540.

4..Crowley, J.L.[James L.], and Parker, A.C.[Alice C.], A Representation for Shape Based on Peaks and Ridges in the Difference of Low-Pass Transform, PAMI(6), No. 2, March, 1984, pp. 156-169.

5..Meer, P., Baugher, E.S., and Rosenfeld, A., Frequency Domain Analysis and Synthesis of Image Pyramid Generating Kernels, PAMI(9), No. 4, July 1987, pp. 512-522.

6..Mallat, S.G., A Theory for Multiresolution Signal Decomposition: The Wavelet Representation, PAMI(11), No. 7, July 1989, pp.674-693.

7..Pentland, A.P., Fractal-Based Description of Natural Scenes, PAMI(6), No. 6, November 1984, pp. 661-674.

8..Peleg, S., Naor, J., Hartley, R.L., and Avnir, D., Multiple Resolution Texture Analysis and classification, PAMI(6), No. 4, July 1984, pp. 518-523.

9..Keller, J.M., Chen, S.S., and Crownover, R.M., Texture Description and Segmentation through Fractal Geometry, CVGIP(45), No.2, February 1989, pp. 150-166.

10..Gagalowicz, A., A New Method for Texture Field Synthesis: Some Applications to the Study of Human Vision, PAMI(3), No. 5, September 1981, pp. 520-533.

11..Julesz, B., and Bergen, R., Textons, The Fundamental Elements in Preattentive Vision and Perception of Textures, Bell System Tech.(62), No. 6, 1983, Part II, pp. 1619-1645.

12..Caelli, T., Three Processing Characteristics of Visual Texture Segmentation, SV(1), No. 1, 1985, pp. 19-30. BibRef 8500

13..Wilson, R., Bergen, A four mechanism model for spatial vision, Vision Research, 1979.

14..DeValois et. al, Responses of the striate cortex cells to gratings and checkerboard patterns, J. Physiology, 1979.

15..Koenderink and van Doorn , Representation of local geometry in the visual system, Biological Cybernetics , 1987.

16..Porat and Zeevi , The generalized Gabor scheme of image representation in biological and machine vision, 1988.

17..Clark, M., Bovik, A.C., Experiments in Segmenting Texton Patterns Using Localized Spatial Filters, PR(22), 1989, pp. 707-717.

18..Watson, A.B., The Cortex Transform: Rapid Computation of Simulated Neural Images, CVGIP(39), No. 3, September 1987, pp. 311-327.

RATE-CONSTRAINED SELF-ORGANIZING NEURAL MAPS AND EFFICIENT PSYCHOVISUAL METHODS FOR LOW BIT RATE VIDEO CODING

Keith L. Ferguson and Nigel M. Allinson
UMIST
Department of Electrical Engineering and Electronics
Manchester, P.O. Box 88, M60 1QD, United Kingdom

Abstract: The video coding problem is essentially an operational distortion-rate issue where the underlying input pixel data, probability distributions and dimensions are discrete, unknown and not smooth. In the low bit rate case the high resolution assumptions for vector quantization are not strictly valid and the problem is exacerbated. However, by considering the rate-constrained operational points on sets of self-organizing neural maps (SOMs), provides a methodology for selecting locally optimal vector quantizers. The learning process of the standard SOM algorithm is modified to minimize the distortion subject to a constraint of entropy approximation. The applied training set is adapted to suit the proposed coding environment. Operating in the discrete wavelet transform (DWT) domain is well suited to the inclusion of a psychovisual model. The spatial frequency response, the multiresolution scene analysis and the central focusing aspects of the visual cortex are incorporated into the model. The resulting video coding algorithm is bit rate scalable from 10k bits per second (bits/s) and provides subjectively acceptable video at a fixed frame rate of 10 frames per second (fps) with a QCIF pixel resolution.

INTRODUCTION

A novel video coding algorithm, developed for videoconferencing environments at data rates less than 64k bits/s, is presented. The important principles of our approach are the use of rate-constrained self-organizing neural maps as vector quantizers and the efficient inclusion of psychovisual factors to maximize perceived image quality for a specified data rate.

The relationship between SOMs and vector quantizers is formulated and followed by a discussion on the bit rate generated by the vector quantizers. The proposed video coding algorithm is described together with the rate-constrained SOM learning algorithm. A distortion-rate characteristic plot of the learned SOMs from one of the DWT sub-bands is presented with a discussion on how an operationally

0-7803-5673-X/99/$10.00 © 1999 IEEE

optimal point may be chosen. The performance results of the application of these points to the video coding algorithm are presented.

BACKGROUND

The relationship between the SOMs [1] and the generalized high resolution vector quantization theory [2] relies on an encoder-decoder noise model [3 pp. 453-461, 4] shown in Figure 1.

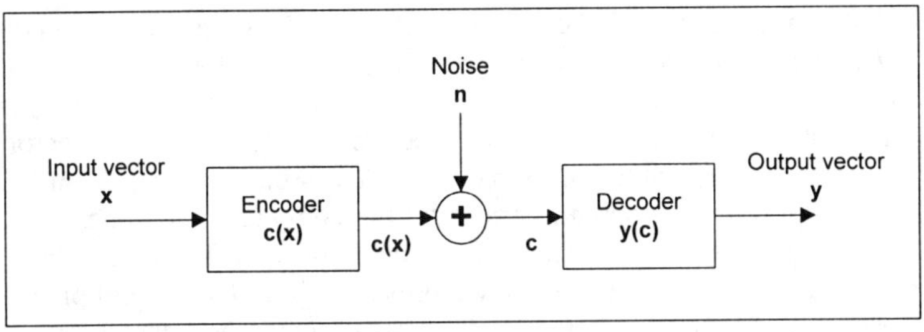

Figure 1: Vector Quantizer Encoder-Decoder Noise Model

The optimal vector quantizer for vectors, $\mathbf{x}$, taken from a sample space defined by an underlying stationary probability density function (pdf), $f(\mathbf{x})$, in a squared error sense, is one that minimizes the long term average distortion defined as:

$$D = \int_{-\infty}^{+\infty} f(\mathbf{x}) \left[\int_{-\infty}^{+\infty} \pi(\mathbf{n}) \|\mathbf{x} - \mathbf{y}(\mathbf{c}(\mathbf{x}) + \mathbf{n})\|^2 \, d\mathbf{n} \right] d\mathbf{x} \tag{1}$$

Here $\pi(\mathbf{n})$ is the pdf of the additive noise in the model and is assumed to be independent of the input signal. The optimal solution requires the joint minimization of the *nearest-neighbor* condition and the *centroid* condition. The *nearest-neighbor* condition provides for selecting the optimal encoder to match a given decoder. For a smooth noise pdf, $\pi(\mathbf{n})$, the minimum condition may be approximated to the best-matching codeword from the vector quantizer codebook, or the best-matching neuron from the SOM. For the gradient decent SOM algorithm with N neurons referenced by an index i such that $i \in \{1 \ldots N\}$, this step is defined as:

$$i(\mathbf{x}) = \arg\min_{j} \left\{ \|\mathbf{x} - \mathbf{y}(\mathbf{c}_j(\mathbf{x}) + \mathbf{n})\| \right\}, \quad j = 1 \ldots N \tag{2}$$

Note that the index, i, completely defines the codeword $\mathbf{c}(\mathbf{x})$ and vise versa. The *centroid* condition provides the selection of the optimal decoder for a given encoder. In the gradient decent SOM algorithm this refers to the neuron weight update step where the neighborhood function is equivalent to the noise pdf.

$$\mathbf{y}_{t+1}(\mathbf{c}_j) = \mathbf{y}_t(\mathbf{c}_j) + \eta_t \pi_t(\mathbf{c}_j - \mathbf{c}(\mathbf{x}_t))[\mathbf{x}_t - \mathbf{y}_t(\mathbf{c}_j)] , \quad j = 1...N \tag{3}$$

Here, η_t is the learning rate parameter and $\mathbf{c}(\mathbf{x}_t)$ is defined by the winning neuron from the *nearest-neighbor* condition at iteration t. If the neighborhood function of the SOM algorithm is defined by the assumed noise model of a zero mean Gaussian pdf, then the SOM algorithm will result in an optimal vector quantizer. The SOM has been shown to perform well as a vector quantizer for INTER frames in a video coding environment [5].

Consider now the application of the learned SOM in a signal compression environment. The vector samples, **x**, are extracted from the signal and the index of the neuron whose weight vector has the lowest squared error distortion with **x**, is transmitted on the channel to the receiver. The receiver decodes the output vector, **y**, in a weight vector look-up table whose neuron index is that received from the channel. Therefore the information conveyed over the communication channel is completely contained in the index. This raises the issue of efficient symbolic representation of the transmitted indices. Since only binary alphabet symbols are considered here, the index is represented by a variable length code, vlc(i), whose bit rate is upper-bounded by the uniform neuron firing distribution case of $R=\log_2 N$ bits per vector. The bit rate or bit length of the code will be denoted by its magnitude, |vlc(i)|.

For any joint pdf $f(\mathbf{x})$ of the input vector, **x** with dimension k, such that $f(\mathbf{x})\rightarrow 0$ as $\mathbf{x}\rightarrow\pm\infty$ defined as $\{x_1\rightarrow\pm\infty , x_2\rightarrow\pm\infty ,\ldots, x_k\rightarrow\pm\infty\}$, there exists an arbitrary low distortion SOM where N is finite and the neuron firing probabilities are described by a probability mass function (pmf), $p(i(\mathbf{x}))$, such that $p(i(\mathbf{x}))\propto f(\mathbf{x})$. This premise is based on the density matching properties of the learned SOM [1, 3]. An entropy coding method is therefore more efficient for transmitting the indices. From an information viewpoint the average entropy, in bits, is defined as:

$$H(i) = -\sum_{i=1}^{N} P_i \log_2 P_i \tag{4}$$

Here P_i is the *a posteriori* probability of index, i being the winning neuron. If a prefix-free variable length binary code is used and allowing non-integer code lengths, then the average length is the index entropy [2, ch. 9]. Therefore the length of the code to represent the index, i is defined as:

$$|\,\mathrm{vlc}(i)\,| = -\log_2 P_i \tag{5}$$

Note that practical Huffman codes must posses integer multiple lengths and therefore (5) acts as an asymptotic lower bound. The long term average bit rate generated from the learned SOM is written as:

$$R = \sum_{i=1}^{N} P_i\,|\,\mathrm{vlc}(i)\,| \tag{6}$$

The rate-constraint is constructed from the trade-off between rate, (required for compression), and distortion (required for image quality). For low bit rate coding we wish to sacrifice image quality for higher compression, but a quantitative analysis is required in order to make a prudent choice of an operational point. An approximate operational distortion-rate function, for a fixed vector dimension and a given vector quantizer, ($\mathbf{c}(\mathbf{x})$, $\mathbf{y}(\mathbf{c})$), may be constructed as a lower bound from the distortion-rate theory underlying entropy-constrained vector quantization [6].

$$D^*(R) = \inf_{(\mathbf{c}(\mathbf{x}),\mathbf{y}(\mathbf{c}))} \left\{ E[\, \|\mathbf{x}-\mathbf{y}\|^2 \,] \mid E[R] \le \log_2 N \right\} \tag{7}$$

Here E[ ] is the statistical expected operator of a stationary process. This equation defines the bounds of the region in the distortion-rate plane within which we are able to operate with the parameter restrictions of our particular coding environment.

Consider the learned SOM where the neighborhood function has decayed to zero and hence, assuming the high resolution case, the noise pdf, $\pi(\mathbf{n})$ in (1) is approximated by the Dirac function. The operational average distortion, D^* is a minimum in the sense that it minimizes the function:

$$D = E[\, \|\mathbf{x}-\mathbf{y}\|^2 \,] = \int_{-\infty}^{+\infty} \|\mathbf{x}-\mathbf{y}\|^2 f(\mathbf{x})\, d\mathbf{x} \tag{8}$$

The *nearest-neighbor* and *centroid* conditions result in an optimal partitioning of the SOM weight space into regions, $\mathbf{R}_i$, [2 pp. 349-358] which may be described as k-dimensional "volumes of influence", such that for each neuron index, i:

$$\mathbf{R}_i \subset \left\{ \mathbf{x} : \|\mathbf{x}-\mathbf{c}_i\|^2 \le \|\mathbf{x}-\mathbf{c}_j\|^2 \right\}, \quad j = 1..N \tag{9}$$

Under this condition the operational distortion, D^* may then be written as:

$$\begin{aligned} D^* &= \sum_{i=1}^{N} \int_{\mathbf{R}_i} \|\mathbf{x}-\mathbf{c}_i\|^2 f(\mathbf{x})\, d\mathbf{x} \\ &= \sum_{i=1}^{N} P_i \int_{-\infty}^{+\infty} \|\mathbf{x}-\mathbf{c}_i\|^2 f(\mathbf{x} \mid \mathbf{x} \in \mathbf{R}_i)\, d\mathbf{x} \end{aligned} \tag{10}$$

Here $f(\mathbf{x} \mid \mathbf{x} \in \mathbf{R}_i)$ is the conditional pdf of $f(\mathbf{x})$ given that $\mathbf{x}$ lies in the region $\mathbf{R}_i$. The rate constraint of (6) is approximated by (4) and introduced with a Lagrange multiplier, λ, to generate the operational distortion-rate function to be minimized.

$$D^*(R) = \sum_{i=1}^{N} P_i \int_{-\infty}^{+\infty} \left(\|\mathbf{x} - \mathbf{c}_i\|^2 - \lambda \log_2 P_i \right) f(\mathbf{x} \,|\, \mathbf{x} \in \mathbf{R}_i)\, d\mathbf{x} \tag{11}$$

Note that the $\lambda \log_2 P_i$ is a constant and therefore affects the *nearest-neighbor* condition but does not affect the *centroid* condition. From (11) the rate-constrained SOM is trained with the gradient decent algorithm to minimize the instantaneous cost function described as:

$$J = \left\| \mathbf{x} - \mathbf{c}_{i(\mathbf{x})} \right\|^2 - \lambda \log_2 P_{i(\mathbf{x})} \tag{12}$$

VIDEO CODING ALGORITHM

The structure of the algorithm is designed to be more suitable than discrete cosine transform (DCT) based algorithms for incorporating the removal of psychovisual redundant information. The DCT-based algorithms generally operate in the spatial domain and it is therefore more difficult to include a psychovisual frequency and multiresolution model. However, DWT-domain processing is similar to the manner in which the visual cortex processes image information and hence inherently facilitates the incorporation of such a model. In the videoconferencing environment, the 'lip synchronization' requirement necessitates complex channel buffering control for algorithms that generate variable length coded bit streams and the dropping of frames to prevent buffer overflow. At low bit rates large frame delays may be required for the buffer control. Our algorithm circumvents these problems by generating fixed length coded bit streams at a constant frame rate. In this way the buffering requirements are minimal and 10fps is subjectively deemed to be suitable for a reasonable amount of lip movement. The complexity is kept low by utilizing the simple structure of neural-based vector quantizers and by excluding INTRA frames and motion compensation. A block diagram of the algorithm is shown in Figure 2.

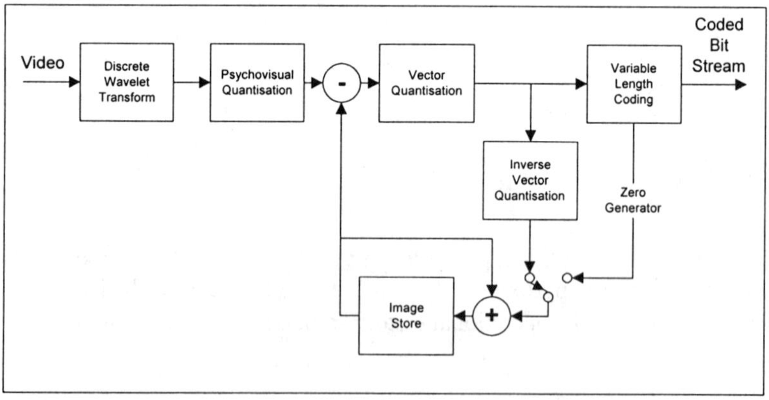

Figure 2: Block Diagram of the Video Coding Algorithm

The coding loop operates in the discrete wavelet transform (DWT) domain. There is evidence to suggest that the visual cortex operates in a multiresolution sub-band manner [7] and although the sub-bands are not octave sub-divisions as in the DWT, they are close approximations. Spatial redundancy is reduced before entering the loop by scalar quantizers that are shaped according to a spatial psychovisual model developed from subjective tests specifically for DWT images [8]. The DWT difference images are coded using vector quantizers within the loop. Entropy codes generate a variable length output bit stream for each coded vector. The coding process is centrally weighted from a multiresolution viewpoint. Coding takes place in stages from the frame center outwards with increasing resolution and color addition up to the specified frame bit limit. All remaining stages are then set to zero. In this way the bit rate of every frame is fixed. The central weighting is justified by assuming that the center of the image is visually the most important. In videoconferencing applications, this is most likely to be true.

RATE-CONSTRAINED TRAINING ALGORITHM

The difference DWT images are vector quantised by rate-constrained SOMs trained for each sub-band and color using a sampled gradient decent method. The zero vectors are run-length encoded and hence the training set consists only of non-zero difference vectors. No INTRA frames are used in the algorithm and therefore the finite size vector quantizers must be able to deal with errors that accumulate over time. The training set is formed from typical image differences and the addition of errors that the vector quantizers will make *after* training. However, knowledge of the final state of the quantizers is unknown during the training process, therefore error approximation is based on the current state. The applied training set is adapted at each iteration during the training process in the following manner. Two random samples, $\mathbf{x}_1$ and $\mathbf{x}_2$, are selected from the difference image training set, T, at iteration t.

$$\{\mathbf{x}_1(t), \mathbf{x}_2(t)\} \in T \quad (13)$$

The update process of the SOM is performed with the first sample, $\mathbf{x}_1$. The error term, $\mathbf{e}_t$, for this sample and the winning neuron weight vector, $\mathbf{c}_1$, is determined. Simulating the effect of this error in the coding loop generates the second training sample, $\mathbf{x}_e$, for the SOM. Therefore the training data, T_t, applied to the SOM at iteration t, may be described as:

$$T_t = \{\mathbf{x}_1(t), \mathbf{x}_e(t)\} \quad (14)$$

Here;

$$\mathbf{x}_e(t) = \mathbf{x}_2(t) - (\mathbf{x}_1(t) - \mathbf{c}_1(t)) \quad (15)$$

The training process continues in this way with a difference image sample and an error adapted sample at each iteration. After initializing the SOM neuron weights with small random values and setting the occurrence frequency, F_j to unity for each neuron, the training proceeds by repeating the steps 1 to 5 defined as follows:

Step 1: Randomly sample the training space to establish $\{\mathbf{x}_1(t), \mathbf{x}_2(t)\}$.

Step 2: Determine the first winning neuron, i_1:

$$i_1 = \min_j^{-1} \left\{ \|\mathbf{x}_1(t) - \mathbf{c}_j(t)\|^2 - \lambda \log_2 P_j \right\}, \quad j = 1..N$$

$$P_j = \frac{F_j}{\sum_{j=1}^{N} F_k}$$

Step 3: Update the neighborhood neuron weight values and the winning neuron occurrence frequency:

$$\mathbf{c}_j(t+1) = \mathbf{c}_j(t) + \eta(t)\pi_{j,i_1}(t)[\mathbf{x}_1(t) - \mathbf{c}_j(t)], \quad j = 1..N$$

$$F_{i_1} = F_{i_1} + 1$$

Here $\eta(t)$ is the exponential decaying learning rate parameter and $\pi_{j,i1}(t)$ is the Guassian exponential neighborhood function with a linearly decaying radius centered on the winning neuron, i_1.

Step 4: Determine $\mathbf{x}_e(t)$ and find second winning neuron, i_2:

$$i_2 = \min_j^{-1} \left\{ \|\mathbf{x}_e(t) - \mathbf{c}_j(t)\|^2 - \lambda \log_2 P_j \right\}, \quad j = 1..N$$

Step 5: Update the second winning neuron weight values and occurrence frequency:

$$\mathbf{c}_j(t+1) = \mathbf{c}_j(t) + \eta(t)\pi_{j,i_2}(t)[\mathbf{x}_e(t) - \mathbf{c}_j(t)], \quad j = 1..N$$

$$F_{i_2} = F_{i_2} + 1$$

Each DWT sub-band exhibits differing statistical properties and probability distributions therefore optimal rate-distortion selections are based on different criteria depending on the sub-band. The sub-band multidimensional probability distributions can not be assumed to be Gaussian nor are infinite resolution approximations necessarily applicable to finite size SOMs. The choice of sub-band vector dimension is limited by the image size. For QCIF image sizes, vector dimensions of 4x4, 2x2, 2x2 and 1x3 are used at sub-band levels of 1, 2, 3 and 4, respectively. The SOM dimensions are restricted by the need for practical variable length codes. The SOM sizes investigated were 8x8, 16x16, 32x32 and 64x64 neurons. Rate-distortion verses λ plots are generated for all sub-bands and color components and used to empirically select generic optimal points for constructing the vector quantizers. The results for the luminance, level 3, low x high y DWT sub-band at four different SOM sizes for a 2x2 dimensional vector, are shown in Figure 3.

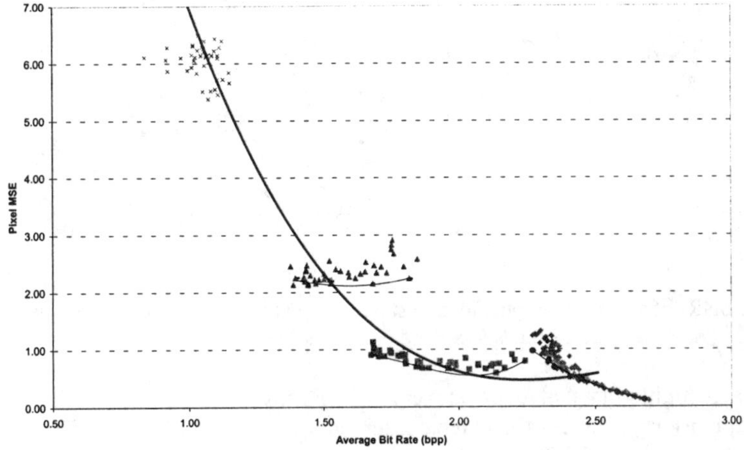

Figure 3: Luminance Level 3 Low X High Y DWT Sub-Band Rate-Distortion. The ✖'s are the distortion-rate operating points for the SOM size, N=8x8 neurons. The ▲'s are for N=16x16, the ■'s are for N=32x32 and the ◆'s are for N=64x64 neurons.

The overall trendline shows the distortion-rate characteristic for the choice of SOM size. From this line, the 32x32 SOM is chosen as the most optimal under these operating parameters. The thinner trendlines indicate the local convex hull for each SOM. The 64x64 neuron SOM has the typical non-increasing convex shape of high resolution smooth pdf vector quantizers. The selected 32x32 SOM shows that the locally optimal operating point is at a minimum. Each sub-band is analyzed in the same way to produce an operational SOM size and Lagrange multiplier, λ.

RESULTS

The rate-constrained SOMs were applied as DWT difference image vector quantizers in the video coding environment described above. A typical videoconferencing test sequence (that was not used in the training process) was used to evaluate the performance. The sequence consisted of color images with a QCIF (176x144) pixel resolution and a frame rate of 10fps. The peak signal-to-noise ratios (PSNR) for a section of the decoded test sequence at bit rates of 10k bits/s, 28.8k bits/s and 64k bits/s are shown in Figure 4. The two minimum regions in the figure show the active part of the sequence where the subject raises her hand to her face and removes it again.

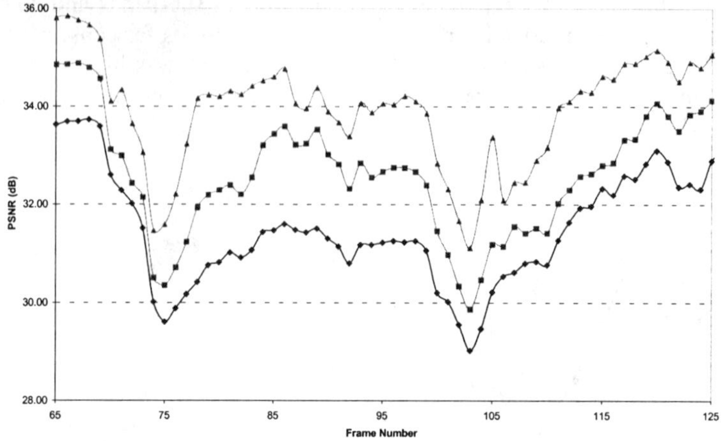

Figure 4: PSNRs of a Section of the Video Test Sequence at bit rates of 10k bits/s (-▲-), 28.8k bits/s (-■-) and 64k bits/s (-◆-).

Our approach highlights that even at very low PSNRs of less than 30dB, the subjective quality is sufficient for video conferencing. Samples of decoded frames from the test sequence at 28.8k bits/s are given in Figure 5.

Figure 5: Decoded Video Test Sequence frames 70, 80, 90 and 100.

CONCLUSION

We have shown that there exist locally optimal operating points for rate-constrained SOMs used as vector quantizers in a DWT domain video coding environment. This is true even if the underlying discrete probability distributions are non-continuous and unknown. If a training set that suits the coding environment is applied to a rate-constrained SOM algorithm, then operating points may be found that perform well in the chosen environment. We have found that a typical videoconferencing sequence of QCIF dimensions at 10fps will produce subjectively acceptable quality at approximately 31dB PSNR for a bit rate of 10k bits/s, 33dB PSNR at 28.8k bits/s and 35dB PSNR at 64k bits/s. The psychovisual model applied to the coding algorithm ensures that even at very low bit rates the subjective quality is satisfactory for practical systems.

REFERENCES

[1] T. Kohonen, Self-Organizing Maps. Springer-Verlag, New York, 2nd Edition, 1997.

[2] A. Gersho and R.M. Gray, VECTOR QUANTIZATION AND SIGNAL COMPRESSION. Kluwer Academic Publishers, Boston, 1992.

[3] S. Haykin, Neural Networks A Comprehensive Foundation. Macmillan College Publishing Company, New York, 1994.

[4] S.P. Luttrell, "Self-organization: A derivation from first principle of a class of learning algorithms," IEEE Conference on Neural Networks, Washington, DC, 1989, pp. 495 – 498.

[5] H. Liu, "Ordered Kohonen vector quantization for very low rate interframe video coding," **SPIE Proceedings on Digital Video Compression: Algorithms and Technologies**, Vol. 2419, pp. 71 – 80, 1995.

[6] P.A. Chou, T. Lookabaugh, and R.M. Gray, "Entropy-Constrained Vector Quantization," **IEEE Transactions on Acoustics, Speech, and Signal Processing**, Vol. 37, pp. 31 – 42, January 1983.

[7] S.G. Mallat, "Multifrequency Channel Decompositions of Images and Wavelet Models," **IEEE Transactions on Acoustics, Speech, and Signal Processing**, Vol. 37, No. 12, pp. 2091 – 2110, December 1989.

[8] A.B. Watson, G.Y. Yang, J.A. Solomon and J. Villasenor, "Visual Thresholds for Wavelet Quantization Error," **SPIE Proceedings on Human Vision and Electronic Imaging**, edited by B. Rogowitz and J. Allebach, Vol. 2657, Paper No. 44, 1996.

SELECTIVE IMAGE SMOOTHING VIA DYADIC WAVELET-BASED CONDUCTION EQUATION

Chwen-Jye Sze, Hong-Yaun Mark Liao *, Shih-Kun Huang, and Chun-Shien Lu

Institute of Information Science, Academia Sinica, Taipei, Taiwan
Email: {liao,arthur}@iis.sinica.edu.tw
Tel: 886-2-27883799 ext. 1811, Fax:886-2-27824814

Abstract. In this paper, we propose a new dyadic wavelet-based conduction approach for selective image smoothing. In our approach, a nonlinear conductivity function is considered in the wavelet-based function decomposition and reconstruction process. Since the proposed approach does not require to solve a PDE, it is therefore more efficient and accurate than the conventional nonlinear diffusion/conduction-based methods. Experimental results using both 1-D synthetic data and a real image demonstrated that the proposed method can efficiently remove noises and preserve real data.

INTRODUCTION

Image smoothing is an important step in low-level computer vision especially when the input images are noisy. An ideal smoothing algorithm is expected to simultaneously remove noises and enhance edges in an image. It is well-known that both noises and edges are high frequency signals. Therefore, it is very difficult to remove one while enhancing (or retaining) the other. Among the existing smoothing algorithms, the approach that bases the smoothing procedure on a diffusion/conduction equation has attracted a lot of attention in recent years[3]. In the literature, signals convoluted with Gaussians of varying width are considered to be the oldest and best-studied representatives of linear diffusion-based smoothing. The first nonlinear diffusion filter was proposed by Perona and Malik[3] . The basic concept behind their work is to replace Gaussian smoothing, which is equivalent to isotropic diffusion, with directional diffusion that preserves edges. In their work, a nonlinear diffusion function is embedded in a nonlinear PDE (Partial Differential Equation), to automatically achieve the smoothing task. Based

* To whom correspondence should be sent.

0-7803-5673-X/99/$10.00 © 1999 IEEE

on computing the solution of the nonlinear PDE, noise reduction and edge enhancement of the input signal can be accomplished simultaneously. However, finding a closed-form solution of a nonlinear PDE is always a difficult task. Usually, only an approximate solution can be found by using numerical methods[3]. In Donoho-Johnstone's approach[1], a special thresholding scheme was introduced to dynamically remove noises using some basic properties of orthogonal wavelet bases. Based on the set of orthogonal wavelet bases, an approximation of the original signal was estimated, and the features such as an L^2-error bound as well as the smoothness capability were presented. However, the above mentioned method is not always correct because the wavelet coefficients which were calculated for approximation can be accurately decided only when the thresholds were pre-determined.

In this paper, we propose a dyadic wavelet-based approach to solve the above mentioned problem. In the wavelet-based decomposition process, a nonlinear conductivity function is used for regularization purpose. Different from Donoho-Johnstone's approach[1], the decomposed wavelet coefficients in our approach are regulated by a continuous function rather than by several crisp thresholds. Our idea is as follows. Basically, a wavelet transform is a linear operator, hence, any nonlinear regulation of wavelet coefficients at a fixed scale will affect the distribution of wavelet coefficients at other scales simultaneously. Therefore, at each iteration, we only regulate the wavelet coefficients at scale 2^1. Under the circumstances, function of the next state can be obtained by executing the inverse wavelet transform on these regulated wavelet coefficients. Moreover, we also convert this nonlinear transform into a nonlinear iterated function system. Here, we call the iterated equation a dyadic wavelet-based conduction equation. Since the proposed approach does not require to solve a PDE, it is therefore more efficient and accurate than the conventional nonlinear diffusion/conduction-based methods.

The rest of this paper is origanized as follows. In the next section, some basic definitions and properties of wavelet theory will be introduced. In the first part of Section 3, we shall discuss some relationships between a diffusion process and a two-band filter system. Then, a wavelet-based conduction approach is proposed. Experimental results using both 1-D synthetic signals and a 2-D real image are reported in Section 4. Finally, a concluding remark is given in Section 5.

WAVELET TRANSFORM

This section introduces briefly some basic knowledge of the dyadic wavelet transform. A wavelet function $\psi(x)$ can be defined as follows:

$$\psi(x) = \frac{\partial\theta(x)}{\partial x}, \tag{1}$$

where $\theta(x)$ is a smoothing function whose integration over the full domain is equal to 1 and converges to 0 at infinity. A wavelet function has to satisfy the following condition:

$$\int_{-\infty}^{\infty} \psi(x)dx = 0. \tag{2}$$

Let $f(x) \in \mathbf{L^2(R)}$. The so-called dyadic wavelet transform[2] of $f(x)$ at scale 2^j can be represented as

$$W_{2^j} f(x) = f * \psi_{2^j}(x), \tag{3}$$

where $\psi_{2^j}(x) = \frac{1}{2^j}\psi(\frac{x}{2^j})$. Let $\phi(x)$ be a scaling function[2] and its integral is equal to one. Basically, $\phi(x)$ is also a smoothing function. A smoothing operator $S^d_{2^j} f(x)$ is defined by

$$S^d_{2^j} f(x) = f * \phi_{2^j}(x), \tag{4}$$

where $\phi_{2^j}(x) = \frac{1}{2^j}\phi(\frac{x}{2^j})$. In general, we can assume that the wavelet function $\psi(x)$ is characterized by the three discrete filters H, G, and K. The three discrete filters satisfy the following relation:

$$|H(\omega)|^2 + G(\omega)K(\omega) = 1. \tag{5}$$

Based on the discrete wavelet transform, a discrete signal $S^d_j f$ at scale 2^j can be decomposed into $S^d_{2^{j+1}} f$ and $W^d_{2^{j+1}} f$, and

$$W^d_{2^{j+1}} f = \frac{1}{\lambda_j} \cdot S^d_{2^j} f * G_j \tag{6}$$

and

$$S^d_{2^{j+1}} f = S^d_{2^j} f * H_j, \tag{7}$$

where H_p and G_p represent the discrete filters obtained by putting $2^p - 1$ zeros between each of the coefficients of the filters H and G.

The inverse wavelet transform algorithm reconstructs $S^d_1 f$ from the discrete dyadic wavelet transform. At each scale 2^j, it reconstructs $S^d_{2^{j-1}}$ from $S^d_{2^j}$ and $W^d_{2^j}$:

$$S^d_{2^{j-1}} f = \lambda_j \cdot W^d_{2^j} f * K_{j-1} + S^d_{2^j} f * \tilde{H}_{j-1}, \tag{8}$$

where K_p represents the discrete filter obtained by putting $2^p - 1$ zeros between each of the coefficients of the filter K and $\tilde{H}_p$ is the conjugate filter of H_p.

DYADIC WAVELET-BASED CONDUCTION

Relation between a diffusion process and a two-band filter bank

Signals convoluted with Gaussian of varying width is considered a linear diffusion-based smoothing or a linear scale-space smoothing. In a linear diffusion smoothing, a signal will be smoothed gradually when the iterated number is increased. This is because the high-frequency components (no mater useful data or noises) in a signal are repeatedly processed without any discrimination.

A linear diffusion-based smoothing (or linear scale-space smoothing) is not an ideal smoothing scheme because most of the time one would like to smooth out the noises but retain the useful data. Nonlinear diffusion-based smoothing[3] was developed to solve the above mentioned problem. Basically, a linear diffusion-based smoothing process repeatedly decomposes the original signal (or the low-frequency component) into the high-frequency and the low-frequency components. The high-frequency component generated at each resolution is then processed by a nonlinear operator which can decrease the magnitude of the noises and retain that of the useful data. The signal reconstruction process can be achieved by adding the processed high-frequency component and the non-processed low-frequency components at the lowest resolution to form the low-frequency component of the second lowest resolution. This low-frequency component is then combined with the processed high-frequency component at the same resolution to form the low-frequency part of the resolution one level up. The process repeats until the orignal signal is reconstructed. Since the noises are selectively smoothed at each resolution, a new signal that contains less noises can be obtained.

A mechanism which can realize the above mentioned process is a two-band filter system shown in Figure 1. The two-band filter system can decompose a signal into a high-frequency component (G in Figure 1) and a low-frequency component (H in Figure 1). Based on the concept descripted in the previous paragraph, a nonlinear regularization function can be applied to the high-frequency component (*i.e.*, G) shown in Figure 1. Figure 2 shows the realization of the above mentioned concept. The dyadic wavelet transformed high-frequency component, $W^d_{2^j} f_n$, will be regularized by a nonlinear function, P_t, before it is used as a component for reconstruction. The nonlinear regularization function used can be any nonlinear function with appropriate characteristics, for example, the nonlinear diffusivity function used in [3]. The objective of regularization is to retain the values of the high-frequency components that corresponds to larger magnitudes and decrease those that corresponds to smaller magnitudes. Experience tells us that the high-frequency components may contain edges and noises simultaneously. Since the magnitude of intensity gradient of edges is usually much larger than that of noises, we shall use the nonlinear diffusivity function adopted in [3] to distinguish them. Besides, since the low-frequency component of a signal is usually irrelevant to the smoothing process, we thus take advantage of the nature of the dyadic wavelet transform[2] to decompose the original signal and then deal with the high-frequency part only. Therefore, the reconstruction process will reconstruct a new signal by integrating the unprocessed low-frequency component and the regularized high-frequency component, iteratively. The dynamic system shown in Figure 2 will turn stable and automatically terminate when there is no more change happened in the high-frequency component.

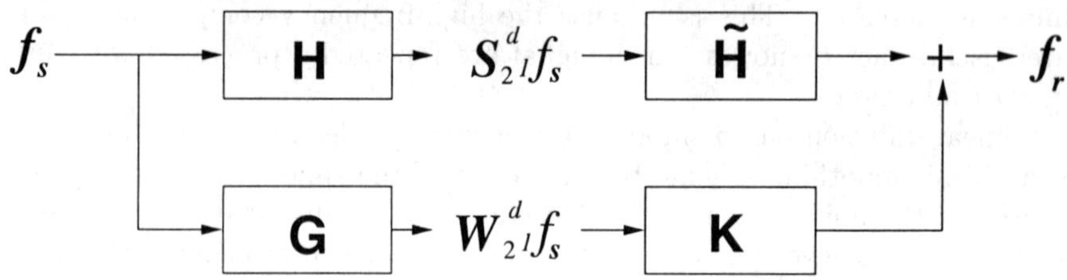

Figure 1: A general two-band filter system.

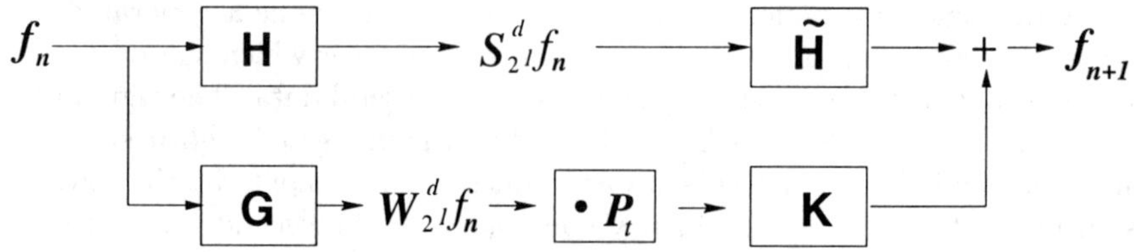

Figure 2: A regularization module.

Wavelet-based Conduction

In what follows, we shall describe how to use the dyadic wavelet transform[2] to realize the mechanism illustrated in Figure 2. Let $f_s(x) \in L^2(R)$. At scale 2^1, the dyadic wavelet transform decomposes f_s into $S^d_{2^1} f_s(x)$ and $W^d_{2^1} f_s(x)$[2] as follows:

$$W^d_{2^1} f_s(x) = (f_s * g_1)(x), \tag{9}$$

and

$$S^d_{2^1} f_s(x) = (f_s * h_1)(x), \tag{10}$$

where g_1 and h_1 are the discrete high-pass filter and low-pass filter at scale 2^1, respectively. The inverse wavelet transform can reconstruct the original signal f_s from $S^d_{2^1} f_s$ and $W^d_{2^1} f_s$ by a procedure shown below:

$$f_s(x) = (W^d_{2^1} f_s * k_1)(x) + (S^d_{2^1} f_s * \tilde{h}_1)(x), \tag{11}$$

where k_1 and $\tilde{h}_1$ ($\tilde{h}_1$ and h_1 are conjugate) are the high-pass filter and low-pass filter for reconstruction at scale 2^1, respectively. Let $p(x)$ be a continuous conductivity function dependent on the absolute value of $W^d_{2^1} f_s(x)$[3]. An approximately reconstructed function $f_{NR}(x)$ of $f_s(x)$ can be defined as follows:

$$f_{NR}(x) = (S^d_{2^1} f_s * \tilde{h}_1)(x) + [p(x) \cdot W^d_{2^1} f_s(x)] * k_1(x). \tag{12}$$

A new iterated function based on Equation (12) can be written as follows:

$$f_{n+1}(x) = (S^d_{2^1} f_n * \tilde{h}_1)(x) + [p_n(x) \cdot W^d_{2^1} f_n(x)] * k_1(x). \tag{13}$$

In order to reduce the complexity for computing $f_{n+1}(x)$, a new dyadic wavelet-based conduction equation is derived. From Equation (13), we can further derive the following:

$$\begin{aligned} f_{n+1}(x) &= (S_{2^1}^d f_n * \tilde{h}_1)(x) + [p_n(x) \cdot W_{2^1}^d f_n(x)] * k_1(x) \\ &= (f_n * h_1 * \tilde{h}_1)(x) + \{[p_n(x) \cdot (f_n * g_1)(x)] * k_1(x)\}. \end{aligned} \tag{14}$$

Taking the Fourier transform of $f_{n+1}(x)$, we obtain

$$\begin{aligned} \hat{f}_{n+1}(\omega) &= F\{(f_n * h_1 * \tilde{h}_1)(x)\} + F\{[p_n(x) \cdot (f_n * g_1)(x)] * k_1(x)\} \\ &= \hat{f}_n(\omega) \cdot H_1(\omega) \cdot \tilde{H}_1(\omega) + F\{p_n(x) \cdot (f_n * g_1)(x)\} \cdot K_1(\omega) \\ &= \hat{f}_n(\omega) \cdot |H_1(\omega)|^2 \\ &\quad + \frac{1}{2\pi} \cdot \{P_n(\omega) * [\hat{f}_n(\omega) \cdot G_1(\omega)]\} \cdot K_1(\omega). \end{aligned} \tag{15}$$

Because $|H_1(\omega)|^2 + G_1(\omega)K_1(\omega) = 1$[2], Equation (15) can be rewritten as follows:

$$\begin{aligned} \hat{f}_{n+1}(\omega) &= \hat{f}_n(\omega)[1 - G_1(\omega) \cdot K_1(\omega)] \\ &\quad + \frac{1}{2\pi} \cdot \{P_n(\omega) * [\hat{f}_n(\omega) \cdot G_1(\omega)]\} \cdot K_1(\omega) \\ &= \hat{f}_n(\omega) - \hat{f}_n(\omega) \cdot G_1(\omega) \cdot K_1(\omega) \\ &\quad + \frac{1}{2\pi} \cdot \{P_n(\omega) * [\hat{f}_n(\omega) \cdot G_1(\omega)]\} \cdot K_1(\omega) \\ &= \hat{f}_n(\omega) + \{\frac{1}{2\pi} \cdot P_n(\omega) * [\hat{f}_n(\omega) \cdot G_1(\omega)] \\ &\quad - \hat{f}_n(\omega) \cdot G_1(\omega)\} \cdot K_1(\omega). \end{aligned} \tag{16}$$

Taking the inverse Fourier transform of Equation (16), we obtain

$$\begin{aligned} f_{n+1}(x) &= f_n(x) + [p_n(x) \cdot [(f_n * g_1)(x)] - (f_n * g_1)(x)] * k_1(x) \\ &= f_n(x) + \{[p_n(x) - 1] \cdot [(f_n * g_1)(x)]\} * k_1(x) \\ &= f_n(x) + \{[p_n(x) - 1] \cdot W_{2^1}^d f_n(x)\} * k_1(x) \\ &= f_n(x) - \{c_n(x) \cdot W_{2^1}^d f_n(x)\} * k_1(x), \end{aligned} \tag{17}$$

where $c_n(x) = 1 - p_n(x)$ and $c_n(x)$ is the traditional diffusivity function[3].

When $c_n(x)$ is set to 1, Equation (17) becomes

$$\begin{aligned} f_{n+1}(x) &= f_n(x) - (W_{2^1}^d f_n * k_1)(x) \\ &= f_n(x) - (f_n * g_1 * k_1)(x). \end{aligned} \tag{18}$$

Again, we take the Fourier transform of Equation(18) and obtain

$$\begin{aligned} \hat{f}_{n+1}(\omega) &= \hat{f}_n(\omega) - \hat{f}_n(\omega) \cdot G_1(\omega) \cdot K_1(\omega) \\ &= \hat{f}_n(\omega)[1 - G_1(\omega) \cdot K_1(\omega)] \\ &= \hat{f}_n(\omega)|H_1(\omega)|^2 \\ &= \hat{f}_n(\omega) \cdot H_1(\omega) \cdot \tilde{H}_1(\omega). \end{aligned} \tag{19}$$

Applying the inverse Fourier transform to both sides of Equation(19), we obtain

$$\begin{aligned} f_{n+1}(x) &= (f_n * h_1 * \tilde{h}_1)(x) \\ &= (S_{21}^d f_n * \tilde{h}_1)(x). \end{aligned} \quad (20)$$

The high-frequency components of $f_n(x)$ can be completely removed in this case because $\tilde{h}_1$ is a low-pass filter. On the other hand, if $c_n(x)$ is set to zero, then Equation (17) becomes

$$f_{n+1}(x) = f_n(x)$$

EXPERIMENTAL RESULTS

In the experiments, two 1-D ideal signals (synthetic data) and a real image were adopted to corroborate the effectiveness of the proposed method. Figure 3(a) showed a 1-D ideal signal of some sinusoidal waves with different periods and amplitudes. Figure 3(b) was a noisy signal (Gaussian white noises) with variance 15. After applying our approach, the results obtained at 25th, 50th and 100th iteration were shown in Figures 3(c), (d) and (e), respectively. It is obvious that when the iteration number increased, noises can be removed more significantly and the signals get closer to the original one. Figure 4(a) showed another 1-D ideal signal with some step edges. Figure 4(b) was a noisy case of Figure 4(a) (Gaussian white noises with variance $\rho = 10$). The results obtained at 30th, 60th, and 120th iteration were shown in Figures 4(c), (d), and (e), respectively. It is also apparent that the results get closer to the original one when the iteration number increased. Figure 5 showed a real test image: Mona Lisa. The result obtained by applying our approach at 10th iteration was shown in Figure 6.

CONCLUSION

In this paper, we have proposed a new dyadic wavelet-based conduction approach for selective image smoothing. In our approach, a nonlinear conductivity function has been considered in the wavelet-based function decomposition and reconstruction process. Since the proposed approach does not require to solve a PDE, it is therefore more efficient and accurate than the conventional nonlinear diffusion/conduction-based methods. Experimental results using 1-D synthetic data and a real image demonstrated that the proposed method can efficiently remove noises and preserve real data.

REFERENCES

[1] D. L. Donoho and I. M. Johnstone, "Ideal spatial adaptation by wavelet

shrinkage," Tech. rep., Department of Statistica, Stanford University, 1992.

[2] S. Mallat and S. Zhong, "Characterization of signal from multi-scale edges," **IEEE Transactions on Pattern Analysis and Machine Intelligence**, vol. 14, no. 7, pp. 710–732, July 1992.

[3] P. Perona and J. Malik, "Scale-space and edge detection using anisotropic diffusion," **IEEE Transactions on Pattern Analysis and Machine Intelligence**, vol. 12, no. 7, pp. 629–639, July 1990.

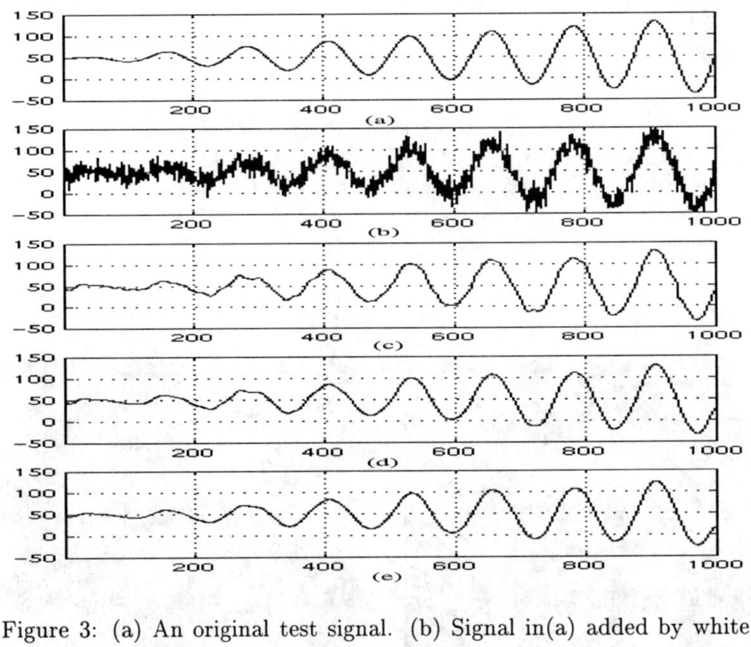

Figure 3: (a) An original test signal. (b) Signal in(a) added by white noise with variance $\rho = 15$. (c) Result obtained by applying our approach at 25th iteration. (d) Result obtained at 50th iteration. (e)Result obtained at 100th iteration.

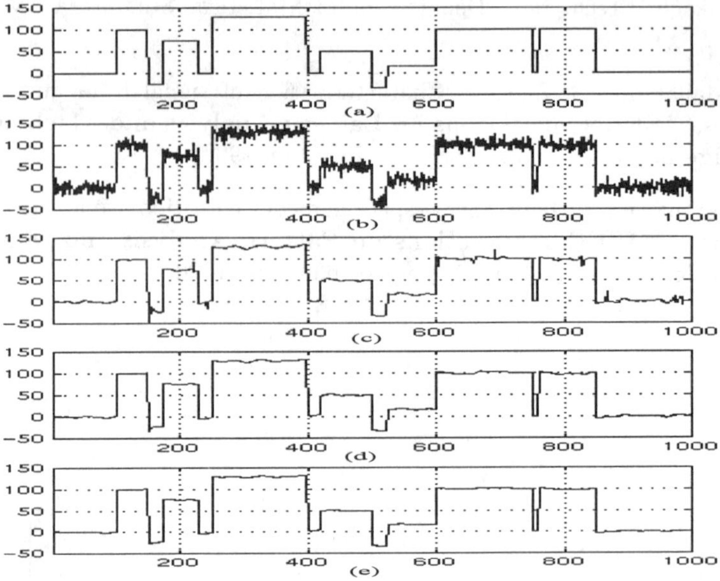

Figure 4: (a) An original test signal. (b) Signal in(a) added by white noise with variance $\rho = 10$. (c) Result obtained by applying our approach at 30th iteration. (d) Result obtained at 60th iteration. (e)Result obtained at 120th iteration.

Figure 5: A test image: Mona Lisa

Figure 6: Results obtained by applying our approach at 10th iteration.

An Adaptive Metric Learning Procedure for Reconfigurable Facial Signature Authentication

Takami Satonaka, Tatsuo Otsuki and Takao Chikamura
Matsushita Electronics Co.
Saiwai-cho 1-1, Takatsuki, Osaka, Japan

Abstract
We present an adaptive metric learning procedure with improved generalization of missing training data for facial signature recognition for use in a smart card system. The conventional learning models suffer from degraded recognition rate due to poor estimation of the margin of a decision boundary. Our model employs an image synthesis method to represent missing patterns of unknown classes by using a mixture distribution. The margin of a decision boundary is dynamically adjusted to input patterns obtained from synthesized images with a time-varying mixing ratio. The metric parameters of mixture distributions have been derived from minimization of the negative log-likelihood probability function. The present method effectively reduces the margin of a class with an improved recognition rate from 81.3% to 100%. Furthermore, we examine the margin structure and select the minimum number of support vectors to represent mixture distributions by using the generalized portrait(GP) method.

1. INTRODUCTION

In this paper, we describe an adaptive metric learning procedure for secure facial signature authentication utilizing a contactless smart card system, which enables dynamic reconfiguration of a neural network[1].

A number of promising approaches assuming probability decision functions have attained good performance in classifying patterns belonging to a finite set of training classes[2][3][4]. In facial signature recognition, facial signature patterns of unknown false users such as intruders are not included in training classes. This makes the estimation of a decision function significantly difficult. The decision function can not separate some patterns belonging to the missing classes of training samples.

For learning from an incomplete set of training samples, we propose an adaptive metric learning procedure based on a mixture model of local prior distributions. In the previous paper[4], we assign a variable number of the mixture classes to each class and determine the mixture number and local metric parameters to minimize the cross entropy between the real distribution and the modeled one.

0-7803-5673-X/99/$10.00

To classify patterns of missing classes correctly, the present model constructs mixture distributions from available training classes by using image synthesis method and the model selection method. The optimal metric parameters of mixture distributions have been derived from minimization of the negative log-likelihood probability function. The metric parameters are dynamically adjusted to input patterns obtained by image synthesis with a time varying mixing ratio. The image synthesis method proposed has an effect to increase the size of training samples enough to fill in missing patterns.

We present two stages of adjusting margin of decision boundary. Firstly, the most likelihood class k_{min} with the maximum margin is selected from a finite set of training classes. Secondly, the margin of class k_{min} is minimized to classify correctly the input patterns of unknown false user classes.

Moreover, we examine margin structure of a decision boundary by using the generalized portrait(GP) method[5][6][7][8]. The GP method determines the trade-off between the number of adjustable parameters and the model complexity in selecting training samples constructed by image synthesis. We can obtain the minimum number of training patterns as supporting vectors which are closest to the decision boundary[6][7]. We present facial signature recognition system using a 13.56MHz contactless smart card, which enables dynamic configuration of a neural network without the need of a large facial database.

2. LEARNING MODEL

2.1 Generalization of missing pattern distributions

For secure facial signature recognition, we present an adaptive metric neural network to represent missing training classes of unknown false users such as intruders. Figure 1 illustrates pattern distributions of correct-user class K_1, known false-user classes K_2 and unknown false-user class K_3 in facial pattern space.

An ideal boundary of correct-user class K_1 is defined by using a complete set of training patterns of class $\overline{K}_1$. In conventional learning, margin of a decision boundary of class K_1 is maximized by using patterns of class K_2 and may include some patterns of class K_3 due to poor generalization of class $\overline{K}_1$, where $\overline{K}_1 = K_2 \cup K_3$. An unknown pattern of class K_3 makes the estimation of a local pattern distribution of $\overline{K}_1$ significantly difficult.

Our model employs two stages of adjusting the margin of a decision boundary by using a mixture distribution. The metric parameters of a mixture distribution are adjusted to an input pattern of a mixture class iteratively. Firstly, the most likelihood class k_{min} ($k_{min} \in K_1$) with maximum margin is derived from mixture distributions of finite training classes K_1 and K_2. Secondly, the margin of a decision boundary of class k_{min} is reduced to separate input patterns of unknown false class K_3. The present image synthesis method constructs the mixture distribution of class $\overline{K}_1$ from patterns of class K_1 and K_2.

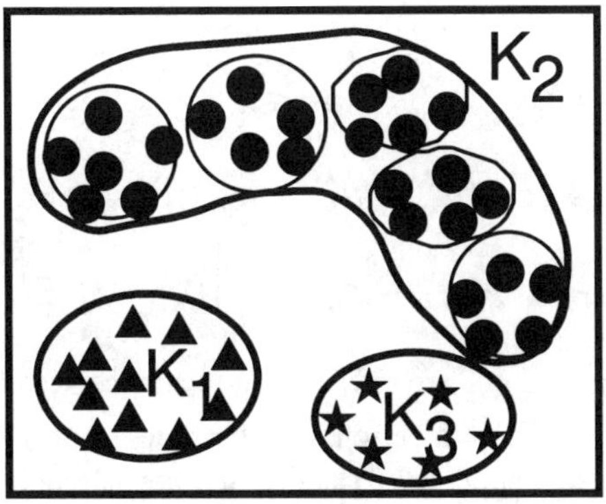

Figure 1. Facial signature pattern space with correct-user class K_1, known false-user classes K_2 and unknown false-user class K_3.

2.2 An adaptive metric learning model

An adaptive metric network is constructed by minimizing the negative log likelihood function of a normal distribution:

$$l(x,\mu^k,\sigma^k) = -\log P(x,\mu^k,\sigma^k)$$
$$= \sum_{i=1}^{N} \frac{1}{2}\left(\frac{x^i - \mu^{k,i}}{\sigma^{k,i}}\right)^2 + \sum_{i=1}^{N} \ln(\sqrt{2\pi}|\sigma^{k,i}|). \qquad (1)$$

Each component of the input vector is assumed to be independently and identically distributed(i.i.d).The first term of equation (1) is the Mahalanobis distance between an input vector and a reference vector $\overline{\mu}^k = (\mu^{k,1},\cdots,\mu^{k,N})$.

To obtain metric parameters from small training samples, we propose a mixture model using the minimization of the cross entropy between an unknown true distribution $P(\mu^k, \sigma^k)$ and a modeled one $Q(\mu^k, \sigma^k)$, which is equal to $-\log P(\mu^k, \sigma^k)$. Small training samples which are close together can give rise to an inappropriate local minima in $-\log P(\mu^k, \sigma^k)$ and may give poor representations of the true distribution. We can obtain optimal metric parameters from the mixture distributions by increasing the size of training samples in cases where the learning parameters have a fixed complexity[4]. We construct a mixture class of prior distributions by using image synthesis:

$$x_{(t)}^{k(j),i} = x_{(t)}^{k,i} + (x_{(t)}^{j,i} - x_{(t)}^{k,i})\zeta \exp(-\frac{t}{\eta}). \qquad (2)$$

The mixture class is allocated to each class k and metric parameters are adjusted to an input pattern of the mixture class. In our simulation, the mixing rate with ζ=0.5 is decreased as a function of learning iteration t and decay factor η=100. For each input vector, we estimate the likelihood function of equation (1) and update components of reference and variance vectors for the first and second most likelihood classes k and l, respectively as depicted in the following equations:

$$\mu_{(t+1)}^{k,i} = \mu_{(t)}^{k,i} + \beta \cdot (x_{(t)}^{j,i} - \mu_{(t)}^{k,i}), \qquad j \in k. \tag{3}$$

$$\mu_{(t+1)}^{l,i} = \mu_{(t)}^{l,i} - \beta \cdot (x_{(t)}^{j,i} - \mu_{(t)}^{l,i}), \qquad j \notin l. \tag{4}$$

$$\sigma_{(t+1)}^{k,i} = \sigma_{(t)}^{k,i}(1 + \alpha \cdot g(|x_{(t)}^{j,i} - \mu_{(t)}^{k,i}| - \sigma_{(t)}^{k,i})), \qquad j \in k. \tag{5}$$

$$\sigma_{(t+1)}^{l,i} = \sigma_{(t)}^{l,i}(1 - \alpha \cdot g(\sigma_{(t)}^{l,i} - |x_{(t)}^{j,i} - \mu_{(t)}^{l,i}|)), \qquad j \notin l. \tag{6}$$

Where $\mu_{(t)}^{k,i}$, $\mu_{(t)}^{l,i}$, $\sigma_{(t)}^{k,i}$ and $\sigma_{(t)}^{l,i}$ are reference and variance vectors for correct class k and false class l . An unit function g(u) becomes 1 or 0 for u ≥ 1 or u< 1, respectively. The parameters α =0.01 and β=0.1 are selected for simulation.

From a fixed number of facial classes in training database, an adaptive metric neural network selects the most likelihood class defined by

$$k_{\min} = argmin \quad l(x, \mu^k, \sigma^k). \tag{7}$$

Our model employs two stages of adjusting the margin of a decision boundary by using a mixture distribution. The first stage is to determine class $\boldsymbol{k}_{\min}$ with the maximum margin among a finite number of classes in the database. The second stage is required to classify an arbitrary pattern of a missing training class. To define the decision boundary of $\boldsymbol{k}_{\min}$, we present a mixture distribution of $\bar{k}_{\min}$ by using image synthesis:

$$x^{\bar{k}_{\min},i} = x^{k_{\min},i} + (x^{j,i} - x^{k_{\min},i})\zeta \quad (x^{\bar{k}_{\min},i} \in \bar{k}_{\min}), \tag{8}$$

where mixing rate ζ =0.5 is selected in this simulation.

2.3 Margin structure and supporting vector

In our model, a decision boundary of class $\boldsymbol{k}_{\min}$ is obtained by assuming a proper mixture distribution of class $\bar{k}_{\min}$. We examine margin structure of a decision boundary by using the GP method[5], which provides the measure to determine the trade-off between the number patterns of a mixture distribution and complexity of classification. The mixture patterns are selected from the training set of (x, y_i) as support vectors. (y_i=1 for $x \in \boldsymbol{k}_{\min}$ and y_i=-1 for $x \in \bar{k}_{\min}$).
The support vector machine constructs a nonlinear decision boundary of the form

$$f(x) = sign(\sum_{i=1}^{N} \alpha_i y_i K(x, x') + b), \tag{9}$$

which is equivalent to an optimal separating hyperplain in the higher dimensional space($\boldsymbol{K}(x,x_1)$, $\boldsymbol{K}(x,x_2)$, $\boldsymbol{K}(x,x_3)$... $\boldsymbol{K}(x,x_N)$), where x_1, x_2, x_3, ..., x_N are the support vectors. The Lagrange functional

$$L(w,b,\alpha) = \frac{1}{2}\|w\|^2 - \sum_{k=1}^{l} \alpha_k \{[(w \cdot \varphi(x) + b]y_k - 1\}, \tag{10}$$

has to be minimized with respect to w and b. The Lagrange duality transforms the primal problem of equation to the following dual representation.

$$\max_{\alpha} W(\alpha) = \max_{\alpha}\{\min_{w,b} L(w,b,\alpha)\},$$
$$= \max_{\alpha}\{-\frac{1}{2}\sum_{i=1}^{l}\sum_{j=1}^{l}\alpha_i\alpha_j y_i y_j k(x_i,x_j) + \sum_{i=1}^{l}\alpha_i\}, \qquad (11)$$

with the constraints, $\alpha_i \geq 0, \quad i = 1,\cdots,l$ and $\sum_{i=1}^{l}\alpha_i y_i = 0$. (12)

Thus, a simple quadratic programming problem has to be solved to obtain an optimal margin of a decision boundary. The resulting decision function is

$$D(x) = w\phi(x) + b = \sum_{k} y_k \alpha_l k(x_k, x_l) + b_0, \qquad (13)$$

where $b_0 = -\frac{1}{2}(w\varphi(x_{k\min}) + w\varphi(x_{\bar{k}\min}))$. (14)

3. RESULTS

3.1 Learning with a finite set of classes

The database consists of 60 persons with 10 distinct facial expressions. The original image of 114x 88 pixels is transformed into an image of 16x16 pixels. The coefficients derived from the 16x16 two dimensional discrete cosine transform(2DCT) are used for learning[4].

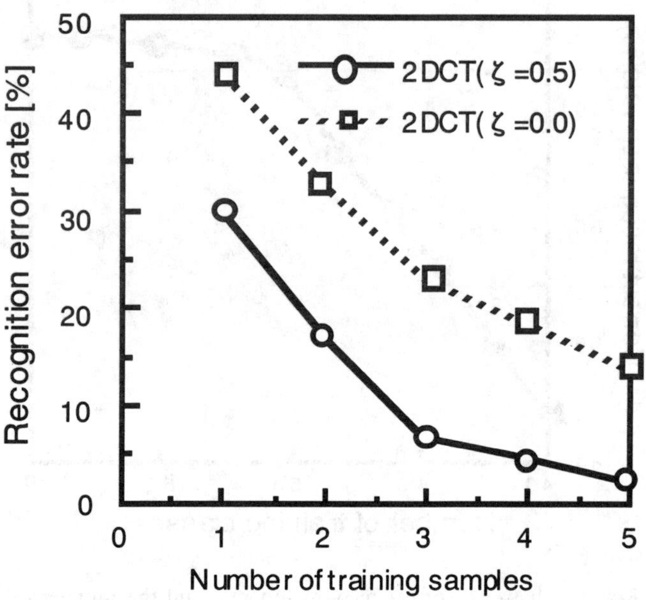

Figure 2. Relation between recognition error rate and varying number of training samples from 1 to 5 with and without image synthesis.

In the adaptive metric learning, we assume a mixture distribution to obtain a decision boundary with the maximum margin among a finite set of training classes. The performance of neural network constructed with image synthesis is compared with that of neural network without it for 40 training classes. Figure 2 shows the relation between average recognition error rate and varying number of training samples from 1 to 5 of each class. The error rates of 2DCT(ζ=0.5) with image synthesis are better than these of 2DCT(ζ=0.0) without it. The image synthesis method proposed improves the recognition error rate by about six times from 13.6% to 2.4% for five training samples. It has an effect to increase the size of training samples enough to fill in missing patterns. We obtain optimal metric parameters from the mixture distributions for a fixed number of training classes by increasing the size of training samples.

3.2 Learning with missing training classes

We present two stages of adjusting the margin of a decision boundary in the learning with unknown false-user class K_3 as shown in figure 1. The first stage is to select the most likelihood class k_{min} with the maximum margin among classes K_1 and K_2. The second one is to adjust the margin of a decision boundary of class k_{min} ($k_{min} \in K_1$) with respect to $\bar{k}_{min}$, where $\bar{K}_1 = K_2 \cup K_3$.

The main issue is to assume a proper distribution of $\bar{k}_{min}$ by using training patterns of class K_2. The choice on the number of classes K_2 is crucial for deriving a decision boundary of classes $\bar{k}_{min}$.

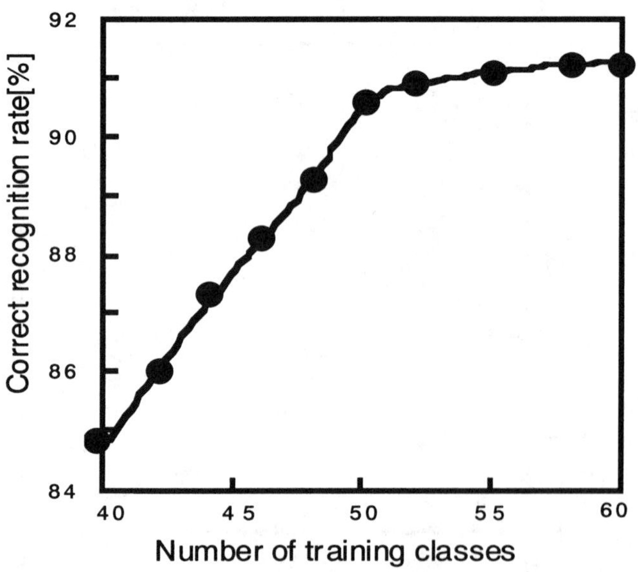

Figure 3. Relation between correct recognition rate and the number of training classes in K_2.

Figure 3 shows the relation between the correct recognition rate and the number of training classes in K_2. We allocate n_1 and n_2 neurons to classify patterns of classes K_1 and K_2. (Where n_1=1 and n_2=40 to 60.) The adaptive metric neural network is trained by ten patterns from each class of K_1 and K_2. The recognition rate to classify test patterns of k_{min} remains 100%. The correct recognition rate to classify test patterns of unknown false user classes K_3 depends on the number of classes in K_2. The correct recognition rate is improved from 87.3% to 91.3% when the number n_2 is increased from 44 to 60. This means the margin of a decision boundary of k_{min} is reduced with an increase of number n_2.

To reduce the margin of classes k_{min} effectively, we assume a mixture class $\bar{k}_{min}$ obtained by using image synthesis in the second stage. Figure 4 shows the relation between correct recognition rate and the number of classes in K_1. The result obtained by assuming mixture class $\bar{k}_{min}$ is compared with these without assuming it. The correct recognition rate at the first stage to classify test patterns in unknown false user classes K_3 is decreased when the number n_1 of classes in K_1 is increased. The recognition rates at the second stage are significantly improved from 81.3% to 100% for five mixture classes. The margin of the maximum likelihood class has been reduced by assuming mixture class $\bar{k}_{min}$.

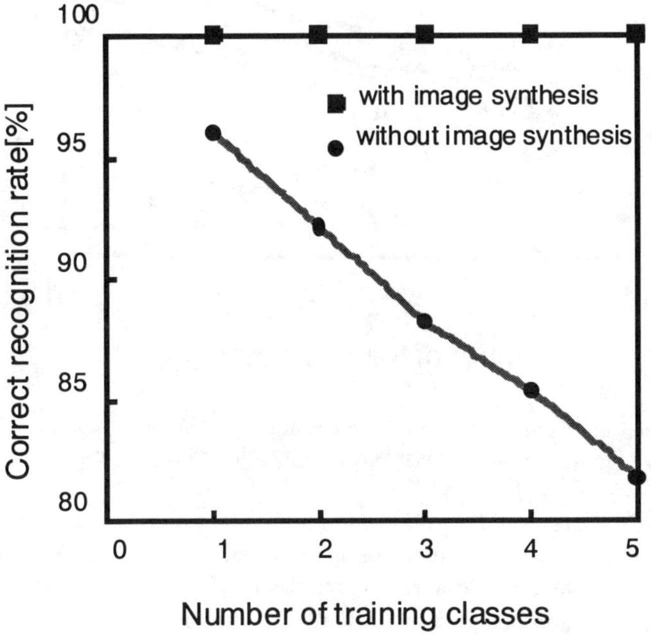

Figure 4. Relation between recognition rate and number of training classes in K_1.

3.3 Margin structure and supporting vectors

In our model, a decision boundary of class k_{min} is derived by assuming a mixture class $\bar{k}_{min}$ to represent unknown false user classes correctly. We examine margin structure of a decision boundary by using the generalized portrait(GP) method. Facial signature patterns of $\bar{k}_{min}$ are constructed by image synthesis as shown in equation (8) with ζ=0.5. Each class in $\bar{k}_{min}$ consists of ten patterns, which are used for image synthesis.

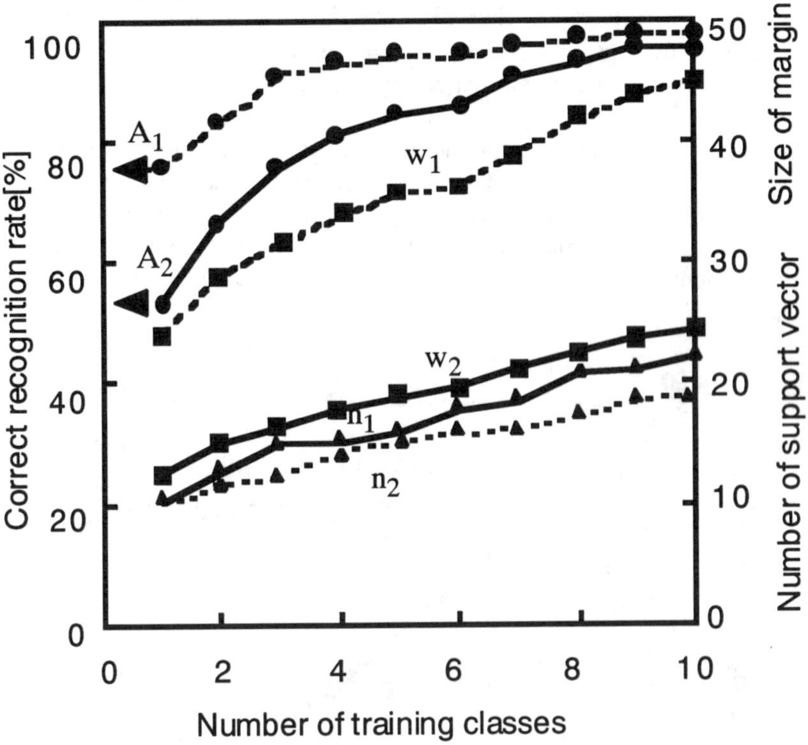

Figure 5. Relation of correct recognition rates, number of support vectors and size of margin to number of training classes with and without image synthesis.

We estimate the validity of assumption for a mixture class $\bar{k}_{min}$ by using recognition rates to classify patterns of unknown false-user classes K_3. Figure 5 shows the relation of recognition rate A_j, number n_j of support vector and margin size $1/w_j$ of a decision boundary to the number of training classes in $\bar{k}_{min}$. In the GP method, a radial basis function is used as a kernel function[6]. The present results (j=1) with image synthesis are compared with these (j=2) without it. The correct recognition rate A_1 with image synthesis is better than that of A_2 without it. The margin $1/w_1$ is smaller than that of $1/w_2$. With an increase of the number of classes, recognition rates A_j are increased and margin sizes $1/w_j$ are decreased .

In our model using image synthesis, a mixture class $\bar{k}_{min}$ reduces the margin of a decision boundary k_{min} significantly and improves the correct recognition rate to classify false patterns of $\bar{k}_{min}$. The GP method determines the trade-offs between the number of adjustable parameters and the complexity of classification. The number n_j of support vectors is not increased. The support vectors near the decision boundary are selected from the training samples obtained by image synthesis.

4. SMART CARD BASED FACE RECOGNITION

Figure 6 illustrates face recognition system utilizing a contactless FeRAM smart card. The smart card LSI integrates 20K byte FeRAM, 64bit encryption circuit, 13.56MHz transceiver and 8 bit RISC processor. FeRAM memory attains about 100ns writing time and 3.0V writing voltage[1].

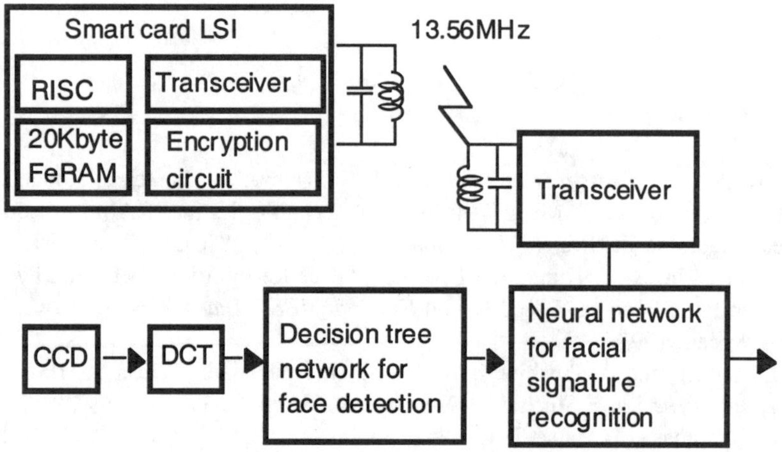

Figure 6. Facial signature recognition system utilizing a contactless FeRAM smart card

By storing supporting vectors of facial signature patterns, the smart card enables facial signature recognition suitable for electronic commerce transactions. This system configures a neural network by loading supporting vectors from the smart card and can identify input facial signature patterns without using large size of facial database. The smart card LSI is powered on by receiving electric energy from 13.56MHz radio frequency field via an antenna circuit. It transmits personal identification number and facial signature patterns to the transceiver. Then, this system starts facial feature extraction and facial recognition.

Face recognition network consists of a tree-structured decision network[9] for face area detection and an adaptive metric network for facial signature recognition. A template image with 32x32 pixels is used to detect a face area with eyes, nose and mouth. The template image is normalized by histogram equalization and transformed by DCT transform. The decision tree network is constructed by using lower frequency parts of DCT coefficients of template images[4]. The adaptive metric network is trained by using DCT features of template facial images with 114x 88 pixels.

5. CONCLUSIONS

We propose an adaptive metric learning employing image synthesis for learning from incomplete training samples, such as facial signature recognition. A mixture distribution utilizing image synthesis is proposed for representing missing training classes. The margin of a decision boundary is dynamically adjusted to input patterns obtained from synthesized images. The optimal metric parameters of mixture distributions have been derived from minimization of the negative log-likelihood probability function. We present a facial signature authentication system using a contactless card without the need of a large facial database.

REFERENCES

[1] K. Asari, Y. Mitsuyama, T. Onoye, I. Shirakawa, T. Hirano, T. Honda, T. Otsuki, T.Baba and T. Meng," Multi-Level Technologies for FRAM Embedded Reconfigurable Hardware, " *In technical digest of ISSCC99*, Feb,1999.

[2] S. H. Lin, S. Y. Kung, and L. J. Lin, "Face Recognition/ Detection by Probabilistic Decision-Based Neural Network,"*IEEE Trans. Neural Networks*, Vol. 8, Jan. 1997.

[3] T. Hastie and R. Tibshirani, "Discriminate Analysis by Gaussian Mixture", *Journal of the Royal Statistical Society,* Vol.58, pp155-176, 1996.

[4] T. Satonaka, T. Baba, T. Otsuki, T. Chikamura and T. Meng, "A DCT-based Adaptive Metric Learning Model Using Asymptotic Local Information Measure", *In Proc. of NNSP97*,Sept.,1997.

[5] B. Boser, I. Guyon,V.Vapnik, "A training algorithm for optimal margin classifiers," *In Proc. of 5th ACM Annual Workshop on Computational Learning Theory.* pp144-152, Pittsburgh, July1992.

[6] V. N. Vapnik, *The Nature of Statistical Learning Theory , Information, Prediction, and Estimation*, NY, Springer-Verlag,1996.

[7] E. Osuna, R. Freund, F. Girosi, "Training Support Vector Machines: An Application to Face Detection," *In Proc. of CVPR'97*, June,Puerto Rico, 1998.

[8] D. Schuurmans, "Improving and generalizing the basic maximum margin algorithm," *In Proc. of Workshop on Support Vector Machines, NIPS97*, Corolado Dec. 1997.

[9]K. Perlmutter, N. Chaddha, J. Buckheit, R. Gray, and R. Olshen, "Text segmentation in mixed-mode images using classification trees and transform tree-structured vector quantization" , *In Proc. of ICASSP 1996.*

GLOBAL COLOR IMAGE SEGMENTATION STRATEGIES: EUCLIDEAN DISTANCE VS. VECTOR ANGLE

Slawo Wesolkowski[1], Robert D. Dony[2] and M. E. Jernigan[1]
[1]Systems Design Engineering
University of Waterloo
Waterloo, Ontario, Canada
[2]School of Engineering
University of Guelph
Guelph, Ontario, Canada
e-mail: s.wesolkowski@ieee.org, rdony@uoguelph.ca, jernigan@uwaterloo.ca

Abstract

In the past few years, researchers have been increasingly interested in color image segmentation. In this paper, we analyze two different global image segmentation algorithms each using its own distance metric: k-means and a mixture of principal components (MPC) neural network. The k-means uses Euclidean distance for color comparisons while the MPC neural network uses vector angles. Two variants of the algorithms will be examined. The first uses the RGB pixel itself for clustering while the second uses a 3x3 neighborhood. Preliminary results on a staged scene image are shown and discussed.

1. INTRODUCTION

The trichromatic model of color representation is fundamental to the human perception of color. Three variables are required to completely describe a point in color space. There exist several equivalent three-dimensional color spaces with varying characteristics [2]. One group of spaces uses Cartesian coordinates to represent points in the space. Examples include the three primary illumination colors RGB, the complementary colors CMY, and the opponent color representation YIQ. An alternative set of spaces employs polar coordinates and includes the HSI and HSV spaces [2,3].

The RGB model corresponds most closely with the physical sensors for colored light such as the cones in the human eye and is implemented as red, green, and blue filters in most color CCD sensors. However, the human perception of color qualities more closely follows the HSI model [14]. The intensity (I) represents the overall brightness which is independent of color and is a linear value. Hue (H) represents the fundamental or dominant color and is measured as an angle on a color circle with the three primary colors spaced 120° apart. Finally, saturation (S) represents the purity of a color, where pastel shades have low saturation values while pure spectral colors are completely saturated. The latter two values specify the chromaticity of a color point.

Because of the added complexity of needing three variables to represent color pixels, the issue of region segmentation in color images is not as well defined as for gray-scale images. In gray-scale images, regions are typically modeled as uniform intensity areas and segmentation algorithms employ some form of

0-7803-5673-X/99/$10.00 © 1999 IEEE

Euclidean distance measure to determine pixel similarity either on a spatially local basis or on a global color basis.

For color images, a number of approaches have been proposed. Most approaches have used the RGB space for their processing [4,8,19,20,22,23] as we do in this paper. Other approaches are based primarily on one color space: HSI [7], HCI [9], CIE Lab [6], and CIE Luv [10]. Some approaches were also designed for non-standard hybrid multidimensional spaces [21]. A few algorithms were designed and tested with a variety of color spaces [17,18].

The color image segmentation field can be roughly divided into four categories: global methods or color clustering [4,6,7,16,18,21], local algorithms [9,23], physical model-based techniques [10,17,20] and hybrid procedures [8,16,22].

We will focus the review on several global image segmentation techniques and some hybrid methods with a color clustering component given that the paper will focus on comparing two global image segmentation methods.

Yang describes a method based on the classification of a vector composed of two angles and the magnitude describing the color pixel RGB triplet [4]. The method uses two thresholds to determine region homogeneity: comparing the Euclidean distance between the angles of two vectors to a "vector direction tolerance" and comparing the magnitude to a second threshold. Woelker implements a color clustering method in CIE Lab space with each cluster being defined as a cube with a Euclidean distance-based tolerance [6]. Weeks and Hague adapted the well known k-means algorithm to perform image segmentation in the HSI space [7].

Park et al. describe an approach based on morphological transforms [18]. They first smooth the 3-D RGB histogram using a difference of Gaussians and then apply closing and dilation operators segment the image. Wu et al. use a fuzzy k-means clustering algorithm for preliminary segmentation of RGB map images for the purpose of extracting lines and text [21]. A neural network is then trained on a set of cluster centers which are used to segment the image with a nearest neighbor approach.

Other methods first use color clustering before more specific region segmentation or merging. For example, Hedley and Yan [8], as well as Zhou et al. [22], use distance metrics which incorporate a spatial component to complement their color clustering approaches in order to mitigate the effects of outliers and anti-aliasing.

The concentration of the research on the RGB color space is logical since it would avoid costly conversions of images into a different color space. Only YUV and YIQ transformations seem to be fast enough to allow processing of images in real time. Transformations to perceptually based spaces as CIE Lab, CIE Luv and various flavors of HSI are very computationally intensive. It is, therefore, desirable to find fast RGB-based methods for real-time applications such as check processing [11].

Euclidean distance and its variants have long been used for image segmentation. However, the Euclidean distance between two colors in RGB does not necessarily reflect the visual separation between these two colors [1,10]. Therefore, a distance measure, which might quantify hue and saturation, might be more suitable for processing color images. The vector angle measure has been recently introduced for that purpose [1]. This distance measure is specifically characterized by its sensitivity to the chromatic information in the image.

This paper is organized in the following manner. The second section will describe the distance measures used in the global image segmentation algorithms: Euclidean distance and vector angle. The third section describes the two global image segmentation algorithms which will be compared (k-means, and a neural network architecture) with two neighborhood variants. The fourth section compares results using both distance measures. The fifth section discusses shortcomings of both methods and possible extensions. The paper concludes with the sixth section.

2. DISTANCE MEASURES

2.1. Euclidean Distance

The Euclidean distance (D_E) metric is usually used to compute distance in N-dimensional vector space. It is defined as

$$D_E(\vec{v}_1,\vec{v}_2) = \left\|\vec{v}_1 - \vec{v}_2\right\| \tag{1}$$

where ||•|| is the L_2 vector norm. For a three plane color space, the distance calculated is

$$D_E(\vec{v}_1,\vec{v}_2) = \sqrt{(v_{1,1}-v_{2,1})^2 + (v_{1,2}-v_{2,2})^2 + (v_{1,3}-v_{2,3})^2} \tag{2}$$

where $\vec{v}_1 = [v_{1,1}\ v_{1,2}\ v_{1,3}]^T$ is a color triplet.

The Euclidean distance does not quantify color similarity in the RGB space as well as in the LUV space [8]. However, in general it can be said that D_E is sensitive to variations in intensity, but not very sensitive to variations in hue and saturation [1].

2.2. Vector Angle

The vector angle (D_A) is an alternate distance metric [1] and is defined as

$$D_A = \cos\theta = \frac{\vec{v}_1^T\vec{v}_2}{\left\|\vec{v}_1\right\|\cdot\left\|\vec{v}_2\right\|} \tag{3}$$

As opposed to Euclidean distance, vector angle is insensitive to intensity differences, but quantifies well hue and saturation differences. In general, there might certain drawbacks to using the angle θ as a similarity distance metric such as the problematic computation of statistics on values in angular coordinates [13] or the complex calculation of the inverse cosine. However, in this case we are only concerned with the calculation of the dot product as will be shown in Section 3.2.

2.3. Euclidean Distance vs. Vector Angle

The definition of how good a distance metric is largely depends on the application for which it is going to be used. Both of these distance measures have advantages and disadvantages in different situations, which could be easily exploited. It has been shown that vector angle generates strong edges in areas of differing hues and very weak edges in areas of high hue similarity [1]. However, hue difference becomes irrelevant at low RGB values since small variations in one of the three values can produce large differences in the angle between two colors. Low intensity areas are not a problem for Euclidean distance, which is unable to assess the similarity of two pixels given their similar hues [1,10].

3. IMAGE SEGMENTATION

There are several general techniques of performing global image segmentation: k-means, k-nearest neighbor, and various neural network architectures. We will examine in detail the k-means approach and the Mixture of Principal Components (MPC) neural network. None of the nearest neighbor approaches can be used due to the very high computational cost per image. We used both the individual pixel and a 3x3 neighborhood around the pixel for clustering. For the 3x3 case, the clustering is done on a 27-element vector since there are three RGB values for each pixel in the neighborhood. This allows for a smoothing effect, which eliminates or mitigates the effect of image noise.

3.1. K-Means

The k-means (KM) algorithm is defined as follows [5]:

1. Choose the number of classes, k.
2. Choose initial cluster centers $\vec{\mu}_1, \vec{\mu}_2, \ldots, \vec{\mu}_k$.
3. Classify each data point.
4. Recompute estimates for $\vec{\mu}_i$ with results of 3.
5. If $\hat{\mu}_i$ are consistent, stop; else go to step 3.

Step 3 is carried out using a simple 1-nearest neighbor approach [5]. The number of classes, k, and choosing initial cluster centers, $\vec{\mu}_1, \vec{\mu}_2, \ldots, \vec{\mu}_k$, are still two challenging problems. In this paper, we will define the number of classes for k-means to operate on and we chose the initial cluster centers randomly.

The Euclidean distance is very suitable for the k-means algorithm since this distance metric can be directly related to the computation of the cluster centers.

3.2. Mixture of Principal Components Neural Network

A new approach to data representation, a mixture of principal components (MPC), has recently been developed [15,24]. Figure 1 illustrates the modular architecture of the network for a single basis vector per class.

Each module consists of a basis vector, which performs a linear transformation on the input data. In addition, each module represents a class of input data and the basis vector of each module defines the class. The input vector, $\vec{x} \in \Re^N$, is linearly transformed by each of the K modules resulting in K coefficient vectors, $\vec{y}_1, \vec{y}_2, \ldots, \vec{y}_K$. Each module contains a single basis vector, $\vec{w}_i$ defining an one-dimensional linear subspace. So, if $\vec{w}_i$ is of dimension N, the coefficient y_i is calculated as

$$y_i = \vec{w}_i^T \vec{x} \tag{4}$$

The classifier then chooses the output of the winning module based on the subspace classifier

$$\arg\{\max_i(\| y_i \|)\} \tag{5}$$

For a network with unit norm basis vectors, this classifier is equivalent to maximizing equation (3), i.e., minimizing the vector angle between the basis vector, $\vec{w}_i$, and the input, $\vec{x}$.

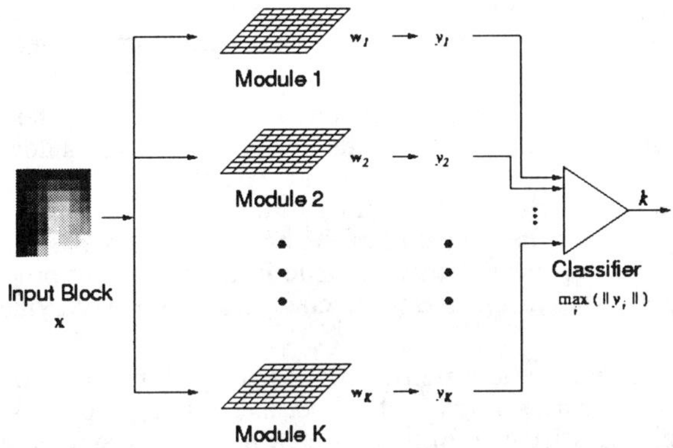

Figure 1: Coding section for MPC network with K classes and M components per subspace.

The basis vector then defines the prototypical angle, which represents the class. Contrast this with k-means, which uses the mean as the class prototype. Therefore, the use of the k-means algorithm to calculate the mean of the class values for the vector angle prototype is inappropriate. Instead, the principal angle of the class is required, or equivalently, the first principal component of the class data (i.e. Euclidean distance is to the class vector mean, what vector angle is to the class first principal component). To compute the principal components while refining the class definitions, an iterative algorithm is employed:

1. Choose the number of classes, K.
2. Initialize the K principal components, $\vec{w}_1, \vec{w}_2, \ldots, \vec{w}_K$, to some appropriate set of values.
3. Classify a data point using the subspace classifier (minimum vector angle) using equations (4) and (5)
4. Modify the basis vector, $\vec{w}_i$, of the winning class using a Hebbian neural network algorithm that extracts the class principal component [12].
5. If no convergence, go to 3, else stop

4. RESULTS

To test the new clustering algorithm, we have constructed a scene and captured its image (see Figure 2[1]). It has several important features: shadows, areas with the same hue, but varying saturation or intensity, areas of differing hue, but similar intensity. The image has been manually segmented (see Figure 3) to show clearly the different regions and for a quantitative performance evaluation (see text below).

Both examined image segmentation methods were applied globally with the cluster number set at eight (we found that less than eight was not practical for this image). We are concerned with examining the effects of the distance measures and are not addressing the cluster validity problem here. The results are shown in Figures 4-7.

For grayscale processing, it can easily be seen that the D_E performs as expected: it groups all areas of similar intensity together (see Figure 5). A 1x1 neighborhood would have given no result for D_A since there is no angle. In a 3x3 case, although there is no hue, the classifier extracts important neighborhood features such as edges.

The impact of using D_A is clearly seen on the color image. The first striking characteristic of the processed images is the disappearance of shadows as separate classes from the objects they fall on. It can be seen that many shadows are included in a variety of classes in Figures 4 and 6. In Figures 5 and 7, these shadows have now mostly become part of the background objects. This is a direct result of using hue as a principal characteristic in the clustering procedure. It can be seen that the hue-based clustering is consistent with a varying number of classes.

The colour image segmentation metrics proposed by Borsotti et al. [16], it was concluded that these metrics would not reflect the results properly since they are based on the Euclidean distance. In this paper, we are proposing an alternate way to measure the performance of the above clustering algorithms.

This method is based on the idea of maximized pixel consistency within a truthed region. To calculate this measure, it is first necessary to manually segment (or truth) the regions of the images that are being examined (see Figure 3 for an illustration). Next, the number of different pixel classes in each separate region is counted. First, the number of pixels which corresponds to the most numerous class in each region is added up. These are considered class-consistent pixels since they best represent the truthed region. Next, the second largest number of pixels for each region is summed, and so forth, up to the number of clusters that the algorithm is trying to classify (in our case either eight).

An example of the class frequency distribution is shown in Figure 8. It can be noticed that the MPC/D_A method has a higher pixel consistency rate than the KM/D_E method. For example, 25% of the KM/D_E results are inconsistent whereas for MPC/D_A this figure is 11%. From both the visual and quantitative performance evaluations, it seems that the D_A measure shows promise. The performance measure, however, does not effectively take into account the possibility that two adjacent regions may be recognized as the same class. This shortcoming is currently under investigation.

Furthermore, there are some problems with the vector angle distance measure. One is its inability to properly process regions of varying intensity. This can be

[1] The image is available at: http://www.uoguelph.ca/~rdony/nnsp99/kzcolour1.gif

noticed especially when comparing the small bottle shapes in the lower part of Figures 4 and 5 or Figures 6 and 7. In this case, the Euclidean distance clustering algorithm performs much better. Another problem is distinguishing between objects of very similar hue. This can be noticed by observing that the green object on the light green background at the bottom of the image is lost in all cases.

The smoothing effect of the 3x3 neighborhood can be noticed by comparing Figures 4 & 6 (i.e. the big blue detergent bottle in the upper half of the image) and by examining Figures 5 and 7 (i.e. various pieces of paper in the upper portion of the image).

5. DISCUSSION

The D_A approach seems to be a good method for obtaining hue and saturation similarity between two pixels in an image. Its major drawback is that it breaks down when the colors are close to 0 (black). In this region, small variations such as noise can cause significant changes in angle. Of course, in the limit, a zero valued vector has no defined angle. The D_E approach is more appropriate for applications where intensity differences are more important. The problem with the latter lies in its inability to determine that the chromaticity of two pixels are identical. It would make sense, then, to try to merge both distance metrics to come up with a method which would use intensity and chromaticity information directly from the RGB image. This was already attempted within various color spaces: in RGB using the three element angle/magnitude representation of a color [4], in HSI [7] and in HCI using a fuzzy logic-based set of operators [9].

Recently, the vector angle and the Euclidean distance metrics have been combined into a unified distance metric and applied to the color edge detection problem [19]. A trade-off parameter based on either the saturation or the intensity component was used to weigh the contribution of both distance metrics. Preliminary results show the elimination of edges caused by shadows usually present when processing gray scale images or color images using Euclidean distance-based metrics while at the same time reinforcing edges between regions of dissimilar saturation and constant hue. We are currently carrying out similar experiments for image segmentation.

As mentioned before, the application is very important when assessing the results of an algorithm. In this case, image segmentation constitutes only a secondary step in an image understanding process. The image segmentation methods shown here should be evaluated in a broader context to verify that their functioning is consistent with our preliminary results. More tests will be carried to fully assess the usefulness of the new distance measure. We are currently investigating differences in performance of both algorithms with respect to other color spaces (e.g., LAB, LUV and HSI).

6. CONCLUSION

The authors have presented a new approach for color image segmentation in the RGB color space. It uses the sine of the vector angle between adjacent pixel values to evaluate their chromatic similarity. Unlike Euclidean distance-based region segmentation, the new method is insensitive to differences in intensity and only produces non-zero values for purely chromatic differences.

7. REFERENCES

[1] R. D. Dony and S. Wesolkowski, "Edge Detection on Color Images Using RGB Vector Angle," Proc. IEEE Canadian Conference on Electrical and Computer Engineering (CCECE), Edmonton, Canada, May 1999.

[2] R. C. Gonzalez and R. E. Woods, Digital Image Processing. Addison-Wesley, 1993.

[3] H. Levkowitz, Color Theory and Modeling for Computer Graphics, Visualization, and Multimedia Applications. Kluwer Academic Publishers, 1997.

[4] Y. Yang, ``Colour edge detection and segmentation using vector analysis," Master's thesis, Electrical and Computer Engineering, University of Toronto, Toronto, Canada, 1995.

[5] R. Schalkoff, and J. Robert, Pattern Recognition: Statistical, Structural and Neural Approaches. John Wiley & Sons, Inc.: New York, 1992.

[6] W. Woelker, "Image segmentation based on an adaptive 3D analysis of the CIE-L*a*b* color space." Visual Comm. and Image Processing '96, Proc. SPIE Vol. 2727, pp. 1197-1203, February 1996.

[7] A. R. Weeks, and G. E. Hague, "Color segmentation in the HSI color space using the K-means algorithm." Nonlinear Image Processing VIII, Proc. SPIE Vol. 3026, pp. 143-154, April 1997.

[8] M. Hedley and H. Yan, ``Segmentation of color images using spatial and color space information," Journal of Electronic Imaging, vol. 1, pp. 374-380, October 1992.

[9] T. Carron and P. Lambert, ``Symbolic Fusion of Hue-Chroma-Intensity Features for Region Segmentation," in Proc. IEEE International Conference on Image Processing, pp. 971-974, October 1996.

[10] L. Shafarenko, M. Petrou, and J. Kittler, ``Automatic watershed segmentation of randomly textured color images," IEEE Trans. on Image Processing, vol. 6, pp. 1530-1544, November 1997.

[11] S. Wesolkowski, P. Bowman, and D. Tunnah, ``A high speed colour cheque image capture and processing system," in Proc. IEEE CCECE, pp. 457-460, 1998.

[12] S. Haykin, Neural Networks: A Comprehensive Foundation. New York, NY: Macmillan, 1994.

[13] G. S. Watson, Statistics on Spheres. Wiley-Interscience, 1983.

[14] Q.-T. Luong, "Color in computer vision," Handbook of Pattern Recognition and Computer Vision, C.H. Chen, L.F. Pau and P.S.P. Wang editors., pp. 311-368. World Scientific Publishing Company, 1993.

[15] R. D. Dony, and S. Haykin, "Neural network approaches to image compression," Proc. IEEE, vol. 83, pp.288-303, February 1995.

[16] M. Borsotti, P. Campadelli, and R. Schettini, "Quantitative evaluation of color image segmentation results," Pattern Recognition Letters, vol. 19, pp. 741-747, 1998.

[17] J. Liu and Y.-H. Yang, "Multiresolution Color Image Segmentation," IEEE Trans. PAMI, vol. 16, no. 7, pp. 689-700, July 1994.

[18] S. H. Park, I. D. Yun, and S.U. Lee, "Color Image Segmentation Based on 3-D Clustering: Morphological Approach," Pattern Recognition, vol. 31, no. 8, pp. 1061-1076, 1998.

[19] S. Wesolkowski and M.E. Jernigan, "Color Edge Detection in RGB Using Jointly Euclidean Distance and Vector Angle," Proc. IAPR Vision Interface 1999, Trois-Rivières, Canada, May 1999.

[20] T. Caelli, and D. Reye, "On the Classification of Image Regions by Colour, Texture and Shape," Pattern Recognition, vol. 26, no. 4, pp. 461-470, 1993.
[21] J. Wu, H. Yan, and A. N. Chalmers, "Color image segmentation using fuzzy clustering and supervised learning," Journal of Electronic Imaging, vol. 3, no. 4, pp. 397-403, October 1994.
[22] J. Zhou, D. Lopresti, and T. Tasdizen, "Finding Text in Color Images," Document Recognition V, Proc. SPIE Vol. 3305, pp. 130-140, 1998.
[23] A. Tremeau, and N. Borel, "A Region Growing and Merging Algorithm to Color Segmentation," Pattern Recognition, vol. 30, no. 7, pp. 1191-1203, 1997.
[24] R. D. Dony, and S. Haykin, "Multi-class maximum entropy coder," in IEEE Int'l Conf. Systems, Man, and Cybernetics, October, 1995.

Figure 2: Gray version of original color scene image.

Figure 3: Hand-segmented truth image.

Figure 4: 1x1 KM/D_E on Figure 3 with 8 classes.

Figure 5: 1x1 MPC/D_A on Figure 3 with 8 classes.

Figure 6: 3x3 KM/D_E on Figure 3 with 8 classes.

Figure 7: 3x3 MPC/D_A on Figure 3 with 8 classes.

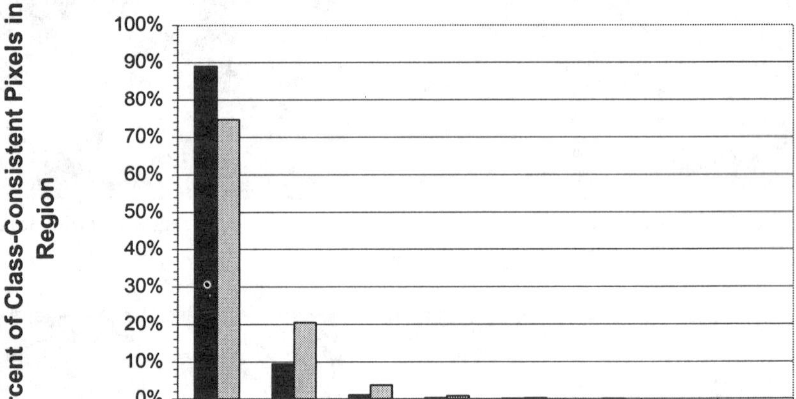

	0	1	2	3	4	5	6	7
■ MPC	89.066	9.481%	1.067%	0.305%	0.069%	0.012%	0.001%	0.000%
▨ k-means	74.709	20.483	3.725%	0.841%	0.202%	0.031%	0.006%	0.003%

Most Consistent Class Within Region

Figure 8: Distribution of most class-consistent pixels

A Topology Independent Active Contour Tracking

Dongxiang Xu, Jenq-Neng Hwang

Information Processing Laboratory
P.O.Box # 352500
Department of Electrical Engineering
University of Washington, Seattle WA 98195
Email: {xdx, hwang}@ee.washington.edu

Abstract

In recent years, active contour (Snake) has become one of the most powerful segmentation algorithms in image processing and computer vision. However, most algorithms based on this model have difficulties in automatic initialization and are hard to handle the problems with topology changes or multiple-objects tracking. In this paper, we propose a novel algorithm, Quad-tree Highest Confidence First (QHCF), for image segmentation first. Based on it, a new framework, called Markov Random Filed (MRF) Based Snake, is then put forward to provide a general purpose image segmentation solution. Since this method combines the most attractive features of MRF and Active Contour model, it provides a more accurate segmentation results. Finally, we further extend this framework to multiple-object scenario and propose a topology independent segmentation algorithm. Experimental results are provided demonstrate its encouraging performance.

1. Introduction

Segmentation is one of the most difficult parts in image processing and computer vision research. In the last decade, researchers have developed numerous approaches for this purpose.[1] These methods are implemented from different perspectives and have shown their successes by applying to various images.

However, most of the existing algorithms have certain limitations when applied in complicated situations. This is mainly due to the models that these algorithms based on is not general enough to cover most of the applications. For example, the active contour model can only process the images with known object topology, and also an initial outline which is close to the object of interest is required to be provided. Therefore, it is sometimes difficult to use them for object tracking in complicated situations with multiple objects.

In this paper, we propose a new framework for image segmentation. It combines the most attractive features of Markov Random Field's reliable region segmentation and active contour model's global optimal boundary tracking. Moreover, it can be easily extended to the case of multiple objects tracking, and hence provides a topology independent object segmentation solution.

The rest of this paper is organized as follows: Section 2 gives an insight review of MRF and active contour model. In Section 3, we propose our algorithm. Section 4 extends it to the multiple-objects scenario and conduct topology-

0-7803-5673-X/99/$10.00 © 1999 IEEE

independent contour tracking. The experimental results and conclusion are given in Section 5.

2. MRF and Active Contour Model

2.1 *MRF Model*

An MRF assumes that the information contained in the local structure of images is sufficient to describe the global image, i.e., the image intensity at a particular location depends only on a neighborhood of pixels and characterized by means of a local, conditional probability distribution. This hypothesis provides a convenient and consistent way for modeling observed images.

Assume X={x} is a realization of random filed when the observed image is Y={y}. At any location s in the image I, the pixel's intensity is y_s with label x_s. The *a posteriori* probability of segmentation result given the observed image can be expressed as:

$$p(\mathrm{X} \mid \mathrm{Y}) \propto p(\mathrm{Y} \mid \mathrm{X}) p(\mathrm{X}), \tag{1}$$

where p(Y|X) is the conditional probability of the observed image given the scene segmentation. In this study, the conditional density is modeled as a Gaussian process, with mean μ_s and variance σ^2_s for the region that s belongs to. Thus, the intensity of each region can be regarded as a signal μ_s plus additive zero mean Gaussian noise with variance σ^2_s, and the conditional density can be expressed as:

$$p(\mathrm{Y} \mid \mathrm{X}) \propto \exp\{-\sum_s \frac{1}{2\sigma_s^2}[y_s - \mu_s]^2\} \cdot \tag{2}$$

Due to the equivalence between MRF and Gibbs Random Field (GRF) from Hammersley-Clifford theorem, the prior probability p(X) follows Gibbs distribution and can be expressed as:

$$p(\mathrm{X}) = \frac{1}{Z}\exp(-\sum_s V_s(x)), \tag{3}$$

where $V_s(x)$ is called the clique energy function at location s with neighborhood G_s. To have a robust segmentation, additional edge constraint is also considered[2] in the model. Therefore, the clique energy function consists of two parts:

$$V_s(x) = V_{sN}(x) + V_{sE}(x), \tag{4}$$

where the $V_{sN}(x)$ is the energy function by considering the spatial constraint of pixel s and its neighbors. $V_{sE}(s)$ is the energy function with edge constraint. The introduction of edge information is very straightforward. Comparing with energy function in traditional MRF models[4,6], this additional constraint can describe the energy function more accurate so that it makes the label updating process more sensitive at the regions boundaries..

From above, we can find the *a posteriori* probability can be obtained as:

$$p(\mathrm{Y} \mid \mathrm{X}) \propto \exp\{\, -\sum_s (y_s - \mu_{x_s})^2 - \sum_s [V_{sN}(x) + V_{sE}(x)] \,\} \cdot \tag{5}$$

and we define the energy function of the whole image:

$$E(\mathrm{X}) = \sum_s (y_s - \mu_{x_s})^2 + \sum_s [V_{sN}(x) + V_{sE}(x)] \cdot \tag{6}$$

2.2 *Active Contour Model*

The original active contour model was first introduced by Kass et al at 1988. It is a deformable curve controlled by two parts, the internal spine energy E_{int} and the external feature energy E_{ext}. The E_{int} imposes the smoothness constraint on the curve and E_{ext} is decided by the distinct features. More specifically, we assume the closed boundary of a region R is expressed as $C(s)=(x(s), y(s))$, where s is the arc length of the contour. The typical energy is defined as:

$$\begin{aligned} E(C) &= \int_{\Omega} \{E_{\text{int}}(s) + E_{ext}(s)\} ds \\ &= \int_{\Omega} \{\frac{1}{2}(\alpha|C_s(s)|^2 + \beta|C_{ss}(s)|^2) + E_{ext}(s)\} ds \end{aligned} \qquad (7)$$

where $C_s(s)$ and $C_{ss}(s)$ are the first order and second order derivatives of $C(s)$.

Although the classical active contour based algorithms have achieved very good performance in boundary tracking, some well known problems are associated with them that limit its applications, such as curve initialization, energy function definition and regularization and the selection of elasticity parameters α and β. Moreover, the topology change scenarios, such as splitting and merging of object contours, also can not be handled by this model.

To overcome some of these problems, a new algorithm, Minimal Path Approach (MPA), was proposed by Laurent et al.[3] It can detect the global minimum of the energy curves given only two end points of an open curve and makes the initialization an easier task. For the case of tracking closed contour, Laurent proposed to provide only one initial point on the search routine and use saddle point searching to accomplish automatic initialization[3]. However, since the saddle points are based on potential field ***P*** which highly depends on the pixels' intensity and the edge detection results, therefore, the whole contour tracking can be easily misled by spurious edges generated by the noise or those very complicated contexts. Hence the MPA has difficulties to reliably extract the closed contour with just one point in most complex scenarios images.

3. Proposed MRF-based Snake Algorithm

To have a better performance in the closed contour detection, this paper presents a new algorithm. It incorporates an enhanced Markov Random Field solution, *Quad-tree Highest Confidence First (QHCF)*, into the active contour based algorithm. Instead of using the saddle point in potential function, the proposed method searches the most confident points on the contour, *control points*, and then use the two-end-point based MPA algorithm to find the curves between each two adjacent control points. By controlling the distribution of these control points, unbalanced errors and unnecessary computations can be avoided and it improves the algorithm's performance.

3.1. *QHCF algorithm*

Many algorithms have been invented to minimize MRF's energy function and they are often classified into two categories. One is stochastic based methods, such as Simulated Annealing (SA)[6] and Gibbs sampler, which are theoretically optimal, but computationally intolerable for practical use. The other category is deterministic methods, of which the Highest Confidence First (HCF)[5] is the most popular.

In the original HCF algorithm, a predefined number of classes $\boldsymbol{K}$ is needed in advance. Then the whole image will be initially segmented (usually by K-means) and each region is specified with a label number $\boldsymbol{k}$, ($1<\boldsymbol{k}<\boldsymbol{K}$). Further segmentation starts based on this initialization. However, as the segmentation result is very sensitive to the choice of $\boldsymbol{K}$, it is very crucial to have a reliable initial estimation.

To overcome this initialization problem, the proposed QHCF algorithm first divides the whole image into four equal blocks by Quad-tree splitting. By checking the value of *Region Criterion* (**RC**) V_{rc}, each block will be evaluated whether it is qualified to be an individual region. The **RC** for each block $\boldsymbol{B_i}$ is defined as:

$$\mathrm{V}_{rc}(i) = \mathrm{var}(B_i)\,. \tag{8}$$

If $\boldsymbol{V_{rc}}$ is smaller than a predefined threshold $\boldsymbol{T}$, the block is assigned a new label. Otherwise, it is divided into smaller blocks by Quad-tree splitting until every small block becomes a region. Comparing with the uniform grid initialization presented in [2], Quad-tree initialization is more accurate and efficient by adaptively finding the labels and hence reducing the number of iterative comparisons in following labeling process. Also, it avoids the "unlabeled small region" problems occurred in [2].

Under this initialization, the confidence of each pixel is defined as:

$$c_s(x) = \begin{cases} \max\{E_s(x_k) - E_s(x_i)\}, & s \textit{ is committed with label } k, \\ E_s(0), & s \textit{ is uncommitted with label } 0; \end{cases} \tag{9}$$

which represents how much we are confident to update the pixel s's label with a new one. Obviously, the more stable of the label-updated image due to the change, the more confident we are that the label at s should be updated.

At the very beginning of the algorithm, all the pixels are assigned label zero (uncommitted) and their confidence are defined as $\boldsymbol{c_s(x)=E_s(x)}$. In each block, the pixel whose intensity is the most close the mean of each region are assigned an unique label. Since the neighbors are uncommitted, they have no contribution to $\boldsymbol{U_N}(s)$ and $\boldsymbol{U_E}(s)$. After labeling these pixels, the energy and confidence of their surrounding pixels will be updated. Selecting the one with largest confidence, we assign it a new label with which the pixel has the lowest energy. If this pixel is uncommitted, then it will become committed. If it has already been committed, it can only change from one label to another label and can not become uncommitted again. The process of confidence updating and label assigning repeats until no more pixels are with positive confidence.

3.2 *Control points searching*

After segmenting the image with QHCF, every pixel is assigned a label with which the pixel has the largest *a posteriori* probability. With this segmentation, the contour of target region can be easily found by searching region's boundary points. However, since MRF model has not any constraint on smoothness of object's contour during the segmentation process, its result is not good enough for some practical applications, such as lumen's contour tracking in medical images. To overcome this problem, a solution is proposed in our algorithm. Basically, it consists of the following steps:

Step 1. Find the boundary points of the region of interest based on the QHCF segmentation;

Step 2. Divide the region's boundary into sections;

Step 3. Select the most reliable boundary point in each section as control point for MPA.

Assume the region of interest is labeled with p and object's boundary is divided into M sections. In section m, $0<m\leq M$, there are totally i_m points. Since these points are on the boundary, they are adjacent to at least one neighboring regions. To simplify the problem formulation, we only consider those points which are with one adjacent region. Given a particular point with label p, suppose its adjacent region's label is q, the a posteriori probability of this point with label p and q can be expressed respectively as:

$$p(x_s = p \mid y) \propto \exp\{ -(y_{i_m} - \mu_p)^2 - \sum_{s\in S_{i_m}} [U_N(x_s = p) + U_E(x_s = p)] \}, \quad (10)$$

$$p(x_s = q \mid y) \propto \exp\{ -(y_{i_m} - \mu_q)^2 - \sum_{s\in S_{i_m}} [U_N(x_s = q) + U_E(x_s = q)] \}.$$

It is obvious that the *a posteriori* probability of the real boundary pixel belonging to label p should usually have bigger value with label p than that with its adjacent region's label q. However, due to the effect from other regions in a noisy image, these two probabilities are generally quite close. For the above two reasons, we can define the reliability of boundary points as:

$$r(s) = 1 - [p(x_s = p \mid y) - p(x_s = q \mid y)]. \quad (11)$$

The value of the reliability is within range [0,1]. If s from the segmented contour is more likely to be a boundary point, its *a posteriori* probability with label p and q will be quite similar. It hence leads to the value of $r(s)$ closer to 1 and makes point s with higher reliability. Therefore, in the control point searching process, we use ***Maximum Reliability*** (MR) as the criterion, which can be expressed as:

$$s = \max_{0<i\leq i_m}\{r(s_i)\} \quad (12)$$

where m is the section number, i_m is the number of boundary points s in the section.

However, the above discussion about the most likely boundary points does not imply that the MR criterion can directly find those control points. This is because the region shape found by QHCF segmentation is already with very high accuracy in terms of its location and shape. To take advantage of this, in Step 2 we divide the closed boundary into sections and build up accurate initial structure with those control points in each section. The splitting of contour generally depends on the image quality and the confidence of the contour points. If the object boundary is severely distorted by the noise, it is inevitable that more sections should be split so as to reach a higher accuracy. This will make the resulting curve to be more noise-resistant and reliable, as well as to reduce the tracking computation.

Step 1 in the control point searching process is more flexible and application dependent in the implementation. In most cases, the exact location of object of interest is unknown and only some prior knowledge of the object can be referenced as additional constraint during the segmentation process. In such a situation, identification function should be designed based on QHCF segmented regions so as to extract the boundary of region-of-interest, which will be used in the further contour fine-tuning.

3.3. *Contour Extraction*

After finding all the control points alone the contour, every two-adjacent can be paired as the end points for minimal-path active contour model. Assume the a pair of stakes are ***p0*** and ***p1***, the energy is along the curve ***C*** connecting them is defined as:

$$E(C) = \int_{\Omega} \{w \left\| \frac{\partial C}{\partial s}(s) \right\|^2 + P(C(s))\} ds = wL(C) + \int_{\Omega} P(C(s)) ds \qquad (12)$$

C(0)=***p0*** and ***C***(***L***)=***p1;*** and ***L*** is the length of the curve; ***s*** is the arc-length parameter. Minimizing above integration [3], we can solve curve ***C*** and use it as the boundary points between ***p0*** and ***p1***. Linking all the curves between every two adjacent stakes, the closed contour can be tracked.

4. Topology Independent Framework

Most existing contour modeling schemes require that the topology of the object to be known before the tracking action can start. Unfortunately, this is very difficult to be satisfied in some scenarios. For example, in the atherosclerosis study with MR imaging, an important concern in lumen tracking is topological change. Along the image sequence, the closed lumen boundary may split because of the blood vessel's bifurcation. However, the position of topological transformation is normally hard to be known in prior.

The basic idea of the proposed algorithm is to use QHCF algorithm to segment it into regions. Based on this, the candidate objects will be searched within region-of-interest (ROI). Then we can apply the MRF-based active contour algorithm introduced in Section 3 to track each object's contour with high accuracy. The basic procedure is as follows:

(1) Load the input image Y and use QHCF to segment image into *N* meaningful regions with labels X={x_1, *x2*, ... x_N}.

(2) Based on the segmented ROI outline, named *c*, use the connected component algorithm on *c* to create a reference image *r*, where the region within *c* is marked as 1 and the rest pixels are with value 0;

(3) Applying "*AND*" operation between each region of $x_i \in \{x_1, x2, \ldots x_N\}$ with *r*. Those regions with non-zero output are candidates of objects of interest. A region filter which is a knowledge-based decision tree will be applied to them so that those regions representing the true objects will be found from candidate list.

(4) To each object region detected in step (3), apply the MRF-based active contour algorithm to track all the objects' boundaries.

It is easy to see from the above processing flow that the accuracy of the first step plays a very important role in the topology detection. This can be guaranteed by the performance of proposed QHCF algorithm. Step 2 is a preparation procedure for object detection. We apply the two-pass connected component algorithm[7], which uses local equivalence table to avoid paging problem. In Step 3, a region filter is used to remove those pseudo-objects in ROI. This is necessary because the strong noise in original image usually can generate some spurious small regions. To remove the unwanted regions in the ROI outline, a region filter is needed. This is where the prior knowledge is applied and it can be modified to identify different types of objects. In our application, the region filter is implemented with a decision tree based on the features such as region area and intensity mean. Other classification algorithms, such as Maximum Likelihood, fuzzy logic decision, etc. can also used to construct region filter. The last step is a further accurate boundary searching in Step 4 on each region with the MRF-based active contour algorithm.

Compared with existing topology independent algorithms, such as those proposed by Malladi[8] and Wang[9], our framework has the following superior features. First, the boundaries of target objects in ROI are tracked individually in our method so that each contour reaches its global optimization in active contour model separately. Secondly, the structure of our framework is very flexible for incorporating additional knowledge or decision criterion. It is hence easy to adapt it to different applications with minor modification. Moreover, it is also possible to identify arbitrary objects within ROI as target region with appropriate region filter designing and then conduct contour fine-tune.

5. Experiments and Conclusion

In this section, we present the experimental results of the proposed algorithm.

First, a typical spin echo MR image is shown in Fig.1(a). Because of the noise or artifacts during imaging process, the intensity of lumen area is not uniform and there are some isolated bright spots inside. QHCF algorithm is applied to this image and segments it into many regions as shown in Fig.1(c). It is clear that the lumen segmentation result is not affected by those bright spots inside lumen area and most of the noise in background have been removed. It is better than the adaptive ICM[4] segmented result shown in Fig.1(b). By tracking the boundary of lumen region based on purely QHCF segmented results, we obtain the contour points of target region as shown in Fig.1(d). It is obvious that some sharp corners on the top-left part of the contour and the bottom part are also not very smooth. Based on this QHCF contour, we split it into six equal sections and searched the control points (seen in Fig.1(e)) by finding the contour points with Maximum Reliability criterion. The MPA algorithm is then used to track the whole contour and the result is shown in Fig.1(f). Compared with the contour in Fig.1(d) and Fig.1(f), the effect of smooth constraint in MPA algorithm is demonstrated. The two rough parts in Fig.1(c) have also been fine-tuned.

Fig.2 shows the framework's ability of tracking multiple objects' boundary with supervision. As shown in Fig.2, the first step uses the QHCF algorithm to reach the optimal region segmentation as shown in Fig.2(b). Then by applying connected-component algorithm followed by a region filtering, we can identify the two the target regions in Fig.2(c). The boundary tracking result with QHCF is shown in Fig.2(e). Fig.2(f) is the result by further contour fine-tuning with MPA. Obviously, the segmentation result in Fig.2(f) is much better than Fig.2(e) in terms of contour smoothness.

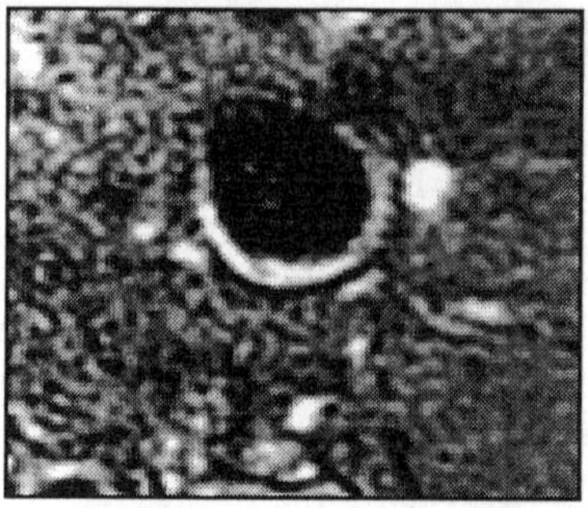

(a) Original MR image with carotid lumen.

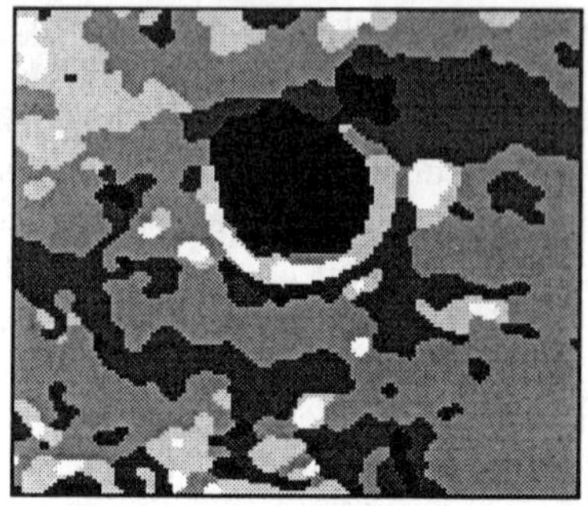

(b) Segmentation by adaptive ICM[4].

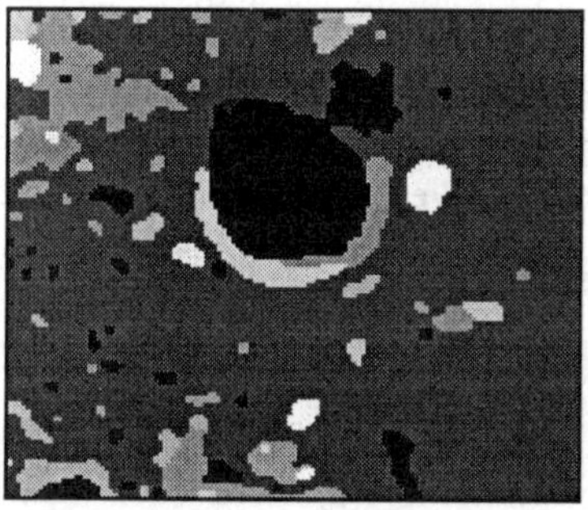

(c) Segmented image by QHCF.

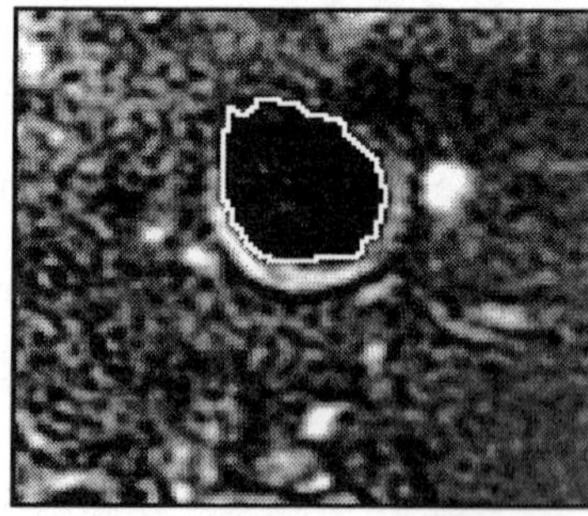

(d) Lumen contour based on QHCF.

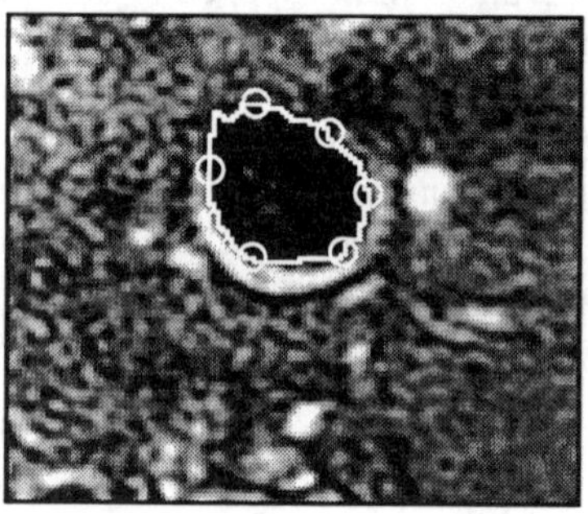

(e) Six control points alone lumen contour based on QHCF segmentation.

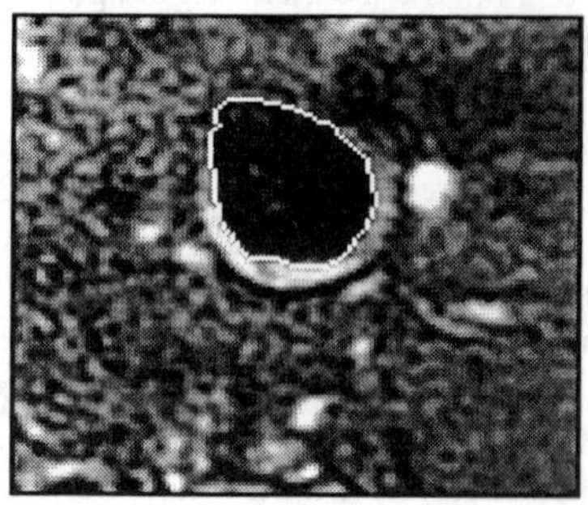

(f) Lumen contour after MPA fine-tune.

Fig1 Lumen segmentation. (a) is the original image; (b) is done by the adaptive ICM algorithm; (c)~(f) shows the results of each procedure in MRF-based Active Contour algorithm.

In this paper, we have proposed algorithm, QHCF and MRF based active contour model as the solutions for image segmentation and automatic multiple-object contour tracking. These algorithms combine some of the most attractive features of Markov Random Field and Active Contour model. We have successfully demonstrated our approach on gray level images and the experimental results demonstrate its encouraging performance.

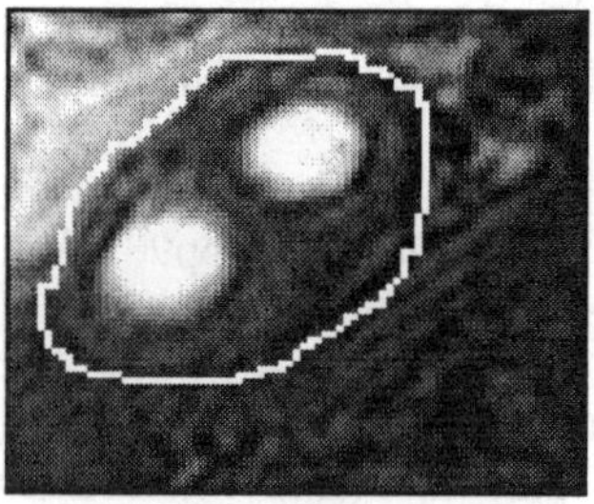

(a) Original MR image. (two lumens for tracking with the ROI outline)

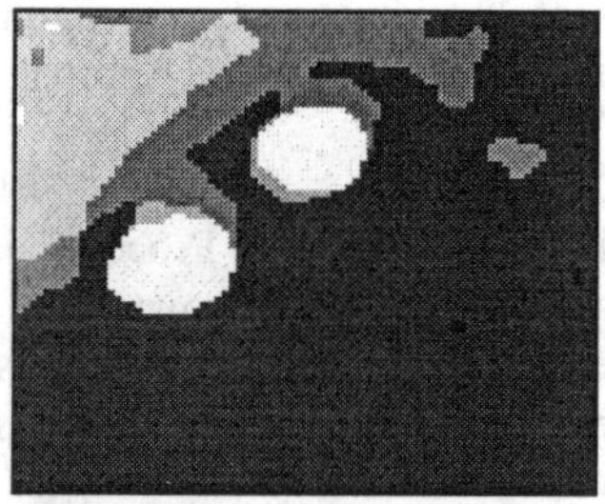

(b) QHCF segmented image.

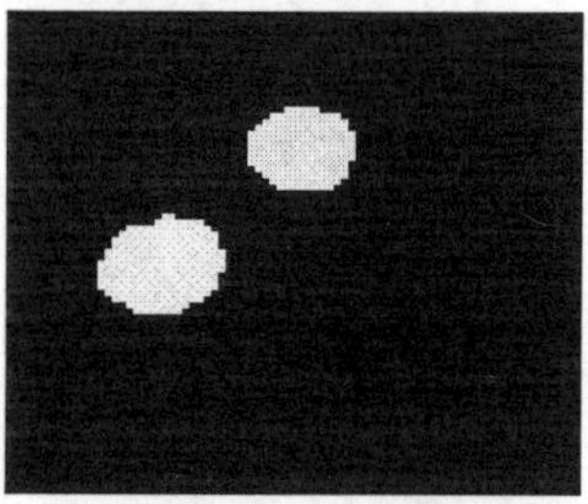

(d) The target region detected by region filter

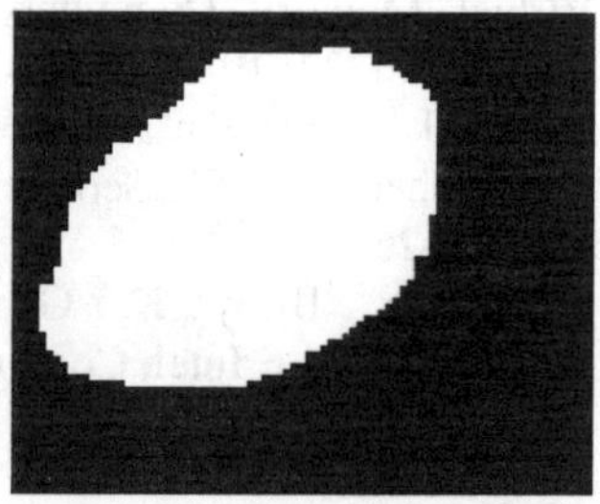

(c) ROI template used for target region searching .

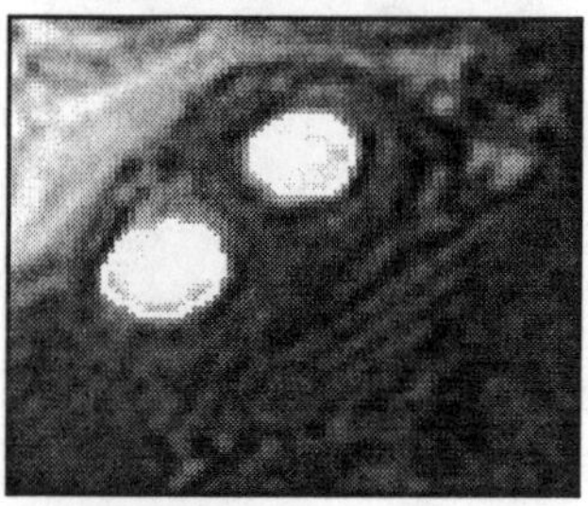

(e) Lumen contour based on QHCF segmentation.

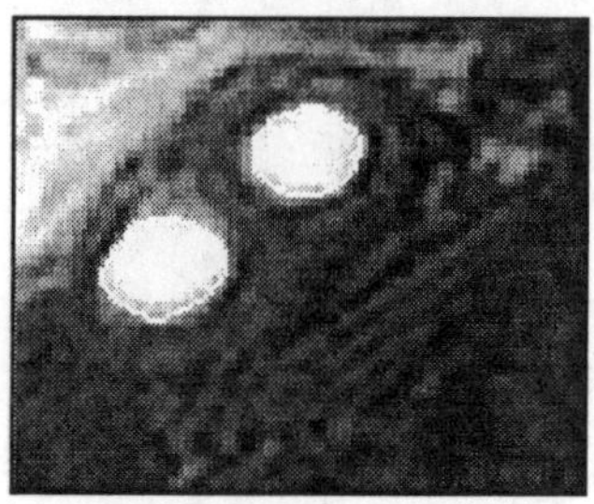

(f) Lumen contour based on the proposed algorithm.

Fig.2 Segmentation of bifurcated carotid lumens. The region-of-interest outline is provided by user.

Reference:

1. N.R.Pal and S.K.Pal, "A review on image segmentatin otechniques", Patter Recognition, vol. 26, no.9, pp 1277-1294, 1995.

2. Thomas Meier, King N. Ngan and Gregory Crebbin, "A Robust Markovian Segmentation Based on Highest Confidence First", ICIP, Santa Barbara, Oct. 1997
3. Laurent D.Chhen and Ron Kimmel, "Global Minimum for Active Contour Model: A Minimal Path Approach," Int. J of Computer Vision, vol. 24, 1997, pp. 57-78.
4. T.N.Pappas, "An Adaptive Clustering Algorithm for Image Segmentation." IEEE Trans. On Signal Processing, vol. 40, no. 4, April 1992, pp901-914.
5. P.B.Chou and C.M.Brown,"The Theory and Practive of Bayesian Image Labeling,", Int. J. of Computer Vision, vol. 4, 1990, pp.185-210
6. S.Geman and D.Geman, "Stochastic Relaxation, Gibbs Distributions, and the Bayesian Restoration of Images," IEEE Trans. on PAMI vol. PAMI-6,no.6, Nov. 1984,pp721-741.
7. R. Lumia, G.Shapiro, and O.Zuniga, "A New Connected Component Algorithm for Virtual Memory Computers," Computer Vision, Graphics and Image Processing, Vol. 22, 1983, pp287-300.
8. R. Malladi, J. A. Sethian and B. C. Vemuri, "A Topology Independent Shape Modeling Scheme", SPIE Geometric Methods in Computer Vision II, Vol.2031, 1993, pp246-258.
9. Haiyan Wang, Bijoy K. Ghosh, "Geometric Deformable Model and Segmentation", IEEE Intel. Conf. On Image Processing, Chicago, USA. 1998.

MODEL COMBINATION AND WEIGHT SELECTION CRITERIA FOR SPEAKER VERIFICATION

Kevin R. Farrell

T-NETIX/SpeakEZ Inc.
67 Inverness Drive East
Englewood, Colorado 80112
kevin.farrell@t-netix.com

Abstract

In this paper, we focus on the score combination for three separate modeling approaches as applied to text-dependent speaker verification. The modeling methods that are evaluated consist of the neural tree network (NTN), hidden Markov model (HMM), and dynamic time warping (DTW). One of the main challenges in combining scores of several models is how to select the weight for each model. One method is to use equal weights for all models used in the combination. Another method is to use the Fisher linear discriminant to select the weights that maximize the ratio of the separation of the inter-class means to the sum of the variances. Both methods are evaluated for three separate databases and the results are compared to the optimal performance as obtained by an exhaustive search over the weight space. Overall, both weight selection methods provide performance close to the optimal point. It is also shown that the optimal combination of three models provides lower error rates than that achievable with two models.

1 INTRODUCTION

Speaker verification consists of determining the authenticity of a person's claimed identity based on a sample of their voice. Speaker verification has numerous applications in telecommunications, facility access, and internet security. Many of these applications pose different implementation scenarios. These include the recording device, channel, noise levels and numerous other characteristics of the environment for the application. For these different applications of speaker verification, one can hardly expect that a single technology will always provide the best solution in every scenario. It is much

0-7803-5673-X/99/$10.00 © 1999 IEEE

more likely that each application will require its own fine-tuned technology to yield the best performance.

It is not practical to develop a new technological approach for every different scenario that can be encountered for speaker verification. We have explored a different option which is to use several different approaches simultaneously and combine their results using data fusion concepts. As long as the errors of the different methods are uncorrelated, improvements can be made with the proper combination. This technique does not only provide robustness for different environments but can additionally improve performance over that obtained when only a single approach is used.

In this paper, we focus on the combination of several different models as applied to speaker verification. The modeling approaches that are used for combination consist of the hidden Markov model (HMM), neural tree network (NTN), and dynamic time warping (DTW). The scores of these different modeling approaches are combined using the linear opinion pool method. This consists of computing a weighted sum of the model scores. Several methods are evaluated for determining the weights to be used for the combination. The model combination is evaluated on data collected from landline telephony, cellular telephony, and multimedia environments.

This paper is organized as follows. The following section provides a brief overview of speaker verification along with a description of the modeling approaches and data fusion method that we consider in this study. This is followed by a description of the equal weight and Fisher discriminant methods as used to determine the weight vector for combining the model scores. Experiment results are provided for several databases collected within landline, cellular, and multimedia environments. A summary of the observations is then given.

2 SPEAKER VERIFICATION

Most commercial applications of speaker verification are text-dependent. Text-dependent systems require that the same password be used for both training and testing. Text-independent systems do not constrain the speech used for testing to be the same as that used for training. Text-dependent speaker verification systems provide an additional level of security over text-independent systems since an impostor must know the password of the person that he or she is attempting to defraud. This paper provides analysis for text-dependent speaker verification.

Text-dependent speaker verification systems typically use modeling approaches that incorporate temporal information within the model. Traditional modeling approaches to text-dependent speaker verification include hidden Markov models (HMMs) [1, 2] and dynamic time warping (DTW) [3] techniques. Neural network approaches, including whole-word neural tree networks (NTNs) [4] and sub-word NTNs [5] have also been evaluated for text-dependent speaker verification.

It has previously been shown that the errors of different modeling approaches, such as the neural tree network (NTN), dynamic time warping (DTW), and hidden Markov model (HMM) are to some degree uncorrelated [6]. Hence, with the proper combination method a system can be developed that provides a lower error rate than any modeling approach being used by itself or even any pair of models. In this paper, we evaluate a three model system for text-dependent speaker verification. The scores of the three models are combined using the linear opinion pool approach for data fusion. The modeling techniques along with the linear opinion pool method of data fusion are briefly described in the following subsections.

2.1 Speaker Verification Models

All models evaluated in this paper use pole-filtered cepstral coefficients [7]. It is also possible to vary the features that are used to create each model, however, this option was not explored in this work. An overview of the HMM, DTW, and NTN is now provided.

The HMM is based on parametric representations of the feature space occupied by the speaker. Typically, it is assumed that the features extracted from a spoken password can be represented by a multi-modal Gaussian distribution. The HMM incorporates transitional information in addition to the distribution parameters of an assumed probability density function. The HMM has been evaluated for speaker verification using both whole-word [8] and sub-word [2] models. The HMM has also been found to perform favorably to DTW [1] for certain speaker verification applications. The HMM is currently one of the most popular modeling approaches for text-dependent speaker verification systems. To simplify the subsequent fusion with other models, the output score of the HMM is normalized to a scale of zero to one by passing the score through a sigmoidal activation function.

The DTW method is based on a distortion measure that is computed between a template representing an average of the enrollment utterances and a given test utterance. For speaker verification, a reference template can be generated by using features extracted from several utterances of the password [3]. Then during testing, a decision can be made to accept or reject the claimed identity based on whether or not the distortion measured to the training template falls below a predetermined threshold. To allow for subsequent fusion with other speaker models, the DTW distortions must be converted to a compatible scale, i.e., from zero to one. We accomplish this by simply raising the scaled negative distortion to an exponential.

The NTN [9] is a hierarchical classifier that uses a tree architecture to implement a sequential linear decision strategy. Specifically, the training data for a NTN consists of feature data from a target speaker, labeled as one, along with feature data from other speakers that are labeled as zero. These other speakers are known as "anti-speakers". The NTN will then learn to distinguish regions of feature space that belong to the target speaker from those that are more likely to belong to an impostor. These regions of feature space will correspond to leaves in the NTN that contain probabilities of the target speaker having generated data that falls within that region of feature space [10]. The NTN has been evaluated for text-independent speaker verification [10], whole-word [4] and subword-based [11] text-dependent speaker verification.

2.2 Linear Opinion Pool

The linear opinion pool method [12] is used for combining the output scores from the NTN, HMM, and DTW models. The linear opinion pool method computes the final score as a weighted sum of the outputs for each model:

$$P_{linear}(x) = \sum_{i=1}^{N} \mathbf{w_i p_i(x)}, \qquad (1)$$

where $P_{linear}(x)$ is the probability of the combined system, $\mathbf{w_i}$ are weights, $p_i(x)$ is the probability output by the ith model, and N is the number of models. The weights $\mathbf{w_i}$ are typically constrained such that their sum is equal to one. A performance comparison of the linear opinion pool, log opinion pool and voting method was previously reported [6] where it was found that the linear opinion pool generally performed the best. The selection of weight values is a key design issue for the linear opinion pool method of data fusion. This issue is addressed in the following section.

3 WEIGHT SELECTION METHODS

There are several methods for selecting weights to use for combining scores in a linear opinion pool. One simple method is to use equal weight values for each score to be combined, or

$$\mathbf{w_i} = \frac{1}{N}, \; 1 \leq i \leq N, \qquad (2)$$

where N is the number of models. For our approach with three models each weight will have a value of $1/3$. In order for this method to be effective, the scores of each model must be on a comparable scale with each other. This normalization procedure occurs automatically when using the Fisher linear discriminant.

The Fisher linear discriminant is a method for reducing the dimensionality of a pattern recognition problem [13]. By projecting all data points onto a line, the dimension of a classification problem can be reduced to one. The Fisher linear discriminant is defined as the linear function that maximizes the criterion function:

$$J(\mathbf{w}) = \frac{|\hat{\mathbf{m}}_1 - \hat{\mathbf{m}}_2|^2}{\hat{s}_1^2 + \hat{s}_2^2} \qquad (3)$$

which represents the ratio of the mean separation to the sum of the variances. This can be accomplished by projecting the observation vectors onto the line defined by:

$$\mathbf{w} = S_w^{-1}(\hat{\mathbf{m}}_1 - \hat{\mathbf{m}}_2) \quad (4)$$

where $\hat{\mathbf{m}}_1$ and $\hat{\mathbf{m}}_2$ are the sample means of the data for class 1 and class 2 and S_w is the scatter matrix defined as:

$$S_w = \hat{\Sigma}_1 + \hat{\Sigma}_2. \quad (5)$$

Here, $\hat{\Sigma}_1$ and $\hat{\Sigma}_2$ are the sample covariance matrices of the data from class 1 and class 2, respectively. The solution for $\mathbf{w}$ in equation 4 will maximize the criterion function of equation 3.

A more generalized method for determining the scatter matrix S_w has also been evaluated as [14]

$$S_w = s\hat{\Sigma}_1 + (1-s)\hat{\Sigma}_2, \quad (6)$$

which allows one to weight the sample covariance matrices for each class through the weight parameter s where $0 \leq s \leq 1$. This method was evaluated for the data in this paper, however, it was found that $s = 0.5$ provided the best overall performance when considering all databases.

4 EXPERIMENTAL RESULTS

The modeling approaches and data fusion methods are evaluated with three speech corpora that were collected in house. These databases were collected within different recording environments with respect to the microphone and channel. The intent is to evaluate the data fusion methods within several diverse implementation scenarios.

The first database was collected over a standard landline telephone and is hereforth referred to as the "landline" database. This database consists of 56 enrolled speakers and 47 development speakers. Each speaker enrolled with the phrase "open sesame". There are a total of 195 true speaker attempts and 11,229 impostor attempts for the landline database.

The second database was collected in a cellular environment and is referred to here as the "cellular" database. The cellular database uses the password "Al Capone" for all speakers. This database was collected using cellular phones and consists of 26 enrolled speakers and and 15 development speakers. There are a total of 273 true speaker attempts and 6,825 impostor attempts for the cellular database.

The third database is known as the "multimedia" database and was collected with a PC microphone. Each person enrolled and tested with the phrase "open the door". The multimedia database consists of 18 enrolled users along with 18 development speakers. There are a total of 90 true speaker trials and 1,638 impostor trials for the multimedia database.

Password text	# speakers devel/eval	# trials true/impostor
Landline	47/56	195/11,229
Cellular	15/26	273/6,825
Multimedia	18/18	90/1,638

Table 1: Specifications for the databases

The statistical aspects of each database are summarized in Table 1. The *development* speakers column in Table 1 refer to the set that are used as "anti-speakers" when training a NTN. To avoid bias in the results, only the *evaluation* speakers are used to measure the actual system performance. The features used for all experiments consist of the pole-filtered cepstral coefficients as described in [7].

For the cellular database, an analysis is performed to determine the correlation between errors for the respective models. In order to evaluate the models with the same performance criteria, we set the thresholds for each model to have zero percent false reject. Hence, all errors are false accept errors. The error correlation between models is shown in Table 2.

The diagonal entries in Table 2 correspond to the false accept error of the model for the case of zero false rejects. In order for data fusion to be successful, models should be selected for combination that have a minimal correlation between errors. For example, if all three models make an error on the same observation then no combination of their scores will rectify the result. The data in Table 2 shows the correlation level between the different models to be relatively low. Hence, for this example we anticipate performance gains through the proper combination of model scores.

In order to determine the potential performance gain by combining the three models, the weights for the linear opinion pool are evaluated exhaustively over the weight space. The weight space is constrained by the sum of the weights being equal to one. The graphical illustration of the weight space for the landline, cellular, and multimedia databases are shown in Figures 1, 2, and 3, respectively.

Model	HMM	NTN	DTW
HMM	0.182	0.044	0.048
NTN	0.044	0.126	0.034
DTW	0.048	0.034	0.106

Table 2: Error correlation between models

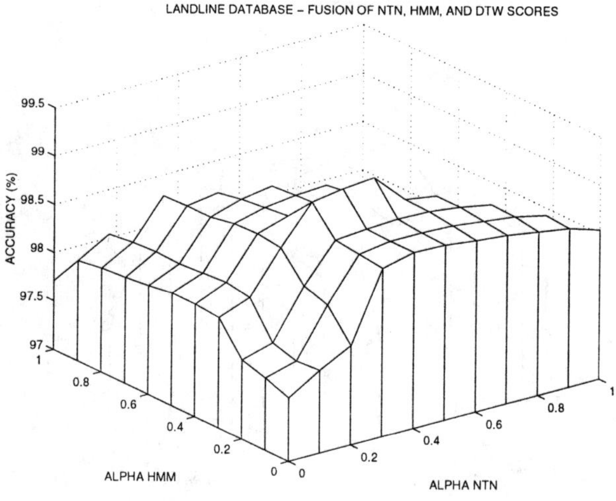

Figure 1: Multi model fusion for landline database

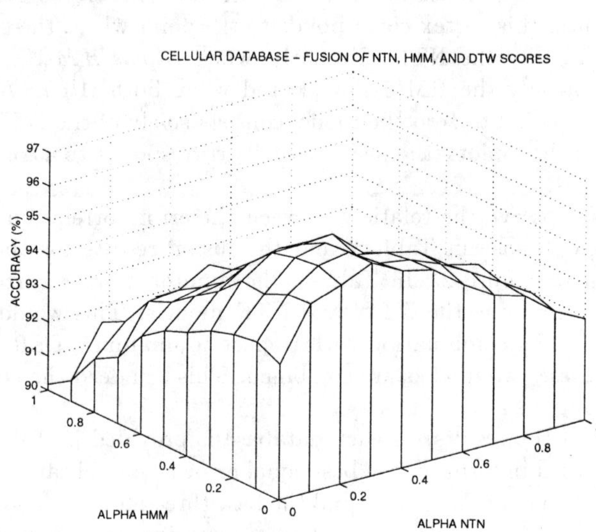

Figure 2: Multi model fusion for cellular database

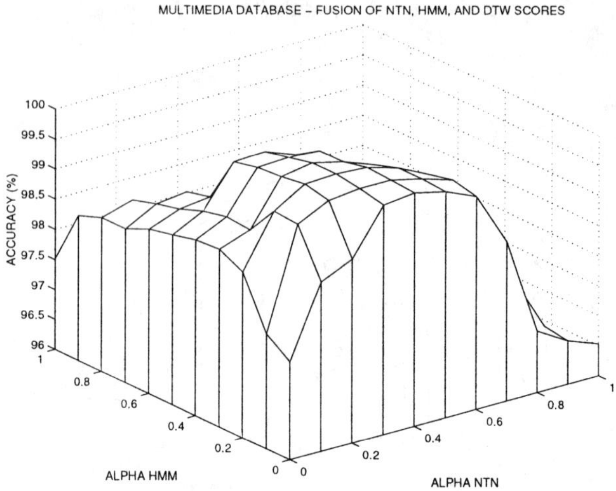

Figure 3: Multi model fusion for multimedia database

For each of these figures, the three vertices represent the performance of the individual models. For example, when *Alpha NTN* = 1, the weights of the two other models must be zero since the weights are constrained to sum to one. Hence, this vertex corresponds to the point where the composite model consists of only the NTN. Similarly, when *Alpha HMM* = 1 the model consists of only the HMM model, and when both *Alpha NTN* and *Alpha HMM* are equal to zero the model consists solely of the DTW model. As can be seen in these plots, the peak typically corresponds to a combination of all three models.

One can also observe the relation between the error correlation provided for the cellular database in Table 2 and the fusion results for the cellular database shown in Figure 2. In Table 2 the minimum error correlation is obtained when combining the DTW and NTN models. This would indicate that the best 2-model combination with respect to minimum EER would be achieved when these two models are combined. This is indeed the case as can be observed in Figure 2.

A summary of the results for each database is provided in Table 3. The results are expressed in terms of the best equal error rate as obtained by using a single model, the best two models, and the best three models as obtained by exhaustive search. The models or model pairs that give the best performance are also shown. As shown in Table 3, a performance improvement over single and dual model approaches is generally achievable by the addition of the third model score. Also, returning to the error correlation shown in Table 2, the minimum error correlation corresponds to the NTN/DTW pair which is consistent with the performance shown in Table 3.

The equal weight method and Fisher linear discriminant approach are

	best single	best dual	best trio
landline	1.46% (NTN)	1.15% (HMM-NTN)	0.84%
cellular	5.99% (DTW)	3.71% (DTW-NTN)	3.71%
multimedia	2.27% (DTW)	0.12% (DTW-NTN)	0.03%

Table 3: Equal Error Rates for Best 1-model, 2-model, and 3-model systems

	landline	cellular	multimedia
equal weights	0.99%	4.27%	0.10%
Fisher	0.95%	3.75%	0.85%
exhaustive	0.84%	3.71%	0.03%

Table 4: Equal error rates for different weight selection methods for 3-model systems

then evaluated to determine the weights to use for combining the scores of the HMM, NTN, and DTW models. The scatter matrix for the Fisher method is computed by using equal weight of the covariance matrix estimates for model scores of the speaker and impostor as represented in equation 5. The results for the different databases are summarized in Table 4.

The equal weight method and Fisher linear discriminant method both provide comparable error rates to that obtained by exhaustive search over the weight space. The Fisher provides closer error rates to the optimal except for the case of the multimedia database. We attribute this to the fact that more data is available for the landline and cellular databases and that the multimedia database is, hence, more susceptible to statistical variation.

5 CONCLUSION

The hidden Markov model (HMM), neural tree network (NTN), and dynamic time warping (DTW) methods are evaluated for model score combination. The linear opinion pool method is used to combine the scores of each model and several approaches for determining the weight values for fusion are considered. The weight selection approaches include the equal weight method and Fisher linear discriminant method. The multiple model fusion analysis is performed on three databases collected within landline telephony, cellular telephony, and multimedia environments. Overall, the error rates are reduced by using three models as opposed to two. Both weight selection methods provide performance that is comparable to the optimum weight vector as found by an exhaustive evaluation over the weight space.

References

[1] J.M. Naik, L.P. Netsch, and G.R. Doddington. Speaker verification over long distance telephone lines. In *Proceedings ICASSP*, pages 524–527, 1989.

[2] A.E. Rosenberg, C.H. Lee, and F.K. Soong. Sub-word unit talker verification using hidden Markov models. In *Proceedings ICASSP*, pages 269–272, 1990.

[3] S. Furui. Cepstral analysis technique for automatic speaker verification. *IEEE Trans. on Acoustics, Speech, and Signal Processing*, ASSP-29:254–272, April 1981.

[4] K.R. Farrell. Text-dependent speaker verification using data fusion. In *Proceedings ICASSP*, 1995.

[5] M. Sharma and R.J. Mammone. Subword-based text-dependent speaker verification system with user selectable passwords. In *Proceedings ICASSP*, 1996.

[6] K.R. Farrell, R.P. Ramachandran, and R.J. Mammone. An analysis of data fusion methods for speaker verification. In *Proceedings ICASSP*, 1998.

[7] D. Naik. Pole-filtered cepstral mean subtraction. In *Proceedings ICASSP*, pages 157–160, 1995.

[8] A.E. Rosenberg, C.H. Lee, and S. Gokeen. Connected word talker recognition using whole word hidden Markov models. In *Proceedings ICASSP*, pages 381–384, 1991.

[9] A. Sankar and R.J. Mammone. Growing and pruning neural tree networks. *IEEE Trans. on Computers*, C-42:221–229, March 1993.

[10] K.R. Farrell, R.J. Mammone, and K.T. Assaleh. Speaker recognition using neural networks and conventional classifiers. *IEEE Trans. Speech and Audio Processing*, 2(1), part 2, 1994.

[11] H. Liou and R.J. Mammone. Text-dependent speaker verification using sub-word neural tree networks. In *Proceedings ICASSP*, 1995.

[12] J.A. Benediktsson and P.H. Swain. Consensus theoretic classification methods. *IEEE Trans. on Systems, Man and Cybernetics*, 22(4):688–704, 1992.

[13] R.O. Duda and P.E. Hart. *Pattern Classification and Scene Analysis.* Wiley, New York, NY, 1973.

[14] K. Fukunaga. *An Introduction to Statistical Pattern Recognition.* Academic Press, San Diego, CA, 1990.

SUBSPACE TECHNIQUES IN SPEECH ENHANCEMENT

Gavin Smith[1], Mahesan Niranjan[2], Tony Robinson[1]
[1]Cambridge University Engineering Dept, Cambridge, CB2 1PZ. UK
Tel: +44 1223 332754, Fax: +44 1223 332662,
Email: {gas1003,ajr}@eng.cam.ac.uk
[2] Computer Science, Sheffield University, Sheffield. S1 4DP. UK
Tel: +44 114 22 21805, Fax: +44 114 22 21810,
Email: M.Niranjan@dcs.shef.ac.uk

Abstract. This paper concerns speech enhancement. Speech is first modelled and then model parameters used to design a Kalman smoother to reduce the background noise. 4SID (subspace state space system identification), DOA (direction-of-arrival) and polynomial techniques are described within a common subspace state space framework, and assumptions for each technique stated. It is shown that 4SID methods make least assumptions about the data, provide detailed spectra for the vocal tract, and can be used effectively for speech enhancement when combined with a Kalman smoother. An iterative technique is also possible.

INTRODUCTION

Speech enhancement estimates clean speech from noisy speech [1] [4]. This paper addresses an iterative two stage approach. In the first stage, a linear speech model is assumed and parameters estimated from the noisy speech. In the second stage, these parameters are used to design a Kalman smoother which is used to filter off the noise from the speech. This filtered speech is fed back into stage one, and the process iterated. The motivation of the paper is to contrast various parameter estimation techniques within a common subspace state space framework. More detail is given in [8].

The structure of the paper is as follows. The theory section presents a state space speech model, various parameter estimation techniques, Kalman smoothing and spectral estimation. Experimental results concerning vocal tract spectral estimation and speech enhancement are then shown. Finally conclusions are made.

0-7803-5673-X/99/$10.00 © 1999 IEEE

THEORY

Speech Modelling

Speech is represented as a combined deterministic-stochastic state space (SS) model [11]. The model is introduced in simple and then block Hankel matrix form. The input $u_t \in \mathbb{R}$ is the glottal waveform and the output $y_t \in \mathbb{R}$ is the speech signal at the microphone. Process and observation noises are modelled as white zero-mean Gaussian stationary noises $w_t \in \mathbb{R} \sim N(0, \sigma_w^2)$ and $v_t \in \mathbb{R} \sim N(0, \sigma_v^2)$ respectively. The system order is p.

$$x_{t+1} = Ax_t + Bu_t + w_t \quad (1)$$

$$y_t = Cx_t + Du_t + v_t \quad (2)$$

$x_t \in \mathbb{R}^{p\times 1}$, $A \in \mathbb{R}^{p\times p}$, $B \in \mathbb{R}^{p\times 1}$, $C \in \mathbb{R}^{1\times p}$, $D \in \mathbb{R}$. The problem definition is to estimate parameters $(A, B, C, D, \sigma_w, \sigma_v)$ from input-output data directly. The state space model can be extended to block matrix form. Time series are of length N and $i > p$ is the block size. Equations 1 and 2 are rewritten as the state sequence and block output equations respectively.

$$X_{i,N} = A^i X_{0,N-i} + \Delta_i^d U_{0,N-1} + \Delta_i^w W_{0,N-1} \quad (3)$$

$$Y_{0,N-1} = \Gamma_i X_{0,N-i} + H_i^d U_{0,N-1} + H_i^w W_{0,N-1} + V_{0,N-1} \quad (4)$$

$X_{i,N}$ is a state sequence matrix; its columns are the state vectors from time i to N. $X_{0,N-i}$ is similarly defined. $U_{0,N-1}$ is a Hankel matrix of inputs from 0 to $(N-1)$. Y, W, V are similarly defined. Δ_i^d and Δ_i^w are the deterministic and full reversed controllability matrices respectively, Γ_i is the observability matrix and H_i^d and H_i^w are Toeplitz matrices. These are all defined below where $I^{p\times 1}$ is a matrix of ones.

$$X_{0,N-i} \stackrel{\text{def}}{=} \begin{bmatrix} x_0 & x_1 & x_2 & \dots & x_{N-i} \end{bmatrix}$$

$$\Delta_i^d \stackrel{\text{def}}{=} \begin{bmatrix} A^{i-1}B & A^{i-2}B & \dots & B \end{bmatrix}$$

$$\Delta_i^w \stackrel{\text{def}}{=} \begin{bmatrix} A^{i-1}I & A^{i-2}I & \dots & I \end{bmatrix}$$

$$\Gamma_i \stackrel{\text{def}}{=} \begin{bmatrix} C \\ CA \\ \vdots \\ CA^{i-1} \end{bmatrix} \qquad U_{0,N-1} \stackrel{\text{def}}{=} \begin{bmatrix} u_0 & u_1 & \dots & u_{N-i} \\ u_1 & u_2 & \dots & u_{N-i+1} \\ \vdots & \vdots & & \vdots \\ u_{i-1} & u_i & \dots & u_{N-1} \end{bmatrix}$$

$$H_i^w \stackrel{\text{def}}{=} \begin{bmatrix} 0 & & & 0 \\ C & 0 & & \\ \vdots & \ddots & \ddots & \\ CA^{i-2} & \dots & C & 0 \end{bmatrix} \qquad H_i^d \stackrel{\text{def}}{=} \begin{bmatrix} D & & & 0 \\ CB & D & & \\ \vdots & \ddots & \ddots & \\ CA^{i-2}B & \dots & CB & D \end{bmatrix}$$

A sequence of outputs can be separated into two block output equations containing *past* and *future* data denoted with subscripts p and f respectively. With $U_p \stackrel{def}{=} U_{0,N-1}$, $U_f \stackrel{def}{=} U_{i,N+i-1}$ and similarly for Y, W and V, and $X_p \stackrel{def}{=} X_{0,N-i}$ and $X_f \stackrel{def}{=} X_{i,N}$, past and future are related by the equations

$$X_f = A^i X_p + \Delta_i^d U_p + \Delta_i^w W_p \tag{5}$$
$$Y_p = \Gamma_i X_p + H_i^d U_p + H_i^w W_p + V_p \tag{6}$$
$$Y_f = \Gamma_i X_f + H_i^d U_f + H_i^w W_f + V_f \tag{7}$$

Model Parameter Estimation Techniques

Speech can be represented as a combined deterministic-stochastic linear state space system. There are various techniques to estimate the parameters. In this section, several parameter estimation techniques are presented together with the assumptions and constraints made by each. Further detail is obtained by considering the state space equations in block matrix form. 4SID techniques, which make least assumptions, are introduced first, and then other techniques as special cases.

Subspace State Space System Identification (4SID) Methods. These include N4SID (numerical algorithms for subspace state space system identification), MOESP (multi-variable output-error state space) and CVA (canonical variate analysis) [11] [12]. The assumptions made are that the process and observation noises are zero-mean white Gaussian. They use a combined deterministic-stochastic state space model as in equations 1 and 2.

4SID algorithms are composed of two stages: low-rank approximation and estimation of the extended observability matrix directly from the input-output data, followed by estimation of system matrices from its column space. Consider the future output block equation from 7

$$Y_f = \Gamma_i X_f + H_i^d U_f + H_i^w W_f + V_f$$

Denote matrix $W_p = (U_p^T Y_p^T)^T$. To remove noise and input effects, matrices undergo an oblique projection onto the row space of W_p along the row space of U_f (reference Appendix 1).

$$Y_f/_{U_f}\mathbf{W_p} = \Gamma_i X_f/_{U_f}\mathbf{W_p} \tag{8}$$

This oblique projection coincides with a minimum squared error between true data Y_f and its linear prediction from W_p and U_f. Greater flexibility is obtained by weighting the oblique projection with weighting matrices W_1 and W_2 and analysing this: $W_1(Y_f/_{U_f}\mathbf{W_p})W_2$. N4SID, MOESP and CVA differ with respect to these weighting matrices, but are otherwise identical. W_1, W_2 relate to weighting the optimization criterion in the frequency domain.

In the absence of a deterministic input signal, the model becomes just stochastic. Thus the oblique becomes an orthogonal projection onto past output samples, and is similar to instrumental variable (IV) methods [12] [13].

$$x_{t+1} = Ax_t + w_t, \quad y_t = Cx_t + v_t \tag{9}$$

Direction-of-Arrival (DOA) Methods.
These algorithms are used in the direction-of-arrival problem, and in estimating damped sinusoids in noisy conditions [5] [6] [10]. Algorithms include ESPRIT and MUSIC. These operate on a state space model with no deterministic input and no process noise. Refer to Appendix 2.

$$x_{t+1} = Ax_t, \quad y_t = Cx_t + v_t \tag{10}$$

From this the correlation matrix can be derived. DOA methods operate on this and assume spatially and temporally white noise.

$$Y_f Y_f^T = \Gamma_i X_f X_f^T \Gamma_i^T + \sigma_v^2 I \tag{11}$$

I denotes identity. These methods yield biased estimates of (A, C) whenever a deterministic input excites the system, or there is process noise or coloured observation noise.

Polynomial methods.
These assume an autoregressive model [9] [10]. They operate on a restricted and modified form of equations 1 and 2. Deterministic inputs and measurement noise are assumed zero, process noise is premultiplied by a matrix F, and (A', C', F) have special structures (reference Appendix 3).

$$x_{t+1} = A'x_t + Fw_t, \quad y_t = C'x_t \tag{12}$$

Consider system order p, with block size $i = p + 1$. In the absence of a deterministic input and noises

$$Y_f = \Gamma_i X_f \tag{13}$$

Consider also a vector $[a - 1]$ which has roots equal to the system poles, where $a \in \mathbb{R}^{1 \times p}$. By the Cayley-Hamilton theorem, the vector $[a - 1]$ (unique to within a scale factor) is orthogonal to the observability matrix $[a - 1]\Gamma_i X_f = 0$. Therefore $[a - 1]Y_f = 0$ also. Thus a becomes a prediction polynomial. Thus the roots of the prediction polynomial only equal the system poles in the absence of a deterministic input, measurement and process noises. Otherwise parameter estimates are biased. A summary of results for all techniques is shown in table 1.

MODEL	STATE SPACE EQUATION	MEASUREMENT EQUATION
Combined SS	$x_{t+1} = Ax_t + Bu_t + w_t$	$y_t = Cx_t + Du_t + v_t$
Stochastic SS	$x_{t+1} = Ax_t + w_t$	$y_t = Cx_t + v_t$
DOA	$x_{t+1} = Ax_t$	$y_t = Cx_t + v_t$
Polynomial	$x_{t+1} = A'x_t + Fw_t$	$y_t = C'x_t$

Table 1: State-space equations for different methods

Filtering the Noisy Speech

Subspace identification methods estimate system matrices using geometry. The optimal state sequence is then estimated using the Rauch-Tung-Striebel formulation of the Kalman smoother [2], and speech resynthesised using the equation $y_t = Cx_t + Du_t$. This estimation-resynthesis can be iterated.

Spectral Estimation

For the combined model, the frequency response $H_c(e^{j\Omega})$ between deterministic input and output is considered. Ω is normalised frequency. For the stochastic SS model, DOA and polynomial models, the transfer function $H_s(e^{j\Omega})$ between process noise and output is considered.

$$H_c(e^{j\Omega}) = C(e^{j\Omega}I - A)^{-1}B + D \quad (14)$$

$$H_s(e^{j\Omega}) = C(e^{j\Omega}I - A)^{-1} \quad (15)$$

EXPERIMENTS

Preparation of the speech and glottal waveforms

Experiments are conducted on the phrase *"a small"*, spoken by an adult male. The speech waveform and laryngograph are obtained from the Eurom 0 database [3] and sampled at 16 kHz. The phrase *"a small"* is chosen because of the proximity of the two lowest frequency formants of the vowel in *"all"* which are only about 300 Hz apart (chapter 2 of [1]). Speech is preemphasised prior to analysis to reduce the effects of lip radiation.

Due to non-stationarity, the speech waveform is divided into fixed-length, overlapping and windowed (hamming) analysis frames 15 ms in duration, shifted 7.5 ms each frame. Speech is assumed stationary within each frame. Overlap gives smoother parameter transitions and better quality reconstruction. Reconstruction employs the overlap-add method. All models are order 8, and the block size for subspace methods is fixed at 20.

Because of the difficulties in modelling the glottal waveform, input is approximated as a series of Rosenberg pulses [7], with closure synchronised to the dominant impulses in the derivative of the laryngographic data. Each pulse is a piecewise trigonometric function. Glottal open and closed phases are T_o and T_c respectively where T is the glottal period. These are also framed and windowed as for the speech. In this analysis $T_o = 0.4T$, $T_c = 0.16T$, $\alpha = 1$.

$$
\begin{aligned}
x(t) &= \tfrac{\alpha}{2}\left[1-\cos(\tfrac{\pi t}{T_o})\right] && 0 \le t \le T_o \\
&= \alpha \cos\left[\tfrac{\pi(t-T_o)}{2T_c}\right] && T_o \le t \le T_o + T_c \\
&= 0 && T_o + T_c < t < T_p
\end{aligned}
$$

Experimental Details

Two experiments are conducted. In the first experiment, the 4SID, DOA and polynomial techniques are compared at estimating the vocal tract frequency response from clean speech data as discussed previously. This is done using the stochastic SS model *sto_pos.m* algorithm of [11], Liu's method [5] and the autocorrelation method [9] as representative of the 4SID, DOA and polynomial techniques respectively. Their spectrograms after one iteration are given in figure 1. The 4SID and DOA methods show better resolution than the polynomial method. DOA tends to produce over-peaked spectra. 4SID methods estimate greater spectral detail because they introduce zeros.

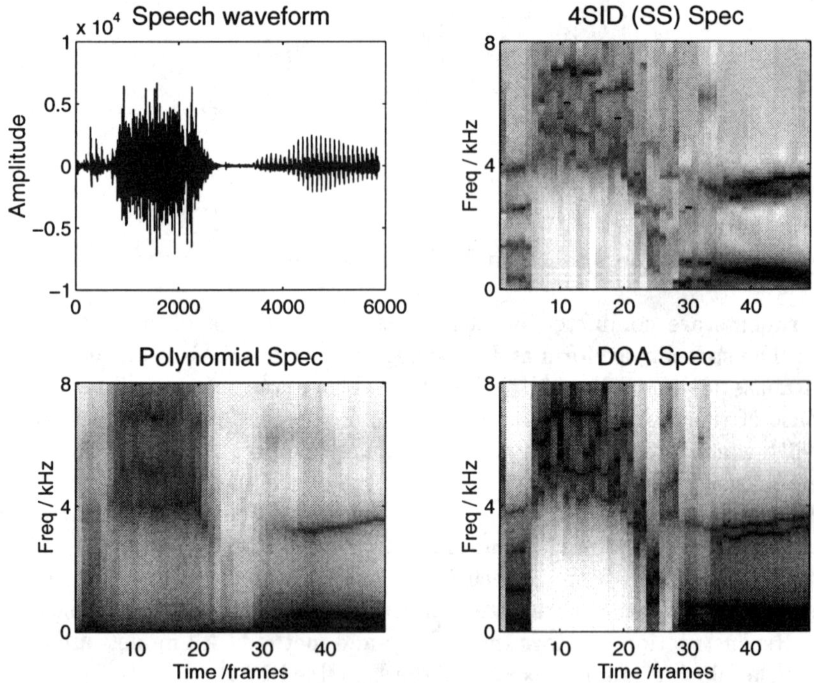

Figure 1: Spectrograms for clean speech

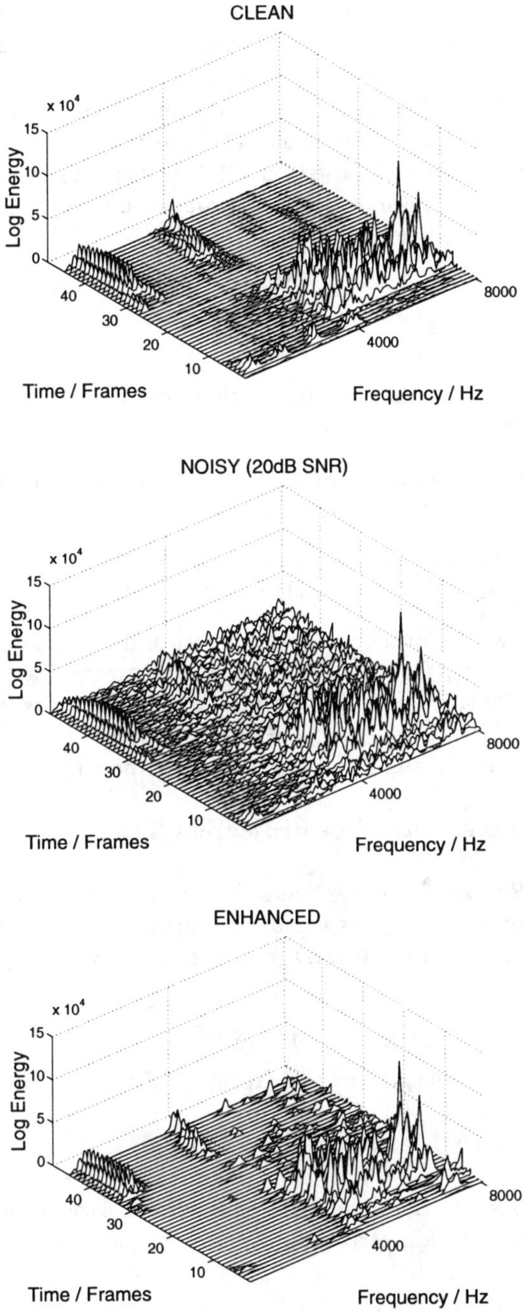

Figure 2: 3D spectrograms showing clean, noisy and enhanced speech

In the second experiment, the stochastic SS model algorithm *sto_pos.m* of [11] is used to illustrate speech enhancement for 4SID methods. Figure 2 shows 3D spectrograms of clean speech, speech corrupted by white Gaussian noise at 20dB SNR, and enhanced speech after 5 iterations. Although observation noise is removed, some musical noise is introduced. The *matlab* default spectrogram plotter (based on DFT principles) is used. Enhancement using a combined SS model and the robust algorithm of [11] produced visually slightly poorer results.

CONCLUSIONS

1. The 4SID, DOA and polynomial methods can be presented in a common subspace state space framework.
2. 4SID and DOA methods resolve close formants better than polynomial methods; DOA produces *peaky* spectra.
3. Iterative state space parameter estimation and Kalman smoothing are shown to be effective for speech enhancement.

Acknowledgements The author is grateful to Nando De Freitas (Cambridge University Eng. Dept.) for valuable discussion regarding Kalman smoothing, for the use of 4SID software supplied with [11] and the smoothing software of Zoubin Ghahramani *www.gatsby.ucl.ac.uk/ zoubin/*. I would like to thank the Schiff Foundation, Cambridge University for their financial support.

Appendix 1 – Linear Algebra Projection Theory

Matrices S and R have row spaces $\mathcal{S}$ and $\mathcal{R}$ respectively. $A/\mathbf{S}$ is the *orthogonal* projection of A onto $\mathcal{S}$. $A/_R\mathbf{S}$ is the *oblique* projection of A onto $\mathcal{S}$ along $\mathcal{R}$. † denotes the Moore-Penrose inverse, and $^\perp$ the orthogonal space.

$$A/\mathbf{S} = AS^T(SS^T)^\dagger S \tag{16}$$

$$A/_R\mathbf{S} = (A/\mathbf{R}^\perp)(S/\mathbf{R}^\perp)^\dagger S \tag{17}$$

Appendix 2 – DOA Methods

These methods assume that the signal is a sum of exponential modes with poles z_i. Using the block matrix definitions as in the theory section

$$Y_f = \begin{bmatrix} 1 & 1 & \dots & 1 \\ e^{z_1} & e^{z_2} & \dots & e^{z_p} \\ e^{2z_1} & e^{2z_2} & \dots & e^{2z_p} \\ \vdots & \vdots & & \vdots \\ e^{i\ z_1} & e^{i\ z_2} & \dots & e^{i\ z_p} \end{bmatrix} X_f + V_f = \Gamma_i X_f + V_f \tag{18}$$

The observability matrix Γ_i has Vandermonde structure and X_f contains the amplitudes of the decaying exponentials. DOA methods consider the correlation matrices (denoted R).

$$Y_f Y_f^T = \Gamma_i X_f X_f^T \Gamma_i^T + V_f V_f^T \quad (19)$$

$$R_{Y_f} = \Gamma_i R_{X_f} \Gamma_i^T + \sigma_v^2 I \quad (20)$$

System poles can be determined through non-linear (or linearly approximated) optimization of one of the criteria in table 2 [8]. These are equivalent and equal to maximum likelihood estimation for spatially and temporally white noise (an assumption which can be only approximated due to the Hankel structure). S_n are vectors which span the noise subspace.

CRITERION	OPTIMIZATION
1	$arg_{\Gamma} \ min\|\|Y - \Gamma X\|\|_F^2$
2	$arg_{\Gamma} \ max \ tr(\Gamma\Gamma^T R)$
3	$arg_{\Gamma} \ min \ tr(\Gamma^T S_n S_n^T \Gamma R)$

Table 2: DOA Maximum Likelihood Estimation Criteria

Defining $A = diag(e^{z_1}, e^{z_2}, \ldots, e^{z_p})$, $C = [1, 1, \ldots, 1] \in \mathbb{R}^{1\times p}$, and x_k as containing amplitude information, then the above equations can be described in state space form

$$x_{k+1} = Ax_k, \quad y_k = Cx_k + v_k \quad (21)$$

Appendix 3 – Polynomial Methods

Polynomial methods use an autoregressive model $y_k = \sum_{i=1}^{p} a_i y_{k-i} + n_k$. Estimation criteria minimise the error n_k. In modified state space form this becomes (modified due to the inclusion of the F matrix)

$$x_{k+1} = A'x_k + Fw_k, \quad y_k = C'x_k \quad (22)$$

where $F \in \mathbb{R}^{p\times p}$ is a matrix of zeros except $F_{p,p} = 1$, $C' \in \mathbb{R}^{1\times p} = [0, 0, \ldots, 0, 1]$, $a \in \mathbb{R}^{1\times p} = [a_p, a_{p-1}, a_{p-2}, \ldots, a_1]$, and

$$A' = \begin{bmatrix} 0 & 1 & 0 & \ldots & \ldots & 0 \\ & 0 & 1 & 0 & \ldots & 0 \\ & & & \ddots & & \\ \hline & & & a & & \end{bmatrix}^{p\times p} \quad (23)$$

REFERENCES

[1] J. Deller, J. Proakis and J. Hansen, **Discrete-Time Processing of Speech Signals**, Englewood Cliffs, NJ: Macmillan Publishing Co, 1993.

[2] A. Gelb, ed., **Applied Optimal Estimation**, Cambridge, MA: MIT Press, 1974.

[3] M. Grice and W. Barry, "Multi-lingual speech input/output: Assessment, methodology and standardization," Tech. rep., University College, London, 1989, ESPRIT Project 1541 (SAM), extension phase final report.

[4] P. Hansen, **Signal Subspace Methods for Speech Enhancement**, Ph.D. thesis, IMM Dept of Mathematical Modelling, Technical University of Denmark, 1997.

[5] K. Liu, "Modal parameter estimation using the state space method," **Journal of Sound and Vibration**, vol. 197, no. 4, pp. 387–402, 1996.

[6] B. Rao and K. Arun, "Model based processing of signals: A state space approach," **Proceedings of the IEEE**, vol. 80, no. 2, 1992.

[7] A. Rosenberg, "Effect of glottal pulse shape on the quality of natural vowels," **Journal of the Acoustical Society of America**, vol. 49, no. 2, pp. 583–590, 1971.

[8] G. Smith, M. Niranjan and A. Robinson, "Subspace techniques in speech enhancement," Tech. Rep. CUED/F-INFENG/TR.345, Cambridge University Engineering Dept, Cambridge, UK, 1999.

[9] C. Therrien, **Discrete Random Signals and Statistical Signal Processing**, Englewood Cliffs, NJ: Prentice-Hall, 1992.

[10] A. Van Der Veen, E. Deprettere and A. Swindlehurst, "Subspace-based signal analysis using singular value decomposition," **Proceedings of the IEEE**, vol. 81, no. 9, 1993.

[11] P. Van Overschee and B. De Moor, **Subspace Identification for Linear Systems: Theory, Implementation, Applications**, Dordrecht, Netherlands: Kluwer Academic Publishers, 1996.

[12] M. Viberg, "Subspace-based methods for the identification of linear time-invariant systems," **Automatica**, vol. 31, no. 12, pp. 1835–1851, 1995.

[13] M. Viberg, B. Wahlberg and B. Ottersten, "Analysis of state space system identification methods based on instrumental variables and subspace fitting," **Automatica**, vol. 33, no. 9, pp. 1603–1616, 1997.

MULTI-CHANNEL PIECEWISE SELECTIVE AVERAGING OF COGNITIVE EVOKED POTENTIALS WITH VARIABLE LATENCY

Arthur Flexer
The Austrian Research Institute for Artificial Intelligence
Schottengasse 3, A-1010 Vienna, Austria
arthur@ai.univie.ac.at
Herbert Bauer
Institute of Psychology, University of Vienna
Liebiggasse 5, A-1010 Vienna, Austria

Abstract.
This work is about the development of an alternative way of averaging evoked potentials (EP) of cognitive activities. Since the main assumption of invariant waveforms time locked to the eliciting events does not hold for cognitive EPs, averaging results in distorted estimates. Our alternative selective averaging finds similar subsequences of fixed length with variable latency which are common to all multi-channel EPs by transforming the multivariate time series to discrete sequences via vector quantization and applying a sequence alignment algorithm. The method yields a significant improvement over common averaging in terms of noise attenuation and is shown to be valid by comparison with results for random data. Results for EP data obtained during a spatial imagination task are reported.

INTRODUCTION

An evoked potential (EP) is the electro cortical potential measurable in the human electro encephalogram (EEG) before, during and after sensoric, motoric or cognitive events. This work is about the development of a new method for the analysis of EPs of cognitive activities for which already existing methods for the simpler sensoric or motoric EPs are not suited.

An EP is defined as the combination of the brain electric activity that occurs in association with the eliciting event and 'noise', which is brain activity not related to the event together with inference from non-neural sources. Since the noise contained in EPs is significantly stronger than the signal, the common approach is to compute an average across several EPs recorded under equal conditions to improve the signal-to-noise ratio. For common averaging of a set of N EPs, the assumed model for the ith EP (given here for only one channel of EP) is the following [6]:

$$x_i(t) = s(t) + n_i(t); \quad i = 1, 2, \ldots N; \quad 0 \leq t < T \tag{1}$$

where $x_i(t)$ is the ith recorded EP, $s(t)$ the underlying signal, $n_i(t)$ the noise associated with the ith EP, and T the duration over which each EP is recorded. The average $\hat{s}(t)$ over the sample of N EPs is used to estimate the underlying signal $s(t)$:

$$\hat{s}(t) = \frac{1}{N}\sum_{i=1}^{N} x_i(t) = s(t) + \frac{1}{N}\sum_{i=1}^{N} n_i(t) \tag{2}$$

Averaging will attenuate the noise $n_i(t)$ and not the signal $s(t)$ given that signal and noise linearly sum together to produce the recorded EPs $x_i(t)$ and the evoked signal $s(t)$ is the same for each recorded EP $x_i(t)$ and the noise contributions $n_i(t)$ can be considered to constitute statistically independent samples of a random process. Averaging is standard for the analysis of exogenous components of motoric or sensoric events like the $N100$ waveform which lie within $100ms$ after the eliciting event, as well as for endogenous components of short duration ($\leq 500ms$). If the above assumptions do not hold, the averaging will result in a biased, distorted estimate of the signal.

Cognitive evoked potentials are endogenous components of the EPs and start about $100ms$ after the onset of the recording and can last several seconds. They do not elicit one specific EP waveform time locked to the onset of the recording, and their analysis is a largely unsolved problem in psychophysiology. Classical methods like [15, 9, 8, 13] and [14] are designed for univariate time series of simpler motoric or sensoric EPs only. They usually assume that the recorded univariate signal is the same for all EPs during the whole duration of the recording but allow variable latencies of the common waveform. Therefore they cannot really cope with the harder problem of analysing cognitive EPs. Other more modern approaches like e.g. Independent Component Analysis [7] use EP data of short duration ($\leq 1sec$) after averaging. Another approach is to average despite improper assumptions thereby loosing all information about faster components and to analyse only positive and negative shifts of very slow potentials [2].

Our hypothesis concerning cognitive EPs is that only pieces (subsequences) of the whole EPs with variable latencies can be expected to be due to the cognitive task. Our approach to discover such subsequences which are common to all EPs across all recording channels is to replace the sequence of the original multi dimensional vectors (our data was measured via 22 electrodes) by a sequence of codebook vectors obtained via vector quantization. The trajectories across codebook vectors are univariate discrete time series to which we can apply a multiple sequence alignment procedure [1] that has originally been designed for molecular biology.

The data was recorded with 22 electrodes with an equidistant matrix montage during a spatial imagination task (Three-dimensional Cube Test 3DC [5]) and consists of 319 EP trials from 10 good female spatializers plus 167 EP trials from 8 poor female spatializers. After appropriate preprocessing (digital low pass filtering to frequencies below $8Hz$ and eliminating the DC-like trend by subtracting a linear fit), each EP trial lasts $8.5sec$ consisting of 2125 samples, each being a 22 dimensional real valued vector.

COMPUTATION OF TRAJECTORIES ACROSS CODEBOOK VECTORS

All EP time series are vector quantized together by using all EP vectors at all sample points as input vectors to a clustering algorithm disregarding their ordering in time. Then the sequence of original vectors x is replaced by the sequence of codebook vectors $\hat{x}$. K-means clustering (see e.g. [3, p.201]) is used for vector quantization using the sum of squared differences $d(x, \hat{x}) = \sum_{i=0}^{k-1} \mid x_i - \hat{x}_i \mid^2$ as measure of distance, where both x and $\hat{x}$ are of dimension k. Since observation of the sum of distances $d(x, \hat{x})$ with growing size of codebooks did not indicate an optimal codebook size, we pragmatically decided to use 64 codebook vectors based on the following consideration: We vector quantized the EP data with increasing numbers of codebook vectors. We then took the original EPs and substituted at each sample point the original real valued vector x_l with the appropriate codebook vector $\hat{x}_i$ (i.e. where $d(x_l, \hat{x}_i) < d(x_l, \hat{x}_s)$ for all s). We then visually inspected both the original EP time series and the coarser codebook time series as series of topographical patterns (spherical spline interpolations of the 22 values at a single point in time) and checked, whether the important features (positive and negative peaks and their development in time) of the topographies still were existent in the coarser codebook approximation. The high number of different discrete symbols (64 code book vectors) did not allow for a more principled information theoretic approach to obtain an optimal codebook size. For the 64 codebook vectors, we calculated a 64×64 distance matrix D_C.

SEQUENCE ALIGNMENT AND SELECTIVE AVERAGING

We chose a so-called *fixed length subsequence* approach for comparison of the sequences made of 64 discrete symbols (corresponding to 64 codebook vectors $\hat{x}$). Given two sequences E and F of length m, all possible overlapping subsequences having a particular window length W from E are compared to all subsequences from F. For each pair of elements the score taken from the distance matrix D_C is recorded and summed up for the comparison of subsequences. The distance between two subsequences of length W from two sequences E and F is therefore:

$$D_{align}(b_e, b_f, W) = \sum_{i=0}^{W-1} d(E_{b_e+i}, F_{b_f+i}) \tag{3}$$

The indices b_e and b_f are the beginning points of the subsequences in the sequences E and F and E_{b_e+i} and F_{b_f+i} are the corresponding codebook vectors. Successive application of this pairwise methods allows for the alignment of more than two sequences. Such a *fixed subsequence* approach that is explicitly designed for *multiple sequence alignment* is given in [1]. It computes a multiple alignment by iteratively comparing sequences to the multiple alignment obtained so far, keeping always just the L best subsequences as an intermediate result.

This approach to multiple sequence alignment is called *progressive alignment.* It works by constructing a succession of pairwise alignments. Initially, two sequences are chosen at random and aligned via the *fixed length subsequence* approach described above. The L best pairwise alignments (i.e. pairs of subsequences) with minimum distances D_{align} are now fixed and stored in a heap. Then a third sequence is chosen at random and aligned to all the L pairwise alignments. All L three-way alignments are now fixed and stored in the heap. This process is iterated until all sequences have been aligned.

When a subsequence is compared to an intermediate "more"-way (let us say p-way) alignment, the resulting score is computed as the sum of the p pairwise comparisons of the subsequences in the intermediate solution with the new subsequence that is to be aligned. The number of all such crosswise comparisons within the final overall alignment is given by $P = \sum_{i=1}^{p-1} i$. The number of all element-wise comparisons within the final overall alignment is given by WP, and its average per element, the average element-wise within alignment distance, by:

$$\bar{D}_{align} = \frac{1}{WP} \sum_{i=1}^{p-1} \sum_{j=i+1}^{p-1} D_{align}(b_i, b_j, W) \tag{4}$$

Desired is a set of beginning points b_i^{min} for which $\bar{D}_{align}$ is minimal. The b_i^{min} are the same for all $d = 22$ channels of the corresponding ith EP. For each channel of EP we can compute an alternative selective average $\hat{s}'(t)$ where the duration T is equal to the length of the subsequences, W, and the beginning points of the averaging are the parameters b_i^{min}:

$$\hat{s}'(t) = \frac{1}{N} \sum_{i=1}^{N} x_i(b_i^{min} + t); \quad 0 \leq t < W \tag{5}$$

This approach guarantees that the obtained multiple alignments contain subsequences that are part of all the original sequences. The number of single element-wise comparisons is $LW(m + 1 - W)P$. For a given L and m, this function is proportional to p^2, in contrast with m^p comparisons in "brute force" searching where not just the L best but all possible alignments are considered. As experiments with L equal 100, 1000 and 10000 showed, it is sufficient to keep 100 intermediate results to avoid the omission of good alignments that are weak in the first few sequences but strong in the later ones. Experiments varying the window length W from 31 to 62, 125 and 187 showed that $W = 125$ (corresponding to $500ms$ of EP) is short enough to yield alignments of satisfactory quality which are still long enough to be significant in terms of their psychophysiological interpretation. For more detail on tuning of the parameters L and W see [4].

RESULTS

In a related study [4] working on a subset of our EP data we have shown the statistical significance of our approach. We verified that our procedure yields better results for real human EPs than for unstructured random input in terms of average element-wise within alignment distance $\bar{D}_{align}$ (see Equ. 4). We compared results obtained from 21 EPs of one test subject with time-shuffled EPs and artificial EPs. The latter consisted of random Gaussian sequences whose power spectrum was changed appropriately to resemble the characteristics of real EPs. A one-way analysis of variance plus additional Duncan t-Tests allowed us to rank the result for real EP as being significantly better than the result for time-shuffled EP, which is again significantly better than the result for random Gaussian EP.

To compare the gain in noise attenuation of the common average and of our selective average, the respective estimated standard deviations of the background noise, $\hat{\sigma}(t)$ and $\hat{\sigma}'(t)$, are being compared.

$$\hat{\sigma}(t) = \left[\frac{\sum_{i=1}^{N} \left[x_i(t) - \hat{s}(t)\right]^2}{N-1} \right]^{\frac{1}{2}} \tag{6}$$

$$\hat{\sigma}'(t) = \left[\frac{\sum_{i=1}^{N} \left[x_i(b_i^{min} + t) - \hat{s}'(t)\right]^2}{N-1} \right]^{\frac{1}{2}} \tag{7}$$

Since the $\hat{\sigma}(t)$ and $\hat{\sigma}'(t)$ are given for each of the $d = 22$ channels and for the duration of $t = m$ or $t = W$ respectively, the following average estimates of the standard deviations of the background noise are being computed:

$$\hat{S} = \frac{1}{dm} \sum_{j=1}^{d} \sum_{t=0}^{m-1} \hat{\sigma}_j(t) \tag{8}$$

$$\hat{S}' = \frac{1}{dW} \sum_{j=1}^{d} \sum_{t=0}^{W-1} \hat{\sigma}'_j(t) \tag{9}$$

$\hat{S}$ is the estimate for the common averaging and $\hat{S}'$ for the selective averaging. An $\hat{\sigma}_j(t)$ is the $\hat{\sigma}(t)$ for channel j given by Equ. 6. An $\hat{\sigma}'_j(t)$ is the $\hat{\sigma}'(t)$

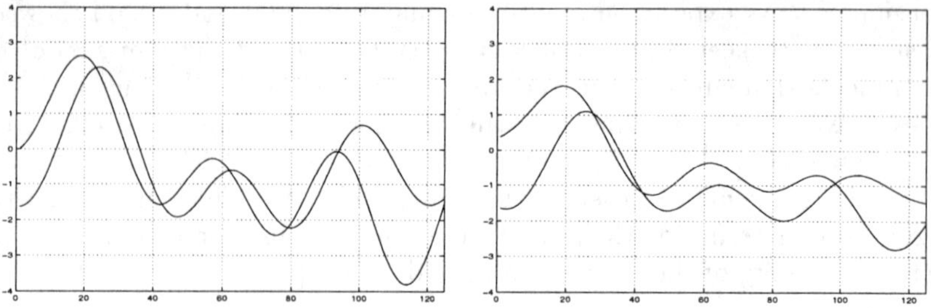

(a) (b)

Figure 1: Comparison of selective averages computed via beginning points b_i^{min5} obtained by minimizing the sum of $\bar{D}_{align}$ from five repeated runs each. (a) is for electrode F1, (b) for P2, both for poor spatializers.

for channel j given by Equ. 7. For all EPs of good and poor spatializers the common average $\hat{s}$ as well as five selective averages $\hat{s}'$ have been computed. Results for the good spatializers were $\hat{S} = 7.68$ vs. mean $\hat{s}' = 4.35 \pm .068$ and for the poor spatializers $\hat{S} = 7.84$ vs. mean $\hat{s}' = 4.37 \pm .048$. Computing Z-values shows the differences in noise attenuation to be significant:

$$Z_{good} = |(4.35 - 7.68)/(.068/\sqrt{5})| = |-109.5| > Z_{99} = 2.58;$$

$$Z_{poor} = |(4.37 - 7.84)/(.048/\sqrt{5})| = |-161.6| > Z_{99} = 2.58.$$

The estimated expected magnitude of the noise residual is now only $\approx$ 0.56 times that of the respective common averages. This is a gain in noise attenuation of more than 40%.

The repeated runs of the fixed subsequence algorithm showed a problem with the reproducibility of the selective averages. Although a majority of the obtained beginning points b_i^{min} are identical, there is a difference in the selective averages due to the different succession in which the sequences are being compared. We are able to overcome this problem by combining five selective averages to one overall selective average: we obtain overall beginning points b_i^{min5} at the points in time where the sums of the respective five $\bar{D}_{align}$ (cf. Equ. 4) of the single selective averages are minimal. This "averaging" across five solutions diminishes their variability. We computed two such overall selective averages by combining 2×5 selective averages to two sets of beginning points b_i^{min5}. The resulting pairs of overall selective averages are now very similar which is corroborated by visual inspection (Fig. 1).

The results of computing selective averages for both good and poor spatializers via beginning points b_i^{min5} are given in Fig. 2 as sequences of topographical patterns. Each topography is a spherical spline interpolation of the 22 values at a single point in time of the selective averaging window. Given are topographies at $40, 80, \cdots, 440, 480 msec$ of the window for poor

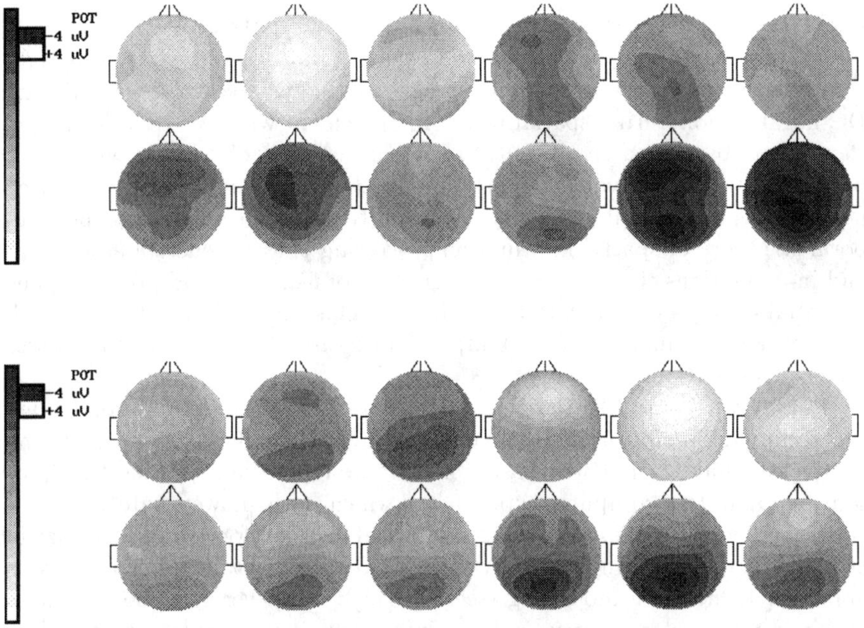

Figure 2: Sequences of topographies for poor spatializers (top two rows) and good spatializers (lower two rows). Scale is from -4 to $+4mV$.

spatializers (top two rows) and good spatializers (lower two rows). We can see that for both good and poor spatializers there is one specific dominant topographical pattern visible, albeit at changing levels of amplitude. It is a pattern of more positive amplitudes at frontal to central regions relative to more negative amplitudes at occipital to parietal regions. This common topographical pattern is generally more negative for poor spatializers.

DISCUSSION

Our selective averaging makes possible the analysis of cognitive EPs by finding common pieces (subsequences) of a set of EPs which have fixed length but variable latencies and are sufficiently similar across all EP channels. We were able to validate our approach both on a statistical basis and in terms of the psychophysiological content of the obtained results. The comparison with artificial random EPs shows the statistical significance of our algorithm. The selective averaging almost doubles the noise attenuation compared to common averaging. The more negative amplitudes at occipital to parietal regions visible in Fig. 2 are well in line with both clinical findings and related EP studies about spatial processing. A study [12] using the exact same experimental design as in this work, but with male subjects and applying common averaging to analyse positive and negative shifts of very slow potentials, confirmed the importance of occipito-parietal regions by showing significant DC-negativation during spatial tasks. In agreement with our results, the authors also observed a significantly higher investment of cortical effort with poorer spatializers investing more activity (i.e. showing more pronounced negative topographical patterns). We additionally observed that the temporal structure of spatial information processing is a sequence of activations and in-activations characterized by sequences of negative- and positive-going potential changes of one dominant pattern. This shows that our method allows for a richer analysis of the temporal structure of the cognitive processes under investigation.

Finally we would like to comment on possible improvements concerning the techniques being used. Our decision to use batch k-means clustering for vector quantization was based on a thorough review of literature plus some comparative computer experiments which both showed batch k-means clustering to be a good choice. Our empirical decision for a codebook size of 64 could be replaced employing an objective measure obtained via a Bayesian approach to mixture modelling (see e.g. [11]). However, codebook centers obtained via clustering serve only as intermediate representations enabling efficient sequence comparison. Computation of selective averages is done using the original signals and not their codebook representations. Errors made during clustering could harm final results only in a very indirect way. Our sequence comparison algorithm finds subsequences which are the same across different subjects both in their topographical *and* timely appearance for the whole length of the analysis window. If we want to allow subsequences to additionally show considerable variation on the time axis, i.e. compression and expansion, algorithms that are able to deal with gaps in the comparison of sequences are needed. Hidden Markov Models (see e.g. [10]) are a promising candidate to solve this problem.

Acknowledgements: The EEG recordings have been made by R. Gstättner, Dept. of Psychology, University of Vienna. Parts of this work were done within the BIOMED-2 BMH4-CT97-2040 project SIESTA, funded by the EC DG XII. The Austrian Research Institute for Artificial Intelligence is supported by the Austrian Federal Ministry of Science and Transport. The author was supported by a doctoral grant of the Austrian Academy of Sciences.

REFERENCES

[1] Bacon D.J., Anderson W.F.: Multiple Sequence Alignment, Journal of Molecular Biology, 191, 153-161, 1986.

[2] Bauer H.: Slow potential topography, Behavior Research Methods, Instruments and Computers, 30(1), 20-33, 1998.

[3] Duda R.O., Hart P.E.: Pattern Classification and Scene Analysis, John Wiley & Sons, N.Y., 1973.

[4] Flexer A., Bauer H.: Discovery of Common Subsequences in Cognitive Evoked Potentials, in Zytkow J.M. & Quafafou M.(eds.), Principles of Data Mining and Knowledge Discovery, Second European Symposium, PKDD '98, Lecture Notes in Artificial Intelligence 1510, p.309-317, Springer, 1998.

[5] Gittler G.: Dreidimensionaler Wuerfeltest (3DW). Ein Rasch-skalierter Test zur Messung des raeumlichen Vorstellungsvermoegens. Theoretische Grundlagen und Manual, Beltz Test, Weinheim, 1990.

[6] Glaser E.M., Ruchkin D.S.: Principles of Neurobiological Signal Processing, Academic Press, 1976.

[7] Makeig S., Jung T-P., Bell A.J., Ghahremani D., Sejnowski T.J.: Blind Separation of Auditory Event-related Brain Responses into Independent Components, Proc. Natl. Acad. Sci. USA, 94:10979-10984, 1997.

[8] McGillem C.D., Aunon J.I.: Measurements of signal components in single visually evoked brain potentials, IEEE Transactions on Biomedical Engineering, 24, 232-241, 1977.

[9] Pfurtscheller G., Cooper R.: Selective averaging of the intracerebral click evoked responses in man: an improved method of measuring latencies and amplitudes, Electroencephalography and Clinical Neurophysiology, 38: 187-190, 1975.

[10] Rabiner L.R., Juang B.H.: An Introduction To Hidden Markov Models, IEEE ASSP Magazine, 3(1):4-16, 1986.

[11] Roberts S., Husmeier D., Rezek I., Penny W.: Bayesian Approaches to Gaussian Mixture Modelling, IEEE Transactions on Pattern Analysis and Machine Intelligence, Vol. 20, No. 11, 1133-1142, 1998.

[12] Vitouch O., Bauer H., Gittler G., Leodolter M., Leodolter U.: Cortical activity of good and poor spatial test performers during spatial and verbal processing studied with Slow Potential Topography, International Journal of Psychophysiology, Volume 27, Issue 3, p.183-199, 1997.

[13] Weerd J.P.C.de, Kap J.I.: A Posteriori Time-Varying Filtering of Averaged Evoked Potentials, Biological Cybernetics, 41, 223-234, 1981.

[14] Westerkamp J., Aunon J.I.: Optimum multielectrode a posteriori estimates of single-response evoked potentials, IEEE Transactions on Biomedical Engineering, 34, 13-22, 1987.

[15] Woody C.D.: Characterization of an adaptive filter for the analysis of variable latency neuroelectric signals, Medical and Biological Engineering, 5, 539-553, 1967.

A COMPARATIVE STUDY OF A HIDDEN MARKOV MODEL DETECTOR FOR ATRIAL FIBRILLATION

Brian Young, Don Brodnick, and Randy Spaulding

GE Marquette Medical Systems
8200 West Tower Avenue
Milwaukee, WI 53223

INTRODUCTION

Atrial fibrillation (AF) is the most common cardiac arrhythmia resulting in hospitalization [1]. It is often characterized as an irregular heart rhythm and is clinically defined as the complete absence of coordinated atrial contractions [2], which causes a random or chaotic heart rate. It is a medical condition that is responsible for significant morbidity and mortality and accounts for significantly large healthcare expenditures [3]. Furthermore, a higher incidence of thromboembolism and stroke in patients with AF adds to the already significant cost of managing and treating AF [2].

In order to manage a patient with AF, it is necessary to quantify various aspects of the occurrence of AF. One common way to do this is through the use of Holter ambulatory electrocardiogram (ECG) recording, which records the ECG of a patient for typically 24 hours on either magnetic tape or solid state media. The data are then reviewed, and a report is generated on an arrhythmia workstation, such as the Marquette Multi-parameter Arrhythmia Review Station (MARS) 8000 system. To facilitate quantification of AF, different computerized AF detection algorithms were investigated for use on the MARS workstation. The MARS is capable of analyzing data from different sources including analog tape and solid state Holter recording devices as well as data acquired on the Marquette Unity network from in-hospital patient monitoring at the bedside or by telemetry transmitters.

AF is very difficult to detect from an ambulatory ECG because of the fact that signals from atrial activity are not reliable and often difficult, if not impossible, to interpret. Although this is clearly a difficult problem, one still might propose the proper solution is to obtain ambulatory ECGs in such a way that reliable atrial activity is clearly and consistently recorded. Because this has been such an elusive problem, the most popular approach has been to detect AF based on the characteristics of ventricular activity in the presence of AF. The ventricular rhythm is usually described as random, chaotic, and irregularly irregular. It is this feature that AF detectors are usually designed to identify. The caveat of such algorithms is that there are many other rhythms that are irregular which are not AF.

Furthermore, the quantity of data for this application is so large and it is

0-7803-5673-X/99/$10.00

an application which requires human interaction such that the speed requirements of the algorithm are critical. Typically AF detection may be a relatively small component of a set of algorithms that will analyze over 20 million data points and 100,000 heart beats for a two channel, 24 hour recording. In addition, the amount of variability from patient to patient, compounded by ECG abnormalities, makes it very difficult to develop algorithms such as AF detection, which will provide reasonable ECG interpretations across such a diverse population.

A comparative study of several AF detection algorithms was done to determine the algorithm best suited for use in real clinical environments to detect AF in ambulatory ECGs. The algorithms that were investigated for this paper are based on the Hidden Markov Model, measures of variance, linear predictive coding, and measurement of approximate entropy.

METHODS

Each of the AF detection algorithms was developed using the same training set of data from the MIT-BIH arrhythmia database [4]. This training set consisted of twelve 30 minute tapes which demonstrated chronic AF, paroxysmal AF, the complete absence of AF, regular non-sinus rhythms, sinus rhythm, and various examples of irregular rhythms (both regularly irregular and irregularly irregular) that were not AF. The training set was used to derive algorithm parameters, thresholds, filters and processing modifications. After development of the algorithms was completed, they were evaluated on an

TABLE 1. DEFINITION OF PERFORMANCE MEASUREMENTS

Sensitivity = TP / (TP + FN)
Specificity = TN / (TN + FP)
Positive Predictive Accuracy = TP / (TP + FP)
Error Rate (FP + FN) / (TP + TN + FP + FN)

Where:

TP (true positive) = count of all beats in AF which are correctly called by the algorithm as being in AF
TN (true negative) = count of all beats NOT in AF which are correctly called by the algorithm as NOT being in AF
FP (false positive) = count of all beats in NOT AF which are incorrectly called by the algorithm as being in AF
FN (false negative) = count of all beats in AF which are incorrectly called by the algorithm as NOT being in AF

independent test set consisting of the MIT-BIH AF database [4], which is composed of 25 recordings approximately 10 hours in length.

The performance of the algorithms is reported in terms of sensitivity, specificity, positive predictive accuracy, and error rate. These values are standard statistics used to measure the performance of ECG arrhythmia analysis algorithms [5]. Sensitivity is a measure of how well the algorithm can identify periods of AF; specificity is a measure of how well the algorithm identifies periods of the signal NOT in AF, and positive predictive accuracy measures how often the algorithm is correct when it calls a time period AF. These are all performance measures that have important meanings to clinicians relying on these algorithms to help them manage patient care. The error rate is a value which is useful as a single value summary of the overall percentage of mistakes made by the algorithm. The definitions for these quantities are listed in Table 1.

HIDDEN MARKOV MODEL

The Hidden Markov Model (HMM) is a statistical model for predicting the state of a state machine based on observations [6][7]. Markov models are signal processing techniques that are popular for particular speech analysis problems but have also been used for ECG arrhythmia analysis [8][9][10]. The observations are not direct observations of the state machine, but observations of events, which provide indirect evidence of the model state. The design of the HMM was divided into three problems, namely: 1) definition of the model parameters; 2) evaluation of probabilities for observation sequences given the specific HMM; and 3) the adjustment of the algorithm parameters so as to best account for the observed signal.

The HMM model is specified by a set of states Q, a set of observations O, a matrix of transition probabilities **A**, and a vector of probabilistic output functions **B**. For this application, a simple two state (N=2) model was empirically chosen, where the two states were defined as AF and non-AF. The state transition probabilities are not known apriori but can be computed from the training set since the underlying states are known from the reference annotations. The observations are sequences of three RR intervals (T=3) between normal and supraventricular beats, which are classified by length relative to the local average RR interval. Figure 1 illustrates this two state HMM model. Thus the model can be described as follows:

(1) $O = [\, o_t \,||\, t = 0, \ldots, T\text{-}1]$
(2) $Q = [\, q_t \,||\, t = 0, \ldots, T\text{-}1]$
(3) $\mathbf{A} = [\, a_{ij} \,||\, i,j = 0, \ldots, N\text{-}1]$
(4) $\mathbf{B} = [b_i(x) \,||\, i = 0, \ldots, N\text{-}1]$
(5) $b_i(x) = \Pr(o_t = x \mid q_t = {}_{Si})$

In the above equations, o_t is the observation at time t, q_t is the hidden

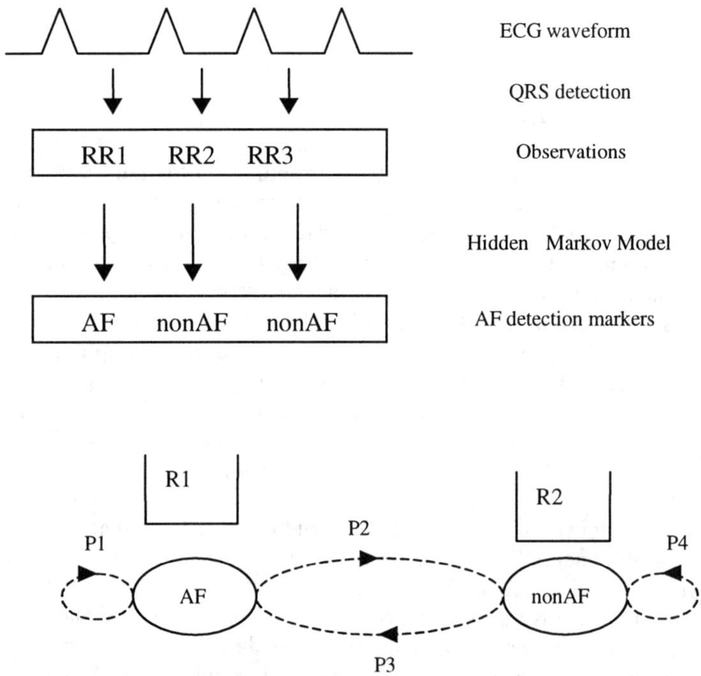

Figure 1. *Diagram illustrating the two state HMM model. The ECG waveform is analyzed for QRS detections, which are used to compute RR interval lengths. The RR intervals are observations used by the HMM to determine the AF/nonAF state of the model. P1, P2, P3, P4 represent state transition probabilities. R1 represents the RR interval observations generated by AF and R2 is the observations from the nonAF state.*

state at time t, **A** is the transition state probability, **B** is the probabilistic output vector, where b_i is the probability that o_t was the result of each of the N states. In this particular HMM, the RR intervals are classified as one of seven possible observations relative to the average RR interval: very short, short, slightly short, normal, slightly long, long, and very long.

Because both the observations and the hidden states are available to the algorithm for the training set, the parameters in (1) through (5) can be directly computed and the observation probability density functions for the states measured.

Finally, algorithm parameters such as the output filtering and the average RR estimator were adjusted to add robustness to the algorithm and minimize the error rate. The HMM output is then low pass filtered to remove jitter at the onset and offset of AF and make delineation of AF episodes easier. The heart rate estimator was designed to be insensitive to isolated ventricular ectopic beats

VARIANCE MEASUREMENTS

Given that AF is characterized by an irregular ventricular rhythm, a simple and common approach to detecting AF is to measure variability in RR intervals. Because normal cardiac rhythms have some variability as well, detection of AF requires measurements that are capable of differentiating AF from other rhythms. Methods of measuring variance have been used in real-time Holter recorders and various types of in-hospital patient monitoring devices. These algorithms have been used in such devices because of their simplicity and low computational cost. The two methods of measuring variance used in this study are 1) an incrementally updated estimate of RR interval variance (RRvar), which is used in real-time Holter applications, and 2) a count of RR interval differences that exceed a fixed threshold (RR100), which is used in various monitoring devices.

The RRvar algorithm measures an estimate of the RR variance, which is essentially an exponentially weighted average of the difference of RR intervals from the local average RR value. This is a measure of the RR variability, but is not a measure of variance as defined in the field of statistics. The average heart rate and heart rate variance is updated with each interval bounded on each end by either a normal or supraventricular beat. When the variance estimate exceeds a specific percentage of the average heart rate, the rhythm is considered AF. The variance and heart rate estimates exclude intervals bounded on either side by ventricular ectopic beats.

The RR100 algorithm is an up/down counter of RR interval differences that exceed 100 milliseconds. Intervals bounded by ventricular ectopic beats are excluded from processing. When the counter exceeds 6, the rhythm is considered AF. The counter is not permitted to exceed a count of 12 or go below a count of zero.

In addition to the simplicity and low computational cost of these algorithms, they are both well suited for real-time applications. The determination of AF is made on a beat-by-beat basis without any time delays. An output of the detector is produced for each new QRS which is not a ventricular ectopic.

LINEAR PREDICTIVE CODING

Linear predictive coding (LPC) is a mathematical method based on the concept that for many signals, past samples can be used to predict future outputs of a system [11]. This is a popular algorithm for some signal compression applications which deal with redundant and periodic signals. The coefficients used for LPC are derived from a parametric spectral estimation.

A spectral estimation method commonly used for LPC is the maximum entropy method (MEM). MEM is based on using a segment of a known autocorrelation function to extrapolate lags which are not known [12]. In other word, we can compute from a segment of signal its autocorrelation signal for lags 0 to p, and then predict lags $p+1$, $p+2$, $p+3$ and so on. MEM computes the extrapolations in such a way as to maximize the entropy of the autocorrelation. The rationale for choice of the maximum entropy criterion is that it imposes the fewest constraints on the unknown time series by maximizing its randomness, thereby producing a solution with minimum bias. The method used to implement the MEM is Burg's method [11], which is the most commonly used MEM algorithm. MEM produces an all pole model, which is not guaranteed to be stable, but is capable of characterizing sharp spectral features. The resulting model is used to generate the linear prediction coefficients. Because the prediction length is relatively short (10 RR intervals), stability is not an issue.

For AF detection, RR intervals between normal and supraventricular beats are used for computing the linear prediction coefficients. Processing is performed to handle isolated ventricular ectopy. Any intervals interrupted by multiple ventricular ectopic beats are excluded from the detector. If available, 20 consecutive intervals are used; otherwise, 10 intervals are used to compute six LPC coefficients. These coefficients are then processed to predict the next 10 intervals. The prediction error is taken as an indicator of AF. If the rhythm is irregularly irregular, such as AF, the prediction error will be high; otherwise, the prediction error should be low.

APPROXIMATE ENTROPY

Approximate entropy (AE) is a measure of signal randomness. In the field of nonlinear dynamics, many complexity measures statistically quantify the evolution of trajectory points in phase space [13]. However, most of these measures also try to quantify chaotic sequences, and consequently require the use of limits as time approaches infinity. AE measures signal randomness without taking limits. The trade off for this is that AE cannot detect chaos. For the purposes of detecting AF this is an acceptable condition.

To compute AE, segments of the signal are used for sliding comparisons with itself. For each comparison at a particular time delay, a count of the number of ``matches" based on a norm measurement and the Heaviside function is made [14]. Or in other words, a correlation sum is computed for different embedding dimensions. This measurement is then converted into a probabilistic value which is the AE measurement. The assumption is that AF will yield significantly higher AE values than other rhythms. Like other algorithms, ectopy will disrupt the heart rate variance and care must be taken to appropriately handle or exclude ectopy.

As with the other algorithms, processing is done to handle isolated ventricular ectopy, and the output of the filtering the output. The RR interval is then adjusted based on this classification. The AE is computed from 10 interval, non-overlapping windows. The AE measurements are then lowpass filtered to detect periods of AF.

RESULTS

Table 2 shows the training data results from algorithms after algorithm implementations had been optimized using the training data set. The training set of data consisted of six hours of data, or 30033 QRS complexes. According to the reference annotations, 41 percent of the QRS complexes in this data set were contained in atrial fibrillation regions.

TABLE 2. RESULTS FROM TRAINING SET

	Sensitivity	Specificity	Positive Pred.	Error Rate
HMM	97.70	88.73	86.77	7.00
AE	94.24	90.26	86.95	9.39
LPC	94.23	78.54	72.95	14.42
RRvar	99.57	56.54	61.27	25.88
RR100	34.72	93.89	79.70	30.28

Table 3 shows the test data set results from the same implementation of the algorithms listed in Table 1, without any further adjustments. The test set of data consisted of approximately 250 hours of data, or 1221578 QRS complexes. According to the reference annotations, 42 percent of the QRS complexes in this data set were contained in atrial fibrillation regions.

TABLE 3. RESULTS FROM TEST SET

	Sensitivity	Specificity	Positive Pred.	Error Rate
HMM	94.75	93.51	91.38	5.97
AE	94.06	92.56	90.41	6.54
LPC	91.68	83.14	79.06	13.16
RR100	60.31	92.47	85.46	21.14
RRvar	97.48	55.96	61.89	26.47

LIMITATIONS OF THE STUDY

This investigational study compares algorithm implementations for a specific real world application. The target computing environment is such that the execution time of the algorithm and its robustness in the presence of noisy or incomplete data are real and significant issues. It is well known that the MIT-BIH database has limited extrapolation of performance in the real clinical world [15]; however, the test set should provide a useful first estimate of the relative performance of these algorithms in the real world. Further testing is planned to more accurately measure performance of the HMM model algorithm in real world clinical settings.

CONCLUSION

Based on the results from the test data set, the HMM algorithm performed best for this application. In general, there is little difference between the performance of the HMM and AE algorithms. However, the implementation of the HMM algorithm is more computationally efficient. Because of the large amount of data that must be analyzed in ambulatory ECG recordings, the computational efficiency must be considered as an issue of practicality.

Review of the data illuminated some of the strengths and weaknesses of the various algorithms. Variance measures performed with either high sensitivity or high positive predictivity, but were not able to achieve a desirable operating point that had both acceptable sensitivity and positive predictivity. Although AE and LPC had acceptable sensitivity and positive predictivity, the HMM performed even better than both of these in terms of overall error rate. It would seem that an observational model such as the HMM, fits the data better than parametric models such as AE and LPC. Finally, as the computing power of medical systems increases, more sophisticated algorithms may be exploited in ways that leads to more accurate computerized ECG interpretation.

REFERENCES

[1] W. Jung, B. Luderitz, "Quality of Life in Patients with Atrial Fibrillation", **Journal of Cardiovascular Electrophysiology** 1998, 9:S177-S186.

[2] S. Levy, G. Breithardt, R.W.F. Campbell, A.J. Camm, et al, "Working Group Report - Atrial Fibrillation: current knowledge and recommendations for management", **European Heart Journal** 1998, 19:1294-1320.

[3] C.R. Kerr, "Guest Editorial - Atrial Fibrillation: The Next Frontier", **PACE** 1994, 17:1203-1207.

[4] G.B. Moody, R.G. Mark, "The MIT-BIH arrhythmia database on CD-ROM and software for use with it", **Computers in Cardiology** 1990, 17:185-188.

[5] Association for the Advancement of Medical Instrumentation (AAMI), "Testing and Reporting Performance Results of Cardiac Rhythm and ST Measurement Algorithm", **EC57** 1999.

[6] L.R. Rabiner, "A Tutorial on Hidden Markov Models and Selected Applications in Speech Recognition", **Proceedings of the IEEE** 1989, 77:257-286.

[7] I.L. MacDonald, W. Zucchini, **Hidden Markov and Other Models for Discrete-valued Time Series**, Chapman and Hall, 1997, ch.2, pp. 53-105.

[8] D.A. Coast, R.M. Stern, G.G. Cano, S.A. Briller, "An Approach to Cardiac Arrhythmia Using Hidden Markov Models", **IEEE Transactions on Biomedical Engineering** 1990 37: 826-836.

[9] D.A. Coast, R.M. Stern, G.G. Cano, S.A. Briller, "Use of Hidden Markov Models for Electrocardiographic Signal Analysis", **Journal of Electrocardiography**, 1990, vol. 23 supplement, pp. 184-191.

[10] P. Gersch, P.M. Eddy, E. Dong Jr., "PVC Detection by the Heart-Beat

Interval Data-Markov Chain Approach", **Computers and Biomedical Research**, 1975, 8:370-378.
[11] W.H. Press, B.P. Flannery, S.A. Teukolsky, W.T. Vetterling, **Numerical Recipes in C**, Cambridge University Press 1988, pp. 447-452, 461-466.
[12] S.M. Kay, S.L. Marple Jr., "Spectrum Analysis - A Modern Perspective", **Proceedings of the IEEE** 1981, 69:1380-1419.
[13] I.A. Rezek, S.J. Roberts, "Stochastic Complexity Measures for Physiological Signal Analysis", **IEEE Transactions on Biomedical Engineering** 1998, 45:1186-1191.
[14] D.T. Kaplan, M.I. Furman, S.M. Pincus, "Techniques for Analyzing Complexity in Heart Rate and Beat-to-Beat Blood Pressure Signals", **Computers in Cardiology** 1991, 18:243-246.
[15] B. Young, D. Brodnick, I. Rowlandson, "Performance Characteristics of a Real-time Arrhythmia Analysis Program on Different Databases", **Journal of Electrocardiography**, 1993, vol. 26 supplement, pp. 25-30.

DISCRIMINATION OF CYLINDERS WITH DIFFERENT WALL THICKNESSES USING NEURAL NETWORKS AND SIMULATED DOLPHIN SONAR SIGNALS

Lars Nonboe Andersen, Whitlow Au (*), Jan Larsen and Lars Kai Hansen

Department of Mathematical Modeling, Building 321
Technical University of Denmark, DK-2800 Lyngby, Denmark
Email: lna,jl,lkhansen@imm.dtu.dk
Web: http://eivind.imm.dtu.dk

(*) Marine Mammal Research Program, Hawaii Institute of Marine Biology
University of Hawaii, P.O. 1106, Kailua, Hawaii, 96734, USA

Abstract. This paper describes a method integrating neural networks into a system for recognizing underwater objects. The system is based on a combination of simulated dolphin sonar signals, simulated auditory filters and artificial neural networks. The system is tested on a cylinder wall thickness difference experiment and demonstrates high accuracy for small wall thickness differences. Results from the experiment are compared with results obtained by a false killer whale (*pseudorca crassidens*).

INTRODUCTION

Dolphins have demonstrated excellent capabilities of detecting, discriminating, recognizing and classifying underwater targets (see e.g. [3], [16]). Comparable target discrimination capabilities have not been achieved with conventional sonar. This has inspired research in new sonar systems based on biological knowledge, i.e. modeling dolphins' discrimination capabilities (see e.g., [15] and [21]).

The fact that the inner ear of the dolphin has many similarities with the human inner ear makes it tempting to use knowledge from simulations of the human auditory system when trying to model the dolphin sonar system. Furthermore, neural networks have proven to be very useful for pattern recognition tasks (see e.g., [21]) and are here applied for classification of features extracted from auditory preprocessing.

Based on earlier work [4], [14], [19] we will describe and present results

0-7803-5673-X/99/$10.00 © 1999 IEEE

from an experiment using a simulated dolphin signal, preprocessing based on auditory modeling and classification using a neural network. The aim is to discriminate between echoes recorded from hollow cylinders with different wall thicknesses.

Results from the experiment will be compared with results from a similar experiment involving a false killer whale (*pseudorca crassidens*) in which the whale was required to discriminate wall thickness differences of the same hollow cylinders used in our artificial experiment.

DATA COLLECTION

Echoes from ten hollow cylinders made of stainless steel were measured with an echo collection system using a planar broadband transducer to project and receive the acoustic signals. All data were collected in Kaneohe Bay, Oahu, Hawaii.

All cylinders had an outer diameter of 38.1 mm, and a height of 12.7 cm. One cylinder had a wall thickness of 6.35 mm and is denoted the standard cylinder. The other nine cylinders differed from the standard by 0 mm, ±0.08 mm, ±0.15 mm, ±0.23 mm and ±0.31 mm within an accuracy tolerance of 0.03 mm and are denoted comparison cylinders. A schematic of the standard and the comparison cylinders is shown in Fig. 1.

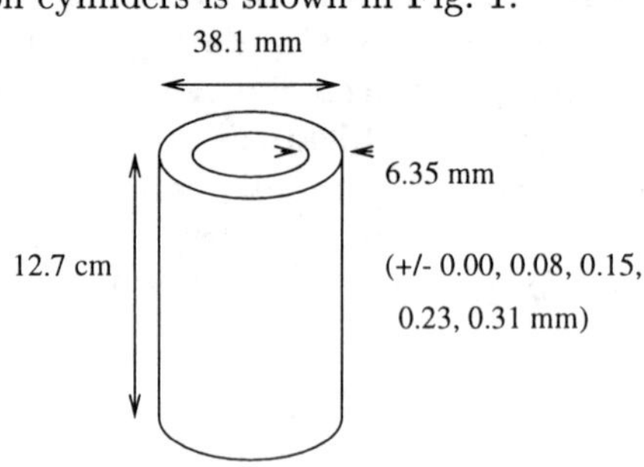

Figure 1: Schematic of the standard and comparison cylinders.

The transducer was mounted on a floating pen located in Kaneohe Bay. The pole supporting the transducer was aligned vertically and attached to the pen, with the transducer pointing directly towards the center of the suspended target. Each target was suspended with a monofilament line in such a manner its longitudinal axis was vertically aligned. With this geometry, the incident signal was nearly perpendicular to the longitudinal axis of the cylinders. The distance between the transducer and the target was approximately 5 m. An illustration of the setup is shown in Fig. 2.

A typical real dolphin sonar signal was recorded, stored and used in the experiments as a simulated dolphin sonar signal. This simulated dolphin sonar signal was projected at the cylinders and a Gage 1012 Data Acquisition Board operating at a sampling frequency of 1 MHz was used to digitize

Figure 2: Illustration of the setup.

the echoes with a resolution of 16 bits. Each echo consisted of 1024 points with the time window placed so the echo from the target was present in the recorded signal. The emitted signal is shown in Fig. 3 and an example of recorded echoes from the standard and the +0.15 mm comparison target is shown in Fig. 4.

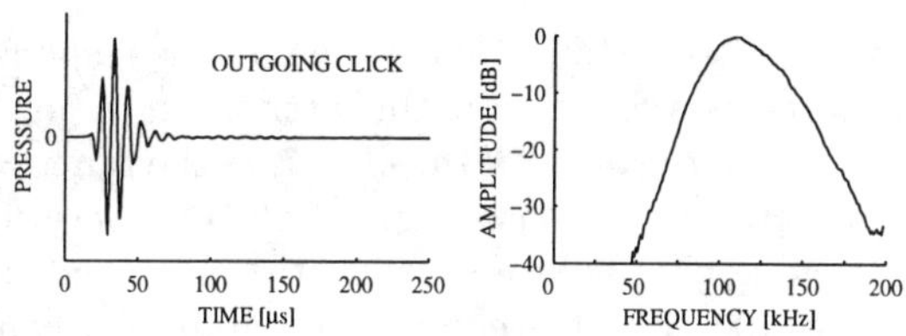

Figure 3: Waveform and spectrum of the projected simulated dolphin sonar signal.

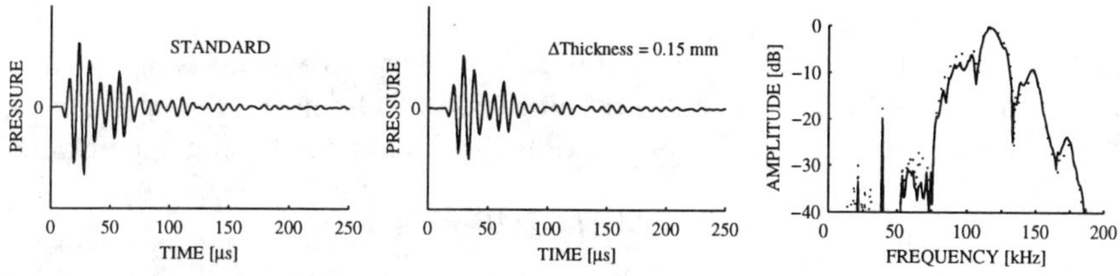

Figure 4: Echoes from the standard and the comparison cylinder that was 0.15 mm thicker than the standard cylinder. The waveforms are shown in the first two plots and the frequency spectra in the last plot. In the spectral plot the solid line is for the standard and the dotted line is for the comparison target.

The procedure for collecting the echoes was as following. the cylinder having the thinnest wall thickness was mounted at the end of the line and suspended in the water. Fifty echoes were then collected at a rate of two echoes per second. The cylinder was then removed from the water and unmounted from the line. This was performed for all ten cylinders. This session involving 500 echoes was repeated ten times giving a data base of 5000 echoes. This procedure was designed to reduce the risk of any potential features which are not related to the cylinders wall thicknesses (e.g. a fish swimming by, a cylinder hanging a little tilted etc.) being essential for classification of the echoes. Therefore, by using this procedure, hopefully only specific information from the cylinders will make it possible, when knowing the echoes in one session, to recognize echoes from the same cylinder from a different session.

PREPROCESSING

Features from the echoes were extracted using a combination of a matched filter, envelope detection, a gammatone filterbank, time integration and principal component analysis.

To find the beginning of each echo a matched filter, which was implemented as the time reversed version of the transmitted simulated dolphin click, was applied. The start of each echo was chosen as the peak of the envelope of the output from the matched filter (see e.g. [9]).

The gammatone filter is defined by its impulse response [1]

$$g(t) = at^{(n-1)} \exp(-2\pi bt) \cos(2\pi f_c t + \varphi),\ t > 0 \tag{1}$$

where b largely determines the duration of the impulse response and thus, the bandwidth of the filter; n is the order of the filter and it largely determines the slope of the skirts. When the order of the filter is in the range 3-5, the shape of the magnitude characteristic of the gammatone filter is very similar to that of the roex(p) filter commonly used to represent the magnitude characteristic of the human auditory filter [18]. The equivalent rectangular bandwidth (ERB) of the auditory filter is given by [11]

$$ERB = 24.75(4.37 f_c/1000 + 1\text{Hz}) \tag{2}$$

where f_c is center frequency of the filter. The center frequencies of the filters has been determined using Fay's modification of Greenwood's equation [10] for estimating cochlea frequency distributions along the basilar membrane

$$f_c(x) = 0.008 f_{\max}(10^{2.1x} - 1.0) \tag{3}$$

where $f_{\max}$ is the maximal frequency perceived by the animal and x is the position of the filter on the basilar membrane expressed as the proportion between the distance from the basal end and the full length of the basilar membrane ($x = 0$ at the basal end and $x = 1$ at the apical end). A high-frequency boundary of 150 kHz for $f_{\max}$ was used to coincide with the bottlenose dolphin upper frequency limit of hearing [13].

If we choose $n = 4$ and f_c/b is large, which is the case here, then b and the 3-dB bandwidth, BW, of the filter are given by [17]

$$b = 1.019 \cdot ERB \tag{4}$$

$$BW = 0.887 \cdot ERB \tag{5}$$

The quality of a filter is defined as $Q = f_c/BW$ and by using the described function for center frequencies all filters have approximately a constant Q-value of 10 which is between the Q-values measured for the bottlenose dolphin using the two different techniques: Critical Bandwidth (CB) ($Q = 2.2$) [5] and Critical Ratio (CR) ($Q = 12.3$) [12].

The remaining constants a and φ in the gammatone filter are chosen as $a = 1$ as the amplification variable and $\varphi = 0$ as the phase variable.

The filterbank consists of N_f such gammatone filters and N_f is limited to 15 for computational reasons. The locations of the filters are chosen to be linearly spaced by the distance, $\mathrm{dx} = 0.0093$ with start and end locations corresponding to center frequencies of 80 kHz ($x = 0.86$) and 150 kHz ($x = 1$). The frequency response characteristic for the 15 gammatone filters are shown in Fig. 5.

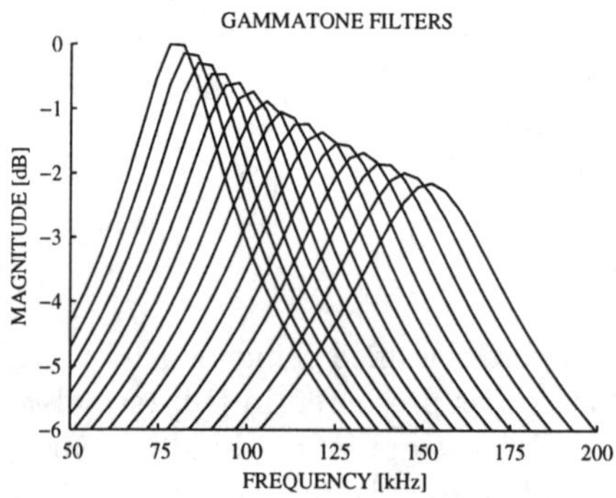

Figure 5: Frequency response for the gammatone filterbank.

The output of the filterbank when filtering the input signal with the bank of filters consists of N_f new signals, $y(n_f, t)$, where n_f is the filter number.

Each of these signals was split up in $N_{\text{bin}}(n_f)$ time bins of length $\delta_{\text{bin}}(n_f) = 1/f_c(n_f \cdot \mathrm{dx})$ and the power was calculated in each time bin, $n_{\text{bin}}(n_f)$ using

$$P(n_f, n_{\text{bin}}) = \frac{1}{\delta_{\text{bin}}} \int_{(n_{\text{bin}}-1)*\delta_{\text{bin}}(n_f)}^{n_{\text{bin}}*\delta_{\text{bin}}(n_f)} y^2(n_f, t)\mathrm{dt} \tag{6}$$

Only power values from time bins within the first 264 μs following the beginning of each echo were used for later analysis. A time window of 264 μs was chosen based on the dolphin's 264 μs integration time (see e.g. [6]). The power calculation in time bins are illustrated in Fig. 6. All power values were then combined into a single data vector P_{all} consisting of 439 values.

The complete data set consisting of all power value vectors were split in three sets: A training set, a validation set and a test set. The combined training-validation set consisting of the combination of the training set and the validation set was used to reduce the high dimensionality of the P_{all} vectors using principal component analysis (PCA) (see e.g. [20]). A singular value decomposition of the combined training-validation set was performed and only the ten highest eigenvalues were retained and used to project all

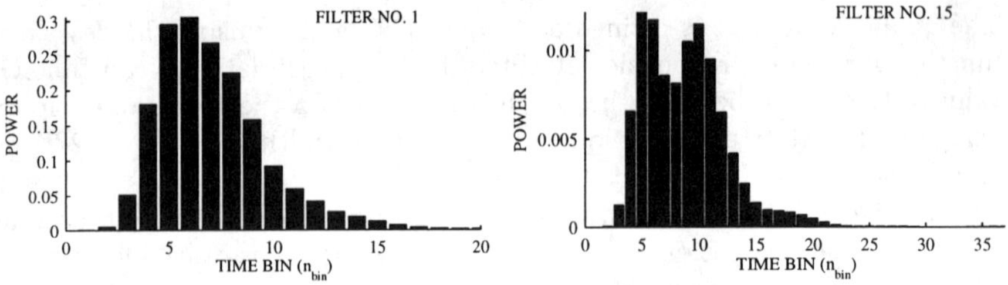

Figure 6: Illustration of power values calculated in time bins for the first filter and the last filter.

data down to a 10-dimensional feature vector, which means 10 features were extracted from each echo. The ten highest eigenvalues contained in average for the nine experiments 85 % of the variance. A limit of ten features was chosen for computational reasons to obtain a reasonable ratio between the number of training data and network complexity which is affected by the chosen number of features.

CLASSIFICATION

The extracted features were classified using a feed-forward net with a modified SoftMax [8] normalization as presented in [14] (see also, [2] [7]). The 2-layer feed-forward network with n_I inputs, n_H hidden neurons and $c-1$ outputs, where c is the number of classes, is defined by:

$$h_j(\mathbf{x}) = \tanh\left(\sum_{\ell=1}^{n_I} w_{j\ell}^I x_\ell + w_{j0}^I\right) \tag{7}$$

$$\phi_i(\mathbf{x}) = \sum_{j=1}^{n_H} w_{ij}^H h_j(\mathbf{x}) + w_{i0}^H \tag{8}$$

where $w_{j\ell}^I$, w_{ij}^H are the input-to-hidden and hidden-to-output weights, respectively. All weights are assembled in the weight vector $\mathbf{w} = \{w_{j\ell}^I, w_{ij}^H\}$. In order to interpret the network outputs as probabilities we used a *modified* normalized exponential transformation [14] similar to SoftMax [8],

$$\widehat{z}_i = \frac{\exp(\phi_i)}{\sum_{j=1}^{c-1}\exp(\phi_j)+1}, \quad 1 \le i \le c-1, \tag{9}$$

$$\text{and } \widehat{z}_c = 1 - \sum_{i=1}^{c-1} \widehat{z}_i, \tag{10}$$

where $\widehat{z}_i$'s are estimates of posterior probabilities.

The network was optimized using the maximum a posteriori technique, i.e., the cost function is the sum of the log-likelihood and a regularization term (prior).

$$C(\mathbf{w}) = S(\mathbf{w}) + R(\mathbf{w}, \kappa) \tag{11}$$

where $R(\mathbf{w}, \kappa)$ is a weight decay parameterized by a set of regularization parameters κ.

The full scheme for optimizing the network was presented at ICASSP 98 [14] but in this work only the weights and the regularization parameters were optimized using a second-order Gauss-Newton scheme based on the training set for the weights and a gradient descend scheme based on the validation set for the regularization parameters as described in [14]. For a more detailed description and use of outlier detection, see [14].

The number of inputs in the network correspond to the number of extracted features ($n_i = 10$), the number of hidden units, n_H was chosen to 5 and the number of classes c was chosen to 2 representing the standard cylinder and a comparison cylinder. Five hidden units was chosen to give an acceptable size of the network compared to the number of training data. Thus, the network initially consisted of $11 \cdot 5 + 6 \cdot 1 = 61$ free weights.

EXPERIMENTS

Ten experiments were performed testing the systems capability of discriminating between the standard cylinder and each of the ten comparison cylinders.

Data for the training set were chosen as all data from the first three sessions, data for the validation set came from the next three sessions while data from the last four sessions were used for the test set. The training, validation and test set for each experiment thus consisted of $50 \cdot 3 \cdot 2 = 300$, $50 \cdot 3 \cdot 2 = 300$ and $50 \cdot 4 \cdot 2 = 400$ data respectively since each set contained fifty echoes from each session from both the standard and a comparison cylinder.

Data from both the training set and the validation set were used for optimizing the neural network as described in [14] whereas data from the test set exclusively were used for testing the system.

RESULTS

Table 1 reports the performance of the system for different wall thickness differences between the standard and the comparison cylinders. The system is capable of discriminating between the standard and the comparison cylinder with a probability of correct classification, *pcc*, higher than 99% when the wall thickness difference is ≥ 0.15 mm. The system performance decreases as the wall thickness difference approaches zero and has a performance close to fifty percent (guessing) for a comparison cylinder expected to have the same wall thickness as the standard cylinder within the given mechanical tolerance. As can be seen in Table 1 retraining on the combined training and validation set using regularization parameters scaled by the increased number of training examples in the combined set improves the performance on the test set in most cases. Although increased performance is often observed using this technique it is not guaranteed. This can be seen in the experiment $\Delta WT = 0.00$ where a small decrease in performance is observed after retraining.

ΔWT [mm]	**Train.** [%]	**Val.** [%]	**Test** [%]	**Test after retrain** [%]
−0.31	100.0	100.0	100.0	100.0
−0.23	100.0	100.0	100.0	100.0
−0.15	100.0	100.0	99.5	99.5
−0.08	100.0	98.7	94.0	95.8
0.00	76.3	78.0	59.0	57.5
+0.08	100.0	93.7	83.5	96.0
+0.15	100.0	100.0	100.0	100.0
+0.23	100.0	100.0	99.8	99.8
+0.31	100.0	100.0	100.0	100.0

Table 1: Results from artificial experiment describing probability of correct classification for different wall thickness differences between the standard and the comparison cylinder.

Results from the artificial experiment are shown in Fig. 7 along with results from a similar experiment involving a false killer whale[1]. The performance of the system is seen to be comparable with the performance of the dolphin with a slightly higher probability of correct classification for the artificial system. The confidence intervals are found using the standard formula for calculating confidence intervals for a binomial distribution.

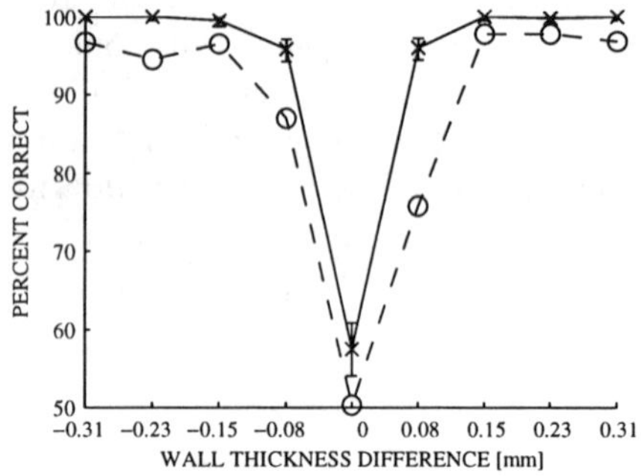

Figure 7: Results from artificial experiment (x-markers connected with full line including 90%-confidence intervals) compared with results from a similar experiment involving a false killer whale (o-markers connected with dashed line).

[1]This experiment has not yet been published.

DISCUSSION AND CONCLUSION

The results from the experiment shows that the described system can discriminate very accurately between the standard cylinder and the comparison cylinders differing more than or equal to 0.15 mm in wall thickness. This small wall thickness difference has been detected under conditions where the targets have been approximately five meters away and subject to wind and wave motions.

Comparing the results from the artificial experiment with results from a similar experiment involving a false killer whale shows that the artificial system can produce performance results which are comparable to results obtained by the whale. It should be emphasized however that although there are similarities in the two experiments differences are also present. In the whale experiment the whale is trained to remember the standard cylinder and during the experiment choose the standard when presented for the standard and a comparison cylinder. This task thus demands the dolphin to remember the standard cylinder over a period of time. The dolphin might also be distracted or simply unconcentrated during the experiment. Such psychological factors are important to have in mind when comparing the results and are possibly some of the reasons why the whale has a slightly lower performance than the artificial system.

Having this in mind the results of this study using a combination of the matched filter technique, the gammatone filterbank, time integration, principal component analysis and neural networks for discriminating cylinders with different wall thicknesses has demonstrated its potential.

ACKNOWLEDGMENTS

The authors would like to express their appreciation to Dr. Robert Gisiner, Office of Naval Research, for allowing us to use data from the experiment involving the *pseudorca crassidens*, to Dr. Herbert Roitblat, University of Hawaii, for many inspiring discussions and to Dr. Paul Nachtigall, University of Hawaii, for making the collaboration possible. Research was supported by the Danish Research Councils through the Danish Computational Neural Network Center (CONNECT) and the THOR Center for Neuroinformatics.

REFERENCES

[1] A. Aertsen and P. Johannesma, "Spectro-temporal receptive fields of auditory neurons in the grassfrog. 1. characterisation of tonal and natural stimuli," **Biol. Cybern.**, vol. 38, pp. 115–135, 1980.

[2] L. Andersen, J. Larsen, L. Hansen and M. Hintz-Madsen, "Adaptive regularization of neural classifiers," in J. Principe, ed., **IEEE Workshop on Neural Networks for Signal Processing VII**, IEEE, Piscataway, NJ, USA, 1997, pp. 24–33.

[3] W. Au, **The Sonar of Dolphins**, New Jersey, USA: Springer Verlag, 1993.

[4] W. Au, L. Andersen, A. Rasmussen, H. Roitblat and P. Nachtigall, "Neural network modeling of a dolphin's sonar discrimination capabilities," **J. Acoust. Soc. Am.**, vol. 98, pp. 43–50, 1995.

[5] W. Au and P. Moore, "Critical ratio and critical bandwidth for the atlantic bottlenose dolphin," **J. Acoust. Soc. Am.**, vol. 88, pp. 1635–1638, 1990.

[6] W. Au, P. Moore and D. Pawloski, "Detection of complex echoes in noise by an echolocating dolphin," **J. Acoust. Soc. Am.**, vol. 83, pp. 662–668, 1988.

[7] C. Bishop, **Neural Networks for Pattern Recognition**, Oxford, UK: Oxford University Press, 1995.

[8] J. Bridle, "Probabilistic interpretation of feedforward classification network outputs with relationships to statistical pattern recognition," in **Neurocomp. - Algor. , Arch. and Appl.**, Berlin, Germany: Springer-Verlag, vol. 6, pp. 227–236, 1990.

[9] W. Burdic, **Underwater Acoustic System Analysis**, New Jersey, USA: Prentice Hall, 1991.

[10] R. Fay, "Structure and function in sound discrimination among vertebrates," in D. Webster, R. Fay and A. Popper, eds., **The Evolutionary Biology of Hearing**, New York, USA: Springer-Verlag, pp. 229–263, 1992.

[11] B. Glasberg and B. Moore, "Derivation of auditory filter shapes from notched-noise data," **Hearing Research**, vol. 47, pp. 103–138, 1990.

[12] C. Johnson, "Masked tonal threshold in the bottlenose porpoise," **J. Acoust. Soc. Am.**, vol. 44, pp. 965–967, 1968.

[13] C. Johnson, "Auditory masking on one pure tone by another in the bottlenose porpoise," **J. Acoust. Soc. Am.**, vol. 49, pp. 1317–1318, 1971.

[14] J. Larsen, L. Andersen, M. Hintz-Madsen and L. Hansen, "Design of robust neural network classifiers," in **ICASSP'98**, Seattle, USA, 1998, vol. 2, pp. 1205–1208.

[15] P. Moore, H. Roitblat, R. Penner and P. Nachtigall, "Recognizing successive dolphin echoes with an integrator gateway network," **Neural Networks**, vol. 4, pp. 701–709, 1991.

[16] P. Nachtigal, "Odontocete echolocation perfomance on object size, shape and material," in R. Busnel and J. Fish, eds., **Animal Sonar**, New York, USA: Plenum, pp. 71–95, 1980.

[17] R. Patterson, J. Holdsworth, I.Nimmo-Smith and P. Rice, "Svos final report: The auditory filterbank," Tech. Rep. 2341, APU, 1988.

[18] R. Patterson and B. Moore, "Auditory filters and excitation patterns as representations of frequency resolution," in B. Moore, ed., **Frequency Selectivity in Hearing**, London, UK: Academic Press Limited, pp. 123–177, 1986.

[19] R. Patterson, K. Robinson, J. Holdsworth, D. McKeown, C. Zhang and M. Allerhand, "Complex sounds and auditory images," in **Advances in the Biosciences**, UK: Pergamon Press, vol. 83, pp. 426–446, 1992.

[20] B. Ripley, **Pattern Recognition and Neural Networks**, Cambridge, UK: Cambridge University Press, 1996.

[21] H. Roitblat, P. Moore, P. Nachtigal, R. Penner and W. Au, "Natural echolocation with an artificial neural network," **Neural Networks**, vol. 1, pp. 239–248, 1989.

STATISTICAL MULTIMODAL INTEGRATION FOR INTELLIGENT HCI

Lizhong Wu, Sharon L. Oviatt & Philip R. Cohen
Center for Human and Computer Communication
Department of Computer Science and Engineering
Oregon Graduate Institute of Science and Technology
P.O.Box 91000, Portland, OR 97291-1000, USA
Phone: 503 748-1248, Fax: 503 690-1548
Email: lwu, oviatt, pcohen@cse.ogi.edu
Web: http://www.cse.ogi.edu/CHCC/

Abstract. This paper presents a statistical approach to developing multimodal recognition systems and, in particular, to integrating the posterior probabilities of parallel input signals involved in the multimodal system. We first derive the performance bounds of multimodal recognition probabilities, and identify the primary factors that influence multimodal recognition performance. We then develop a technique, a Members-Teams-Committee (MTC) recognition approach, designed to optimize accurate recognition during the multimodal integration process. We evaluate these methods using Quickset, a speech/gesture multimodal system, and report evaluation results based on an empirical corpus collected with Quickset. From an architectural perspective, the integration technique presented here offers enhanced robustness. It also is premised on more realistic assumptions than previous multimodal systems using semantic fusion. From a methodological standpoint, the evaluation techniques that we describe provide a valuable tool for evaluating multimodal systems. [1]

INTRODUCTION

There are two main types of multimodal systems, one of which integrates signals at the feature level and the other at a semantic level. Systems that utilize feature fusion generally are based on multiple HMMs or temporal neural networks. In a feature fusion architecture, the correlation structure between modes can be taken into account automatically via learning. Feature fusion generally is considered more appropriate for closely coupled and synchronized

[1]This research is supported in part by DARPA under contract number DABT63-95-C-007, in part by ONR grant number N00014-95-1-1164.

modalities, such as speech and lip movements. However, such a system tends not to generalize as well if it consists of modes that differ substantially in the time scale characteristics of their features as is the case with speech and gesture input. Modeling complexity, computational intensity, and training difficulty typically are other problems associated with the feature fusion integration approach. Due to the high dimensionality of input features and high degree of freedom of system models, a large amount of training data also is required for building this type of system. Of course, multimodal corpora rarely have been collected and labeled for training purposes, and they tend not to be publicly available so therefore are at a high premium.

Generally, multimodal systems using semantic fusion include individual recognizers and a sequential integration process. These individual recognizers can be trained using unimodal data, which are relatively easy to collect or are already publicly available for modalities like speech and handwriting. The architecture of this type of system also can leverage from existing and relatively mature unimodal recognition techniques. Such unimodal systems can be integrated directly without re-training. Compared with systems based on feature fusion, in this respect systems using semantic fusion scale up easier, whether in number of modes or size of command set.

Multimodal systems with fusion at the semantic level include Bolt's seminal work "Put-That-There" [1], ShopTalk [3], CUBRICON [8], Virtual World [2], Finger-Pointer [5], VisualMan [13], Jeanie [12], and others as described in [7, 11]. All these previous efforts on multimodal integration have concentrated primarily on semantic representations and incorporation of new input technologies rather than on the statistical integration process that defines a multimodal system architecture. Such systems typically also have assumed that the individual modes in a multimodal interaction function independently of each other. As a result, a multimodal command's posterior probability has been the cross product of the posterior probabilities of the associated constituents. Although this independence assumption has provided a starting point and it simplifies the integration process, it nonetheless is a naive assumption since modalities like speech and lip movements or speech and manual gestures are known to be highly correlated [10].

It also is known that a constituent in one mode typically associates with only a limited number of constituents in another mode and that input modes differ in both their information content and recognition accuracy. An additional problem with past multimodal architectures is that the overall recognition accuracy of different input modes has been assumed to be equally reliable, although this is rarely the case. Even within the same mode, recognition accuracy varies considerably from one constituent to another. By refining the multimodal integration process so that different weights are assigned to different modes and different constituents, recognition errors potentially could be avoided so that overall system robustness is enhanced.

In theory, the optimal weights for combining the posterior probabilities from different modes can be determined by the mode-conditional input feature density functions [15]. In practice, it is difficult or even impossible to evaluate these conditional density functions because of the high dimensional input features. In this paper, we have developed a modeling technique, referred to as Members-Teams-Committee (MTC), to approximate these conditional densities and to obtain the class-dependent weighting parameters for the posterior probabilities.

Another critical issue involved in the multimodal integration process is how to evaluate and estimate system recognition performance. Given a set of individual recognizers with known accuracies, is the multimodal system's performance bounded? If so, what are the theoretical lower and upper performance bounds? By estimating performance bounds, it becomes possible to evaluate the performance of alternative integration techniques in comparison with a theoretical optimum. From a diagnostic perspective, it also becomes possible to identify key factors that influence multimodal performance. To our knowledge, past research has not estimated performance bounds for guiding the development of multimodal systems. In this paper, we have derived a lower and upper bound of multimodal recognition performance.

The outline of this paper is as follows. Section 2 briefly introduces **Quickset** [4], a multimodal system developed at OGI that has been a testbed for our research. Section 3 derives the multimodal recognition probability, the multimodal performance lower and upper bounds and their boundary conditions. It also identifies the primary factors that influence multimodal recognition performance. Section 4 provides a theoretical solution for combining the multimodal posterior probabilities. Section 5 develops an MTC technique to integrate and optimize the multimodal posterior probabilities. Empirical evaluation and performance comparisons are reported in Section 6. We summarize our findings in Section 7.

A MULTIMODAL SYSTEM

Our multimodal system, called Quickset [4], consists of parallel recognizers for the speech and gesture input modalities, both of which fuse at the semantic level. A command in this multimodal system is represented by constituents that are joined from the two different input modes. Each constituent is a target of an individual mode's recognizer. During the recognition process, an individual recognizer analyzes a set of input features and then produces the constituents as an N-best list of alternatives, along with posterior probabilities. Figure 1 illustrates **Quickset**'s architecture. The gesture recognition for **Quickset** was developed at OGI, and the speech recognition is Microsoft's Whisper. **Quickset** integrates multimodal input in the following three sequential steps:

1. Temporally, **Quickset** combines speech and gesture input that is overlapped, or that falls within a certain lag relation when signals arrive sequentially. The temporal constraints of **Quickset**'s integration were determined by empirical research with users [10]. It was found that when users speak and gesture in a sequential manner, they gesture first, then speak within a rela-

tively short time window; speech rarely precedes gesture. As a consequence, the multimodal synchronizer in **Quickset** prefers to integrate a gesture with speech that follows within a 4-second interval, rather than integrating it with preceding speech. If speech arrives after that interval, the gesture is interpreted unimodally. The precise lag threshold adopted when signals arrive sequentially can be learned by the system using training data, or pre-set by the system developer for a particular domain.

2. Statistically, **Quickset** integrates the posterior probabilities of constituents from individual modes, and then generates an N-best list for a multimodal command that includes posterior probabilities for each final interpretation. The original version of **Quickset** relied on the independence assumption, and takes the cross product of the probabilities of individual modes as the multimodal probability for each item in the final multimodal N-best list. One goal of the present work is to supersede the independence assumption by developing a more powerful statistical integrator based on the realities of empirical data.

3. Semantically, **Quickset** determines whether a given gestural element and a spoken element in the N-best lists can be combined legally into a coherent multimodal interpretation that is executable by the system. The semantic information contained within the two modes in **Quickset** are represented as typed feature structures that can be unified if the elements are compatible semantically. The unification of typed feature structures in **Quickset** has been detailed elsewhere [6].

Quickset has been applied to various map-based applications that enable users to set up and control distributed interactive simulations. Further information and videotape examples of **Quickset** can be found at http://www.cse.ogi.edu/CHCC.

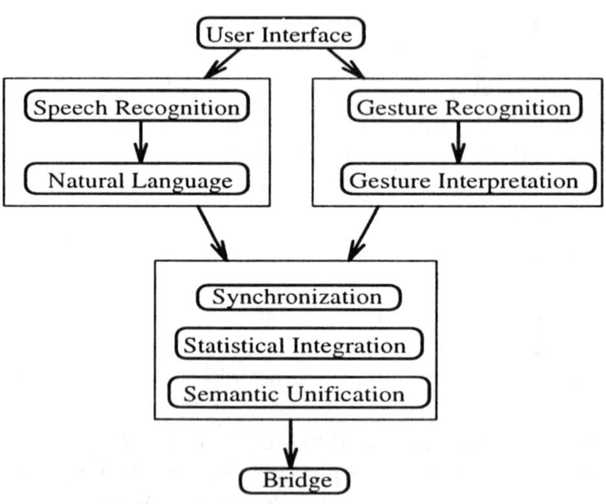

Figure 1: Architecture of **Quickset**, a collaborative, handheld, agent-based, speech/gesture system developed at OGI [4].

MULTIMODAL RECOGNITION PERFORMANCE BOUNDS

In a speech/gesture multimodal system, assume that $S_i, i = 1, \ldots, M$ is the output from the speech mode, and $G_j, j = 1, \ldots, N$ is the output from the gesture mode. The system is designed to recognize $C_k, k = 1, \ldots, K$ multimodal classes. The number of multimodal classes K cannot be larger than the number of attainable integrated classes $(M+1)(N+1)-1$, but will at least equal the larger number of the two output modes $max[M, N]$.

We define the projection between the index of multimodal classes and the indices of the individual modal output as a *multimodal associative map*. For a given corpus, the associative map defines all meaningful relations that exist between the set of speech constituents and the set of gesture constituents for each multimodal command. During multimodal recognition, the defined associative map supports a simple process of table lookups. This table can be defined directly by a user, or it can be built automatically using labeled data.

Assume that X represents a multimodal input feature vector, which is a combination of gesture input feature X^G and and speech input feature X^S. A system designed to recognize K classes of commands will partition the input feature space into K disjoint decision regions $R_k, k = 1, \ldots, K$. The probability of correct recognition for X is thus:

$$P(correct) = \sum_{k=1}^{K} P(X \in R_k, C_k) = \sum_{k=1}^{K} P_{X_k} P(C_k) \qquad (1)$$

where

$$P_{X_k} = P(X \in R_k | C_k) = \int_{R_k} p(X|C_k) dX \qquad (2)$$

is the probability of correct recognition for the kth class, $p(X|C_k)$ is its class-conditional density and $P(C_k)$ is its prior probability.

By expressing the multimodal recognition probability using the recognition probabilities of its associated modes, we have obtained the following multimodal recognition probability bound [15]:

$$\sum_{k=1}^{K} P_{X_j^G} P_{X_i^S} P(C_k) \leq P(correct) \leq \sum_{k=1}^{K} max[P_{X_j^G}, P_{X_i^S}] P(C_k) \qquad (3)$$

where $P_{X_j^G}$ and $P_{X_i^S}$ are respectively the correct recognition probabilities of the associated gesture and speech constituents for the kth multimodal class.

Equation 3 has revealed that a multimodal system performs at its lower bound if individual modes are assumed to be independent and a simple joint probability estimate is calculated during integration. Furthermore, a multimodal system performs at the upper bound when the information in one mode is completely redundant with that in the other mode. In general, we would summarize that multimodal recognition performance is determined by the following factors: (1) recognition accuracy of the individual modes, (2) the structure of the associative map, (3) the manner of combining posterior probabilities, and (4) the prior distribution of multimodal commands.

INTEGRATION OF MULTIMODAL POSTERIOR PROBABILITIES

During recognition, our goal is to evaluate the posterior probabilities of all multimodal classes given an unknown input feature set. Combining the posterior probabilities involves combining the class-conditional density function $p(X|C_k)$. In [15], we have found:

$$p(X|C_k) = \frac{1}{2}\left[p(X^S|C_k, X^G)p(X^G|C_k) + p(X^G|C_k, X^S)p(X^S|C_k)\right] \quad (4)$$

where $p(X^S|C_k)$ and $p(X^G|C_k)$ are the class-conditional densities estimated by the speech recognizer and gesture recognizer. $p(X^G|C_k, X^S)$ and $p(X^S|C_k, X^G)$ are the mode-conditional input feature densities for the kth class.

Equation 4 provides a theoretical solution for integrating multimodal class-conditional density functions and their corresponding posterior probabilities. However, due to the high dimensional input features, a large amount of training data is required to evaluate the mode-conditional input feature densities directly. Instead, an approximation would be

$$\alpha_k = \frac{1}{2}p(X^S|C_k, X^G) \text{ and } \beta_k = \frac{1}{2}p(X^G|C_k, X^S). \quad (5)$$

α_k and β_k are class-conditional, but independent of individual input features. Equation 4 can then be rewritten as

$$p(X|C_k) = \alpha_k p(X^G|C_k) + \beta_k p(X^S|C_k). \quad (6)$$

α_k and β_k are the weighting parameters to the modal class-conditional densities. In the next section, we develop a technique to evaluate these parameters.

MTC TECHNIQUE

MTC Architecture

The MTC architecture consists of three layers. The bottom layer is formed by multiple recognizer members. Each member is a local posterior estimator with an assigned input variable subset, a specified model type and complexity, and a given training and validation data set. The members cooperate with each other via the multiple teams built at the mid-layer. Different teams observe different training data, and are initialized and trained differently. The team integrates the members. Multiple teams are built to reduce integration uncertainty. Output from the teams forms an empirical posterior distribution that then is sent to the committee at the upper layer. The committee makes a final decision after comparing the empirical posterior distributions of different targets.

MTC Recognition Algorithm

In general, we define the input feature set $I = \{I_1, I_2, \ldots, I_n\}$ and the recognition target set $T = \{T_1, T_2, \ldots, T_m\}$. The input feature I is formed by n-streams, whose dimensions may differ. The target T consists of m different classes, for example of different multimodal commands. The MTC recognition algorithm goes through three bottom-up steps:

1. Estimating the local posteriors of members: Each member computes a local posterior estimate under the specified modeling condition. The modeling specifications include the model type, the model complexity, the extraction of input features, the training and validation data, and the learning algorithm. If there is a total of M combinations of modeling specifications in which we are interested, then we would compute M local posterior estimates from the M members as follows:

$$\hat{P}(T_k|I, S_i), \text{ with } k = 1, \ldots, m \text{ and } i = 1, \ldots, M \tag{7}$$

where S_i stands for the ith combination of modeling specifications.

2. Coordinating the local posteriors into teams: The team integrates the local posterior estimates of different specifications. We have

$$\hat{P}(T_k|I) = \sum_{i=1}^{M} \hat{P}(T_k|I, S_i)P_k(S_i), \text{ for } k = 1, \ldots, m \tag{8}$$

where $P_k(S_i)$ is the mode probability of the kth target associated by the ith combination of modeling specifications. The team is trained to learn the mode probability matrix. Different training data and approaches will result in different mode probability estimates. Subsequently, the multiple team posteriors are obtained:

$$\hat{P}^{(l)}(T_k|I), \text{ with } k = 1, \ldots, m \text{ and } l = 1, \ldots, L \tag{9}$$

where l is the index of ways of estimating the mode probability, and L is the total number of ways that we are interested in.

3. Making a recognition decision via committee: The output from multiple teams forms an empirical distribution of posterior $P(T_k|I)$, which is approximated by a normal or t-student distribution, depending on the size of the samples. Given a confidence level, the committee runs through a series of pair-by-pair hypothesis tests and obtains a significance matrix H. H is an $m \times m$ square matrix, where m is the number of recognition targets. The element of H, h_{ij}, is either 1 when the posterior estimate $\{\hat{P}^{(l)}(T_i|I), l =$

$1, \ldots, L\}$ is significantly greater than $\{\hat{P}^{(l)}(T_j|I), l = 1, \ldots, L\}$ within the given confidence level, -1 when it is significantly less, or otherwise 0. By definition, the diagonal elements of H are all zeros, if $h_{ij} = 1$ then $h_{ji} = -1$, if $h_{ij} = 0$ then $h_{ji} = 0$, and $\sum_i \sum_j h_{ij} = 0$. The recognition targets then are ranked by summing over each row in H, and this summary value is called a significance number. All significance numbers form an m-dimensional significance vector V. The maximal significance number is $m - 1$. If there is a significance number that equals $m - 1$, the input is recognized as the target corresponding to the row index of this maximal significance number. If all significance numbers are smaller than $m - 1$, then the current input cannot be recognized with confidence, and further external information is required.

MTC Training Procedure

The goal of the first tier involving the MTC's members is to learn a set of local posterior estimates, as indicated in equation 7. The key of this first layer is to identify the modeling specifications and their combinations. The members within the MTC can represent different types of models, with the training algorithm for the members being model-dependent. Among the various modeling specifications, the most important one is the extraction of input features via exploratory data analyses. Once the input features have been extracted, the model type is selected to fit the characteristics of input features. In the MTC, a variety of input features can be extracted, and different types of models can be selected to fit different types of input features.

With respect to the training procedure for teams, the goal of this second layer is to learn the mode probability matrix in equation 8. By adjusting the mode probability matrix $\{P_k(S_i)\}$, we maximize (i.e., reward) the kth posterior $\hat{P}(T_k|I)$ and simultaneously reduce (i.e., penalize) the other posteriors $\hat{P}(T_j|I)$, for $j = 1, \ldots, m$ and $j \neq k$, when the kth-class pattern is applied. In order to meet the constraint that the sum of all posteriors must equal one, we impose a softmax function on the output. The detailed learning algorithm is given in [14].

The team integrates the members' posterior estimates. Multiple teams are built to reduce integration uncertainty. The goal of the committee is to compare the empirical posterior distribution formed by the teams and make a final recognition decision. To train the committee, no free system parameter is needed. The confidence level for recognition is pre-determined, and different confidence levels will result in a different system error rate/rejection rate tradeoff. The higher the confidence level, the lower the error rate but the higher the rejection rate.

MTC Multimodal Statistical Integration

Our proposed MTC technique is well suited to experimenting with ways to integrate multiple modes on the basis of posterior probabilities and other factors. Using this technique, the recognizers of different modes become the members of an MTC statistical integrator. Multiple teams built in the MTC integrator are trained to coordinate and weight the output from different modes. Each team establishes the posterior estimate for a multimodal command, given the current multimodal input received. The committee of the MTC integrator analyzes the empirical distribution of the posteriors and establishes the N-best ranking for each multimodal command.

EMPIRICAL RESULTS

Our Quickset system was the testbed used to formulate and evaluate the derived performance bounds and the MTC integration concepts. The data corpus used in the present work was collected using Quickset while users performed community fire and flood management tasks, as described in [9]. All commands were multimodal, involving both speech and gesture input. For example, to control the system's map display, one can say "zoom in" and simultaneously use the pen to circle an area of the map. The corpus consisted of 1539 labeled commands collected from sixteen users, eight native speakers of English and eight accented non-native speakers. We randomly assigned the data from the first eight users for development, and the rest for test purposes. With this corpus of commands, the recognition performance ranged from 78.91% correct at the lower bound, to 96.65% correct at the upper bound. The new integration technique outlined above achieved 95.26% correct test performance or within 1.4% of the upper bound.

CONCLUDING REMARKS

Developing an architecture for integrating multimodal input modes and for evaluating multimodal recognition performance are two critical issues in developing a multimodal system. We have derived a lower and upper bound of multimodal recognition probability in this paper. We also have developed a technique to integrate the multimodal posterior probabilities. The analytic tools and integration techniques presented in this paper could be extended to multimodal systems involving other modes or more than two modes.

REFERENCES

[1] R. Bolt, "Put that there: Voice and gesture at the graphics interface," **Computer Graphics**, vol. 14, no. 3, pp. 262–270, 1980.

[2] C. Codella, R. Jalili, L. Koved, J. Lewis, D. Ling, J. Lipscomb, D. Rabenhorst, C. Wang, A. Norton, P. Sweeney and C. Turk, "Interactive simulation in a multi-person virtual world," in **ACM Conference on Human Factors in Computing Systems (CHI'92)**, 1992, pp. 329–334.

[3] P. Cohen, M. Dalrymple, D. Moran, F. Pereira, J. Sullivan, R. Gargan, J. Schlossberg and S. Tyler, "Synergistic use of direct manipulation and natural language," in **Human Factors in Computing Systems: CHI'89 Conference Proceedings, ACM, Addison-Wesley Publishing Co.**, New York, April 1989, pp. 227–234.

[4] P. Cohen, M. Johnston, D. McGee, S. Oviatt, J. Pittman, I. Smith, L. Chen and J. Clow, "Quickset: Multimodal interaction for distributed applications," in **Proceedings of the Fifth ACM International Multimedia Conference**, New York: ACM Press, 1997, pp. 31–40.

[5] M. Fukumoto, Y. Suenaga and K. Mase, "Finger-pointer: pointing interface by image processing," **Computer Graphics**, vol. 18, no. 5, pp. 633–642, 1994.

[6] M. Johnston, P. Cohen, D. McGee, S. Oviatt, J. Pittman and I. Smith, "Unification-based multimodal integration," in **Proceedings of the 35th Annual Meeting of the Association for Computational Linguistics**, San Francisco, CA: Morgan Kaufman, 1997, pp. 281–288.

[7] D. Koons, C. Sparrell and K. Thorisson, "Integrating simultaneous input from speech, gaze and hand gestures," in M. Maybury, ed., **Intelligent Multimedia Interfaces**, Menlo Park, CA: MIT Press, pp. 257–276, 1993.

[8] J. Neal and S. Shapiro, "Intelligent multimedia interface technology," in J.Sullivan and S.Tyler, eds., **Intelligent User Interfaces**, New York: ACM Press, pp. 11–43, 1991.

[9] S. Oviatt, "Mutual disambiguation of recognition errors in a multimodal architecture," in **Proceedings of the Conference on Human Factors in Computing Systems: CHI'99 (in press)**, Pittsburgh, PA: ACM Press, 1999.

[10] S. Oviatt, A. DeAngeli and K. Kuhn, "Integration and synchronization of input modes during multimodal human-computer interaction," in **Proceedings of the Conference on Human Factors in Computing Systems: CHI'97**, Atlanta, Georgia: ACM Press, pp. 415–422, 1997.

[11] R. Sharma, T. Huang, V. Pavlović, Y.Zhao, Z.Lo, S.Chu, K.Schulten, A.Dalke, J.Phillips, M.Zeller and W.Humphrey, "Speech/gesture interface to a visual computing environment for molecular biologists," in **Proceedings of International Conference on Pattern Recognition**, Aug. 1996, pp. 964–968.

[12] M. Vo and C. Wood, "Building an application framework for speech and pen input integration in multimodal learning interfaces," in **Proceedings of IEEE International Conference of Acoustic, Speech and Signal Processing**, Atlanta, GA, 1996, pp. 3545–3548.

[13] J. Wang, "Integration of eye-gaze, voice and manual response in multimodal user interface," in **Proceedings of IEEE International Conference on Systems, Man and Cybernetics**, 1995, pp. 3938–3942.

[14] L. Wu, S. Oviatt and P. Cohen, "From members to teams to committee - a robust approach to gestural and multimodal recognition," journal manuscript in submission.

[15] L. Wu, S. Oviatt and P. Cohen, "Multimodal integration - a statistical view," journal manuscript in submission.

HIERARCHY OF PROBABILISTIC PRINCIPAL COMPONENT SUBSPACES FOR DATA MINING

Lan Luo Yue Wang* Sun Yuan Kung
Department of Electrical Engineering and Computer Science
Catholic University of America, Washington, DC 20064, USA
Department of Electrical Engineering
Princeton University, Princeton, NJ 08544, USA

Abstract. Visual exploration has proven to be a powerful tool for multivariate data mining. Most visualization algorithms aim to find a projection from the data space down to a visually perceivable rendering space. To reveal all of the interesting aspects of complex data sets living in a high-dimensional space, a hierarchical visualization algorithm is introduced which allows the complete data set to be visualized at the top level, with clusters and subclusters of data points visualized at deeper levels. The methods involve multiple use of standard finite normal mixture models and probabilistic principal component projections, whose parameters are estimated using the expectation-maximization and principal conponent neural networks under the information theoretic criteria. We demonstrate the principle of the approach on two three-dimensional synthetic data sets.

INTRODUCTION

As a step toward understanding complex data sets, structural information reveals insight that may prove useful in data interpretation [1, 2]. To convey more structural information, new visualization methods have been progressively developed to model and display the contents of the data sets [3, 4]. However, most of such algorithms are based on a projection of the data onto a two-dimensional visualization space, which often prove inadequate in the face of more complex and high dimensional data sets [1, 5]. For example, a single projection of the data onto a visualization space may not be able to capture all of the interesting aspects of the data set. This motivates the consideration of a hierarchical visualization paradigm involving multiple statistical models and visualization spaces.

Once we explore the possibility of using many complementary visualization projections, structure decomposition and dimensionality reduction are

*Further author information: Send correspondence to wang@pluto.ee.cua.edu.

0-7803-5673-X/99/$10.00 © 1999 IEEE

the two natural strategies. Structure decomposition permits the use of relatively simple models for each of the local structures, offering greater ease of interpretation as well as the benefits of analytical and computational simplification. This philosophy for modeling complexity is similar in spirit to the "mixture of experts" approach following *divide-and-conquer* principle [6]. On the other hand, dimensionality reduction allows better visual interpretation and less computational demand. The price to be paid however is the possible loss of structural resolution and explicitness. Many researchers have recently proposed various methods to improve data visualization [2, 5]. The most close work related to our methodology was reported by Bishop and Tipping in [1, 7]. They introduce a hierarchical visualization algorithm based on a two-dimensional hierarchical mixture of latent variable models, whose parameters are estimated using the expectation-maximization (EM) algorithm. The construction of the hierarchical tree proceeds top down in which the structure decomposition is driven interactively by the user and optimal projection is determined by the maximum likelihood principle. There are two major potential limitations associated with the present approach. First, although a probability density is defined in the data space through a latent variable model, the prior and order of the mixture model are heuristically selected that may misrepresent the true data structures. The second important limitation with the present model is that the parameters including optimal projections are determined by maximum likelihood, and this criterion need not always lead to the most interesting or interpretable visualization plots. For example, alternative models may be those which optimize other criteria such as the separation of clusters.

In this paper, we propose using multiple finite normal mixture models and hierarchical visualization spaces to develop new techniques for data visualization and interpretation. The strategy is that the top-level model and projection should explain the entire data set, best revealing the presence of clusters and relationships, while lower-level models and projections should display internal structure within individual clusters, such as the presence of subclusters, which might not be apparent in the higher-level models and projections. With many complementary mixture models and visualization projections, each level will be relatively simple while the complete hierarchy maintains overall flexibility yet still conveys considerable structural information. Based on the concept of combining minimax entropy modeling and principal component analysis to optimize structure decomposition and dimensionality reduction, the particular advantages of our algorithm are: 1) At each level, probabilistic principle component analysis is performed to project the data space down to a desired two-dimensional visualization space, leads to a optimal dimensionality reduction allowing best separation of local clusters; 2) Learning from the data directly, minimax entropy principle is used to select model structures and estimate its parameter values, results in the standard finite normal mixture models with minimum conditional bias and variance; 3) By performing principle component analysis and minimax entropy modeling alternatively, a complete hierarchy of complementary projections and models can be generated automatically, leads to a statistical model best fit to the data.

THEORY AND METHOD

One of the difficulties inherent in data visualization is the problem of visualizing multi-dimensionality. When there are more than three variables, it stretches the imagination to visualize their relationships. Fortunately in data set with many variables, groups of variables often form clusters. One reason for this is that more than one variable may be measuring the same driving principle governing the behavior of the system. In many systems, there are only a few such driving forces. We can reduce data dimensionality by visualizing the principal component projections of the data. By examining plots of these few variables, researchers often develop a deeper understanding of the driving forces that generated the original data.

Principal component analysis (PCA) is an optimal method for achieving this simplification [7]. For a set of observed d-dimensional data vectors $\{\mathbf{t}_i\}$, $i \in \{1, ..., N\}$, the q principal axes $\mathbf{w}_m$, $m \in \{1, ..., q\}$, are those orthogonal axes onto which the retained variance under projection is maximal. It can be shown that the principal axes $\mathbf{w}_m$ are given by the q dominant eigenvectors of the sample covariance matrix $\mathbf{C} = \sum_i(\mathbf{t}_i - \boldsymbol{\mu}_\mathbf{t})(\mathbf{t}_i - \boldsymbol{\mu}_\mathbf{t})^\mathrm{T}/N$ such that $\mathbf{C}\mathbf{w}_m = \lambda_m \mathbf{w}_m$ and where $\boldsymbol{\mu}_\mathbf{t}$ is the sample mean. The vector $\mathbf{x}_i = \mathbf{W}^\mathrm{T}(\mathbf{t}_i - \boldsymbol{\mu}_\mathbf{t})$, where $\mathbf{W} = (\mathbf{w}_1, \mathbf{w}_2, ..., \mathbf{w}_q)$, is thus a q dimensional reduced representation of the observed vector $\mathbf{t}_i$. The optimality of PCA are twofold, the projection onto the principal subspace minimizes the squared reconstruction error and maximizes the separation of data clusters. This second advantage is particularly significant. However, because of its global linearity, the application of PCA is necessarily somewhat limited [7]. For example, the inherent multimodal nature of the data set may be completely obscured when it is projected onto the lower dimensional principal subspace, i.e., a global linear PCA may not separate clustered data effectively or sufficiently. Thus, it is important to note that although the cluster structure of the data set may be evident from the higher dimensional plot of the raw data, it is quite conceivable to have the intrinsic cluster structure of the data concealed after a projection in the more general case of high-dimensional data sets [8]. A naturally motivated alternative paradigm is to model multimodal structure with a collection of local linear subspaces [7, 9]. The method is a two-stage procedure: a partitioning of the data space followed by estimation of the principal subspace within each partition. In this section, we discuss the concept of standard finite normal mixture (SFNM) models. We then introduce probabilistic principal component analysis (PPCA), showing how the principal subspace of a set of data points can be obtained within a minimax entropy framework.

Suppose the data space is d-dimensional. Now consider a two-dimensional projection space $\mathbf{x} = (x_1, x_2)^\mathrm{T}$ together with a linear function which maps the projection space to the data space by $\mathbf{t} = \mathbf{W}\mathbf{x} + \boldsymbol{\mu}_\mathbf{t}$ where $\mathbf{W}$ is a $d \times 2$ matrix. If we introduce a probability distribution $p(\mathbf{t})$ over the data space, then a similar reduced dimension probability distribution of the new variables $\{\mathbf{x}_i\}$ in the projection space can be defined by the Radon transform of $p(\mathbf{t})$, resulting in a form of [10]

$$f(\mathbf{x}) = \mathcal{R}p(\mathbf{t}) = \int p(\mathbf{t})\delta(\mathbf{x} - \mathbf{W}^\mathrm{T}\mathbf{t} + \mathbf{W}^\mathrm{T}\boldsymbol{\mu}_\mathbf{t})d\mathbf{t} \tag{1}$$

where $\mathcal{R}$ is the Radon operator and $\delta(.)$ is the delta function. Recently there has been considerable success in using the SFNM to model the distribution of a multimodal data set and to guide further split of the data [11]. Assume that the data points $\{\mathbf{t}_i\}$ in the data space come from K_0 clusters $\{\boldsymbol{\theta}_{\mathbf{t}1}, ..., \boldsymbol{\theta}_{\mathbf{t}k}, ..., \boldsymbol{\theta}_{\mathbf{t}K_0}\}$, where $\boldsymbol{\theta}_{\mathbf{t}k}$ is the Gaussian kernel parameter vector of cluster k in the model. The data distribution takes a sum of the following general form:

$$p(\mathbf{t}) = \sum_{k=1}^{K_0} \pi_k g(\mathbf{t}|\boldsymbol{\theta}_{\mathbf{t}k}) \tag{2}$$

where π_k is the corresponding mixing proportion, with $\pi_k \geq 1$ and $\sum \pi_k = 1$. The log-likelihood of observing the data under this model is $\mathcal{L} = \sum_{i=1}^N \log p(\mathbf{t}_i)$. According to the linear superposition property of Radon transform and the projection invariant property of normal distribution, from Eq. (1) the data distribution in the projection space has a similar reduced dimension form as Eq. (2)

$$f(\mathbf{x}) = \sum_{k=1}^{K_0} \pi_k g(\mathbf{x}|\boldsymbol{\theta}_{\mathbf{x}k}) \tag{3}$$

where $g(\mathbf{x}|\boldsymbol{\theta}_{\mathbf{x}k}) = \int g(\mathbf{t}|\boldsymbol{\theta}_{\mathbf{t}k})\delta(\mathbf{x} - \mathbf{W}^\mathrm{T}\mathbf{t} + \mathbf{W}^\mathrm{T}\boldsymbol{\mu}_\mathbf{t})d\mathbf{t}$. The corresponding log-likelihood of projecting the data under Radon transform is $\mathcal{L} = \sum_{i=1}^N \log f(\mathbf{x}_i)$.

The problem of the SFNM modeling addresses the combined estimation of regional parameters $(\pi_k, \boldsymbol{\theta}_{\mathbf{t}k})$ and detection of the structural parameter K_0 in Eq. (2) based on the observations $\mathbf{t}$. It has been shown that when K_0 is given, the maximum likelihood (ML) estimate of the regional parameters can be obtained using the EM algorithm. There are two practical problems associated with the current approach. First, when the dimension of the data space is high, the implementation of the EM algorithm is very complex and time consuming. Second, the initialization of the EM algorithm is in general heuristically chosen which often leads to only a local optimal solution. Therefore, it is reasonable to consider the model parameter values being estimated firstly in the projection space and then further fine tuned in the data space. One natural criterion used for learning the optimal parameter values is to minimize the distance between the SFNM distribution $f(\mathbf{x})$ and the data histogram $f_\mathbf{x}$. Suggested by information theory, relative entropy (Kullback-Leibler distance) is a suitable measure of the reconstruction error, given by $D(f_\mathbf{x}\|f) = \sum_\mathcal{X} f_\mathbf{x}(\mathbf{x}) \log[f_\mathbf{x}(\mathbf{x})/f(\mathbf{x})]$. It can be shown that when relative entropy is used as a distance measure, distance minimization is equivalent to the ML estimation, summarized by $\mathcal{L} = \exp(-N[H(f_{\mathbf{x}}) + D(f_{\mathbf{x}}||f)])$, where H is the entropy calculator [3].

An important issue concerning unsupervised distribution learning is the detection of the structural parameter K_0, called model selection [3, 8, 11]. The objective is to propose a systematic procedure for determining the optimal number of local clusters when the prior knowledge about the true data structure is not available. This is indeed the case particularly critical in real-world applications where the structure of the data patterns may be arbitrarily complex. To avoid the subjective nature of the previous methods such as a sequence of hypothesis testing [8, 11], we propose to use two information theoretic criteria, i.e., the Akaike information criterion (AIC) [12] and minimum description length (MDL) [13], to guide model selection. The major thrust of this approach has been the formulation of a model fitting procedure in which an optimal model is selected from the several competing candidates such that the selected model best fits the observed data. AIC tries to reformulate the problem explicitly as an *approximation* of the true structure by the model, implies that AIC will select the model that gives the minimum value defined by

$$AIC(K_a) = -2\log(\mathcal{L}_{ML}) + 2K_a \tag{4}$$

where $\mathcal{L}_{ML}$ is the maximum likelihood of the model and K_a is the number of free adjustable parameters in the model. From a quite different point of view, MDL reformulates the problem explicitly as an information coding problem in which the best model fit is measured such that it assigns high probabilities to the observed data while at the same time the model itself is not too complex to describe [13]. A model is selected by minimizing the total description length defined by

$$MDL(K_a) = -\log(\mathcal{L}_{ML}) + 0.5K_a \log N. \tag{5}$$

Note that, different from AIC, the penalty term in MDL takes into account the number of observations. Further interpretation on model selection based on the minimax entropy principle can be found in [3, 14].

The association of a probability model with PCA offers the promising prospect of being able to visualize complex data structures with a combination of local PCA projections through the mechanism of a mixture of probabilistic principal component subspaces. By a simple extension of the maximum a posterior (MAP) formulation for data classification in the standard K_0-ary Bayes hypothesis testing, we can obtain a principal component projection along the desired axes onto which a particular portion of the data set is highlighted, by weighting all of the data points in the whole data set with their posterior probabilities of belonging to that portion. This involves a *soft* clustering algorithm where instead of any given data point being assigned exclusively to one principal component subspace, the responsibility for its generation is shared among all of the subspaces.

Under the SFNM model defined by Eq. (2), the posterior Bayesian probability z_{ik} of a given data point $\mathbf{t}_i$ belonging to cluster k is

$$z_{ik} = \frac{\pi_k g(\mathbf{t}_i|\boldsymbol{\theta}_{\mathbf{t}k})}{p(\mathbf{t}_i)}. \tag{6}$$

where $k = 1, 2, ..., K_0$ and $\sum_k z_{ik} = 1$. These posterior probabilities, together with the computational simplicity of performing PCA (involving no more than finding the top q eigenvectors of the covariance matrix of the data points) make it a good candidate for the linear subspace in the mixture. The q principal components define the local subspace assumed for the multimodal. The contributions of the input to the k subspace are the activities of the weighted data points $\{\mathbf{t}_{ik}\}$ for input cluster k. This can be obtained by $\mathbf{t}_{ik} = z_{ik}(\mathbf{t}_i - \boldsymbol{\mu}_{\mathbf{t}k})$, where $\boldsymbol{\mu}_{\mathbf{t}k}$ is the weighted sample mean of cluster k. The subspaces for the highlighted clusters are generated by a localized linear PCA with $\mathbf{C}_k = \sum_i z_{ik}(\mathbf{t}_i - \boldsymbol{\mu}_{\mathbf{t}k})(\mathbf{t}_i - \boldsymbol{\mu}_{\mathbf{t}k})^{\mathrm{T}} / \sum_i z_{ik}$ such that $\mathbf{C}_k \mathbf{w}_{mk} = \lambda_{mk} \mathbf{w}_{mk}$. The vector $\mathbf{x}_{ik} = \mathbf{W}_k^{\mathrm{T}} z_{ik}(\mathbf{t}_i - \boldsymbol{\mu}_{\mathbf{t}k})$, where $\mathbf{W}_k = (\mathbf{w}_{1k}, \mathbf{w}_{2k}, ..., \mathbf{w}_{qk})$, is thus a q dimensional reduced representation of the k-cluster focused vector $\mathbf{t}_{ik}$. Then, Eq. (3) may be modified as $\{f(\mathbf{x}_k)\} = \{\pi_k g(\mathbf{x}_k|\boldsymbol{\theta}_{\mathbf{x}k})\}_{k=1}^{K_0}$, where $\mathbf{x}_k = \mathbf{W}_k^{\mathrm{T}} z_k(\mathbf{t} - \boldsymbol{\mu}_{\mathbf{t}k})$, $z_k = \pi_k g(\mathbf{t}|\boldsymbol{\theta}_{\mathbf{t}k})/p(\mathbf{t})$, and $g(\mathbf{x}_k|\boldsymbol{\theta}_{\mathbf{x}k}) = \int g(\mathbf{t}|\boldsymbol{\theta}_{\mathbf{t}k})\delta(\mathbf{x}_k - \mathbf{W}_k^{\mathrm{T}} z_k \mathbf{t} + \mathbf{W}_k^{\mathrm{T}} z_k \boldsymbol{\mu}_{\mathbf{t}k})d\mathbf{t}$. It is important to understand that each component in Eq. (2) now corresponds to an independent subspace model with parameters $\boldsymbol{\theta}_{\mathbf{t}k}$ and $\mathbf{W}_k$.

As discussed above, the initialization of the SFNM model identification will be performed over $\mathbf{x}$-space, but a mapping from $\mathbf{t}$-space to $\mathbf{x}$-space may have the intrinsic cluster structure of the data concealed, leading to an incorrect correspondence between Eq. (3) and Eq. (2). Once again using *divide-and-conquer* principle, we now extend the mixture representation of Eq. (2) to form a hierarchical mixture model generally enough to be applicable to mixtures of any parametric density model. With Eq. (1) and Eq. (2) we have discussed a two-level system consisting of a single Radon transform at the top level and a mixture of K_0 normal distributions at the second level. We can now reformulate the hierarchy to a third level by associating a group $\mathcal{G}_k$ of SFNM models with each model k in the second level. The resulting probability distribution can be written as

$$p(\mathbf{t}) = \sum_{k=1}^{L_0} \pi_k \sum_{j=1}^{L_k} \pi_{j|k} g(\mathbf{t}|\boldsymbol{\theta}_{\mathbf{t}(k,j)}) \tag{7a}$$

where $\pi_{j|k}$ again correspond to a set of mixing proportions, one for each k, with $\sum_j \pi_{j|k} = 1$, and $K_0 = \sum_{k=1}^{L_0} L_k$. This reformulation will be guided by the model selections over $\mathbf{x}$-subspaces such that $\mathcal{G}(\mathbf{x}_k|\boldsymbol{\theta}_{\mathbf{x}k}) = \int \sum_{j=1}^{L_k} \pi_{j|k} g(\mathbf{t}|\boldsymbol{\theta}_{\mathbf{t}(k,j)})\delta(\mathbf{x}_k - \mathbf{W}_k^{\mathrm{T}} z_k z_{j|k} \mathbf{t} + \mathbf{W}_k^{\mathrm{T}} z_k z_{j|k} \boldsymbol{\mu}_{\mathbf{t}k})d\mathbf{t}$, where $z_{j|k} = \pi_{j|k} g(\mathbf{t}|\boldsymbol{\theta}_{\mathbf{t}(k,j)})/\mathcal{G}(\mathbf{t}|\boldsymbol{\theta}_{\mathbf{t}k})$. Thus, each level of the hierarchy corresponds to a generic model, with lower levels giving more focused and interpretable representations. Determination of the parameters of the subspaces at the third level can again be performed similarly as at the second level.

ALGORITHMS

Based on the theory behind hierarchical mixtures of probabilistic principal component subspaces we have discussed above, we now present the description of our algorithm involving major steps of the visual hierarchy construction. Although the tree structure of the hierarchy may be empirically defined [1], a more interesting effort, with greater scientific applicability, is to build the tree *automatically and interactively.* Guided by the two information theoretic criteria, our algorithm progressively proceeds by fitting a series of submodels to the clusters of the data set, in which model order is selected automatically and algorithm initialization is conducted interactively.

For low dimensional data sets, we directly evaluate the covariance matrix to find $\mathbf{W}_k$. For high dimensional cases, since only the top two eigenvectors of the covariance matrix of the data points are of the interest, it may be computationally more efficient to apply our previously developed adaptive principal components extraction (APEX) algorithm [4]. With a soft partitioning of the data set via the EM algorithm, data points will now effectively belong to more than one cluster at any given level. This step is automatically available in our approach since the estimation of parent subspace involves the calculation of posterior probabilities z_{ik}. Thus, the effective input values are $\mathbf{t}_{ik} = z_{ik}(\mathbf{t}_i - \boldsymbol{\mu}_{\mathbf{t}k})$ for an independent visualization space k in the hierarchy. Since the determination of $\mathbf{W}_k$ is now based on neural network learning with posterior probability weighted inputs, new algorithm will be termed adaptive probabilistic principal component extraction (APPEX). Further projections $\mathbf{W}_{j|k}$ can again be performed by using the effective input values $\mathbf{t}_{i(k,j)} = z_{ik} z_{i,j|k}(\mathbf{t}_i - \boldsymbol{\mu}_{\mathbf{t}(k,j)})$ for the visualization subspace j. Formal version of the APPEX algorithm is the following.

1. Initialize the feedforward weight vector $\mathbf{w}_{mk}$ and the feedback weight vector a_k to small random values at time $i = 1$. Assign a small positive value to the learning rate parameter η.

2. Set $m = 1$, and for $i = 1, 2, ...,$ compute $\mathbf{w}_{1k}(i+1) = \mathbf{w}_{1k}(i) + \eta[y_{1k}(i)\mathbf{t}_{ik} - y_{1k}^2(i)\mathbf{w}_{1k}(i)]$ where $y_{1k}(i) = \mathbf{w}_{1k}^{\mathrm{T}}(i)\mathbf{t}_{ik}$. For large i we have $\mathbf{w}_{1k}(i) \longrightarrow \mathbf{w}_{1k}$ where $\mathbf{w}_{1k}$ is the eigenvector associated with the largest eigenvalue of the covariance matrix $\mathbf{C}_k$.

3. Set $m = 2$, and for $i = 1, 2, ...,$ compute $\mathbf{w}_{2k}(i+1) = \mathbf{w}_{2k}(i) + \eta[y_{2k}(i)\mathbf{t}_{ik} - y_{2k}^2(i)\mathbf{w}_{2k}(i)]$ and $a_k(i+1) = a_k(i) - \eta[y_{2k}(i)y_{1k}(i) + y_{2k}^2(i)a_k(i)]$ where $y_{2k}(i) = \mathbf{w}_{2k}^{\mathrm{T}}(i)\mathbf{t}_{ik} + a_k(i)y_{1k}(i)$. For large i we have $\mathbf{w}_{2k}(i) \longrightarrow \mathbf{w}_{2k}$ where $\mathbf{w}_{2k}$ is the eigenvector associated with the second largest eigenvalue of the covariance matrix $\mathbf{C}_k$.

Once $\mathbf{W}_k$ is estimated, the visualization subspace is generated by plotting each data point $\mathbf{t}_i$ at the corresponding $\mathbf{x}_{ik} = \mathbf{W}_k^{\mathrm{T}} z_{ik}(\mathbf{t}_i - \boldsymbol{\mu}_{\mathbf{t}k})$ in k-subspace. Based on this plot, model selection is performed to automatically determine the optimal order of models to fit at the next level down, and the user then selects points $\boldsymbol{\mu}_{\mathbf{x}(k,j)}$ on the plot representing the centers of local clusters.

To calculate the posterior probabilities $z_{i,j|k}$ through further soft clustering, we first apply the EM algorithm in k-subspace based on $\{\mathbf{x}_{ik}\}$ to estimate the parameters of $\mathcal{G}(\mathbf{x}_k|\boldsymbol{\theta}_{\mathbf{x}k})$ where $z_{\mathbf{x}i,j|k}$ is simultaneously determined and it can be shown that the mixing proportion $\pi_{j|k}$ are projection-invariant [15]. The resulting points $\boldsymbol{\mu}_{\mathbf{t}(k,j)} = \mathbf{W}_k\boldsymbol{\mu}_{\mathbf{x}(k,j)} + \boldsymbol{\mu}_{\mathbf{t}k}$ in data space are then used to initialize the means of the respective submodels. Including the important step of model selection, the formal version of the adaptive EM (AEM) algorithm is the following.

1. For $L_{\mathbf{x}k} = 1, 2, ..., L_{\mathbf{x}k\mathrm{MAX}}$ and $j = 1, ..., L_{\mathbf{x}k}$, select $\boldsymbol{\mu}_{\mathbf{x}(k,j)}$ and $\pi_{j|k}$. Assign a 2×2 unit matrix to the covariance matrix $\mathbf{C}_{\mathbf{x}(k,j)}$. Perform the EM algorithm in $\mathbf{x}$-space. Once the EM algorithm has converged, for $i = 1, 2, ..., N$, compute $z_{\mathbf{x}i,j|k} = \pi_{j|k}g(\mathbf{x}_{ik}|\boldsymbol{\theta}_{\mathbf{x}(k,j)})/\mathcal{G}(\mathbf{x}_{ik}|\boldsymbol{\theta}_{\mathbf{x}k})$, where $\pi_{j|k} = \sum_i z_{\mathbf{x}i,j|k}/\sum_i z_{\mathbf{x}ik}$, $\boldsymbol{\mu}_{\mathbf{x}(k,j)} = \sum_i z_{\mathbf{x}i,j|k}\mathbf{x}_{ik}/\sum_i z_{\mathbf{x}i,j|k}$, $\mathbf{C}_{\mathbf{x}(k,j)} = \sum_i z_{\mathbf{x}i,j|k}(\mathbf{x}_{ik} - \boldsymbol{\mu}_{\mathbf{x}(k,j)})(\mathbf{x}_{ik} - \boldsymbol{\mu}_{\mathbf{x}(k,j)})^{\mathrm{T}}/\sum_i z_{\mathbf{x}i,j|k}$. Calculate the corresponding AIC and MDL values, where $L_{\mathbf{x}k,a} = 7L_{\mathbf{x}k} - 1$, and select a $\mathcal{G}(\mathbf{x}_{ik}|\boldsymbol{\theta}_{\mathbf{x}k})$ with the optimum order $L_{\mathbf{x}k,0}$ corresponding to the minimum values of the AIC and MDL criteria.

2. Set $\boldsymbol{\mu}_{\mathbf{t}(k,j)} = \mathbf{W}_k\boldsymbol{\mu}_{\mathbf{x}(k,j)} + \boldsymbol{\mu}_{\mathbf{t}k}$, $L_{\mathbf{t}k,0} = L_{\mathbf{x}k,0}$, and $\pi_{j|k}$. Assign a 2×2 unit matrix to the covariance matrix $\mathbf{C}_{\mathbf{t}(k,j)}$. Perform the EM algorithm in $\mathbf{t}$-space. Once the EM algorithm has converged, for $i = 1, 2, ..., N$, compute $z_{i,j|k} = \pi_{j|k}g(\mathbf{t}_i|\boldsymbol{\theta}_{\mathbf{t}(k,j)})/\mathcal{G}(\mathbf{t}_i|\boldsymbol{\theta}_{\mathbf{t}k})$, where $z_{ij} = z_{ik}z_{i,j|k}$, $\pi_{j|k} = \sum_i z_{ij}/\sum_i z_{ik}$, $\boldsymbol{\mu}_{\mathbf{t}(k,j)} = \sum_i z_{ij}\mathbf{t}_i/\sum_i z_{ij}$, $\mathbf{C}_{\mathbf{x}(k,j)} = \sum_i z_{ij}(\mathbf{t}_i - \boldsymbol{\mu}_{\mathbf{t}(k,j)})(\mathbf{t}_i - \boldsymbol{\mu}_{\mathbf{t}(k,j)})^{\mathrm{T}}/\sum_i z_{ij}$.

With the resulting $z_{i,j|k}$ in $\mathbf{t}$-space, we apply the APPEX algorithm to estimate $\mathbf{W}_{j|k}$. The next level visualization subspace is generated by plotting each data point $\mathbf{t}_i$ at the corresponding $\mathbf{x}_{i,j|k} = \mathbf{W}_{j|k}^{\mathrm{T}}z_{ik}z_{i,j|k}(\mathbf{t}_i - \boldsymbol{\mu}_{\mathbf{t}(k,j)})$ in $j|k$-subspace. The construction of the tree structure hierarchy is automatically completed when no further data split is recommended by the information theoretic criteria in all of the subspaces.

ILLUSTRATION AND DISCUSSION

We first illustrate the application of our algorithm to a simple synthetic data set generated from a mixture of two Gaussians in a three-dimension space. Each Guassian is consisted of 150 data points, and has its distinct mean and covariance. These two cloud-like clusters are relatively overlapped which was chosen to illustrate the global and local principal axes. To explore the data, we first perform a single global PCA, and visualize the data points in the projection space. Two information theoretic criteria clearly suggest the presence of two distinct clusters within the data. The user then selects two initial cluster centers and the EM algorithm is applied to conduct a soft clustering of the data points. This leads to a mixture of two probabilistic principal component subspaces as shown in Figure 1.

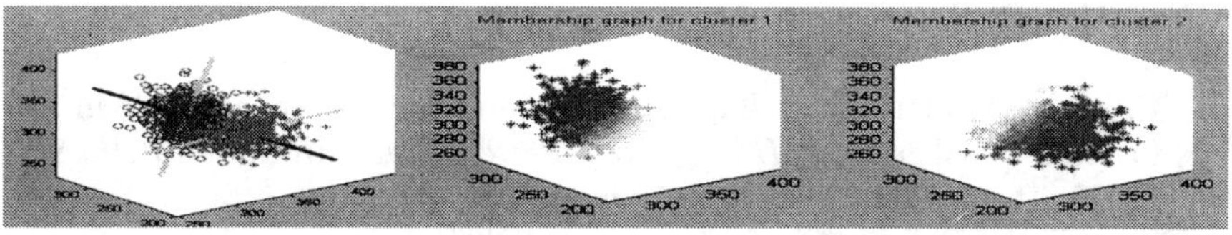

Figure 1: A summary of the final results from the data set 1. Both global and local principal axes are indicated. Each subspace is plotted with weighted data inputs.

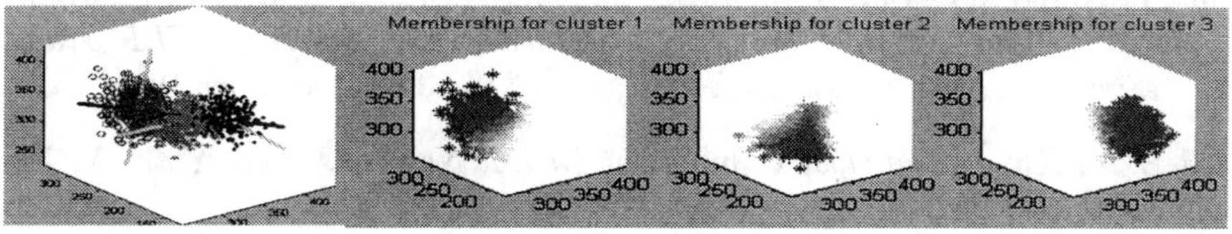

Figure 2: Results of fitting the data set 2 containing three local clusters. A three level hierarchy model is therefore adopted reflecting the true data structure.

As an example of a more complex problem, we consider a data set arising from a mixture of three Gaussians consisting of 300 data points. Two of these cloud-like clusters are well separated while the third is spaced in between. By performing the same operation as above, the second level space is generated with a mixture of two local PCA subspaces. We then apply the two information theoretic criteria to examine this two plots, that on the "right" shows evidence of further split. At this third level data mining, a hierarchical model is adopted, which illustrates that there are indeed total three clusters within the data set.

We have presented a novel approach to data mining which is both statistically principled and visually effective. This method, as illustrated by the well-planned simulations, can be very capable of revealing hidden structure within data. It is important to emphasize that in relation to previous work [1, 7], one interesting consideration with the present algorithm is that the models are determined by minimax entropy, and this criterion can not only estimate the parameters but also select the structure. This approach allows a self-consistent fitting of the whole tree, so that the an automated procedure for generating the hierarchy is available. Other possible advantages include the determination of data projection by maximum the separation of clusters which in turn optimizes the other crucial operations such as model selection and parameter initialization, and data rendering algorithms which permit user or hypothesis driven nature of the data projection. While the optimality of these new techniques is often highly data-dependent, we would expect the hierarchical visualization model to be a very effective tool for the data visualization and exploration in many applications. We are currently investigating applications to computer-aided diagnosis for breast cancer detection and intelligent mapping of gene expression profiles.

REFERENCES

[1] C. M. Bishop and M. E. Tipping, "A hierarchical latent variable model for data visualization," *IEEE Trans. Pattern Anal. Machine Intell.*, Vol. 20, No. 3, pp. 282-293, March 1998.

[2] E. R. Tufte, *Visual Explanation: Images and Quantities, Evidence and Narrative*, Graphics Press, Cheshire 1996.

[3] Y. Wang, S. H. Lin, H. Li, and S, Y. Kung, "Data mapping by probabilistic modular networks and information theoretic criteria," *IEEE Trans. Signal Processing*, Vol. 46, No.12, pp. 3378-3397, December 1998.

[4] S. Y. Kung, *Principal Component Neural Networks*, New York: Wiley, 1996.

[5] G. M. Nielson, "Challenges in visualization research," *IEEE Trans. Visualization and Computer Graphics*, vol. 2, no. 2, pp. 97-99, June 1996.

[6] M. I. Jordan and R. A. Jacobs, "Hierarchical mixture of experts and the EM algorithm," *Neural Computation,* Vol. 6, pp. 181-214, 1994.

[7] M. E. Tipping and C. M. Bishop, "Mixtures of probabilistic principal component analyzers," *Neural Computation*, Vol. 11, pp. 443-482, 1999.

[8] S. Haykin, *Neural Networks: A Comprehensive Foundation,* MacMillan College Publishing Company, Inc., 1994.

[9] G. E. Hinton, P. Dayan, and M. Revow, "Modeling the manifolds of images of handwritten digits," *IEEE Trans. Neural Net.*, Vol. 8, No. 1, pp. 65-74, January 1997.

[10] R. N. Bracewell, *Two-Dimensional Imaging,* Prentice-Hall, Inc., 1995.

[11] D. M. Titterington, A. F. M. Smith, and U. E. Markov, *Statistical analysis of finite mixture distributions.* New York: John Wiley, 1985.

[12] H. Akaike, "A new look at the statistical model identification," *IEEE Trans. Automatic Control*, Vol. 19, No. 6, pp. 716-723, 1974.

[13] J. Rissanen, "Modeling by shortest data description," *Automat.*, Vol. 14, pp. 465-471, 1978.

[14] E. T. Jaynes, "Information theory and statistical mechanics," *Physical Review*, Vol. 108, No. 2, pp. 620-630/171-190, May 1957.

[15] L. I. Perlovsky and M. M. McManus, "Maximum likelihood neural networks for sensor fusion and adaptive classification," *Neural Networks*, Vol. 4, pp. 89-102, 1991.

A NEW TOOL FOR THE MARKET RESEARCH USING A MODIFIED AUTO-ASSOCIATIVE MEMORY*

V. Giménez-Martínez, J. Castellanos
Dept. Matemática Aplicada
Fac. Informática - UPM
Campus de Montegancedo
28660 Boadilla del Monte
Madrid - SPAIN
E-mail: vgimenez@fi.upm.es

L. F. Mingo
Dept. Lenguajes, Proyectos
y Sistemas Informáticos
EU. de Informática - UPM
Crta. Valencia Km, 7
Madrid - SPAIN
E-mail: lfmingo@eui.upm.es

Abstract. This paper describes a new application of the *Hopfield* auto-associative memory. This kind of networks have been used as very good tools for solving optimization problems but not as tools for the investigation of public opinion poll, market researches or stock market predictions We have considered the training set as a sample of public-opinion polls. The aim of this research is to design the net as a tool for getting statistical information of the general opinion of the population represented for that sample. For getting this goal, and based on preliminary researches, we are used a modified version of the *Hopfield* auto-associative memory. Once obtained the weights that control the net, we may, trough a set of "external parameters" control the number of fixed points of the net. Moreover, the state vector space may be classified in as many classes as the dimension of the space and we can control the number of fixed points in any one of those classes. So, not only we have the degree of correlation between a given element of the population and the sample of polls, but also we will have the degree of correlation between that element and any one of the classes of the space. Associated to any element of the population, we will also have a probability frequency distribution, which may be used to obtain the ratio of elements of the population whose answer to a certain question is similar to the one given by the element to be tested. We present, in this paper, the general features of the theory, so as an instance that proves the performance of the algorithm.

*THIS RESEARCH IS CARRIED OUT UNDER GRANTS NUMBER TIC95 0122.

INTRODUCTION

In the classical approach a *Hopfield* auto-associative memory is a discrete time, discretevalued dynamic system which at any given instant of time t is characterized by a *bipolar* binary state vector. For a recurrent network with n neurons, the state vector at time t is:

$$x(t) = [x_1(t), \cdots, x_i(t), \cdots, x_n(t)] \in \{-1, 1\}^n \tag{1}$$

The behavior of the system along the time axis is described by a dynamic equation of the type:

$$x_i(t+1) = Sgn\left[\sum_{j=1}^{n} w_{ij}x_j(t) - \theta_i\right], \quad i = 1, 2, \cdots, n \tag{2}$$

The classical solution to impose fixed points by means of the synaptic matrix W is the ***Hebb's*** law, which state that the synaptic weight w_{ij} should increase whenever neurons i and j have simultaneously the same activity level and it should decrease in the opposite case [6]. As it was pointed out above, the prototype vector components must belong to the set $\{-1, 1\}$. This fact is the cornerstone of the ***Hebb's*** law mathematical interpretation. The reason is that when the prototype ξ^μ is stored, neurons i and j may receive activation from $\{-1, 1\}$. Then, there are four different ways for neurons i and j to be simultaneously activated. But if only the sign of the received activation was taken into account, the four possibilities could be reduced to two possibilities: neurons i and j may receive a similar sign or not. The mathematical advantage of this interpretation lies in the fact that when the prototype ξ^μ is acquired, the synaptic weight w_{ij} should increase if neurons i and j receive a similar sign; in other words if $\xi_i^\mu \xi_j^\mu$ is positive. And w_{ij} should decrease if $\xi_i^\mu \xi_j^\mu$ is negative. The updating of the weights may be then expressed by $\nabla w_{ij} = \eta \xi_i^\mu \xi_j^\mu$ (where η is a positive learning factor); which means that, $\{-\eta, \eta\}$. The synaptic matrix W should be interpreted as a sort of sign correlation matrix of the prototypes. The weak point in the classical deterministic approach comes from the fact that it does not have the appropriate tools to deal efficiently with the correlation between the state vectors and the prototype vectors. The capacity of the net is very poor because we can only know if one given vector is adequately correlated with the prototypes or not and we are not able to know what is the exact correlation degree between one given component vector and the prototypes [7]. The interest of our approach lies precisely in the fact that it provides these tools.

At section 2 we present a new *training algorithm* and R^2 as the *energy* landscape of the net. Each of the energy points associated with state-vectors having the same *Hamming* distance to the zero vector is located over the same energy line in R^2. The state-vector space may be then classified in n classes according to the n different possible distances from any of the state-vector to the zero vector. At section 3, in the dynamics stage, a parameter vector θ is introduced, this parameter is used for controlling the capacity of the net[2]. It will be proved that the bigger the θ_i component is, the lower the number of fixed points located in the r_i energy line. We will also define the relative weight vector $w(t)$ associated to any state vector $x(t)$ of the space, the components of this vector may be interpreted as a sort of frequency distribution of probabilities, which may be used to obtain the ratio of prototypes whose i component is similar to $x_i(t)$. It is clear that the special features of the training will allow to obtain statistical information about the population represented by the training set. A new performance of the Hopfield algorithm is explained in this paper, in such a way that not only the correlation degree between a given pattern and the training prototypes may be measured but also the correlation degree between a prototype and the class to which it belongs. At section 4 we present the results and conclusions of the work so as an instance that prove the performance of the algorithm.

TRAINING AND ENERGY

In our approach $x(t) \in \{0,1\}$ and associated to the network there will be a complete graph G[2], with n vertices $\{v_1, \cdots, v_n\}$, and one bi-directional edge a_{ij} for every possible pair of different vertices. Initially, at the training stage a null value w_{ij} is assigned to every edge a_{ij} in the graph; afterwards, when a pattern ξ^μ belonging to $\{0,1\}^n$, is presented to the net, this value is modified as follows:

$$\nabla w_{ij} = \begin{cases} +1 & if \quad \xi_i^\mu = \xi_i^\mu = 1, \quad i \neq j \\ -1 & if \quad \xi_i^\mu = \xi_i^\mu = 0, \quad i \neq j \\ 0 & if \quad otherwise \end{cases} \tag{3}$$

This procedure, as in the classical approach, is also based on Hebb's law, but with a significant difference: now there are three candidates for updating the weights, since $\nabla w_{ij} \in \{-1,0,1\}$ (or $\nabla w_{ij} \in \{-\eta, 0, \eta\}$ if a learning factor were used). Moreover, the final weight w_{ij} is not going to represent any sort of sign correlation. We are, in some way, separating the null and unit components of the pattern to be acquired. This idea may be much more easily understood if we use the next graphical interpretation of the training algorithm: At the first step, a null value is assigned to all the edges a_{ij}, of the complete graph G, then, when a learning pattern ξ^μ is acquired by the net, it is superposed over the graph G. That means that their components $\{\xi_1^\mu, \cdots, \xi_n^\mu\}$, are going to be mapped over the vertices $\{v_1, \cdots, v_n\}$, in the complete graph G. This mapping may be interpreted as a coloring of the edges in G [1], in such a way that, if $\xi_i^\mu = \xi_j^\mu = 1$, the edge a_{ij} (whose ending vertices are v_i and v_j) are going to be colored with a certain color, for example red. On the other hand, if $\xi_i^\mu = \xi_j^\mu = 0$, then a_{ij} will be colored with a different color, as for example blue. The rest of the edges in G remain uncolored. Once this coloring has been done, the value assigned over the red edges are positively reinforced and the value assigned over the blue ones are negatively reinforced. The value over the rest of the edges remains unchanged. Once the pattern ξ^μ is acquired, the colors are erased and we repeat the same color assignation with the next pattern to be acquired by the net, and so on until the last pattern is acquired[3]. It is clear that the value of w_{ij} will represent the rate of coincidence between unit and null coincidences at the i and j components of the prototypes. After applying our training algorithm, all the matrices obtained as weight matrices W, the resulting graph G, will fulfill this property:

$$w_{ij} + w_{rs} = w_{is} + w_{rj} \quad \forall\, i,j,r,s \quad \text{from} \quad 1 \rightarrow n \quad \text{and} \quad (i \neq j \neq r \neq s) \tag{4}$$

The set of constrains makes possible to reduce the weight-matrix to a vector $p = \{p_1, \cdots, p_i, \cdots, p_n\}$ where $w_{ij} = p_i + p_j$ [5]. The state vector x at time t could also be interpreted as a coloring of the edges in G, but now this coloring is going to be used to retrieve the stored data. If the graph G is colored with the coloring associated with the pattern $x(t)$, it is easy to understand (taking into account how the training algorithm was designed), that the bigger the summation of all the edges in the red graph and the lower the summation of all the edges in the blue graph are, then the more correlated the pattern $x(t)$ must be, with those that were used during the training stage. So, if W is the weight matrix of G, the energy point *EP* of the net is defined as a pair of numbers. The first of them represents the summation of all the values on the edges of the red graph, and the second one represents the same summation, but on the blue ones. So if G is colored with the color associated to $x(t)$, then $\{I(t), O(t)\}$ may be defined as the pair of quadratic forms:

$$\begin{cases} I(t) = \frac{1}{2} x(t) W x(t)^t \\ O(t) = \frac{1}{2} \bar{x}(t) W \bar{x}(t)^t \end{cases} \tag{5}$$

Which may be expressed as a function of p as:

$$I(t) = (n_1 - 1) p x(t) \quad \text{and} \quad O(t) = (n_0 - 1) p \bar{x}(t) \tag{6}$$

In other researches [5], we have proved that if n_1 and n_0 is the number of unit and null components of $x(t)$, then

$$\frac{I(t)}{n_1 - 1} + \frac{O(t)}{n_0 - 1} = K \quad \text{where} \quad K = p_1 + \cdots + p_n \tag{7}$$

As all state vectors $x(t)$ with the same Hamming distance i to the zero vector contains the same numbers n_1 and n_0 of unit and null components, the energy points $\{I(t), O(t)\}$, associated to all of them, will be placed in the same line r_i, whose equation expressed in (x, y) is:

$$r_i : (n_0 - 1)x + (n_1 - 1)y - (n_0 - 1)(n_1 - 1)K = 0 \tag{8}$$

In other words, all the *EP's* associated with vector-states with the same number of unit components are placed in the same line of the energy field, and the equation of this line is the one represented in (8). In this way the state vector space is classified in as many classes as the dimension n of the space. We have also proved that if we do a constant translation of value m (where m is equal to minus the arithmetic mean of the components of p), to the vector parameter p, we obtain a new vector q, which may be expressed as: $q = p + mI = p - \frac{I}{n}\sum_{i=1}^{n} p_i$. It is clear that now the sum of all the components in q must be zero, which means that if the weight vector is q, then $K = 0$ and the equations of the energy lines s_i will be homogeneous equations associated to the equations obtained before in (9), in other words:

$$s_i : (n_0 - 1)x + (n_1 - 1)y = 0 \tag{9}$$

We have also proved that the distribution of the *EP's* in line r_i will be similar to the distribution of the *EP's* in the translated line s_i [5]. From this very point we are going to consider as parameters of the net the weight vector p, which we suppose has already been transferred to the origin. To understand how the dynamic equation is designed, we must always keep in mind that associated to every possible state vector $x(t)$ of the net there is a coloring of the graph G as well as a energy point $\{I(t), O(t)\}$, whose coordinates are equal to:

$$I(t) = (n_1 - 1)px(t) \quad \text{and} \quad O(t) = (n_0 - 1)p\bar{x}(t) \tag{10}$$

Where n_1 is the total number of unit components of $x(t)$ and n_0 is the total number of the null ones. The energy pair number distribution of the state space over the energy field is represented as a set of lines (all of them pass through the origin), and the equations of these lines are

$$r_i : (n_0 - 1)x + (n_1 - 1)y = 0 \tag{11}$$

where $i = n_1$ (r_1 is the line $x = 0$ and r_n is the line $y = 0$).

By other way, and as we said in the introduction, the nature of the algorithm we propose let to know how the value of $x_i(t)$ affects the whole energy of the state $x(t)$. We may define the relative weight of the neuron i when the net is in state $x(t)$ as the contribution of this neuron to the component $I(t)$, if $x_i(t) = 1$; or as the contribution of this neuron to the component $O(t)$, if $x_i(t) = 0$. So, if $x_i(t) = 1$, we define the relative weight $w_i(t)$ of the neuron i when the net is in state $x(t)$ as:

$$w_i(t) = \frac{\sum_{j=1}^{n} x_j(t)w_{ij}}{I(t)} \tag{12}$$

Once the resulting graph G is colored with the coloring associated to $x(t)$, if $x_i(t) = 1$, the value of $w_i(t)$ represents the sum of the values over all the red edges (with one end in vertex v_i) divided by the sum of the values over all the red edges of the Graph. The same, but in the complementary space (blue graph) could be interpreted when $x_i(t) = 0$. The problem now is how to express the relative weight $w_i(t)$ as a function of the transferred to the origin weight vector p. Taking into account that $w_{ij} = p_i + p_j$, for any $(i \neq j)$, and that $w_{ii} = 0$, then

$$w_i(t) = \frac{1}{x(t)x(t) - 1} + \frac{x(t)x(t) - 2}{x(t)x(t) - 1} \frac{p_i}{px(t)} \tag{13}$$

when $x_i(t) = 1$. In case that $x_i(t) = 0$, then $x(t)$ must be replaced by $\bar{x}(t)$ (If the value of $w_i(t)$ in the above expression were not defined, we define $w_i(t) = 0$).

DYNAMICS

Defining the energy associated to $x(t)$ as $E(t) = I(t)O(t)$. If in time t the state vector $x(t)$ is in class $[j]$, we have proved [3], that given any i from 1 to n, if the relative weight $w_i(t)$ fulfill that:

$$w_i(t) < 1 - \sqrt{\frac{\alpha_j}{\alpha_{j-1}}} \tag{14}$$

(being α_j and α_{j-1} the slopes of the lines r_j and r_{j-1} where the *EP's* of $x(t)$ and $x(t+1)$ are located), then if $x_i(t)$ change its state, then $E(t+1) < E(t)$. If in time t the state vector $x(t)$ is in class $[j]$, then for any i from 1 to n, the dynamic equation is defined as:

$$x_i(t+1) = f_h\left[(f_b(x_i(t))(w_i(t) - \theta_j)\right] \tag{15}$$

Where f_b is the function defined as $f_b(x) = 2x - 1$, which achieves the transformation from the domain $\{0, 1\}$ to the domain $\{-1, 1\}$ and f_h is the Heaviside step function. In other words, if $x(t)$ is in class $[j]$ then $x_i(t)$ must change its state if the relative weight $w_i(t)$ is less than the parameter θ_j, whose value must be less than the bound defined in (14). It is clear that the lower we set the value of θ_j, the more states in class $[j]$ will have theirs relatives weights greater than θ_j which means that more fixed points the class $[j]$ will have.

RESULTS

The form v_i to be filled by 100 persons in the sample for the investigation of a public opinion poll contains 12 questions. Each of the questions may be answered with a yes (value 0) or a not (value 1). The data have been obtained from the *El País* (Spanish newspaper) and are a sample from the Spanish population. The vector of questions is, (man or woman, older than 45 or not, middle or high class, high level education or not, married or not, smoker or not, left or right politic ideas, worker or unemployed, watch frequently T.V. or not, listen frequently radio or not, read frequently books or not). The sample has been realized by the Montecarlo method.

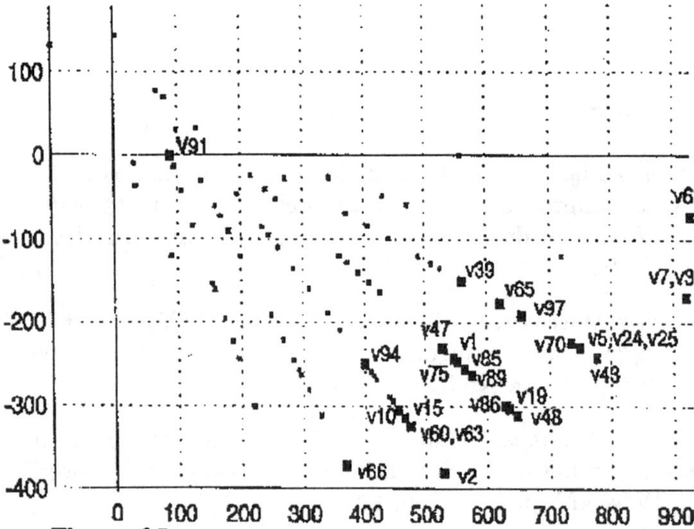

Figure 1: Fixed points when $\theta_i = 0.12$, for all i from 1 to 12.

The big points in the figure are the fixed points of the sample, the distribution of the sample *EP's* in the energy lines may also be observed in the figure. The more extreme points of the lines, correspond to those elements of the population that belonging to a certain class are the more correlated with the prototypes. We have also proved [4], that if, $w(t) = \{w_1(t), \cdots, w_n(t)\}$ is the vector of the relative weights of the components of the state vector $x(t)$, then.

$$\frac{1}{4}\sum_{i=1}^{n} w_i(t) = 1 \tag{16}$$

which means, that associated to any element of the sample, there is a probability frequency distribution [4]. So, taken as an example the element of the population $(0,0,1,0,1,1,1,0,1,1,1,0)$ its associated relative weight vector is $(0.19, 0.26, 0.21, 0.68, 0.31, 0.32, 0.23, 0.49, 0.35, 0, 35, 0.24, 0.36)$. Each component of this vector represents the proportion of the population that share the same component or feature with that person. Many other conclusions, about the population, could be deduced from the energy lines distribution in the figure, as for example: How much is the deviation of a given element in the population to the standard of that population? What is the standard prototype of the smoker woman?, etc. The classical Hopfield algorithm may not be used for such proposes, because the stored data are not separated in classes and it doesn't exist a probability frequency distribution associated to any prototype either. The conclusions, which could be derived from the present research, can be seen under the point of view of the reinterpretation of classical Neural Network Theory as Graph Formalisms. A parameter vector θ is introduced, this parameter is used for controlling the capacity of the net and for studying the degree of correlation among the components of any vector of the population with the sample of prototypes.

REFERENCES

[1] V. Giménez-Martínez, P. Gómez, M. Pérez-Castellanos and V. Rodellar, "A new approach for finding the weights in a neural network using graphs," **Proc. Midwest Symposium on Circuits and Systems**, pp. 113–116, 1993.

[2] V. Giménez-Martínez, P. Gómez, M. Pérez-Castellanos and E. Torrano, "A new approach for controlling the capacity of a recursive neural network," **Proc. of the AMS**, pp. 90–93, 1994.

[3] V. Giménez-Martínez, P. Gómez, E. Torrano and M. Pérez-Castellanos, "A new algorithm for implementing a recursive neural network," **Proc. of the IWANN**, pp. 252–259, 1995.

[4] V. Giménez-Martínez, M. Pérez-Castellanos, J. Castellanos and J. Rios, "Using the statistical deviation of the prototypes for eliminating the parasitic fixed points in a recursive neural network," **Proc. Computer Science and Information Technologies**, pp. 182–195, 1997.

[5] V. Giménez-Martínez, E. Torrano, P. Gómez and M. Pérez-Castellanos, "A class of recursive neural networks based on analytic geometry," **Proc. Intl. Conference on Brain Processes: Theory and Models**, pp. 330–339, 1995.

[6] Y. Kamp and M. Hasler, "Recursive neural networks for associative memory," **Interscience Series in Systems and Optimization**, pp. 10–34, 1990.

[7] W. Kinzel and Z. Phis, "B-condensed matter," **Learning and Pattern Recognition in Spin Glass Models**, vol. 60, pp. 90–93, 1985.

Adaptive Web Caching using Logistic Regression

Annie P. Foong, Yu-Hen Hu and Dennis M. Heisey[1]
Department of Electrical and Computer Engineering
University of Wisconsin
1415 Engineering Drive Madison, WI 53706, USA
Phone: +1 608.262.2445
E-mail: foong, hu@ece.wisc.edu, heisey@surgery.wisc.edu

Abstract. A fundamental understanding of web access patterns is necessary before we can tackle cache performance issues. Due to the extremely dynamic nature of the Web, any techniques we employ must be able to adapt to changing conditions on the Web. We introduce a near-optimal cache algorithm, based on complete *a priori* knowledge of future web access. In reality, such knowledge is unavailable. To enable us to predict future accesses, we have adopted a logistic regression model. Using this model, a web caching agent can acquire knowledge about the objects it encounters and deduce an effective strategy dynamically. Preliminary results (based on object hit and byte hit ratios) on trace-driven simulations of five different servers are very promising.

1 INTRODUCTION

The introduction of World Wide Web (WWW) has made the Internet the world's fastest growing communication medium. Ironically, the WWW has become a victim of its own success. While web traffic increases exponentially, the available end-to-end bandwidth grows at a much slower pace. A survey based on text-only download of home pages of 20 popular sites (from 27 cities over a T1 connection) revealed download times ranging from 3 to 15 seconds [1] . The lower end of such performance is obviously not acceptable.

Improving the performance of such a diverse network is a complex and multi-faceted problem. A common method of improving performance is to implement caches on the Web. They do so by bringing web objects closer to the end users. It alleviates congestion of the Internet by keeping traffic local. The wave of commercial cache products, which have recently hit the market, is testimony of their importance [2]. Most engage in brute force pre-fetching in an attempt to improve response times. Others depend on manual intervention to tune performance. Although this had allowed rapid development of products, such approaches aim to find quick cures for symptoms, and not to understand the underlying issues.

---

[1] Heisey is at the Department of Surgery, University of Wisconsin.

0-7803-5673-X/99/$10.00 © 1999 IEEE

In this paper, we make use of users' access patterns to guide our cache strategy. The remainder of the paper is organized as follows. Section 2 discusses more details of web caching and compares the motivation of this paper with previous work. Section 3 summarizes the methodologies used in our work. Section 4 discusses implementation details and performance results from our prototype. We conclude in Section 5 with suggestions for future work.

2 BACKGROUND

Caching had long been an effective technique used in traditional systems to improve performance. However, there are enough differences between traditional and web caches that we cannot adopt well-known algorithms verbatim. The most important being that the basic entity seen by traditional caches is a fix-sized page, whereas a web cache deals with complete objects (text files, images or video clips). The variable-sized web object demands a different approach in caching strategy. In addition, the response times of web accesses are in the order of seconds (versus milliseconds for system access), which give us latitude to implement more elaborate caching algorithms.

The main cache design issue is in its replacement algorithm. It is a set of rules that determines the object to replace when the cache runs out of space. If *a priori* knowledge is available, the optimal replacement algorithm for a two-level system cache, dealing with fixed-size pages, is the well-known Longest Forward Distance (LFD) algorithm [3]. It seeks to replace the page whose next reference is furthest in the future. More recently, [4] extends this work and deduce a near-optimal algorithm for three-level caches in a global memory system. However, when pages (or objects) are of variable size, the problem becomes NP-hard [5]. They proposed the use of dynamic programming techniques to approximate the optimal solution. All these techniques had been developed based on the assumption that future access patterns are known. In reality, such is usually not the case. As a result, a large number of heuristic-based replacement algorithms exists, each to varying degrees of success. They include variations of the least-recently-used (LRU) and least-frequently-used (LFU) algorithms, with size and last accessed times as secondary considerations [6, 7]. Nine different algorithms were identified in [8], and another proposed that is based on a user-determined cost.

We believe that any heuristic-based algorithms should be an attempt to approximate the optimal algorithm. We therefore develop our strategy as follows:

1. Our first task is to develop a simple heuristic to approximate the optimal algorithm. This is done assuming that we have complete *a priori* knowledge of future accesses,
2. Once this is achieved, we need a way of predicting such knowledge. A logistic regression (LR) model is used to learn from a training set, which is available as a server or proxy statistics log file. This also enables the algorithm to adapt to the site it is learning from.

3. Finally, we test the viability of our algorithm by comparing its performance to existing algorithms on real world traces.

3 METHODOLOGY: THE ADAPTIVE CACHE STRATEGY

Approximating the Optimal Cache Algorithm (OPT)

Since Internet access times are of several orders of magnitude larger than disk cache access, the easiest way to reduce user response time is to reduce cache faults. There are 3 types of cache faults: compulsory, consistency and capacity. A compulsory fault occurs when an object is accessed for the very first time. A consistency fault occurs when an object previously accessed has been changed. A capacity fault occurs when an object previously accessed is not found, due to the finite size of the cache. In this paper, we consider compulsory faults to be unavoidable (since we do not engage in pre-fetching). Also, we consider any object that has changed to be a completely new object. Consistency faults are therefore included in compulsory faults. We will focus only on reducing capacity faults. We have used the term cache "admittance" (in place of "replacement") to refer to the more general policy whereby the set of objects to be replaced includes the latest arrival to cache.

As mentioned before, a full optimization is potentially NP-hard. To simplify analysis, we have defined a simple re-access model where we are interested only in the next access of an object. Therefore, each time an object is removed from cache, it will create exactly one fault. The optimal cache algorithm becomes one that minimizes the total number of removals, subjected to a cache size constraint. We focus only on the set of events, E_D, where the total size of objects exceeds the given cache size. We then seek to optimize each disjoint subset individually. We define the *lifetime,* L(k), of object k within E_D, to be the time between successive accesses to the same object. Let s(k) be the size of object k.

The heuristics for a near-optimal admittance algorithm is as follows:

1. On the *i*th access, identify E_D and separate into disjoint sets.
2. For each object k (including the *i*th object), identify its corresponding L(k). If L(i) = L(k), remove the object with the larger size. If s(i) = s(k), then remove the object with the larger lifetime.
3. Repeat until $E_D = \emptyset$.

As a further approximation, we used the concept of a *weighted lifetime* in our implementation. The OPT algorithm admits objects with the smallest weighted lifetime. Weighted lifetime of object k = E[Time till next access of k] * s(k), where E[X] is the expected value of X. Since accesses are fully deterministic, E[Time till next access of k] = Time of next access of object k – now.

∴ *Weighted Lifetime* of object k = (Time of next access of object k – now) * s(k)

The Logistic Regression Model

The Logistic Regression (LR) model can be used model the probability of an event as a function of the factors (predictors) known or suspected to affect the event [9]. Our goal is to express the outcome of the dependent variable Y, in terms of its predictors, $(1, X_1, \ldots, X_k)$ and their respective coefficients $(\beta_0, \beta_1, \ldots, \beta_k)$. The LR probability is given by:

$$P_{LR} = P(Y{=}g \mid 1, X_1, \ldots, X_k) = \frac{1}{1+\exp(-z)} \qquad \text{Equation (1)}$$

$$\text{where } z = \sum_{j=0}^{k} \beta_j X_j \qquad -\infty < \mathbf{z} < +\infty$$

Given a set of observed data, the coefficients can be estimated by a suitable method (learning phase). Once the coefficients are determined, the LR probability can be calculated for other objects (prediction phase). In the simple case, Y takes on the value one or zero, according to whether or not a specific event occurs during the defined study period (time t_0 to t). In our context, the event of interest is whether a document is re-accessed at least once, in the next *N* accesses. Predictors include size and type of object, number of times it had been accessed before and time since last access.

Let Y_i be the binary outcome (1 for an event, 0 for a nonevent) observed on the i-th access. Let $\mathbf{X}_i$ be the associated vector of predictors. Assuming $Y_1, Y_2, \ldots, Y_m$ are independent, the joint probability of the sequence $Y_1, Y_2, \ldots, Y_m$ is proportional to the product

$$\prod_{i=1}^{m} P(Y{=}Y_i \mid \mathbf{X}_i)$$

This is the likelihood equation. The optimal coefficients β*'s* are found by maximizing the likelihood equation. Iterative algorithms, such as Newton-Raphson, are usually employed to perform the maximization. . The p-value is commonly used to measure the strength of evidence that the estimates of β are statistically significantly different from zero [9]. For our purposes, we considered any predictors with p-value ≤ 0.1 to be significant.

Relating LR probability to OPT heuristics (LROPT)

Our OPT algorithm requires the lifetime of an object to be known. Instead of using the time dimension, we do a count the number of trials between 2 successive accesses to the same object. We call this the *lifespan,* L'(k) of the object. Consider a series of N independent trials. The probability of at least 1 success (i.e.

accessing the same object) out of N trials, is given by the LR probability, P_{LR}. If p is the probability of a success on the next trial, then

$$
\begin{aligned}
&P \text{ (no success on the next trial)} &&= 1 - p \\
&P \text{ (no success in the next N trials)} &&= (1 - p)^{N} = 1 - P_{LR} \\
&\therefore p = 1 - (1 - P_{LR})^{(1/N)}
\end{aligned}
$$

The expected number of trials up to and including the next success is the mean of a geometric distribution [10].

$$E[L'(k)] = 1/p = 1 / (1 - (1 - P_{LR})^{(1/N)})$$

Weighted Lifespan of object k = E[L'(k)] * s(k) Equation (2)

We call our new algorithm LROPT since it uses heuristics from OPT and prediction from the LR model. We associate a weighted lifespan with every object the cache encounters. The LROPT heuristics admits objects with the smallest weighted lifespan into cache.

4 RESULTS

We have restricted our predictor variables to those easily available from server log files. We are capturing the access patterns on a particular server, which are misses from client and proxy caches. The problem description becomes:

$$Y = \begin{cases} 1 & \text{document re-accessed in the next N accesses} \\ 0 & \text{otherwise} \end{cases}$$

X_1 = SIZE, size of document
X_2 = TYPE, type of document (coded as a categorical variable)

1: HTML & Text files e.g. html; 2: Special format files e.g. ps,pdf;
3: Image files e.g. jpeg, gif; 4: Audio files e.g. aud, snd, wav
5: Video files e.g. avi; 6: Executables e.g. pl, exe
0: All others

X_3 = BHITS, number of hits in a backward-looking widow, W_B
X_4 = SINCE, time since last access

We used traces from five web server sites, available at the Internet Traffic Archive [11], which we have named EPA, CAL, NASA, CLAR and SASK traces. A typical trace contains the following information: IP address of requesting host, time of request, requested Uniform Resource Locator (URL), bytes in reply (file size) and reply status. We preprocessed the traces to remove badly formed lines and extraneous characters. Files with the same URL but different sizes are considered to be modified versions. Some accesses result in a "Not modified 304" response from the server, indicating that the file can be served directly from a cache elsewhere in the web. These files contain no size information, and we substitute it with the last known size from its trace history. Files without extensions are considered to be type 1 (HTML/Text) objects.

We used a commercially available statistics package, SAS, to perform the LR modeling and coefficient extraction. The learning was done on N_L accesses and prediction on the next N_P accesses. For simplicity, we performed only one pass of learning and prediction. We based our learning on the number of accesses, instead of time, to circumvent the problem of possible down time on the servers. Furthermore, we ensure that learning is based on an equal number of samples for each trace. For our prototype, we have arbitrarily chosen the following values for our experiments: $N_L = 500$, $N_P = 5000$, $N = 50$, $W_B = 50$. N_L traces were submitted as the learning data to obtain the ML estimates of the coefficients. The coefficients for the 5 traces are given in (Table 1). The coefficients when substituted into (1) give us P_{LR}. The weighted lifespan is given by (2).

TABLE 1 SIGNIFICANT PREDICTORS OF RE-ACCESS

Trace	Coefficients associated with predictors	p-value
EPA	0.58 x BHITS	0.0001
	-7.2e-7 x SINCE	0.0001
NASA	0.78 x BHITS	0.0001
	(-, 17.5, -5.7, 18.4, -, -, -) x TYPE	0.02
	-4e-5 x SIZE	0.06
	-1.4e-5 x SINCE	0.0007
SASK	0.45 x BHITS	0.0003
	-7.8e-6 x SINCE	0.01
CLAR	-1.29 x BHITS	0.1
	- 6e-6 x SINCE	0.0001
CAL	0.66 x BHITS	0.0001
	-9.3e-7 x SINCE	0.0001

Two commonly used metrics for measuring web cache performance are:

$$\text{Byte hit ratio} = \frac{\text{Number of bytes served from cache}}{\text{Total number of bytes}}$$

$$\text{Object hit ratio} = \frac{\text{Number of objects served from cache}}{\text{Total number of objects}}$$

The byte hit ratio gives us a direct measure of raw bandwidth requirements. A high object hit ratio reduces server load by reducing the number of requests a server must process. We first determine the minimum cache size needed for emulating an infinite cache for each of the traces. We ran simulation on cache sizes that were

1%, 5%, 10% and 20% relative to the "infinite" cache size. Hence, we ensured that each trace had ample chance to "warm" the cache and comparisons are standardized across the different traces. We also determined what the "infinite" cache hit ratios were, and present results relative to this ideal. A trace-driven cache simulator is used to measure these metrics. We compared the performance of our LROPT policy to the more traditional policies — LRU, FIFO, LFU and Random (Table 2). Since the size of an object factor so heavily in LROPT, we have also implemented an algorithm based only on size. Finally, we also compare its performance to the two algorithms requiring a priori knowledge, OPT and LFD. The motivation is two-fold: to determine how well OPT and LFD approaches the infinite cache ideal; and how well the more realistic algorithms fare against them.

TABLE 2 ADMITTANCE POLICIES

Policy	**Criteria**	**Extent of knowledge**
OPT	Replace object with largest weighted-lifetime. Admit only if smaller	Perfect future knowledge
Longest Forward Distance (LFD)	Replace object that is accessed furthest in the future. Admit only if smaller	Perfect future knowledge
Logistic Regression-based OPT (LROPT)	Replace object with the largest predicted weighted-lifespan. Admit only if smaller	Last W_B accesses
Size (SIZE)	Replace object with the largest size. Admit only if smaller	Objects in cache
Least Recently Used (LRU)	Replace object with the largest time since last access	Objects in cache
Least Frequently Used (LFU)	Replace object with the smallest number of accesses	Objects in cache
First In First Out (FIFO)	Replace object with the earliest time of entry to cache	Objects in cache
Random	Randomly chosen	None

The results from the LR model are intuitively appealing. "Number of hits in W_B" and "Time since last access" (Table 1) are found to be predictors of future re-accesses. It indicates that more hits in W_B will result in a higher probability of re-access (basis of LFU). It also shows that the longer the "Time since last access", the lower the probability of re-access (basis of LRU). It is important to note that in the CLAR trace, the model indicates that more hits in the past actually lead to lower probability of re-access. In the NASA trace, all 4 predictors are significant. We, therefore, reiterate the need for adaptive algorithms that will not only learn the nuances of each server, but also adapt over time as user behavior changes.

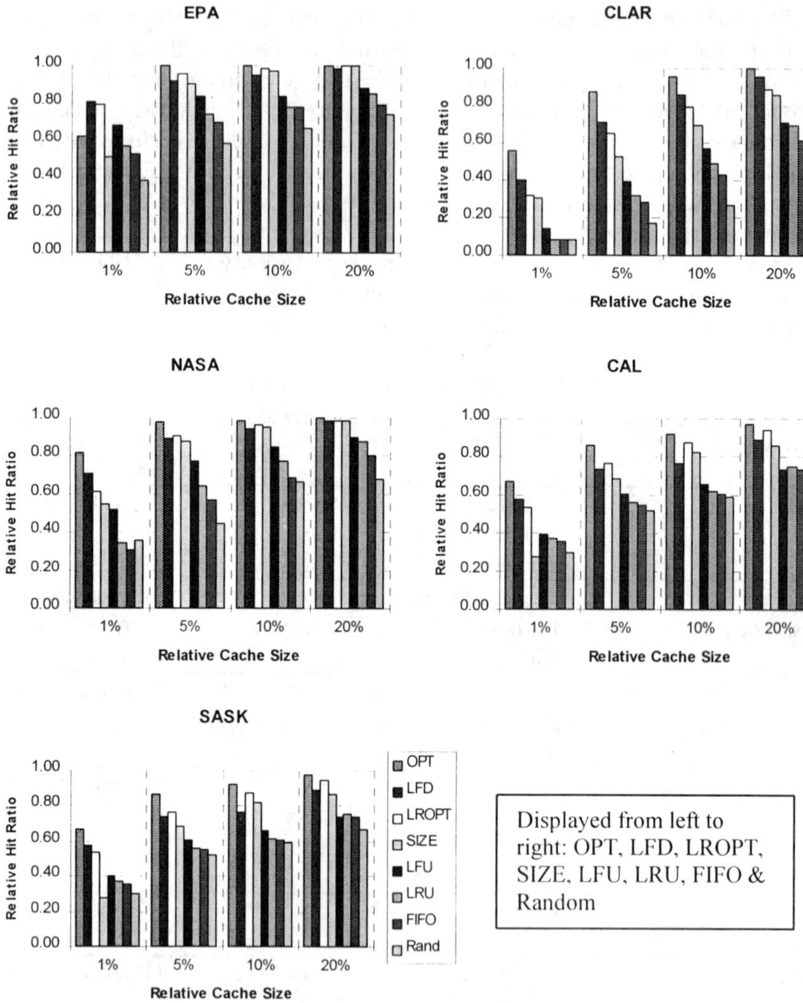

FIGURE 1 OBJECT HIT RATIOS

OPT and LFD performs extremely well as they should, with OPT performing much better in object hit ratios (Figure 1). Our LROPT is the best performer in object hit ratios among the implementable algorithms. In many cases, it performed better than LFD. It also performs better than SIZE. This confirms that both lifespan and size must be taken into consideration simultaneously. LROPT has an average performance in terms of byte hit ratios (Figure 2). This is expected as LROPT is obtained by optimizing only for object hit ratios. We will require a different formulation if our goal is to maximize byte hit ratio. In addition, our LR learning was done in batch mode and did not have the opportunity to continuously adapt itself as it encounters more objects. We acknowledge this oversight and hope to remedy it in future implementation. LROPT has the added benefit of having a low turnover of cache objects (results not reported here). The selective nature of the algorithm reduces the need for constant cache maintenance and may be useful for applications in remote browsing.

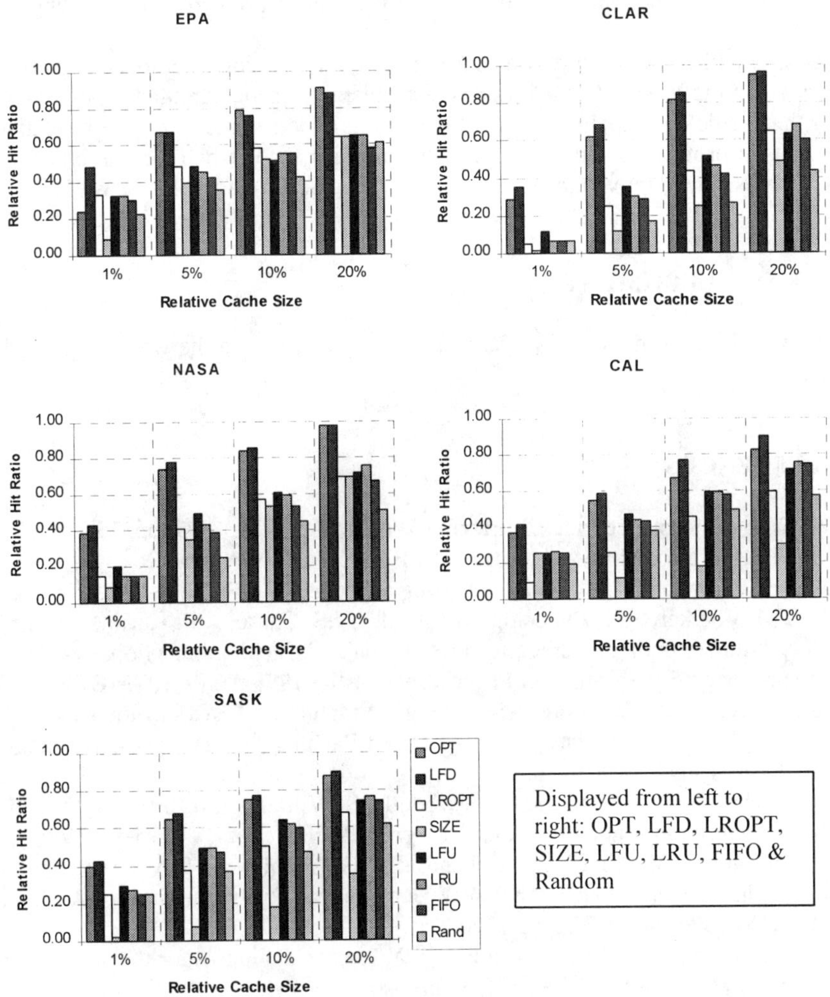

FIGURE 2 BYTE HIT RATIOS

5 FUTURE WORK AND CONCLUSION

Some of the predictors used in this paper, such as window sizes and length of learning, have been arbitrarily chosen. Work is in progress to allow window sizes to be dependent on cache sizes and dynamically adapted. We can also improve our

LR analysis by using a continuous learning-prediction model. We wish to investigate the significance of other predictors, such as number of links and keywords on a page. We need to replay web accesses online and determine more contextual information of each web object. With an online simulator, dynamic metrics such as response times and consistency, can also be measured. Finally, we hope to apply the learning abilities of the LR model in predictive pre-fetching.

Caching will continue to play a central role in maintaining a reasonable quality of service on the Web. We have proposed an adaptive cache admittance policy based on the prediction capabilities of a logistic regression model. Despite a relatively simplistic prototype, our results are very promising. We believe this bodes well for harnessing adaptive models to improve cache performance on the WWW.

ACKNOWLEDGMENT

We would like to thank the researchers [11] who collected the web traces used in this study.

REFERENCES

[1] "Web performance Index," **Internet World,** April 26, 1999, pp. 25.
[2] B.L. Georgia, "More web, less wait," **BYTE**, July 1998, pp. 126.
[3] A.S. Tanenbaum, **Modern Operating Systems**, NJ: Prentice-Hall, Inc, 1992.
[4] G.M. Voelker, T.K. Anderson, M.J. Feeley, J.S. Chase, A.R. Karlin, and H.M. Levy, "Implementing Cooperative Prefetching and Caching in a Globally-Managed Memory System," in **Proceedings of the 1998 ACM SIGMETRICS Conference on Performance Measuring, Modeling and Evaluation**, June 1998.
[5] S. Hosseini, "Investigation of Generalized Caching," **Ph.D. thesis**, Washington University, 1997.
[6] M. Abrams, C.R. Standridge, G. Abdulla, S. Williams and E.A. Fox, "Caching Proxies: Limitations and Potentials," **Virginia Tech Report**, TR-95-12, 1995.
[7] S. Williams, M. Abrams, C.R. Standridge, G. Abdulla and E.A. Fox, "Removal Policies in Network Caches for WWW Documents," in **Proceedings of ACM SIGCOMM 96**, Stanford, Aug 1996.
[8] P. Cao and S. Irani, "Cost-aware WWW Proxy Caching Algorithms," in **Proceedings of the 1997 USENIX Symposium on Internet Technology & Systems**, 1997, pp. 193-206.
[9] D.G. Kleinbaum, L.L. Kupper, and H. Morgenstern, **Epidemiologic Research: Principles and Quantitative methods**, NY: Van Nostrand Reinhold Co, 1982.
[10] K.S. Trivedi, **Probability & Statistics with Reliability, Queuing, and Computer Science Applications**, NJ: Prentice-Hall, 1982, pp. 70-73.
[11] L. Bottomley, R. Fridman, S.Balbach, J. Dumoulin and E. Fogel, **Internet Traffic Archive**, http://ita.ee.lbl.gov/html/traces.html, 1995.

PATTERN CLASSIFICATION USING A MIXTURE OF FACTOR ANALYZERS

Naonori Ueda†, Ryohei Nakano†
Zoubin Ghahramani‡, and Geoffrey Hinton‡
† NTT Communication Science Laboratories
Hikaridai Seika-cho Soraku-gun Kyoto 619-0237 Japan
E-mail: {ueda,nakano}@cslab.kecl.ntt.co.jp
‡ Gatsby Computational Neuroscience Unit, University College London
17 Queen Square, London WC1N 3AR, UK
E-mail: {zoubin,hinton}@gatsby.ucl.ac.uk

Abstract. This paper describes a practical application of a mixture of factor analyzers (MFA) to pattern recognition. The MFA extracts *locally* linear manifolds underlying a given high dimensional data. In this respect, the MFA-based approach is similar to the conventional *subspace methods* that approximate the data space with low dimensional linear subspaces. However, the MFA-based classifier, unlike the conventional subspace methods, can perform classification based on the Bayes decision rule due to its probabilistic formulation. Experimental results show that the MFA-based approach can obtain better classification performance than the conventional subspace methods.

INTRODUCTION

A mixture of factor analyzers (MFA), recently developed in the field of neural networks [1], is an extension of a single FA [2] and can be thought of as a reduced dimensional mixture of Gaussians. The MFA model performes clustering and dimensionality reduction simultaneously. More precisely, the MFA model extracts low dimensional *locally* linear manifolds underlying a given high dimensional data. The MFA model fitting to each class data allows us to estimate the probability density of the class and therefore it can perform pattern classification based on the relative density approach.

The MFA-based classification is related to the *subspace methods* [3]-[5] that approximate the data space with low dimensional subspaces. However, it has an essential advantage over the subspace methods that it can perform the Bayesian classification due to its probabilistic formulation. In the previous paper, we proposed the split and merge EM algorithm [6] to ovecome the local maxima problem when estimating MFA model parameters. The aim of

0-7803-5673-X/99/$10.00 © 1999 IEEE

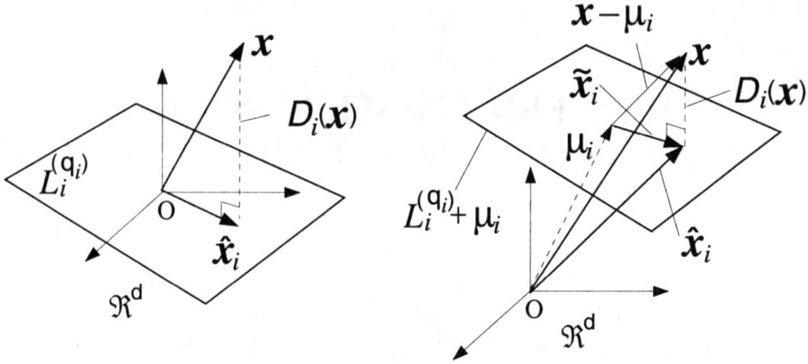

(a) CLAFIC (b) PD

Figure 1: CLAFIC and Projection distance methods.

this paper is to evaluate the practical effectiveness of the MFA model as a classifier.

CONVENTIONAL SUBSPACE METHODS

A well known pattern classification method based on dimensionality reduction is the CLAFIC method has been well known [3]. In the case of a C-class problem, in the CLAFIC method, first, a q_i dimensional linear subspace $L_i^{(q_i)}$ is extracted from the training sample space $\mathcal{R}^d$ for each class ω_i, $i = 1, \ldots, C$. Then the Euclidean distance $D_i(\boldsymbol{x})$ between an input vector $\boldsymbol{x}$ to be classified and its reconstructed vector $\hat{\boldsymbol{x}}_i$ in the ith subspace $L_i^{(q_i)}$ is computed for each class. Next, as shown in Fig. 1(a), the input vector is classified as the class ω_i^* where $i^* = \arg\min_j D_j(\boldsymbol{x})$ or equivalently $i^* = \arg\max ||\hat{\boldsymbol{x}}_i||$.

The subspace $L_i^{(q_i)}$ is spanned by q_i eigenvectors $\boldsymbol{u}_{i1}, \ldots, \boldsymbol{u}_{iq_i}$ corresponding to the largest q_i eigenvalues $\lambda_{i1} > \cdots > \lambda_{iq_i}$ of the sample autocorrelation matrix:

$$S_i = \frac{1}{N_i} \sum_{n=1}^{N_i} \boldsymbol{x}_n \boldsymbol{x}_n^T, \tag{1}$$

where N_i is the training data size in class ω_i. This is the same procedure as PCA. Note that setting $\mathbf{U}_i = [\boldsymbol{u}_{i1}, \ldots, \boldsymbol{u}_{iq_i}]$, we can easily compute the reconstructed vector as $\hat{\boldsymbol{x}} = \mathbf{U}_i \mathbf{U}_i^T \boldsymbol{x}$.

The dimensionality q_i is determined in an ad hoc manner: using some threshold value κ, the q_i that satisfies the following condition is chosen:

$$a(q_i) \leq \kappa \leq a(q_i + 1). \tag{2}$$

Here $a(q_i)$ is the accumulated ratio of the largest q_i eigenvalues and is computed from

$$a(q_i) = \sum_{j=1}^{q_i} \lambda_{ij} / \sum_{j=1}^{d} \lambda_{ij}. \tag{3}$$

Another variant, called the Projection Distance (PD) method [4], has been proposed. In this method, the mean vector $\boldsymbol{\mu}_i$ is taken into account and therefore $L_i^{(q_i)}$ is constructed by using the sample covariance matrix:

$$\boldsymbol{\Sigma}_i = \frac{1}{N_i} \sum_{n=1}^{N_i} (\boldsymbol{x}_n - \boldsymbol{\mu}_i)(\boldsymbol{x}_n - \boldsymbol{\mu}_i)^T \tag{4}$$

In the PD method, as shown in Fig. 1(b), an input vector $\boldsymbol{x}$ is decomposed into $\boldsymbol{\mu}_i$ and $\boldsymbol{x} - \boldsymbol{\mu}_i$. Then, the reconstructed vector $\hat{\boldsymbol{x}}_i$ is obtained by orthogonally projecting $\boldsymbol{x} - \boldsymbol{\mu}_i$ onto $L_i^{(q_i)}$ and adding $\boldsymbol{\mu}_i$.

In the SS method, a single subspace is extracted for each class. A Multiple Subspace (MSS) method has also been proposed [5]. In this method, first, the data is divided into several clusters by using the k-means algorithm. Then, for each cluster the CLAFIC or PD method is performed. The MSS method can be expected to improve the classification performance of the CLAFIC method, but it as well as the CLAFIC and PD methods do not perform *Bayesian classification* due to the absence of a probability density model.

MFA-BASED APPROACH

Mixture of Factor Analyzers

A single factor analyzer (FA) [2] assumes that an observed d-dimensional variable $\boldsymbol{x}$ is generated as a linear transformation of some lower q-dimensional *latent* variable $\boldsymbol{z} \sim \mathcal{N}(\mathbf{0}, \mathbf{I})$ plus additive Gaussian noise $\boldsymbol{v} \sim \mathcal{N}(\mathbf{0}, \Psi)$. Ψ is a diagonal matrix. That is, the generative model can be written as

$$\boldsymbol{x} = \mathbf{W}\boldsymbol{z} + \boldsymbol{v} + \boldsymbol{\mu}. \tag{5}$$

Here, $\mathbf{W} \in \mathcal{R}^{d \times q}$ is a transformation matrix and is called a *factor loading matrix*. $\boldsymbol{\mu}$ is a mean vector. Then, from a simple calculation, the probability density function (pdf) of the observed data by an FA model can be obtained by

$$p(\boldsymbol{x}; \Theta) = \mathcal{N}(\boldsymbol{\mu}, \mathbf{W}\mathbf{W}^T + \Psi). \tag{6}$$

Noting that $\mathbf{W}$ can be represented by linearly independent column vectors as $\mathbf{W} = [\boldsymbol{w}_1, \ldots, \boldsymbol{w}_q]$ where $\boldsymbol{w}_i \in \mathcal{R}^{d \times 1}$, we can rewrite (5) as

$$\boldsymbol{x} = \sum_{i=1}^{q} z_i \boldsymbol{w}_i + \boldsymbol{v} + \boldsymbol{\mu}, \tag{7}$$

where z_i is the ith component of $\boldsymbol{z}$. Clearly, (7) shows that $\boldsymbol{w}_1, \ldots, \boldsymbol{w}_q$ form the basis in a latent space. Hence, FA can be interpreted as a dimensionality reduction model extracting a linear manifold (affine subspace) $\mathcal{M} = L^{(q)} + \boldsymbol{\mu}$ underlying the given observed data space, where $L^{(q)}$ denotes the linear subspace of $\mathcal{R}^d$ spanned by the basis $\boldsymbol{w}_1, \ldots, \boldsymbol{w}_q$.

A mixture of factor analyzers (MFA) is an extension of a single FA [1]. That is, MFA is defined by the linear combination of M factor analyzers and can be thought of as a reduced dimension mixture of Gaussians. The MFA model extracts q-dimensional *locally* linear manifolds $\mathcal{M}_m = L_m^{(q)} + \boldsymbol{\mu}_m$ for $m = 1, \ldots, M$ underlying the given high dimensional data. Since the MFA model extracts a *globally* nonlinear manifold, it is more flexible than the single FA model.

The pdf of the observed data by a mixture of M FAs is given by

$$p(\boldsymbol{x};\Theta) = \sum_{m=1}^{M} P_m \mathcal{N}(\boldsymbol{\mu}_m, \mathbf{W}_m \mathbf{W}_m^T + \Psi_m). \tag{8}$$

Here P_m is a mixing proportion ($P_m > 0$ and $\sum_{m=1}^{M} P_m = 1$). Note that (8) is a natural extension of (6). See [1] for the details.

The maximum likelihood (ML) estimate of the unknown parameters $\Theta = \{P_m, \boldsymbol{\mu}_m, \mathbf{W}_m | m = 1, \ldots, M\}$ of the MFA model is a value of Θ that maximizes the log-likelihood function:

$$\begin{aligned} \mathcal{L}(\Theta;\mathcal{X}) &= \log \prod_{\boldsymbol{x}\in\mathcal{X}} p(\boldsymbol{x};\Theta) \\ &= \sum_{\boldsymbol{x}\in\mathcal{X}} \log \sum_{m=1}^{M} P_m \mathcal{N}(\boldsymbol{\mu}_m, \mathbf{W}_m \mathbf{W}_m^T + \Psi_m), \end{aligned} \tag{9}$$

where $\mathcal{X}$ is a set of sample data. The parameter estimation for a MFA model was derived based on the EM algorithm [1]. In practice, however, it suffers from the local optima problem associated with the EM algorithm.

Split and Merge EM Algorithm

To overcome this problem, we proposed the split and merge EM (SMEM) algorithm and demonstrated its usefulness [6]. In the case of mixture models, the local maxima often arise when there is too many components in one part of the space and there are few in another. It is not possible to move a component in an overpopulated region to an underpopulated region without passing through positions that give a lower likelihood. We therefore introduce a discrete move that simultaneously merges two components in an overpopulated region and splits a component in an underpopulated region. This is the basic idea of the SMEM algorithm for mixture models.

We will breifly explain how this split and merge is used in the EM algorithm below. The probability density function (pdf) of a mixture of M

density models is in general given by

$$p(\boldsymbol{x};\Theta) = \sum_{m=1}^{M} P_m p_m(\boldsymbol{x};\theta_m), \tag{10}$$

where P_m is the mixing proportion mentioned before. The $p_m(\boldsymbol{x};\theta_m)$ is a d-dimensional density model corresponding to the mth model component. $\Theta = \{(P_m, \theta_m),\ m = 1, \dots, M\}$ is an unknown parameter set.

In the case of mixture models, the model index $m \in \{1, \dots, M\}$ is unknown for an observed data $\boldsymbol{x}_n$ and therefore m corresponds to the unobserved data. Thus, the pdf of the complete data is $P_m p_m(\boldsymbol{x};\theta_m)$ in this case. Then, the EM algorithm maximizes the following Q function:

$$Q(\Theta|\Theta^{(t)}) = \sum_{n=1}^{N}\sum_{m=1}^{M}\{\log P_m p_m(\boldsymbol{x}_n;\theta_m)\}P(m|\boldsymbol{x}_n;\Theta^{(t)}). \tag{11}$$

Here, $P(m|\boldsymbol{x}_n;\Theta^{(t)})$ is a posterior probability and is computed as

$$P(m|\boldsymbol{x}_n;\Theta^{(t)}) = \frac{P_m^{(t)} p_m(\boldsymbol{x}_n;\theta_m^{(t)})}{\sum_{l=1}^{M} P_l^{(t)} p_l(\boldsymbol{x}_n;\theta_l^{(t)})}. \tag{12}$$

N is the number of observed data points.

Looking at (11) carefully, one can see that holding the posterior probabilities fixed, the Q function can be represented in the form of a direct sum:

$$Q(\Theta|\Theta^{(t)}) = \sum_{m=1}^{M} q_m(\theta_m|\Theta^{(t)}), \tag{13}$$

where

$$q_m(\theta_m|\Theta^{(t)}) = \sum_{n=1}^{N} P(m|\boldsymbol{x}_n;\Theta^{(t)}) \log P_m p_m(\boldsymbol{x}_n;\theta_m) \tag{14}$$

and depends only on P_m and θ_m. Let Θ^* denote the parameter values estimated by the usual EM algorithm. Then, after the EM algorithm has converged, the Q function can be rewritten as

$$Q^* = q_i^* + q_j^* + q_k^* + \sum_{m, m \neq i,j,k} q_m^*. \tag{15}$$

We then try to increase the first three terms of the right-hand side of (15) by merging model components i and j to produce a model component i', and splitting the model component k into two model components j' and k'. Since there are so many candidates, we introduced reasonable criteria for ordering the split and merge candidates to make the algorithm efficient.

The parameters of these newly generated model components are reestimated by using EM steps, but note that instead of (12) we use the following modified posterior probability:

$$P(m'|\boldsymbol{x};\Theta^{(t)}) = \frac{P_{m'}^{(t)} p_{m'}(\boldsymbol{x};\theta_{m'}^{(t)})}{\sum_{l=i',j',k'} P_l^{(t)} p_l(\boldsymbol{x};\theta_l^{(t)})} \times \sum_{m=i,j,k} P(m|\boldsymbol{x};\Theta^*), \quad \text{for } m' = i', j', k'. \tag{16}$$

Using (16), the sum of posterior probabilities for models i', j' and k' becomes equal to the sum of posterior probabilities for models i, j, and k just before split and merge. That is,

$$\sum_{m'=i',j',k'} P(m'|\boldsymbol{x};\Theta^{(t)}) = \sum_{m=i,j,k} P(m|\boldsymbol{x};\Theta^*) \tag{17}$$

always holds during the reestimation process. By this, we can reestimate the parameters for models i', j', and k' consistently without affecting the other models. We call these EM steps *partial EM steps.* These partial EM steps make the total algorithm efficient.

After the partial EM steps, the usual EM steps, called the *full EM steps*, are performed as a post processing operation. After these steps, if Q is improved, then we accept the new estimate and repeat the above after setting the new parameters to Θ^*. Otherwise, we reject it and go back to Θ^* and try another candidate. We summarize these as the following SMEM algorithm:

1. Perform the usual EM updates from some initial parameter value Θ. Let Θ^* and Q^* denote the estimated parameters and corresponding Q function value, respectively.

2. Sort the split and merge candidates by computing split and merge criteria (described in the next section) based on Θ^*. Let $\{i, j, k\}_c$ denote the cth candidate.

3. For $c = 1, \ldots, C_{max}$, perform the following: After making the initial parameter settings based on Θ^*, perform the *partial EM steps* for $\{i, j, k\}_c$ and then perform the *full EM steps.* Let Θ^{**} be the obtained parameters and Q^{**} be the corresponding Q function value. If $Q^{**} > Q^*$, then set $Q^* \leftarrow Q^{**}$, $\Theta^* \leftarrow \Theta^{**}$ and go to Step 2.

4. Halt with Θ^* as the final parameters.

Note that when a certain split and merge candidate which improves the Q function value is found in Step 3, the other successive candidates are ignored. There is no guarantee therefore that the split and merge candidates that are chosen will give the largest possible improvement in Q. This is not a major problem, however, because the split and merge operations are performed repeatedly. If there were no heuristics for ordering potential split and merge operations, we would have to consider them all $C_{max} = M(M-1)(M-2)/2$,

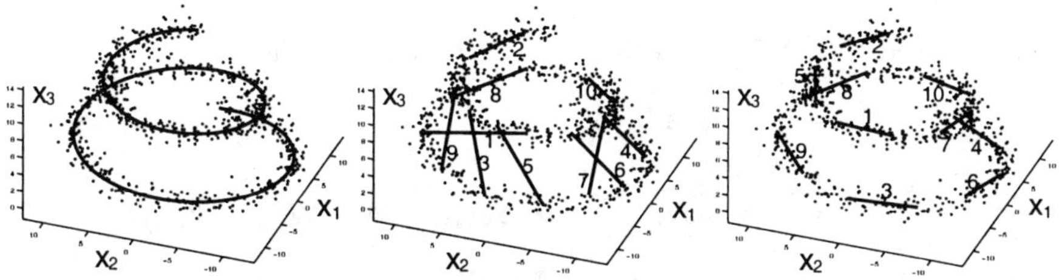

(a) Data and true manifold (b) Result by EM (c) Result by SMEM

Figure 2: Manifold extraction by MFA

but experimentally we have confirmed that $C_{max} \simeq 5$ may be enough because the split and merge criteria do work well.

Figure 2 shows results of extracting a one-dimensional manifold from three-dimensional data (noisy shrinking spiral) using the EM and SMEM algorithms. In this case, each factor loading matrix $\mathbf{W}_m$ becomes a three-dimensional column vector corresponding to each thick line in Figure 2. More precisely, the center position and the direction of each thick line are $\boldsymbol{\mu}_m$ and $\mathbf{W}_m$, respectively. In addition, the length of each thick line is $2||\mathbf{W}_m||$. Although the EM algorithm converged to poor local maxima as shown in Figure 2(b), the SMEM algorithm successfully extracted the locally linear manifolds shown in Figure 2(c).

MFA-based Classifier

The MFA model is also applicable to pattern recognition tasks since once an MFA model is fitted to each class, we can compute the posterior probabilities for each data point. More specifically, the pdf of class ω_i by the MFA model is given by

$$p_i(\boldsymbol{x};\Theta_i) = \sum_{m=1}^{M} P_{im}\mathcal{N}(\boldsymbol{\mu}_{im}, \mathbf{W}_{im}\mathbf{W}_{im}^T + \boldsymbol{\Psi}_{im}), \tag{18}$$

where Θ_i is a set of the MFA model parameters for class ω_i. That is, $\Theta_i = \{P_{im}, \boldsymbol{\mu}_{im}, \mathbf{W}_{im}, \boldsymbol{\Psi}_{im} | \; m = 1, \ldots, M\}$. P_{im} is the mixing proportion of the mth model for class ω_i.

The posterior probability of class ω_i given $\boldsymbol{x}$ is computed from

$$P(\omega_i|\boldsymbol{x}) = \frac{P_i \; p_i(\boldsymbol{x};\Theta_i)}{\sum_{j=1}^{C} P_j \; p_j(\boldsymbol{x};\Theta_j)}. \tag{19}$$

Here C is the number of classes and P_i is a priori probability of class ω_i and is estimated from $P_i = N_i/N$, where N_i is the training sample size of class ω_i and N is the total sample size of all classes. Then, the optimum class i^*

for x based on the *Bayes decision rule* is written as follows:

$$\begin{aligned} i^* &= \arg\max_i P(\omega_i|\boldsymbol{x}) \\ &= \arg\max_i N_i \sum_{m=1}^{M} P_{im}\mathcal{N}(\boldsymbol{\mu}_{im}, \mathbf{W}_{im}\mathbf{W}_{im}^T + \boldsymbol{\Psi}_{im}) \end{aligned} \quad (20)$$

EXPERIMENTS

We compared the MFA-based classifier with the conventional subspace methods (CLAFIC, PD, and MSS) mentioned in this paper using the following three real data sets:

- Hand-written digits [8]: 10 classes, Glucksman's feature (16 dimensional data), 2000 (2000) for training (test),
- Heart1 [9]: two classes, 35 dimensional data, 690 (230) for training (test),
- Diabetes1 [9]: two classes, 8 dimensional data, 576 (192) for training (test).

The recognition results for Hand-written digits are shown in Figure 3. In Figure 3(a), the best results for the test data are presented after varying the threshold value κ. In Figure 3(b),(c), q denotes the dimensionality of latent variable z and M is the number of components in the mixture. Note that the CLAFIC and PD methods correspond to $M = 1$ in the MSS method shown in Figure 3(a). For training data, the larger M was, the higher the classification rates became. On the other hand, for test data, the classification rates started to decrease from some M value. This is clearly due to *overfitting*.

One can see that the MFA-based classifier with the SMEM algorithm drastically exceeded the MSS methods on recognition rate for test data when q=3 and $M = 5$ and it provided the best result among all methods. The recognition rate by the 3-nearest neighbor classifier was 88.3%. It is interesting that the MFA approach by the SMEM algorithm could outperform the nearest neighbor approach when $q = 3$ and $M = 5$. This suggests that the intrinsic dimensionality of the data would be three or so. Clearly, it is quite important to optimally determine the q and M values. *Cross-validation* could be used for this purpose.

Figures 4 and 5 show the results for Heart1 and Diabetes1 in a similar way. For both data sets the MFA-based classifier with the SMEM algorithm outperformed the other methods on the average. According to [9], the best results by the multilayer neural network were 80% and 75% for Heart1 and Diabetes1, respectively, which means that these data sets seem to be hard to classify. For Heart1, the results by the MSS methods were less than that by the neural network, while the MFA-based classifier with the SMEM algorithm obtained better results when $M > 5$. As for Diabetes1, the results by the

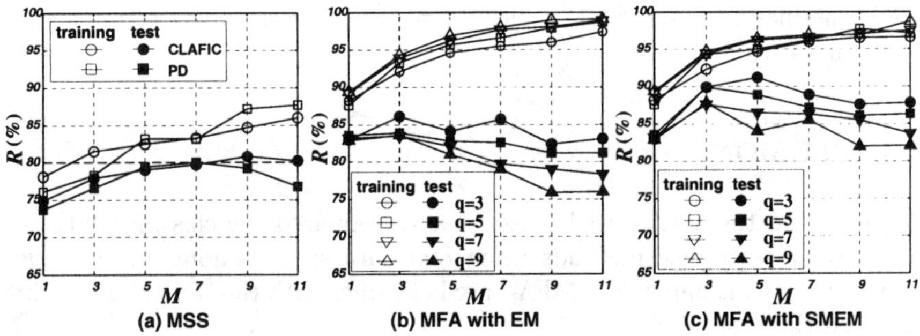

Figure 3: Classification rate for Hand-written digits.

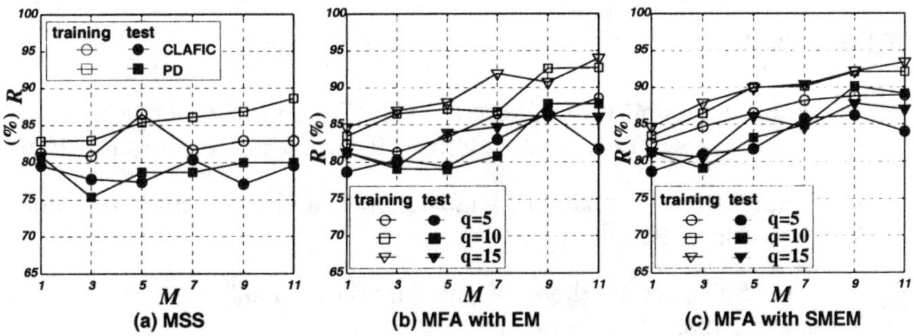

Figure 4: Classification rate for Heart1.

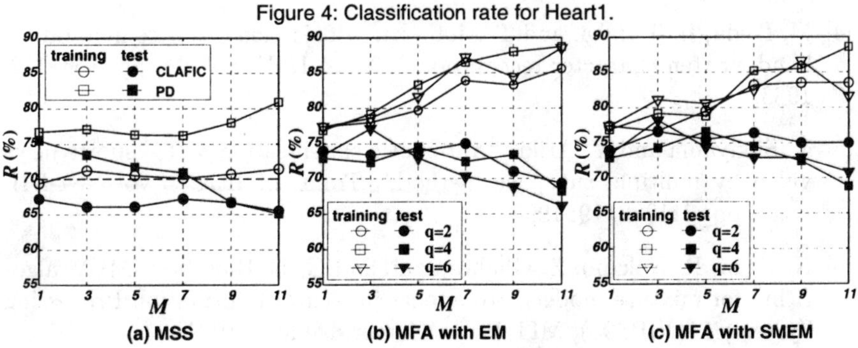

Figure 5: Classification rate for Diabetes1.

MFA-classifier with the SMEM algorithm were similar to the result by the neural network.

CONCLUSION

We described the MFA-based classifier and compared the classifier with the conventional subspace methods using real data sets. Among the methods considered in this paper, the MFA-based classifier with the SMEM algorithm provided the best results.

REFERENCES

[1] Z. Ghahramani and G. E. Hinton, "The EM algorithm for mixtures of factor analyzers," Tech. Report CRG-TR-96-1, Univ. of Toronto, 1997.

[2] D. J. Bartholomew, "Latent variable models and factor analysis," John Wiley & Sons, 1983.

[3] E. Oja, "Subspace methods of pattern recognition," Research Studies Press Ltd., 1983.

[4] M. Ikeda, H. Tanaka, and T. Motooka, "Projection distance method for hand-written character recognition," Trans. IPSJ, vol. 24, pp. 106–112, 1983.

[5] Y. Sugiyama and Y. Ariki, "Automatic classification of TV sports news video by multiple subspace method," Trans. of IEICE, vol. J81-DII, no. 9, pp. 2112–2119, 1998.

[6] N. Ueda, R. Nakano, Z. Ghahramani, and G. E. Hinton, "SMEM algorithm for mixture models," to appear in Neural Information Processing Systems 11(NIPS98), MIT Press, Cambridge MA, 1999.

[7] A. P. Dempster, N. M. Laird and D. B. Rubin, "Maximum likelihood from incomplete data via the EM algorithm," Journal of Royal Statistical Society B, vol. 39, pp. 1–38, 1977.

[8] K. Ishii, "Design of a recognition dictionary using artificially distorted characters," Systems and computers in Japan, vol. 21, no. 9, pp. 669–677, 1989.

[9] L. Prechelt, "Proben1 – a set of neural network benchmark problem and benchmarking rules," Tech. Report 21/94, Fakultat fur Informatic, Universitat Karlsruhe, 1994.

NEAR-OPTIMAL FLIGHT LOAD SYNTHESIS USING NEURAL NETS

Michael T. Manry, Cheng-Hsiung Hsieh, and Hema Chandrasekaran
Department of Electrical Engineering
University of Texas at Arlington
Arlington, Texas 76019
e-mail : manry@uta.edu

Abstract

This paper describes the use of neural networks for near-optimal helicopter flight load synthesis (FLS), which is the process of estimating mechanical loads during helicopter flight, using cockpit measurements. First, modular neural networks are used to develop statistical signal models of the cockpit measurements as a function of the loads. Then Cramer-Rao maximum a-posteriori bounds on the mean-squared error are calculated. Then, multilayer perceptrons for FLS are designed which approximately attain the bounds. It is shown that all of the FLS networks have good generalization.

I. INTRODUCTION

Rotor systems in helicopters suffer extreme vibrations. As a result, system components accumulate amounts of fatigue damage which vary considerably with different types of flight. Since it is necessary to avoid rotor system component failure, a conservative usage spectrum is used which estimates the amount of time spent in each type of flight. Unfortunately, the usage spectrum (1) leads to early retirement of some components and a resulting waste of money, and (2) is difficult to keep track of for individual aircraft.

Flight Load Synthesis (FLS) is one approach to solving these problems. In FLS, cockpit measurements are processed into estimates of the mechanical loads encountered by critical parts. Accumulated loads, rather than a usage spectrum, can then be used to determine component retirement times. The loads estimated from FLS should allow for longer part lifetimes where appropriate. Here, we develop near-optimal neural networks for FLS using data from Bell Helicopter Textron in Fort Worth.

II. DESCRIPTION OF DATA FILES

In FLS, the goal is to process cockpit measurements called features into useful

0-7803-5673-X/99/$10.00 © 1999 IEEE

loads. In this section, we describe the features and loads found in data provided by Bell Helicopter. We then describe the corresponding training data files produced for neural networks.

The first set of FLS experiments uses parameters that are available in the basic health usage monitoring system (HUMS), plus some others. The data was obtained from the M430 flight load level survey conducted in Mirabel Canada in early 1995. The input features include: (1) CG F/A load factor, (2) CG lateral load factor, (3) CG normal load factor, (4) pitch attitude, (5) pitch rate, (6) roll attitude, (7) roll rate, (8) yaw rate, (9) corrected airspeed, (10) rate of climb, (11) longitudinal cyclic stick position, (12) pedal position, (13) collective stick position, (14) lateral cyclic stick position, (15) main rotor mast torque, (16) main rotor mast rpm, (17) density ratio, (18) F/A acceleration, transmission, (19) lateral acceleration, transmission, (20) vertical acceleration, transmission, (21) left hand forward pylon link, (22) right hand forward pylon link, (23) left hand aft pylon link, and (24) right hand aft pylon link.

The desired output loads are as follows: (1) fore/aft cyclic boost tube oscillatory axial load (OAL), (2) lateral cyclic boost tube OAL, (3) collective boost tube OAL, (4) main rotor (MR) pitch link OAL, (5) MR mast oscillatory perpendicular bending st., (6) MR yoke oscillatory beam bending sta., (7) MR blade oscillatory beam bending sta., (8) MR yoke oscillatory chord bending sta., and (9) resultant mast bending, sta.position

Three training data files were produced. File F17.dat used features 1 through 17 and all nine output loads. The features correspond to those used on the basic HUMS. File F20.dat used features 1 through 20 and all nine output loads. File F24.dat used features 1 through 24 and all nine output loads.

III. OBTAINING BOUNDS ON ESTIMATION ERROR VARIANCES

Cramer-Rao maximum a-posteriori (CRMAP) bounds [1] have been shown to provide lower bounds on neural network testing error variances [2,3]. Specifically, let the neural network error E be defined as

$$E \equiv \sum_{i=1}^{M} MSE_i \tag{1}$$

where MSE_i denotes the error for the ith output parameter (load) and M denotes the number of outputs (number of loads). E and MSE_i denote training error or testing error, depending upon the situation. The CRMAP bound on $var(\theta_i'-\theta_i)$, the variance of the error between the ith output (load) θ_i and the ith output's estimate θ_i' is denoted by B_i. Therefore,

$$\begin{aligned} MSE_i &\geq B_i \\ E &\geq \sum_{i=1}^{M} B_i \end{aligned} \tag{2}$$

When we have equality in (2), the estimates are usually optimal. In this section, we show the details of how CRMAP bounds are calculated from training data.

A. Getting Bounds

In this subsection, we describe how to calculate the M by M MAP Fisher information matrix [3] (FIM), given statistics on the training data. Let $\{\mathbf{x}_p, \boldsymbol{\theta}_p\}$ for $1 \leq p \leq N_v$ represent the training set for a neural network. Here, $\mathbf{x}_p$ represents the pth example of the N-dimensional random input vector $\mathbf{x}$ and $\boldsymbol{\theta}_p$ represents the pth example of the M-dimensional random parameter vector $\boldsymbol{\theta}$.

Elements of the M by M MAP FIM, $\mathbf{J}^{MAP}$ are defined as

$$J_{ij}^{MAP} = E_\theta[J_{ij}^{MLE}] + E_\theta[\frac{\partial\Lambda^{AP}}{\partial\theta_i}\frac{\partial\Lambda^{AP}}{\partial\theta_j}],$$
$$J_{ij}^{MLE} \equiv E_n[\frac{\partial\Lambda^{MLE}}{\partial\theta_i}\frac{\partial\Lambda^{MLE}}{\partial\theta_j}] \tag{3}$$

where $E_\theta[\cdot]$ denotes expected value over the parameter vector $\boldsymbol{\theta}$ and $E_n[\cdot]$ denotes expected value over the noise, where $\Lambda^{MAP} = \ell n(p_{\theta|x}(\boldsymbol{\theta}|\mathbf{x}))$, $\Lambda^{MLE} = \ell n(p_{\mathbf{x}|\theta}(\mathbf{x}|\boldsymbol{\theta})$, and $\Lambda^{AP} = \ell n(p_\theta(\boldsymbol{\theta}))$. Assume that the elements x(k) of $\mathbf{x}$ are modeled as

$$x(k) = s(k) + n(k) \tag{4}$$

where the elements s(n) and n(n) are respectively elements of the signal and noise vectors $\mathbf{s}$ and $\mathbf{n}$. $\mathbf{s}$ is a deterministic function of the parameter vector $\boldsymbol{\theta}$. The N by N covariance matrix of $\mathbf{x}$ or $\mathbf{n}$ is denoted by $\mathbf{C}_{nn}$. The elements of $\mathbf{J}^{MAP}$ in (3) can now be evaluated as

$$E_\theta[J_{ij}^{MLE}] = E_\theta[(\frac{\partial \mathbf{s}}{\partial\theta_i})^T C_{nn}^{-1}(\frac{\partial \mathbf{s}}{\partial\theta_j})],$$
$$E_\theta[\frac{\partial\Lambda^{AP}}{\partial\theta_i}\frac{\partial\Lambda^{AP}}{\partial\theta_j}] = d_\theta(i,j) \tag{5}$$

where $\mathbf{C}_\theta$ denotes the M by M covariance matrix of the M-dimensional parameter vector $\boldsymbol{\theta}$ and $d_\theta(i,j)$ denotes an element of $\mathbf{C}_\theta^{-1}$. Let $(J^{MAP})^{ij}$ denote an element of $(\mathbf{J}^{MAP})^{-1}$. Then [3],

$$var(\theta_i' - \theta_i) \geq (J^{MAP})^{ii} \tag{6}$$

where θ_i' can be any estimate of θ_i. The B_i in (2) are the $(J^{MAP})^{ii}$ of (6).

In order to calculate the log-likelihood functions and the CRMAP lower bounds on the variance of the parameter estimates, a statistical model of the input vector $\mathbf{x}$ is required. This model consists of a deterministic expression for the signal vector $\mathbf{s}$ in terms of the parameter vector $\boldsymbol{\theta}$, the joint probability density of the additive noise vector $\mathbf{n}$. In most applications however, the signal model is unknown and bound calculation is not possible. One approach to this problem is to create an approximate signal model from the given training data.

B. Signal Modeling

Given the training patterns for our estimator, we want to find the signal

$$x_p = s_p + n_p \tag{7}$$

component model and noise probability density function (pdf). We make the following assumptions, which make this problem solvable.

(A1) The unknown, exact signal model of (4) can be written in vector form as where $\mathbf{x}_p$ and $\boldsymbol{\theta}_p$ are defined as before, $\mathbf{s}_p$ denotes the pth example of $\mathbf{s}$, and $\mathbf{n}_p$ denotes the noise component of $\mathbf{x}_p$.

(A2) The elements $\theta(k)$ of $\boldsymbol{\theta}$ are statistically independent of $\mathbf{n}$.

(A3) The noise vector $\mathbf{n}$ has independent elements with a jointly Gaussian pdf.

(A4) The mapping $\mathbf{s}(\boldsymbol{\theta})$ is bijective.

1. Basic Approach

From (5) and the assumptions, we need to find a differentiable deterministic input signal model $\mathbf{s}(\boldsymbol{\theta})$ and the statistics of $\mathbf{n}$ and $\boldsymbol{\theta}$. Our first step is to rewrite the signal model of (7) as

$$x_p = s_p' + n_p' \tag{8}$$

where $\mathbf{s}_p'$ and $\mathbf{n}_p'$ denote approximations to $\mathbf{s}_p$ and $\mathbf{n}_p$ respectively. Noting that $\mathbf{s}$ is a function of the desired output $\boldsymbol{\theta}$, we propose to approximate the nth element of $\mathbf{s}_p$ with an inverse neural net,

$$s_p'(n,w) = \sum_{k=1}^{N_u} w_o(n,k) \cdot f_p(k,w_i) \tag{9}$$

for $1 \leq n \leq N$ where $w_o(n,k)$ denotes the coefficient of $f_p(k,\mathbf{w}_i)$ in the approximation to $s_p(n)$, $f_p(k,\mathbf{w}_i)$ is the kth input or hidden unit in the network, $\mathbf{w}_i$ a vector of weights connecting the input layer to a single hidden layer, and N_u is the number of units feeding the output layer. The vector $\mathbf{w}$ is the concatenation of the input weight vector $\mathbf{w}_i$ with the output weight vector whose elements are $w_o(n,k)$. Note that $f_p(k,\mathbf{w}_i)$ can represent a multinomial function of parameter vector $\boldsymbol{\theta}$ in a functional link network [4] or a hidden unit output in a MLP or radial basis function network [5]. In practice, because of the capabilities of the MLP for approximating derivatives [6,7], a MLP neural network would be the first choice for generating $\mathbf{s}'$.

Before we can find the neural net weight vector $\mathbf{w}$ in (9), we must have an error function to minimize during training, which includes desired output vectors. Since we want $\mathbf{s}_p'$ to approximate $\mathbf{s}_p$, the natural choice for the desired output is $\mathbf{s}_p$. Because $\mathbf{s}_p$ is unavailable, we can try using $\mathbf{x}_p$ as the desired output vector, which yields the error function for the nth output node,

$$E_x(n) = \frac{1}{N_v} \sum_{p=1}^{N_v} [x_p(n) - s_p'(n,w)]^2 \tag{10}$$

The hope is that by minimizing $E_x(n)$ with respect to **w**, we are simultaneously approximately minimizing

$$E_s(n) = \frac{1}{N_v} \sum_{p=1}^{N_v} [s_p(n) - s_p'(n,w)]^2 \tag{11}$$

The signal model is determined from the noisy data by using a gradient approach, such as backpropagation (BP) or output weight optimization (OWO) [8], to minimize $E_x(n)$ with respect to **w**, whose elements are denoted by w(m). The gradient of $E_x(n)$ is

$$\frac{\partial E_x(n)}{\partial w(m)} = \frac{-2}{N_v} \sum_{p=1}^{N_v} [x_p(n) - s_p'(n,w)] \frac{\partial s_p'(n,w)}{\partial w(m)} \tag{12}$$

A MLP network for obtaining the signal model is shown in Fig. 1.

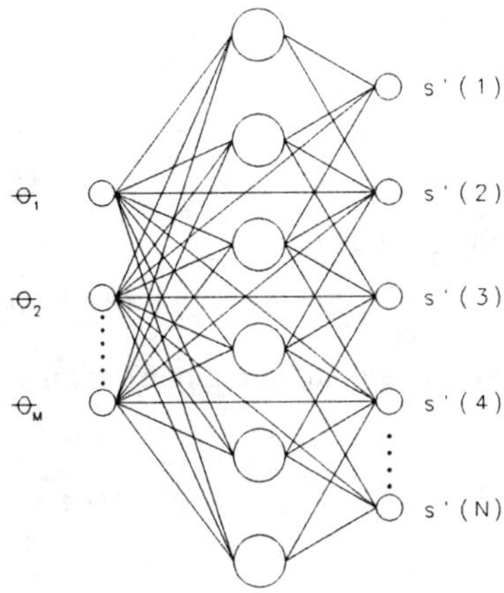

Figure 1. Inverse Network for Signal Modeling

Given a model $\mathbf{s}_p'$ for the deterministic signal component and given the noisy input vectors $\mathbf{x}_p$, we get $\mathbf{n}_p'$ from (8) as

$$\boldsymbol{n}_p' = \boldsymbol{x}_p - \boldsymbol{s}_p'$$

The mean vector and covariance matrix of the noise component are then estimated as

$$\boldsymbol{m}_n' = \frac{1}{N_v} \sum_{p=1}^{N_v} \boldsymbol{n}_p' ,$$

$$C_{nn}' = \frac{1}{N_v} \sum_{p=1}^{N_v} (\boldsymbol{n}_p' - \boldsymbol{m}_n')(\boldsymbol{n}_p' - \boldsymbol{m}_n')^T \tag{13}$$

In addition to the statistics of **n**, calculation of the CRMAP bounds in (3) requires knowledge of the statistics of **θ**, which are

$$\boldsymbol{m}_\theta' = \frac{1}{N_v} \sum_{p=1}^{N_v} \boldsymbol{\theta}_p ,$$

$$C_\theta' = \frac{1}{N_v} \sum_{p=1}^{N_v} (\boldsymbol{\theta}_p - \boldsymbol{m}_\theta')(\boldsymbol{\theta}_p - \boldsymbol{m}_\theta')^T \tag{14}$$

When **θ** is non-Gaussian it is still possible to calculate good bounds using our approach. Theoretical justification of this can be found in [9,10].

C. Convergence of the Method

Since we are minimizing $E_x(n)$ rather than $E_s(n)$ in the previous subsection, it is not clear that $\mathbf{s}_p'$ is a good model for $\mathbf{s}_p$. In this subsection, we analyze the convergence of the gradient of $E_x(n)$ to the gradient of $E_s(n)$ with the following theorem.

Theorem: Assume that the training of the input weight vector $\mathbf{w}_i$ is stopped after iteration number N_{it} which is fixed and bounded, and that training of the output weights $w_o(n,k)$ is allowed to proceed. Then, in the limit as N_v approaches infinity, $\partial E_x(n)/\partial w(m) = \partial E_s(n)/\partial w(m)$.

Proof: The proof is divided into three parts.

1. Energy of the Gradient Noise

Continuing (12) we get

$$\frac{\partial E_x(n)}{\partial w(m)} = \frac{-2}{N_v}\sum_{p=1}^{N_v}[s_p(n) - s_p^{/}(n,w)]\frac{\partial s_p^{/}(n,w)}{\partial w(m)} + e_{nm}$$
$$= \frac{\partial E_s(n)}{\partial w(m)} + e_{nm}, \qquad (15)$$

$$e_{nm} = \frac{2}{N_v}\sum_{p=1}^{N_v} n_p(n)\frac{\partial s_p^{/}(n,w)}{\partial w(m)}$$

Using the fact that

$$E[n_p(n)n_q(n)] = \sigma_n^2 \cdot \delta(p-q)$$

the mean-square of the noise term e_{nm} is evaluated as

$$E[e_{nm}^2] = \frac{4}{N_v^2}\sum_{p=1}^{N_v}\sum_{q=1}^{N_v} E[n_p(n)n_q(n)]\frac{\partial s_p^{/}(n,w)}{\partial w(m)}\frac{\partial s_q^{/}(n,w)}{\partial w(m)}$$
$$= \frac{4\sigma_n^2}{N_v^2}\sum_{p=1}^{N_v}\left(\frac{\partial s_p^{/}(n,w)}{\partial w(m)}\right)^2$$
$$= \frac{4\sigma_n^2 \cdot E_m}{N_v}$$

where E_m is the average energy of the partial derivative of $s_p'(n,\mathbf{w})$ with respect to $w(m)$ and σ_n^2 is the variance of $n(n)$. It remains for us to show that E_m is bounded.

2. E_m for the Output Weight Case

First, assume that $w(m)$ corresponds to an output weight $w_o(n,j)$. Then from (9),

$$\frac{\partial s_p^{/}(n,w)}{\partial w(m)} = f_p(j,w_i),$$

$$E_m = \frac{1}{N_v}\sum_{p=1}^{N_v} f_p^2(j,w_i)$$

Here, the terms in E_m are input or hidden unit activations, or multinomial combinations of inputs for the functional link case. These terms in E_m are bounded if the inverse network inputs (θ_i) are bounded and if the activation functions are bounded. If (9) represents a functional link net, bounding of the inputs produces bounding of the multinomials $f_p(j,\mathbf{w}_i)$ and E_m.

3. E_m for the Input Weight Case

Assume that w(m) corresponds to a weight $w_i(u)$ which feeds unit number j (a hidden unit). Then

$$\frac{\partial s_p'(n,w)}{\partial w(m)} = w_o(n,j)\frac{\partial f_p(j,w_i)}{\partial w_i(u)},$$

$$E_m = \frac{w_o^2(n,j)}{N_v}\cdot\sum_{p=1}^{N_v}\left(\frac{\partial f_p(j,w_i)}{\partial w_i(u)}\right)^2 \tag{16}$$

Functional link nets have no input weight vector and have $E_m = 0$ for the input weight case. For MLP and RBF networks, consider the two factors in E_m separately. In the second factor, partials of $f_p(j,\mathbf{w}_i)$ are bounded for bounded activations. Unfortunately, the $w_o^2(n,j)$ term in the first factor in (16) is not bounded in general. Its value depends upon the initial weight values, the training method chosen, and the learning parameters. This can be solved by stopping the training of the input weight vector $\mathbf{w}_i$ after a bounded number of iterations, N_{it}. After this iteration, the vector $\mathbf{w}$ consists only of elements $w_o(n,k)$, and we have only the output weight case.

In the limit, $E_x(n)$ and $E_s(n)$ have equal derivatives and the same local and global minima.

We have developed CRMAP lower bounds on FLS error variance, given the data file described in the previous section, and using the approach described in [11].

IV. Designing Networks for Flight Load Synthesis

After the analyses summarized in section III, we should expect the training errors FLS networks designed for 17, 20, and 24 inputs to be around $.30 \times 10^8$, $.29 \times 10^8$, and $.28 \times 10^8$ respectively. In this section we design MLPs for the FLS data.

A. MLP Training Results

For all data files three-layer MLP structures are employed. The structures 17-13-9, 20-13-9, and 24-11-9 are for data files F17.DAT, F20.DAT, and F24.DAT, respectively. The number of training iterations for each case is 50. Figures 1, 2, and 3 show training MSE versus number of iterations for all data files. Each figure also has CRMAP bounds obtained from an MLP signal model and from a Modular net [12,13] signal model. The average computation time per iteration, on 486DX 50MHz machine, for data files F17.DAT, F20.DAT, and F24.DAT are 15.43 seconds, 17.57 seconds, and 17.28 seconds, respectively.

From figures 1-3, we see that bounds from modular net signal models are smaller than those from MLP signal models. It remains to be seen which bounds are more correct for testing errors.

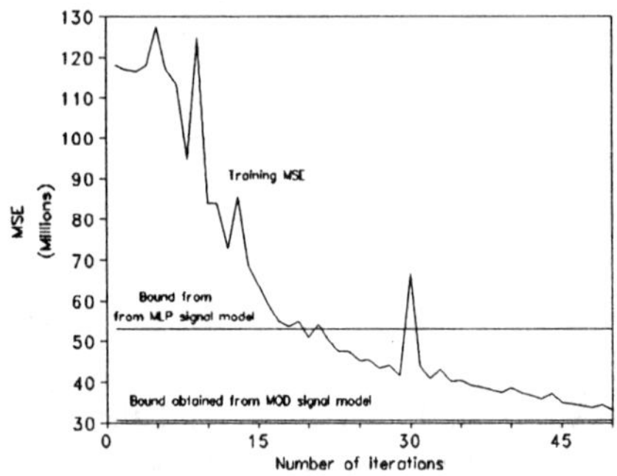

Fig. 1. MLP Training Results and Total Bound for File F17.DAT

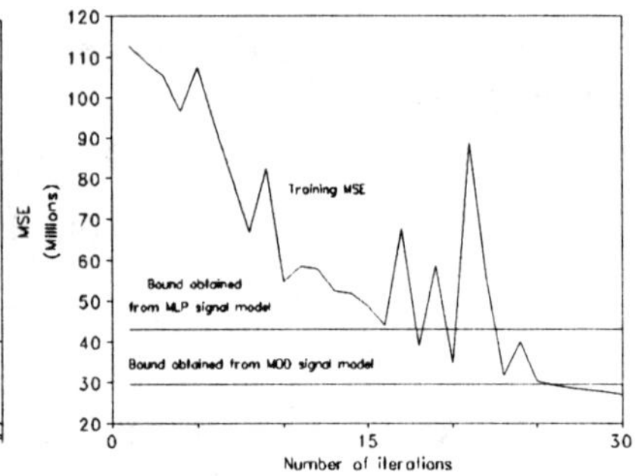

Fig. 2. MLP Training Results and Total Bound for File F20.DAT

B. Testing Results for Designed FLS Networks

Lacking testing data for the FLS problem, we split each original FLS data set into two subsets: one for training and the other for testing. By doing so, we can evaluate the generalization ability of the designed FLS networks. The ratio of the training set to testing set sizes is 3 to 2. The number of training patterns is 2823 and the number of testing patterns is 1922. For the new training data sets, the structures 17-18-9, 20-12-9, and 24-2-9 are used for 17-input, 20-input, and 24-input files, respectively. All networks were trained for 50 iterations. The testing MSEs are obtained when the designed FLS networks are trained after 10, 20, 30, 40, and 50 iterations. The training and testing results are shown in figures 4 to 6, which are respectively for files F17.DAT, F20.DAT, and F24.DAT.

Comparing figures 1 and 4, 2 and 5, and 3 and 6, we see that the modular net signal models provide good lower bounds on testing error. We also see that the FLS networks are close to optimal.

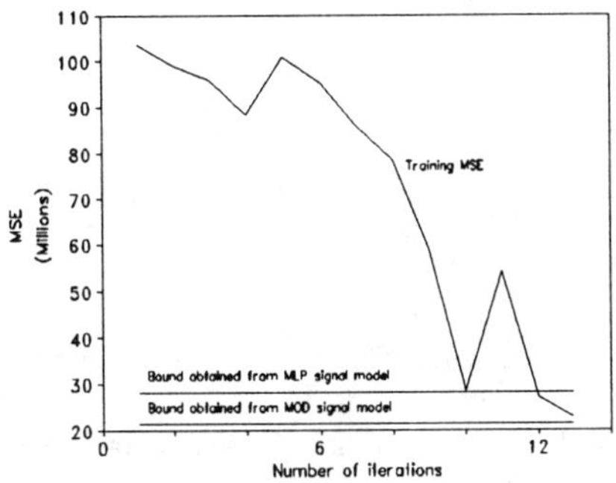

Fig. 3. MLP Training Results and Total Bound for File F24.DAT

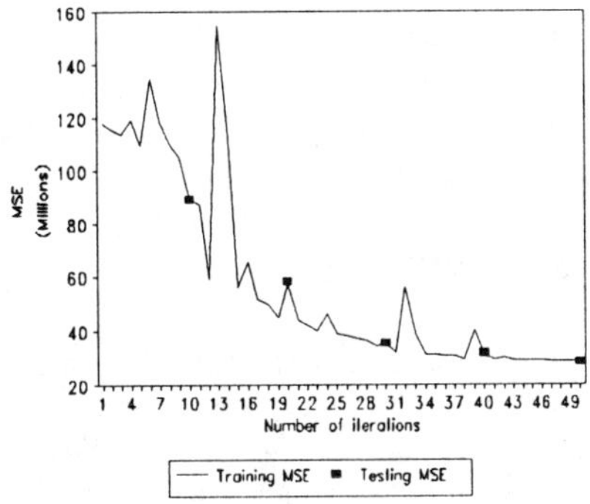

Fig. 4. MLP Training and Testing Results (F17.DAT)

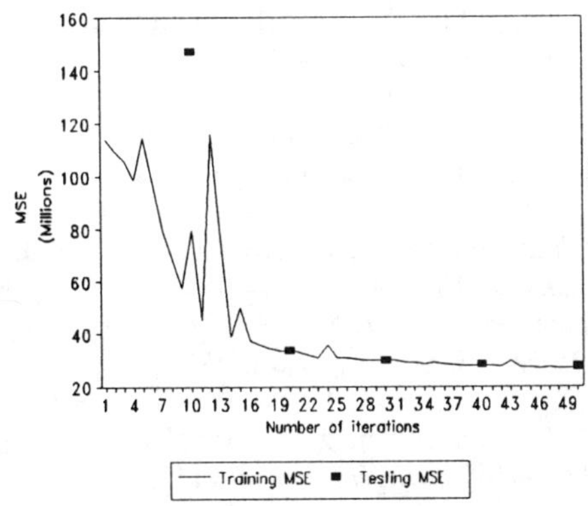

Fig. 5. MLP Training and Testing Results (F20.DAT)

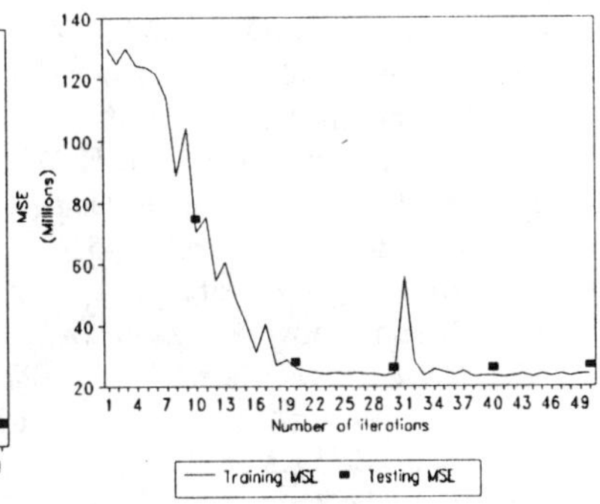

Fig. 6. MLP Training and Testing Results (F24.DAT)

V. CONCLUSIONS

In this paper, a method for calculating CRMAP bounds on neural network testing error variance has been described. The method has been successfully applied to the problem of flight load synthesis in helicopters. MLP neural nets have been trained that approximately attain optimal performance, as indicated by CRMAP bounds. Much work remains to be done. We need to size the inverse networks reliably , in order to produce better bounds. We also need to determine the bijectivity of mappings directly from the training data.

Acknowledgement

This work was funded by the Advanced Technology Program of the state of Texas under grant 003656-063.

VI. REFERENCES

[1] H. L. Van Trees, *Detection, Estimation, and Modulation Theory -Part I*, New York, NY: John Wiley and Sons, 1968.

[2] S.J. Apollo, M.T. Manry, L.S. Allen, and W.D. Lyle, "Optimality of Transforms for Parameter Estimation," *Conference Record of the Twenty-Sixth Annual Asilomar Conference on Signals, Systems, and Computers*, Oct. 1992, vol. 1, pp. 294-298.

[3] M.T. Manry, S.J. Apollo, and Q. Yu, "Minimum Mean Square Estimation and Neural Networks," *Neurocomputing*, vol. 13, September 1996, pp. 59-74.

[4] Y-H Pao, *Adaptive Pattern Recognition and Neural Networks*, Addison-Wesley Publishing Company, Inc., New York, 1989.

[5] S. Chen, C.F.N. Cowan, and P.M. Grant, "Orthogonal Least Squares Learning Algorithm for Radial Basis Function Networks," *IEEE Trans. on Neural Networks*, vol. 2, no. 2, March 1991, pp. 302-309.

[6] K. Hornik, M. Stinchcombe, and H. White, "Universal Approximation of an Unknown Mapping and its Derivatives Using Multilayer Feedforward Networks," *Neural Networks*, Vol. 3, 1990, pp. 551-560.

[7] Y. Ito, "Approximations of Differentiable Functions and Their Derivatives on Compact Sets by Neural Networks," *Math. Scientist*, Vol. 18, 1993, pp. 11-19.

[8] H-H.Chen, M.T. Manry and Hema Chandrasekaran , "A Neural Network Training Algorithm Utilizing Multiple sets of Linear Equations", to appear in *Neurocomputing*.

[9] M.T. Manry, C-H Hsieh, M.S. Dawson, A.K. Fung, and S.J. Apollo, "Cramer Rao Maximum A-Posteriori Bounds on Neural Network training Error for Non-Gaussian Signals and parameters", *International Journal of Intelligent Control and Systems*, Vol. 1, No. 3, 1996, pp. 381-391.

[10] C-H Hsieh, M. T. Manry, and H-H.Chen,, "Cramer Rao Maximum A-Posteriori Bounds for a Finite Number of Non-Gaussian Parameters", *Conference Record of the 30th Asilomar Conference on Signals, Systems, and Computers*, vol. 2, Nov. 1996, pp.1161-1165.

[11] W. Liang, M.T. Manry, Q. Yu, S.J. Apollo, M.S. Dawson, and A.K. Fung, "Bounding the Performance of Neural Network Estimators, Given Only a Set of Training Data," *Conference Record of the Twenty-Eighth Annual Asilomar Conference on Signals, Systems, and Computers*, vol. 2, Nov. 1994, pp.912-916.

[12] Hema Chandrasekaran and Michael T. Manry, "Convergent Design of a Piecewise Linear Neural Network", to appear in *International Joint Conference on Neural Networks (IJCNN'99)*, July 1999.

[13] S.Subbarayan, K.Kim, M.T.Manry, V.Devarajan and H.Chen, "Modular Neural Network Architecture using Piecewise Linear Mapping", *Thirtieth Asilomer Conference on Signals, Systems & Computers,* Vol. 2, pp 1171-1175, Nov 1996.

TWO-DIMENSIONAL JOINT PROCESS ADAPTIVE FILTERING VIA PRINCIPAL COMPONENT SUPPORT REGION

Dai I. Kim, and P. De Wilde
Department of Electrical and Electronic Engineering,
Imperial College of Science, Technology and Medicine,
the University of London, London SW7 2BT, UK
E-mail: d.i.kim@ic.ac.uk

Abstract. This paper describes a 2-D joint process adaptive filtering algorithm using the orthogonal property of the principal component support region in order to speed up the convergence rate. We also introduce the symmetrical sparse support region (SSSR) to reduce the computational burden of the duplicated support region (DSR). Computer simulation results are given to verify the performance of the proposed model.

INTRODUCTION

Image restoration techniques are very important in a variety of fields, such as astronomical imaging, medical imaging, satellite photography, and machine vision. The convergence speed of the stochastic gradient based adaptive filtering algorithms depends on the eigenvalue disparity of the autocorrelation matrix. To overcome this drawback, many two-dimensional (2-D) joint process adaptive filtering algorithms [1], [2], [3], [4] have been proposed to recover or estimate the original scene from degraded observations.

In this paper, a 2-D joint process adaptive filtering algorithm is proposed in order to speed up the convergence rate using the orthogonality of the principal component support region. We also introduce the symmetrical sparse support region (SSSR) to reduce the computational burden of the duplicated support region (DSR) used as the filter input of the 2-D least mean square (LMS). Moreover, the SSSR is decomposed into successive filter stages. Thereby, the joint process filter coefficients are determined independently at each stage and their adaptation is updated by the residual error in the corresponding stage (see Figure 2).

0-7803-5673-X/99/$10.00 © 1999 IEEE

2-D ADAPTIVE FILTER ALGORITHM OF THE SPARSE SUPPORT REGION (SSR)

Sparse Support Region (SSR)

The adjustment of 2-D filter coefficients is restricted by the region of support as shown in Figure 1. Most of 2-D adaptive algorithms [5] utilize the successive support region using the sliding window method. These methods have the high computational complexity. To minimize the computational burden of the DSR, the SSSR is proposed. The size of DSR for the conventional 2-D LMS algorithm is $(N-1) \times (N-1) + N - 1$, while the proposed algorithm for the SSSR requires $(q-1) \times (q-1) + q - 1$ in the case of the first filter stage between adjacent support frames, where N and q denote the filter order and the size of support region, respectively. However, an overlapping support region does not exist for the second filter stage in the case of the SSSR.

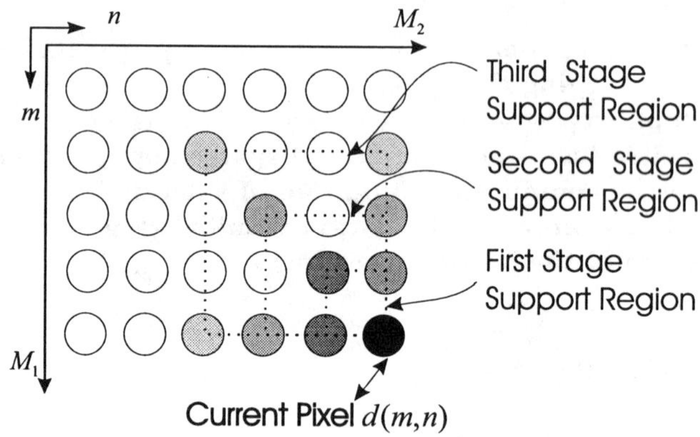

Figure 1: Symmetrical sparse support region (SSSR).

The SSSR shown in Figure 1 are generated by the 2-D delay operator,

$$\mathcal{Z}_{SSSR} = diag[1 \;\; z_1^{-1} \;\; z_1^{-1}z_2^{-1} \;\; z_2^{-1}]. \tag{1}$$

The input-output relationship for the pth filter stage can be defined as follows:

$$\mathbf{x}_k^{(p)}(m,n) = \mathcal{Z}_{SSSR}\mathbf{x}_k^{(p-1)}(m,n) \tag{2}$$

where $\mathbf{x}_k^{(0)}(m,n)$ is the input data vector defined as $[x_k^{(0)}(m,n) \; x_k^{(0)}(m,n) \; x_k^{(0)}(m,n) \; x_k^{(0)}(m,n)]$ and $0 \leq m \leq M_1$, $0 \leq n \leq M_2$ and the filter input signal is generated by the principal component of the neural model in the pre-layer of the adaptive filter.

2-D joint process adaptive filtering algorithm via the principal component support region

Here, we derive the 2-D joint process LMS adaptive filtering algorithm based on the SSSR. The 2-D joint process adaptive filtering algorithm is shown in Figure 2. This model is very similar to the 2-D joint process adaptive lattice model developed by Youla [2]. In [2], the input data of the transversal filter is provided by prediction error fields generated by the quarter plane (QP). We propose the principal component based 2-D joint process structure as shown in Figure 2. The neural model is employed as pre-layer/transform layer to provide the decorrelated signals. This neural model is trained by the generalized Hebbian algorithm (GHA). We derive the Wiener equation. The

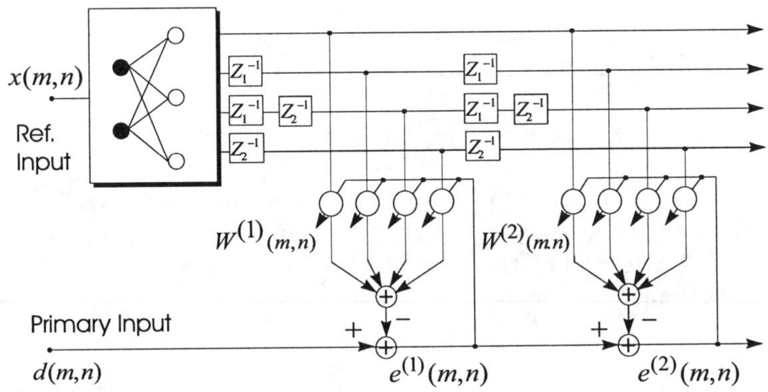

Figure 2: Block diagram of 2-D adaptive filter (2 stages) by the SSSR.

output $y_k^{(p)}(m,n)$ in each filter stage of Figure 2 is given by

$$y_k^{(p)}(m,n) = \sum_{i=0}^{q-1}\sum_{j=0}^{q-1} w_k^{(p)}(i,j)x_k^{(p)}(m-pi, n-pj). \quad (3)$$

The error signal $e_k^{(p)}(m,n)$ in each stage is defined by

$$\begin{aligned} e_k^{(p)}(m,n) &= e_k^{(p-1)}(m,n) - y_k^{(p)}(m,n), \\ e_k^{(p)}(m,n) &= e_k^{(p-1)}(m,n) - \sum_{i=0}^{q-1}\sum_{j=0}^{q-1} w_k^{(p)}(i,j)x_k^{(p)}(m-pi, n-pj) \quad (4) \end{aligned}$$

where q denotes the the size of support region, $d_k^{(p)}(m,n) = e_k^{(p-1)}(m,n)$ and $e_k^{(0)}(m,n)$ is equivalent to $d_k(m,n)$ of the desired signal. Thus, the restored image of the final stage corresponds to

$$\hat{d}_k(m,n) = e_k^{(L)}(m,n)$$

$$= d_k(m,n) - \sum_{p=1}^{L}\sum_{i=0}^{q-1}\sum_{j=0}^{q-1} w_k^{(p)}(i,j)x_k^{(p)}(m-pi,n-pj) \quad (5)$$

The Wiener solution is to obtain a set of optimum weights such that the mean square error (MSE) in (5) is minimized. The MSE is defined as

$$J = E\{[e_k^{(L)}(m,n)]^2\} \quad (6)$$

From (5) the squared error can be expressed as

$$\begin{aligned}
&[e_k^{(L)}(m,n)]^2 \\
&= d_k^2(m,n) - 2\sum_{p=1}^{L}\sum_{i=0}^{q-1}\sum_{j=0}^{q-1} w_k^{(p)}(i,j)d_k(m,n)x_k^{(p)}(m-pi,n-pj) \\
&\quad + \sum_{u=1}^{L}\sum_{v=1}^{L}\sum_{l=0}^{q-1}\sum_{r=0}^{q-1}\sum_{i=0}^{q-1}\sum_{j=0}^{q-1} w_k^{(u)}(l,r)w_k^{(v)}(i,j) \\
&\quad \cdot x_k^{(u)}(m-ul,n-ur)x_k^{(v)}(m-vi,n-vj). \quad (7)
\end{aligned}$$

The input data vector, $\mathbf{x}_k^{(p)}(m,n)$ and the weight vector, $\mathbf{w}_k^{(p)T}(m,n)$ restricted by the size, $q \times q$ of the support region can be defined as a $q^2 \times 1$ column vector to reduce the computational complexity and to use the 1-D LMS algorithm,

$$\begin{aligned}
&\mathbf{x}_k^{(p)}(m,n) \\
&= [x^{(p)}(m,n)\ x^{(p)}(m,n-p)\ \cdots\ x^{(p)}(m,n-p(q-1)), \\
&\quad x^{(p)}(m-p,n)\ x^{(p)}(m-p,n-p)\ x^{(p)} \cdots\ x^{(p)}(m-p,n-p(q-1)), \\
&\quad \vdots \\
&\quad x^{(p)}(m-p(q-1),n)\ x^{(p)}(m-p(q-1),n-p)\ \cdots \\
&\quad x^{(p)}(m-p(q-1),n-p(q-1))]^T \quad (8)
\end{aligned}$$

$$\begin{aligned}
\mathbf{w}_k^{(p)}(m,n) &= [w^{(p)}(0,0)\ w^{(p)}(0,1)\ \cdots\ w^{(p)}(0,q-1), \\
&\quad w^{(p)}(1,0)\ w^{(p)}(1,1)\ \cdots\ w^{(p)}(1,q-1), \\
&\quad \vdots \\
&\quad w^{(p)}(q-1,0)\ w^{(p)}(q-1,1)\ \cdots\ w^{(p)}(q-1,q-1)]^T \quad (9)
\end{aligned}$$

Equation (3), (5) and (7) can be rewritten by the vector form using the (8) and (9),

$$y_k^{(p)}(m,n) = \mathbf{w}_k^{(p)T}(m,n)\mathbf{x}_k^{(p)}(m,n), \quad (10)$$

$$\hat{d}_k(m,n) = e_k^{(L)}(m,n) = d_k(m,n) - \sum_{p=1}^{L}\mathbf{w}_k^{(p)T}(m,n)\mathbf{x}_k^{(p)}(m,n), \quad (11)$$

$$
\begin{aligned}
[e_k^{(L)}(m,n)]^2 &= d_k^2(m,n) - 2\sum_{p=1}^{L} \mathbf{w}_k^{(p)T}(m,n) d_k(m,n)\mathbf{x}_k^{(p)}(m,n) \\
&\quad + \sum_{u=1}^{L}\sum_{v=1}^{L} \mathbf{w}_k^{(u)T}(m,n)\mathbf{x}_k^{(u)}(m,n)\mathbf{x}_k^{(v)T}(m,n)\mathbf{w}_k^{(v)}(m,n)
\end{aligned}
\tag{12}
$$

where $(\cdot)^T$ denotes the transpose operator. By differentiating with respect to $\mathbf{w}_k^{(p)T}(m,n)$ for each filter stage p and taking expectation in (12), we can obtain the normal equation in the independent filter stage

$$
\nabla_J = -2E[d_k(m,n)\mathbf{x}_k^{(p)}(m,n)] + 2\mathbf{w}_k^{(p)}(m,n)E[\mathbf{x}_k^{(p)}(m,n)\mathbf{x}_k^{(p)T}(m,n)]. \tag{13}
$$

Hence, the optimum weight vector of the SSSR algorithm can obtained by nullifying (13)

$$
\mathbf{w}_{opt}^{(p)} = \mathcal{R}^{(p)^{-1}}\mathcal{P}^{(p)} \tag{14}
$$

The matrix $\mathcal{P}^{(p)}$ of (14) is defined as the crosscorrelation matrix, $E[d_k^{(p)}(m,n)\mathbf{x}_k^{(p)}(m,n)]$ between the desired response $d_k^{(p)}(m,n)$ and the reference input such that

$$
\begin{aligned}
&\mathcal{P}^{(p)} \\
&= [d_k^{(p)}(m,n)x^{(p)}(m,n)\ d_k^{(p)}(m,n)x^{(p)}(m,n-p)\cdots \\
&\quad d_k^{(p)}(m,n)x^{(p)}(m,n-p(q-1)), \\
&\quad d_k^{(p)}(m,n)x^{(p)}(m-p,n)\ d_k^{(p)}(m,n)x^{(p)}(m-p,n-p)\cdots \\
&\quad d_k^{(p)}(m,n)x^{(p)}(m-p,n-p(q-1)), \\
&\quad \vdots \\
&\quad d_k^{(p)}(m,n)x^{(p)}(m-p(q-1),n)\ d_k^{(p)}(m,n)x^{(p)}(m-p(q-1),n-p)\cdots \\
&\quad d_k^{(p)}(m,n)x^{(p)}(m-p(q-1),n-p(q-1))]^T.
\end{aligned}
\tag{15}
$$

The matrix $\mathcal{R}^{(p)}$ of (14) as the autocorrelation matrix has the form

$$
\mathcal{R}^{(\mathbf{p})} = \begin{bmatrix}
r_{0,0}^{(p)} & r_{0,p}^{(p)} & \cdots & r_{p(q-1),p(q-1)}^{(p)} \\
r_{0,p}^{(p)} & r_{0,0}^{(p)} & \cdots & r_{p(q-1),pq}^{(p)} \\
\vdots & \vdots & \ddots & \vdots \\
r_{p(q-1),p(q-1)}^{(p)} & r_{p(q-1),pq}^{(p)} & \cdots & r_{0,0}^{(p)}
\end{bmatrix}. \tag{16}
$$

where the matrix element $r_{\alpha\beta} = E[x(m-l,n-r)x(m-u,n-v)]$, $\alpha = u-l$ and $\beta = v-r$ and the $\mathcal{R}^{(p)}$ is Toeplitz matrix. The Wiener-Hopf equation of the proposed SSSR algorithm is completed through (14) to (16).

Applying the method of stochastic gradient descent to the instantaneous squared error, we can obtain a recursive weight update equation of LMS algorithm by SSSR in each filter stage

$$\mathbf{w}_{k+1}^{(p)}(m,n) = \mathbf{w}_{k}^{(p)}(m,n) + \frac{\mathcal{U}_s}{\rho_k^{(p)}(m,n)}\mathbf{x}_k^{(p)}(m,n)e_k^{(p)}(m,n), \quad (17)$$

$$\rho_k^{(p)}(m,n) = \beta\rho_{k-1}^{(p)}(m,n) + (1-\beta)\mathbf{x}_k^{(p)T}(m,n)\mathbf{x}_k^{(p)}(m,n) \quad (18)$$

where $\mathcal{U}_s$ is the fixed step size with $diag[\mu_1, \mu_2, \mu_3, \cdots, \mu_p]$ and $\rho_k^{(p)}(m,n)$ is the power estimation in each filter stage. The recursion of (18) proceeds using the raster scan type of the images, i.e., from left to right and from top to bottom.

COMPUTER SIMULATION RESULTS FOR THE IMAGE RESTORATION

The experimental configuration is shown in Figure 3 for the noise cancellation. The adaptive filtering structure employs the structure of Figure 2. The reference input $x(m,n)$ is a colored signal generated by passing white Gaussian noise $w(m,n)$ through an all pole filter $1/A(z_1,z_2)$, the autoregressive (AR) model. The primary input $d(m,n)$ consists of a signal $s(m,n)$ plus a noise signal $v(m,n)$, which is correlated to $x(m,n)$, where m and n take all values from 0 to M_1-1, M_2-1, respectively.

The signal-to-noise ratio (SNR) and MSE are used to check the performance of the proposed algorithm as follows:

$$SNR = 10log_{10}\left\{\frac{\sum_{m=0}^{M_1-1}\sum_{n=0}^{M_2-1}s^2(m,n)}{\sum_{m=0}^{M-1}\sum_{m=0}^{N-1}[s(m,n)-\hat{d}(m,n)]^2}\right\}, \quad (19)$$

$$MSE = \frac{\sum_{m=0}^{M_1-1}\sum_{m=0}^{M_2-1}[s(m,n)-\hat{d}(m,n)]^2}{M_1M_2}. \quad (20)$$

The MSE and the enhancement of SNR are shown in Table 1. The MSEs were estimated by averaging between $\hat{d}(200,1)$ and $\hat{d}(256,256)$ pixels to verify the misadjustment in the steady state. As shown in Table 1, the superior performance is presented comparing with the conventional 2-D LMS algorithm as well as 2-D lattice model developed by Youla. Simulation results are also given in Figure 4. Figure 4 (f) was restored from the blurred image of SNR=5.52 dB.

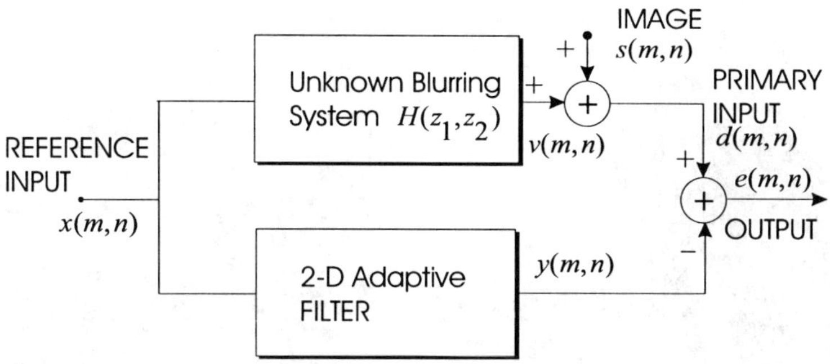

Figure 3: Joint process estimation using the 2-D adaptive filter.

Table 1: Signal-to-Noise Ratio (SNR) Results and Mean Square Error (MSE) of the Image Restoration

Algorithm	Blurred Image SNR(dB)	Restored Image SNR(dB)	MSE
2-D LMS	1.08	7.14	0.0128
	5.52	11.46	0.0036
Youla Model[2]	1.08	8.20	0.0102
(Lattice Model)	5.52	12.11	0.0038
2-D LMS	1.08	10.74	0.0075
(PCA-SSSR)	5.52	14.03	0.0027

CONCLUSIONS

In this paper, a 2-D adaptive filtering structure was developed based on the principal component support region. The superior performance was presented comparing with the conventional 2-D LMS algorithm as well as a 2-D lattice model developed by Youla for image restoration. However, the proposed model has a disadvantage which is the computational complexity due to a neural model.

REFERENCES

[1] R. R. Gharieb, T. I. Haweel and H. M. Ibrahim, "A two-dimensional adaptive lattice (TDAL) structure for noise cancelling," *Int. J. Electronics*, vol. 78, No. 3, pp. 483-492, 1995.

[2] H. Youlal, M. Janati-I and M. Najim, "Two-dimensional joint process lattice for adaptive restoration of image," *IEEE Trans. on Image Processing*, vol. 1, No. 3, pp. 366-378, July. 1992.

Figure.4:Re restored images: (a): original image, (b):blurred image (SNR 1.08 dB), (c): 2-D LMS (SNR 7.14 dB), (d): 2-D lattice model (SNR 8.20 dB), (e): 2-D LMS by SSSR (SNR 10.74 dB), (f): 2-D LMS by SSSR (SNR 14.03 dB)

[3] A. H. Kayran, "Improved 2D joint process lattice for adaptive restoration of images," *Electronics Letters*, vol. 34, No. 4, pp. 350-351, 1998.

[4] T. Nakachi, K. Yamashita and N. Hamada, "A 2-D adaptive joint-process estimator for image restoration," *IEICE Trans. Fundamentals*, vol. E80-A, No. 1, pp. 140-147, 1997.

[5] M. M. Hadhoud and D. W. Thomas, "The two-dimensional adaptive LMS (TDLMS) algorithm," *IEEE Trans. on Circuit and Systems*, vol. 35, No. 5, pp. 485-494, May 1988.

ON CONDITION MONITORING OF EXHAUST VALVES IN MARINE DIESEL ENGINES

T. L. Fog[§] L. K. Hansen[‡], J. Larsen[‡], H. S. Hansen[§], L. B. Madsen[§], P. Sørensen[§], E. R. Hansen[§], and P. S. Pedersen[§]

§ : Research & Development, MAN B&W Diesel A/S
Teglholmsgade 41, DK-2450, Copenhagen SV, Denmark.
E-mail: tof, hsh, lbm, pcs, er, psp@manbw.dk. Website: www.manbw.dk

‡ : THOR Center for Neuroinformatics, Department of Mathematical Modeling, Bldg. 321, Technical University of Denmark, DK-2800 Lyngby, Denmark
E-mail: lkhansen, jl@imm.dtu.dk. Website: eivind.imm.dtu.dk/thor

Abstract. The feasibility of non-invasive characterisation of exhaust valve conditions in large marine diesel engines, were experimentally investigated on a four cylinder 500 mm bore 2-stroke marine diesel engine at MAN B&W Diesel's Research Center in Copenhagen, Denmark. The experiments comprised three different valve conditions, where two were concerned with artificially induced valve burn-through situations. The primary potential monitoring measurements were vibration and structure-borne stress waves, also known as Acoustic Emission (AE). Results have shown that the AE signals have a major advantage over other involved sensors, indicating sensitivity to both mechanical and fluid-mechanical combustion related activity. Recorded data has been preprocessed and features extracted using Principal Component Analysis (PCA). From a number of applied heuristics and statistics, a search for the optimal sub-space of principal components to use, have been carried out. The chosen feature-space have been used for classification of involved exhaust valve conditions by applying both regularized feed-forward neural classifiers and linear discriminators. The complexity of the neural networks have further been optimized by Optimal Brain Damage (OBD) pruning, leading to increased generalization.

INTRODUCTION

The exhaust valve is an important part of a combustion engine. Its main purpose is to seal during the compression stroke, thereby providing maximum

0-7803-5673-X/99/$10.00 © 1999 IEEE

combustion pressure for maximum engine performance. Further, during the expansion stroke at engine-builder specific timing, to lift its seat and thereby provide scavenge (from scavenge air system) of the polluted air in the cylinder and fresh air intake for the following compression and combustion.

Valve burn-through, or leakage, can be caused by dent marks from deposits (combustion particles) on the sealing face or material damage from for instance knocking [14]. Leakage's causes reductions of both pressure and power (i.e. performance) in the cylinder(s). Moreover, the high temperatures and gas pressure impulses during the combustion phase, can further aggravate the exhaust valve condition. In case of valve leakage, the valve can in some cases be reconditioned and re-installed for further use. But in severe cases it has to be replaced. In both situations, it is a time consuming operation and it might cause serious transport delay. Different attempts have been made to investigate possibilities to monitor the exhaust valve condition by means of exciting vibrations from the exhaust valve housing [1, 16].

Having a condition monitoring system which could offer industrial users the ability to carry out planned maintenance by indicating the level of degradation before a fault occurs or efficiency is reduced to unacceptable levels, is of great interest. Further, integration of an adaptive condition monitoring system with other electronic engine control and maintenance systems, is an important progress towards fully automated maintenance scheduling and the concept of the *Intelligent Engine* [13].

To investigate possibilities for detection of exhaust valve burn-through, experiments were carried out on MAN B&W Diesel's Research Engine in Copenhagen. The aim of the experiments was to establish whether it was possible to derive any information's of exhaust valve leakage under different engine and valve conditions. Three different valve conditions were considered; normal valve, a small valve leakage, and a large valve leakage. These were evaluated over a broad range of engine operating conditions.

Since engine signals are quasi-periodic the recorded time-series have been preprocessed for improved stationarity. Extraction of features of importance have been investigated using principal component analysis (PCA) and transformations carried out for investigation of the optimal feature sub-space. From a number of applied heuristics and statistic methods as suggested in literature [7], a search for the optimal set of principal components have been carried out. Two types of models are investigated for automatic classification of involved exhaust valve conditions, based on features derived via PCA. These comprises a linear discriminant, similar to Fisher's discriminant for two classes, and a Bayes optimal classifier using a neural network which outputs estimates of the posterior probabilities of class membership. The regularized neural classifier has been trained via second order learning and model complexity optimized using Optimal Brain Damage Pruning (OBD).

EXPERIMENTAL INVESTIGATIONS

The Research Engine is a 2-stroke 4-cylinder marine diesel engine with an approximate max. output on 10,000 BHP[1]. The aim of the experiment was to establish whether it was possible to derive informations of exhaust valve leakage under different engine and valve conditions. Three different valve conditions were considered: 1) normal valve, 2) relatively small grinded valve leak, with a cross-sectional area of approximately $4mm^2$, 3) large grinded valve leak, with a cross-sectional area of approximately $20mm^2$.

For each valve condition, measurements were obtained from 8 different engine operating conditions, varying both speed and load. As a side experiment, misfire was simulated by taking off combustion on the cylinder unit, which thereby simply pumped air. During the experiments, "snapshots" of time-series over a number of piston revolutions were captured by data acquisition systems for all involved engine and valve conditions. The signals obtained comprised the primary potential monitoring measurements, which were vibration and structure-borne stress waves, also known as Acoustic Emission (AE) (see e.g. [12, 10]). Further, shaft timing signals revealing 2048 angle

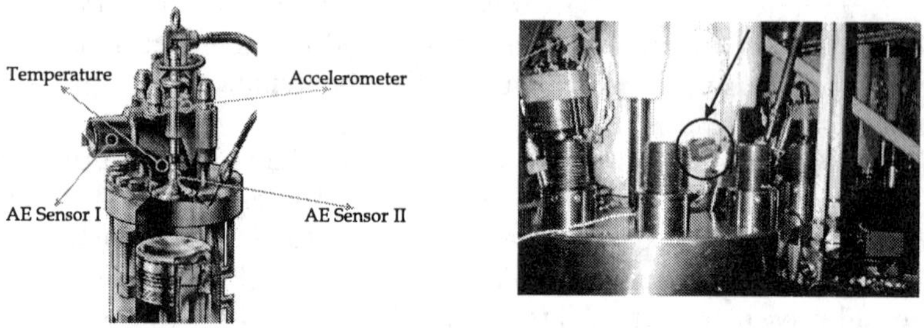

Figure 1: Left panel : Approximate positions of the two AE sensors on the outside surface of the valve housing, the placement of an accelerometer, and the approximate position of a thermocouple. Right panel : AE sensor mounted in a magnetic hold-down on the valve housing.

specific pulses per piston cycle and a TDC[2] signal, together with cylinder pressure, exhaust gas temperature and other pressure measurements. All recordings were obtained from cylinder 1 only which was the unit used for the experiments. The sensors were positioned on the exhaust valve housing as indicated in Fig.1.

The 8 involved engine operating conditions where measurements were recorded for each valve condition comprised:

1.	25% load, propeller curve	2.	25% load, propeller curve, misfire
3.	50% load, propeller curve	4.	50% load, propeller curve, misfire
5.	75% load, propeller curve	6.	100% load, propeller curve
7.	25% load, generator curve	8.	25% load, generator curve, misfire

[1] Brake Horse Power

[2] Top Dead Center

Here, the term, propeller curve, refers to engine propulsion usage with a fixed pitch propeller while generator curve refers to constant speed as using the engine for stationary power plant purposes. Since the Research Engine is a stationary plant the propeller was simulated by a water brake.

The measurements recorded, initiated the ability to investigate data from different engine conditions and with the same load and speed values. Hence, in this experiment, engine condition 1, 2 and 7, 8 have the same engine load (i.e., 25% load), while engine condition 6, 7, and 8 have the same engine speed values. Conditions 3, 4, and 5, further yielded possibilities to broaden the investigations over a larger range of different engine operating conditions.

While vibrations and pressure measurements are traditional widely used techniques for monitoring reciprocating machinery such as engines and compressors, AE has only recently been applied for engine monitoring. Some recent results can be found in [3, 4, 15]. AE has further proven advantageous to the other measurements obtained, indicating sensitivity to mechanical and fluid-mechanical events, where acceleration is sensitive only to the mechanical activity of involved machine parts. AE has proven capability to monitor valve activity and condition and fuel injection. Interference of AE events from adjacent cylinders have been found to be minimal, hence allowing easy isolation of mechanical events related to the monitored cylinder. Another important result is that detection of AE events seems possible anywhere on the valve housing from which the signals were acquired. This makes AE measurements relatively easy to carry out, since the non-intrusive measurements are not too sensitive to the positioning of the sensors. The results obtained by recording RMS AE signals, shows good discrimination between both normal valve condition and the small valve leak, and further the visual "difference" is also clearly observed in the time-series between the small and the large valve leak (cf. Fig. 2).

FEATURE EXTRACTION

Since engine signals in general are quasi-periodic the RMS AE time-series were trigger sampled by use of a shaft timing signal obtained from an angle encoder pulse signal yielding 2048 angle positions per piston cycle. Hence, stationarity has been obtained revealing time-series consisting of 2048 samples per piston cycle. While the trigger sampling process corresponds to a huge dimensionality reduction, we want to explore whether it is possible to represent, for the problem, relevant features in yet a lower dimension of *feature-space*. In situations where the input dimension of data is large and the data vectors are highly correlated, PCA might prove useful as a dimensionality reduction method. The main objective of PCA is to provide a lower dimensional data description by projection of the data vectors onto the eigendirections corresponding to the *largest* eigenvalues of the sample covariance matrix $\widehat{\mathbf{\Sigma}}$. Consider the *centered* data matrix, $\widetilde{\mathbf{X}}$, holding data observations of preprocessed RMS AE time-series, as column vectors. To

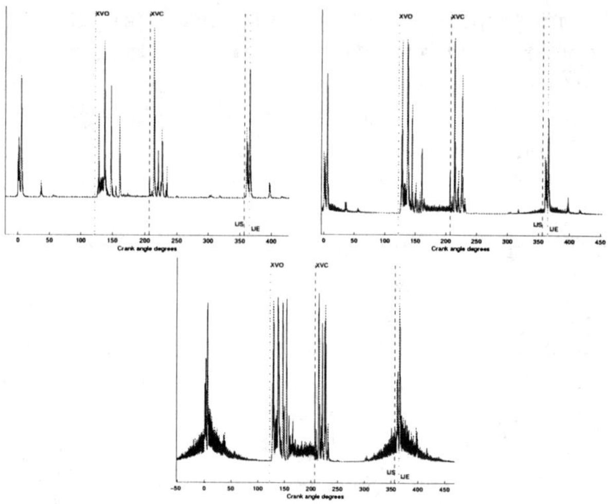

Figure 2: RMS AE time-series from 25% engine load (propeller curve). Time-series obtained from AE sensor II (see Fig.1) for the large valve leakage. Dotted vertical lines indicate timing of mechanical combustion related events, such as exhaust valve open (XVO), exhaust valve close (XVC), injection start (IJS) and injection end (IJE). Left panel demonstrates RMS AE for normal valve condition while the middle and right panel illustrates RMS AE results for the small and large valve leak, respectively.

carry out PCA, we use single value decomposition (SVD)

$$\widetilde{\mathbf{X}} = \mathbf{U}\mathbf{\Lambda}^{1/2}\mathbf{V}^T \tag{1}$$

decomposing the centered data matrix, $\widetilde{\mathbf{X}}$, into a product of the eigenvectors of the constellation, $\widetilde{\mathbf{X}}\widetilde{\mathbf{X}}^T$, which is represented as column vectors by $\mathbf{U}$, $\widetilde{\mathbf{X}}^T\widetilde{\mathbf{X}}$, represented as column vectors by $\mathbf{V}$ and a diagonal matrix, $\mathbf{\Lambda}^{1/2}$ holding the standard deviations, i.e. the square root of the eigenvalues. The principal components (the PC's) can now be computed from

$$\mathbf{PC} = \mathbf{U}^T\widetilde{\mathbf{X}} = \mathbf{\Lambda}^{1/2}\mathbf{V}^T \tag{2}$$

The PC transformation from SVD produces transformed data vectors which holds uncorrelated components, ordered according to their variance level. To find a representative sub-space with dimension $k < p$, the k largest eigenvalues from the eigendecomposition of the covariance matrix is sought, as a choice for mapping a p-dimensional data space into a k-dimensional space spanning the new feature-space. From SVD, the eigenvalues ordered according to their magnitude is found in the diagonal of $\mathbf{\Lambda}$. These are illustrated in Fig.3. The main objective of applying feature extraction in the form of a dimensionality reduction technique (here PCA), is to preserve as much *relevant* information as possible in the new transformed data. Selection of the optimal sub-space of features, constitutes though a non-trivial problem.

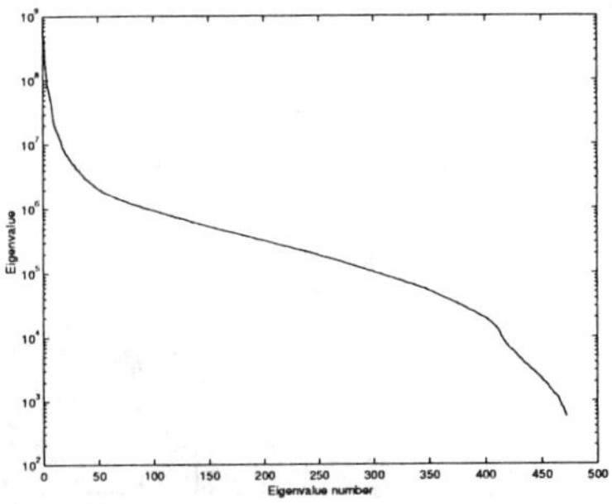

Method	No. of PC's retained
Broken stick	17
Average root	36, 46
SCREE-plot	70
Bartlett's test	66
Velicer's method	76
Neural network	61
Linear model	65

Figure 3: Left panel : $n-1=472$ ordered eigenvalues of the $\widetilde{\mathbf{X}}^T\widetilde{\mathbf{X}}$ covariance matrix with rank$(\widehat{\mathbf{\Sigma}}_{min}) \leq (n-1)$ from the synchronized RMS AE data. Right panel : The table indicates various methods used as selection criterion for retaining PC's. Note that the neural net and linear models are supervised schemes (by valve conditions) whereas the remaining are unsupervised.

To evaluate selection criteria's for retaining a number of PC's, we have employed various heuristics and statistics as suggested in literature (see e.g. [7, 8]). The suggested number of PC's to retain for the applied selection methods, is given in the table in Fig.3. For comparison, the number of PC's to select have been investigated in the context of minimizing a generalization term for involved neural network classifiers and linear discriminators, similar to Fisher's discriminant for 2 classes. This is similar to the work reported in [5]. Some ambiguity among the retained number of PC's is present. We recommend that the number of PC's to retain is 60–65, as suggested by the *supervised* selection methods (linear model, neural network), see Fig. 4. The PC subspace selected by these methods is optimal with respect to the overall objective; minimizing the classification error. But in the following we will base the classifiers on a 76 dimensional subspace corresponding to the largest subspace suggested by any of the methods.

PATTERN RECOGNITION

Our classification tasks aim at classifying levels of degradation of the three involved exhaust valve conditions. The data set, consists of 473 observations each holding 76 chosen features, corresponding to the largest suggested PC subspace to choose.

For performance comparison, in sense of misclassification rate, we employ two different classifiers; a linear discriminant, which for 2 classes is similar to Fisher's discriminant, and a regularized softmax [2] neural classifier. The linear model which has been chosen is the so-called *best linear rule* [11]. A discriminant function can be expressed as [11]

$$F_i(\mathbf{x}) = \left[-2\widehat{\mu}_i^T\widehat{\mathbf{\Sigma}}^{-1}\mathbf{x} + \widehat{\mu}_i^T\widehat{\mathbf{\Sigma}}^{-1}\widehat{\mu}_i - 2\log\pi_i\right], \tag{3}$$

and the class membership is given by $C_i = \arg\min_i F_i(\mathbf{x}^n)$. In Eq.(3) the class mean is found as $\widehat{\mu}_i = N_{\text{train},i}^{-1} \sum_{n_i=1}^{N_{\text{train},i}} \mathbf{x}^{n_i}$ with $N_{\text{train},i}$ denoting the number of training data examples in each class and $\mathbf{x}^{n_i}$ denoting data example n belonging to class i, drawn from the pool of data. The covariance matrix, $\widehat{\boldsymbol{\Sigma}}$ is estimated as

$$\widehat{\boldsymbol{\Sigma}} = \sum_{i=1}^{c} \frac{n_i}{N_{\text{train,tot}}} \widehat{\boldsymbol{\Sigma}}_i = \frac{1}{N_{\text{train,tot}}} \sum_{i=1}^{c} \sum_{n_i=1}^{N_{\text{train},i}} (\mathbf{x}^{n_i} - \widehat{\mu}_i)(\mathbf{x}^{n_i} - \widehat{\mu}_i)^T \tag{4}$$

under the assumption of equal covariances. $N_{\text{train,tot}}$ is here the sum of data examples for each class involved, i.e. $\sum_{i=1}^{c} N_{\text{train},i}$. Further the prior, $\widehat{\pi}_i$, is estimated from data as [11] $\widehat{\pi}_i = N_{\text{train},i}/N_{\text{train,tot}}$.

Following [6] for the construction of the neural classifier, we aim at modeling the posterior probabilities, $p(C_i|\mathbf{x}), i = 1 \cdots c$, of class membership by assigning class label C_i to $\mathbf{x}$ if $i = \arg\max_j p(C_j|\mathbf{x})$. Above, C_i refers to the i'th class, c to the number of classes, and $\mathbf{x}$ is a PC feature vector. Since the neural network outputs, $\widehat{y}_i$, are estimates of the posterior probabilities which sum to 1, we need only to estimate $c-1$ posterior probabilities. Our initial 2-layer feed-forward network consists of $n_I = 76$ inputs, corresponding to our principal components, $n_H = 4$ hidden neurons, and $n_O = c - 1 = 2$ outputs reflecting the 3 classes of exhaust valve conditions involved. We next define our network as : $V_j(\mathbf{x}^n) = \tanh\left(\sum_{k=1}^{n_I} w_{jk} x_k^n + w_{j0}\right)$ and $\phi_i(\mathbf{x}^n) = \sum_{j=1}^{n_H} W_{ij} V_j(\mathbf{x}^n) + W_{i0}$ where V_j denotes the activity of the hidden neurons and ϕ_i the output activity. The network parameters are represented by the input-hidden weight-space, w_{jk}, and the hidden-output weight-space, W_{ij}. W_{i0} and w_{j0} are thresholds in the respective network layers. To be able to interpret the outputs as posterior probabilities, we use a modified normalized exponential activation function, $\widehat{y}_i^n = \exp(\phi_i(\mathbf{x}^n)) / \sum_{j=1}^{c-1} \exp(\phi_j(\mathbf{x}^n)) + 1$ and $\widehat{y}_c^n = 1 - \sum_{i=1}^{c-1} \widehat{y}_i^n$, similar to softmax [2]. All the input vectors, $\mathbf{x}^n$, in the entire data set has been assigned to targets, y_i^n, labeled for class membership as

$$y_i^n = \begin{cases} 1 & \text{if} \quad \mathbf{x}^n \in C_i \\ 0 & \text{if} \quad \mathbf{x}^n \notin C_i \end{cases} \tag{5}$$

with i denoting the ith class. When training the network, we divide the data into a training set consisting $N_{\text{train}} = 70\%$ of available observations, and the remaining 30% kept in an independent evaluation set. Training is carried out by minimizing the negative log-likelihood, where the likelihood parameters is given by $P\left(\mathcal{D}_{\text{train}}|\mathbf{w}\right) = \prod_{n=1}^{N_{\text{train}}} \prod_{i=1}^{c} (\widehat{y}_i^n)^{y_i^n}$. The training error is thus found by $E_{\mathcal{D}_{\text{train}}}(\mathbf{w}) = -N_{\text{train}}^{-1} \log P\left(\mathcal{D}_{\text{train}}|\mathbf{w}\right) = -N_{\text{train}}^{-1} \sum_{n=1}^{N_{\text{train}}} \epsilon(\mathbf{y}^n, \widehat{\mathbf{y}}^n, \mathbf{w})$ with $\epsilon(\mathbf{y}^n, \widehat{\mathbf{y}}^n, \mathbf{w}) = \log\left(1 + \sum_{j=1}^{c-1} \exp(\phi_j(\mathbf{x}))\right) - \sum_{i=1}^{c-1} y_i \phi_i(\mathbf{x})$. In order to reduce overfitting, the networks are trained with a regularized cost function consisting of the training error augmented by a weight decay regularization term. Weight decay parameters have been chosen separately for each layer, and both have been optimized during initial investigations. The weight decay parameters were found to be $\alpha_H = 0.21$ for the input-hidden layer and

$\alpha_O = 1.6 \cdot 10^{-5}$ for the hidden-output layer. Next, using the optimal weight decay parameters, 10 fully connected networks were pruned by Optimal Brain Damage (OBD) [9], by removing two weights in turn. The training scheme consisted of a hybrid first-order/second-order approach, initialized by a few gradient descent steps and followed by pseudo Gauss-Newton diagonal approximation to the Hessian in the remaining iterations.

The two types of classifiers, i.e. the non-linear neural network classifier and the linear discriminant, are not directly comparable since they are derived under different assumptions and with different cost functions. One way to compare performance is to investigate each of the classifiers generalization performance on the independent test set in the form of the number of misclassifications. We find the "best" network around 104 parameters, where the initial network consisted of 323 parameters. Further, among the ensemble of the 10 pruned networks, 4 nets had pruned away 29 inputs, while the remaining had removed 21-28 inputs. More than 50% of the networks further agree to delete the *same* 28 inputs.

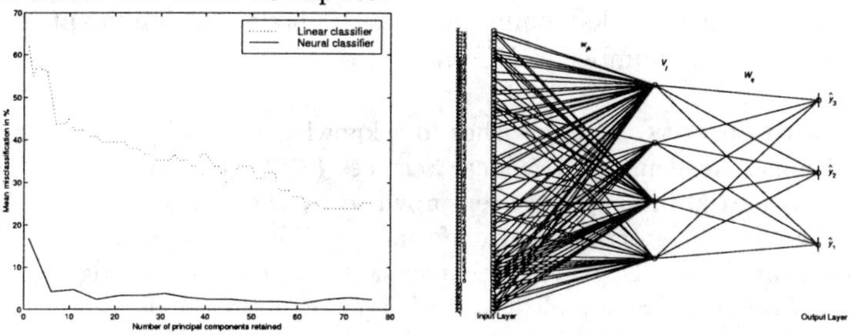

Figure 4: Left: Comparison of generalization performance for the linear model and neural networks, evaluated as the mean misclassification rate on the independent test set vs. the number of principal components utilized for the models. Right: Typical net among the 10 best nets found via OBD pruning. The number of parameters has been reduced from 323 to only 104 parameters. Further, 29 inputs are unused.

CONCLUSIONS

We have in this work carried out experimental investigations of exhaust valve burn-through on a large marine diesel engine situated in Copenhagen, Denmark. The engine involved in the experiments is MAN B&W Diesel's Research Engine, which is a 2-stroke 500 mm bore 4-cylinder marine diesel engine with an approximate output on 10.000 BHP. Measurements have been recorded for 3 different exhaust valve conditions (normal valve, small valve leak, large valve leak) over a broad range of engine operating conditions, varying speed and load and simulating misfire. Results from the measurements show that Acoustic Emissions (AE) have a major advantage over other involved more traditional used sensors (pressures, temperatures, vibrations) for monitoring reciprocating machinery.

Attention in this work, has been given to classification of different degradation levels of exhaust valve leakages as normal condition, acceptable fault, and unacceptable fault. Two different classifiers have been involved, namely a linear classifier, similar to Fisher's discriminant for two classes, and a neural classifier. Results from using time-series obtained by non-intrusive AE sensors, show that the neural classifier had a largely better performance than the linear discriminant (cf. Fig. 4). The neural classifier had a misclassification rate as low as 2%, while the linear discriminant yielded a lowest misclassification rate on 25%. By OBD pruning of the neural classifier, the number of parameters were reduced from initially 323 to 104. This reduction have reduced the computational burden, but the performance of the classifier, measured in terms of misclassification rate, seems not to deviate significantly even with smaller networks. This suggestion is further supported by generalization based investigations of the PC feature sub-space to choose. Despite the various heuristics and statistics involved to indicate a suitable PC subspace (cf. table in right panel of Fig. 3), the generalization performance based PCA investigations (cf. left panel in Fig. 4), shows no significant difference by choosing a lower number of PC's.

Acknowledgements. T. Fog wishes to acknowledge the European Commission and the Danish Academy of Technical Sciences (ATV) for supporting this work. Additionally, partial funding has been provided by the Danish Research Councils through the THOR Center for Neuroinformatics. We also want to address special thanks to Ewan Brown, Dept. of Mechanical & Chemical Engineering, Heriot-Watt University, Edinburgh, Scotland, UK, for his participation in the involved exhaust valve experiments.

REFERENCES

[1] O. Bardou, M. Watremetz and M. Sidahmed, "Early Detection of Leakage in Outlet Valves of High-power Diesel Engine using Vibration Analysis," in **Proceedings of the Third Internalional Machinery Monitoring and Diagnostics Conference 1991**, Las Vegas, NV., 1991, pp. 249–253.

[2] J. Bridle, "Probabilistic Interpretation of Feedforward Classification Network Outputs with Relation-ships to Statistical Pattern Recognition," **Neurocomputing Algorithms, Architectures and Applications**, vol. 6, pp. 227–236, 1990.

[3] T. L. Fog, E. R. Brown, H. S. Hansen, L. B. Madsen, P. Sørensen, E. R. Hansen, J. A. Steel, R. L. Reuben and P. S. Pedersen, "Exhaust Valve Leakage Detection in Large Marine Diesel Engines," in **In Proceedings of the 11th International Conference and Exhibition on Condition Monitoring and Diagnostic Engineering Management (COMADEM '98)**, Launceston, Tasmania, Australia, 1998, vol. 1, pp. 269–278.

[4] J. D. Gill, R. L. Reuben, M. Scaife, E. R. Brown and J. A. Steel, "Detection of Diesel Engine Faults using Acoustic Emission," in **Proceedings of the Planned Maintenance, Reliability and Quality, 2nd International**

Conference, 1998, Oxford, England, 2nd–3rd April 1998.

[5] L. K. Hansen and J. Larsen, "Unsupervised Learning and Generalization," in **Proceedings of the 1996 IEEE International Conference on Neural Networks**, Washington DC, USA, 1996, pp. 25–30, Vol. 1.

[6] M. Hintz-Madsen, L. Hansen, J. Larsen, M. Pedersen and M. Larsen, "Neural Classifier Construction using Regularization, Pruning and Test Error Estimation," **Neural Networks**, vol. 11, no. 9, pp. 1659–1670, Dec. 1998.

[7] J. E. Jackson, **A User's Guide To Principal Components**, Wiley Series in probability and Mathematical Statistics, USA: John Wiley & Sons, INC., 1991.

[8] I. T. Jolliffe, "Discarding Variables in Principal Component Analysis. I: Artificial Data," **Applied Statistics**, vol. 21, pp. 160–172, 1972.

[9] Y. Le Cun, J. Denker and S. Solla, "Optimal Brain Damage," in **Advances in Neural Information Processing Systems**, 1990, vol. 1, pp. 598–605.

[10] Paul McIntyre and R. K. Miller, **Acoustic Emission Testing**, vol. 5 of **Nondestructive Testing Handbook**, American Society for Nondestructive Testing, 2nd edn., 1987.

[11] B. D. Ripley, **Pattern Recognition and Neural Networks**, New York, USA: Cambridge University Press, 1996.

[12] L. M. Rogers, "Use of acoustic emission methods for crack growth detection in offshore and other structures," **The Institute of Marine Engineers : Transactions**, vol. 110, no. 3, pp. 171 – 180, 1998.

[13] P. Sørensen and P. S. Pedersen, "The Intelligent Engine and Electronic Products - A Development Status," in **Proceedings of the 22nd CIMAC International Congress on Combustion Engines**, Copenhagen, May 1998, pp. 551 – 564.

[14] P. Storegård Jensen, **Skibsdieselmotorer: Konstruktion Drift Vedligehold**, Copenhagen, Denmark: G-E-C Gads Forlag, 1994, In Danish.

[15] G. T. Venkatesan, D. West, K. M. Buckley, A. H. Tewfik and M. Kaveh, "Automatic Fault Monitoring Using Acoustic Emissions," in **Proceedings of the IEEE 1997 International Conference on Acouistics, Speech, and Signal Processing (ICASSP '97)**, Munich, Germany, 1997, pp. 1893–1896.

[16] J. Yang and Y. Zhou, "Diagnosis of Moving Components of Internal Combustion Engines by Analysing Vibration Signals," **Machine Vibration**, vol. 5, no. 3, pp. 163–169, 1996.